$4295
D0072477

ELEMENTS
OF
GASDYNAMICS

GALCIT AERONAUTICAL SERIES

AERODYNAMICS OF THE AIRPLANE. By *Clark B. Millikan.*

MATRIX AND TENSOR CALCULUS WITH APPLICATIONS TO MECHANICS, ELASTICITY, AND AERONAUTICS. By *Aristotle D. Michal.*

ELASTICITY IN ENGINEERING. By *Ernest E. Sechler.*

AN INTRODUCTION TO THE THEORY OF AEROELASTICITY. By *Y. C. Fung.*

ELEMENTS OF GASDYNAMICS. By *H. W. Liepmann and A. Roshko.*

Other volumes in preparation

ELEMENTS
OF
GASDYNAMICS

BY

H. W. LIEPMANN

A. ROSHKO

California Institute of Technology

GALCIT AERONAUTICAL SERIES

JOHN WILEY & SONS, INC.

New York • Chichester • Brisbane • Toronto • Singapore

26 25 24 23 22

ISBN 0 471 53460 9

Library of Congress Catalog Card Number: 56-9823

PRINTED IN THE UNITED STATES OF AMERICA

Editors' Preface

The aim of the present volume is to modernize and extend the treatment of compressible fluid aerodynamics which appeared in 1947 in the GALCIT† series, under the authorship of Hans W. Liepmann and Allen E. Puckett. The new volume includes a review of many problems of high-speed aerodynamics which have received increased attention by engineers and scientists in the last decade. The editors believe that in addition to such extensions in the treatment of aerodynamic problems, the reader will welcome the inclusion of the last chapter, containing a short presentation of the fundamental concepts of the kinetic theory of gases, and also the references in Chapter 1 to some problems in aerothermochemistry, such as the law of mass action and dissociation. It appears that certain fundamentals of physics and chemical kinetics are gaining more and more importance in the field of aeronautical engineering.

This book is intended mainly for the use of students in aeronautics, but it is hoped that it will also be useful to practicing engineers and scientists who work on problems involving the aerodynamics of compressible fluids.

THEODORE VON KÁRMÁN
CLARK B. MILLIKAN

November, 1956

†Guggenheim Aeronautical Laboratory, California Institute of Technology.

Authors' Preface

Problems of flow of a compressible fluid have been studied for a long time. For instance, shock waves were investigated in the last century, some of the hodograph methods were studied around 1900, and many features of nozzle flow, supersonic jets, etc., were known at about the same time. The fundamental formula for the theory of thin supersonic wings could have been found in Lord Rayleigh's "Theory of Sound"! The interest in this field, however, was restricted to a very small group of people. On the one hand, the unsolved problems interested only a few physicists; the majority considered the subject closed and uninteresting. On the other hand, the applications of gasdynamics were almost entirely restricted to ballistics and steam turbine design and thus interested only a small number of engineers.

This situation changed radically during World War II with the development of fast aircraft, of missiles, and of explosives with large energy release. Concepts from compressible fluid flow theory became suddenly essential for a very large number of mathematicians, physicists, and engineers; the field has been developing and expanding at a very rapid rate ever since then.

The *Introduction to Aerodynamics of a Compressible Fluid* by A. E. Puckett and the senior author appeared in 1947. That book, developed from war training courses taught by the authors, was intended to furnish a coherent account of the topics from gasdynamics which were of prime interest for aeronautical application at that time. For a number of years now, it has been apparent that a new edition, with revision and extension of that material, was needed. At the time of publication, the book was practically the only English textbook on the subject, but a number of excellent books have appeared since then, and some rudiments of compressible fluid flow can now be found in many elementary fluid mechanics texts.

It was therefore decided to abandon a revision of the original text and instead to rewrite the book completely. The choice of material today is much more difficult than it was at the time the *Introduction* appeared. After some consideration, it was decided to split the material into two independent volumes. One of these, the present one, includes the fundamental material of gasdynamics but goes no further into applications than what is necessary to illustrate the theory. We hope this will be followed eventually by another volume in the GALCIT† series, more advanced and specialized, with particular emphasis on the aeronautical and missile fields.

Thus the present volume is intended to cover the fundamentals of gas-dynamics. Even so the choice of material was not easy. For the book to be up-to-date without being excessively larger and more expensive than the original one, it was necessary to omit some topics and methods which are less interesting now than they were ten years ago. These are now mostly of classical and historical interest, or of special interest in some research problems. Their omission does not alter the aim of the book, to provide a working understanding of the essentials of gas flow.

Compared to the *Introduction* the present volume is a little more advanced. However, we feel that this is completely in line with the general trend in physics and engineering education and that the present volume fits into the educational program at about the same point that the *Introduction* fitted at the time it appeared. The book is again intended to form the foundation for a study of the specialized literature and should give the necessary background for reading original papers on the subject. It is *not* intended to be a handbook in any sense of the word. We have not attempted to include all possible approaches and methods; neither have we attempted to include complete material in tables, graphs, and charts for specific use in engineering design. It is often stated that the inclusion of such material will make a book "practical" for engineering use. We believe that this is not so but that an inclusion of such material would make the book very "impractical" and very rapidly dated. The choice of a specific configuration, such as an optimum wing, a wind tunnel, or a shock tube, is governed by a large number of stringent constraints. General principles and essentials can be given in a textbook such as the present one but the application of these to a specific design must be left to the designer. Thus the omission of specific design data reflects a high opinion of engineering design, not the contrary!

The exercises which are given at the end of the book are mainly intended to demonstrate the use of the material in the text and to outline additional subjects, results and equations. Simple numerical problems have been almost entirely omitted. We feel that these problems are best made up by the instructor or even by the student himself. In the text they would have taken up much space with comparatively little help to either instructor or student. In other words, the function and competence of the instructor is regarded in quite the same way as that of the designer.

The general prerequisites of the book are such that it can be used as a textbook in senior and first graduate courses. At the California Institute of Technology, part of the subject matter is given in an introductory course and part in a more advanced, graduate course. A working knowledge of

†Guggenheim Aeronautical Laboratory, California Institute of Technology.

calculus and elementary concepts of physics is assumed. Here and there a "starred" article of more advanced scope is inserted, but these do not seriously affect the continuity of the text, and can be omitted at first reading.

References in the text are usually made only where a specific, recent paper has been used. References to work that has become "classical" have not been systematically given. A list of suggested reading is included at the end of the book. We apologize for the unsystematic way in which references are given in the text and for the obvious omissions and inconsistencies in which this results. Any author who is part of an active and closely knit research group tends toward the outlook and interests of the group. Since we are no exceptions, there is a certain predominance of GALCIT material in some parts of the book.

During the work on the manuscript we have benefited from contact with many colleagues. Specifically we wish to express our appreciation to Z. Bleviss, J. D. Cole, E. W. Graham, P. A. Lagerstrom, and C. B. Millikan. Much of the material in Chapter 5 developed from discussions with Dr. P. Wegener. We are also indebted to Dr. W. D. Hayes for his critical and constructive review of an early manuscript and to Mr. Bradford Sturtevant for his careful and competent proofreading and checking. Mrs. Beverly Cottingham and Mrs. Alrae Tingley have contributed greatly to the preparation and completion of the manuscript.

H. W. LIEPMANN
A. ROSHKO

Pasadena, California
November, 1956

Contents

CHAPTER 13. EFFECTS OF VISCOSITY AND CONDUCTIVITY

CHAPTER 14. CONCEPTS FROM GASKINETICS

CHAPTER I

Concepts from Thermodynamics

1·1 Introduction

The basis of any physical theory is a set of experimental results. From these special primary observations, general principles are abstracted, which can be formulated in words or in mathematical equations. These principles are then applied to correlate and explain a group of physical phenomena and to predict new ones.

The experimental basis of thermodynamics is formalized in the so-called principal laws. The law of conservation of energy, which thermodynamics shares with mechanics, electrodynamics, etc., is one of these principal laws. It introduces the concept of internal energy of a system. The other principal laws of thermodynamics introduce and define the properties of entropy and temperature, the two concepts which are particular and fundamental for thermodynamics.

The principles laid down in these fundamental laws apply to the relations between equilibrium states of matter in bulk. For instance, thermodynamics yields the relation between the specific heats at constant pressure and at constant volume; it relates the temperature dependence of the vapor pressure to the latent heat of evaporization; it gives upper bounds for the efficiency of cyclic processes, etc.

Fluid mechanics of perfect fluids, i.e., fluids without viscosity and heat conductivity, is an extension of equilibrium thermodynamics to moving fluids. The kinetic energy of the fluid has now to be considered in addition to the internal energy which the fluid possesses when at rest. The ratio of this kinetic energy per unit mass to the internal energy per unit mass is a characteristic dimensionless quantity of the flow problem and in the simplest cases is directly proportional to the square of the Mach number. Thermodynamic results are taken over to perfect fluid flow almost directly.

Fluid mechanics of real fluids goes beyond classical thermodynamics. The transport processes of momentum and heat are of primary interest here, and a system through which momentum, heat, matter, etc., are being transported is not in a state of thermodynamic equilibrium, except in some rather trivial cases, such as uniform flow of matter through a fixed system.

But, even though thermodynamics is not fully and directly applicable to all phases of real fluid flow, it is often extremely helpful in relating the initial and final conditions. This complex of problems is best illustrated with a simple example. Assume a closed, heat-insulating container divided into two compartments by a diaphragm. The compartments contain the same gas but at different pressures p_1 and p_2, and different temperatures T_1 and T_2. If the diaphragm is removed suddenly, a complicated system of shock and expansion waves occurs, and finally subsides due to viscous damping. Thermodynamics predicts the pressure and temperature in this final state easily. Fluid mechanics of a real fluid should tackle the far more difficult task of computing the pressure, temperature, etc., as a function of time and location within the container. For large times, pressure and temperature will approach the thermodynamically given values. Sometimes we need only these final, equilibrium values and hence can make very good use of thermodynamic reasoning even for problems that involve real fluid flow.

In fluid mechanics of low-speed flow, thermodynamic considerations are not needed: the heat content of the fluid is then so large compared to the kinetic energy of the flow that the temperature remains nearly constant even if the whole kinetic energy is transformed into heat.

In modern high-speed flow problems, the opposite can be true. The kinetic energy can be large compared to the heat content of the moving gas, and the variations in temperature can become very large indeed. Consequently the importance of thermodynamic concepts has become steadily greater. The chapter therefore includes material that is more advanced and not needed for the bulk of the later chapters. Articles that are starred can be omitted at first reading without loss of continuity.

1·2 Thermodynamic Systems

A thermodynamic *system* is a quantity of matter separated from the "*surroundings*" or the "*environment*" by an *enclosure.* The system is studied with the help of measurements carried out and recorded in the surroundings. Thus a thermometer inserted into a system forms part of the surroundings. Work done by moving a piston is measured by, say, the extension of a spring or the movement of a weight in the surroundings. Heat transferred to the system is measured also by changes in the surroundings, e.g., heat may be transferred by an electrical heating coil. The electric power is measured in the surroundings.

The enclosure does not necessarily consist of a solid boundary like the walls of a vessel. It is only necessary that the enclosure forms a *closed surface* and that its properties are defined everywhere. An enclosure may transmit heat or be a heat insulator. It may be deformable and thus capable of transmitting work to the system. It may also be capable of transmitting mass. Every real wall has any one of these properties to a certain degree.

There do not exist perfectly rigid walls, for example, and similarly there is no perfect heat insulator. However, it is convenient to use an idealized enclosure, consisting of parts which have well-defined properties such as complete heat insulation, etc.

For our purposes it is sufficient to deal with fluids only. The systems that we shall consider here are:

(*a*) A simple, homogeneous system composed of a single gas or liquid.

(*b*) A homogeneous mixture of gases.

(*c*) A heterogeneous system composed of the liquid and gaseous phase of a single substance.

1·3 Variables of State

If a system is left alone for a sufficiently long time, that is, if no heat and no mass is transferred to it and no work is done on it during this time, it will reach a state of equilibrium. All macroscopically measurable quantities will become independent of time. For example, the pressure p, the volume V, and the temperature θ can be measured, and in equilibrium do not depend upon time.

Variables that depend only upon the state of the system are called *variables of state.* p and V are evidently such variables, and these two are already familiar from mechanics. For a complete thermodynamic description of a system, we need new variables of state, foreign to mechanics. Thus it is a result of experience that the pressure of a system is not only a function of its volume. A new variable of state, θ, the temperature, has to be introduced. For a simple system,

$$p = p(V, \theta) \tag{1·1}$$

Following R. H. Fowler, one states the "zeroth law of thermodynamics":

> There exists a variable of state, the temperature θ. Two systems that are in thermal contact, i.e., separated by an enclosure that transmits heat, are in equilibrium only if θ is the same in both.

Consequently, with the help of Eq. 1·1, we can use the pressure and the volume of an arbitrary system as a thermometer.

When we discuss the exchange of work or heat between a system and its surroundings, we find the need for a variable of state E, the internal energy, which measures the energy stored in the system. The *first law of thermodynamics* introduces E, as will be seen later.

Furthermore we shall find it necessary to introduce a variable of state S, the entropy, which, for example, is needed to decide whether a state is in stable equilibrium. The *second law of thermodynamics* introduces S and defines its properties.

For a simple system E and S are functions of p, V, θ. But, since p can be expressed by V and θ, using Eq. 1·1, it is sufficient to write:

$$E = E(V, \theta) \tag{1·2}$$

$$S = S(V, \theta) \tag{1·3}$$

Relations like Eqs. 1·1, 1·2, and 1·3 are called *equations of state*. Specifically Eq. 1·1 is called the "thermal equation of state"; Eq. 1·2, the "caloric equation of state." A specific substance is characterized by its equations of state. The forms of these equations cannot be obtained from thermodynamics but are obtained from measurements or else, for a particular molecular model, from statistical mechanics or kinetic theory.

Any variable of state is uniquely defined for any equilibrium state of the system. For example, if a system changes from one state of equilibrium, say A, to another state B, then $E_B - E_A$ is independent of the process by which the change occurred. The important consequences of this property of the variables of state will become evident later.

One distinguishes between *intensive* and *extensive* variables of state. A variable is called *extensive* if its value depends on the mass of the system. The mass M of a system is thus an *extensive* quantity, and so are E, V, and S. For example, the internal energy E of a certain mass of a gas is doubled if the mass is doubled; the energy of a system that consists of several parts is equal to the sum of the energies of the parts.

Variables of state that do not depend upon the total mass of the system are called *intensive* variables. p and θ are typical intensive variables. For every extensive variable like E we can introduce an intensive variable e, the energy per unit mass or specific energy. Similarly we can define a specific volume v, specific entropy s, etc. *Specific* quantities will be denoted by lower-case letters.

1·4 The First Principal Law

Consider a fluid contained in a heat-insulating enclosure, which also contains a paddle wheel that can be set into motion by a falling weight. The pressure of the system is kept constant. The temperature θ and the volume V are measured initially (state A). The weight is allowed to drop a known distance, and θ and V are measured again after the motions in the system have died down and a new state of equilibrium B has been established.

In this way a certain amount of work W, equal to the decrease in potential energy of the weight, has been done on the system. Conservation of energy requires that this work is stored within the system. Hence there exists a function $E(V, \theta)$ such that

$$E_B - E_A = W \tag{1·4}$$

It is also possible to use work to produce an electric current and to supply this work to the system in the form of heat given off by a heating coil. Both of these experiments were performed by Joule in his classical studies on the mechanical equivalent of heat. A given amount of work done on the system yields the same difference in internal energy regardless of the rate at which the work is done and regardless of how it is transmitted.

One can furthermore relax the condition of complete heat insulation and allow also the passage of a certain amount of heat Q through the enclosure. Q can be defined calorically by the change in temperature of a given mass of water, or one can use Joule's experiments to define Q entirely in mechanical terms. It is important, however, to define Q and W in terms of changes measured in the surroundings.

We can thus formulate the first law:

There exists a variable of state E, the internal energy. If a system is transformed from a state of equilibrium A to another one, B, by a process in which a certain amount of work W is done by the surroundings and a certain quantity of heat Q leaves the surroundings, the difference in the internal energy of the system is equal to the sum of Q and W,

$$E_B - E_A = Q + W \tag{1·5}$$

It is often convenient to discuss a simple idealized enclosure, the cylinder-piston arrangement of Fig. 1·1. The cylinder walls are assumed rigid. We

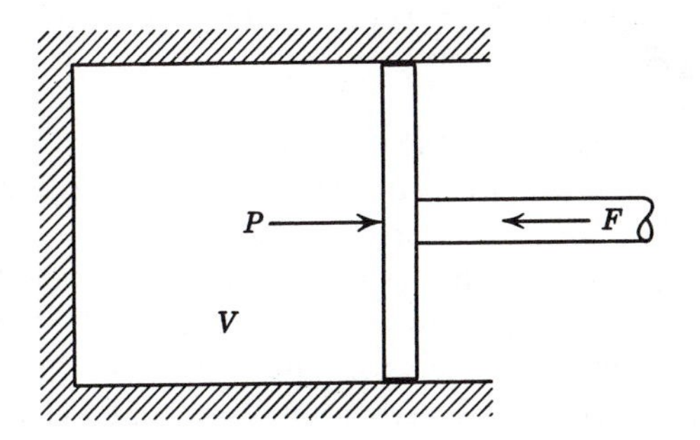

FIG. 1·1 Cylinder-piston arrangement.

can assume them to be heat-insulating or capable of heat transmission, depending on the process that we wish to study. Work can be done by the surroundings only by the displacement of the piston. W is defined as in mechanics in terms of a force vector $\mathbf{F}$ and a displacement $d\mathbf{r}$,

$$W = \int \mathbf{F} \cdot d\mathbf{r} \tag{1·6}$$

The force acting on the piston is parallel to the displacement; thus, introducing the pressure p and the piston surface area A, we have

$$W = \int pA\,dr = -\int p\,dV \tag{1·7}$$

with the convention that dV is positive if the volume of the system *increases*. It is not difficult to show that Eq. 1·6 leads to Eq. 1·7 even in the case of pressures acting on a deformable enclosure of any shape. (Shear forces can be introduced also; this is done later in discussing the equations of motion of a real fluid in Chapter 13.) For a small change of state, we can write Eq. 1·5 in differential form,

$$dE = dQ + dW \tag{1·8}$$

or, using Eq. 1·7,

$$dE = dQ - p\,dV \tag{1·8a}$$

Equation 1·8*a* can also be written for unit mass:

$$de = dq - p\,dv \tag{1·8b}$$

Now E is a variable of state, whereas Q and W depend on the process followed in changing the state. This is sometimes indicated by writing δW and δQ instead of dW and dQ. We shall not follow this custom here.

1·5 Irreversible and Reversible Processes

A change of state of a system is possible only by a process for which

$$\Delta E = Q + W$$

The first law does not restrict the possible processes any further.

Now in the paddle wheel experiment of Joule it is evidently impossible to reverse the direction of the process. One cannot induce the wheel to extract the energy ΔE from the system and to lift the weight. The process is *irreversible.* It is very easy to find other similar situations, and indeed *all natural or "spontaneous" processes are irreversible.* If one scrutinizes these irreversible processes, it becomes evident that the deviation of the system from equilibrium *during* the process is of primary importance. A motion like the stirring of a fluid, sudden heating, etc., induces *currents* in the system. The term *current* refers to the flux of a quantity like heat, mass, momentum, etc. A current of heat flows, if there exists a finite temperature difference; a current of mass flows, if there exist differences in concentration of one component; a current of momentum flows, if there exist differences in velocity.

A system is in a state of equilibrium if it is free of currents. A process leading from one state to another is *reversible* if the system remains during the whole process in equilibrium; i.e., if the work W and the heat Q are added in such a way that no currents are produced. Such an ideal reversible process can actually be closely approximated in an experiment. For example, instead of using the paddle wheel, W could be transferred to an insulated system by a slow displacement of a piston, so that the pressure and temperature remain uniform within the system during the whole process. (Exercise 1·9 gives a simple and instructive example of an irreversible process.)

The changes of state discussed here lead from one static condition of the system to another. It is often much more convenient to consider processes that proceed at a steady rate. This is true for many measurements in thermodynamics and is essential for fluid mechanics. Thus, instead of dealing with a paddle wheel in a closed "calorimeter," as in Joule's experiment, we may consider a heat-insulated duct in which a fluid flows at a steady rate through a turbine wheel or fan. The system consists now of a certain mass of fluid which passes through the fan. Instead of dealing with a system before and after the motion of the paddle wheel, we now deal with the fluid upstream and downstream of the fan. Our definitions of thermodynamic equilibrium can be extended to this case easily. For direct comparison with thermodynamic processes like Joule's experiment, we have to require the fluid to flow very slowly so that its kinetic energy is negligible. In the next chapter we shall drop this restriction and extend the same considerations to high-speed fluid flow.

1·6 Perfect Gases

It is convenient to introduce at this stage the concept of a perfect gas. A perfect gas is the simplest working fluid in thermodynamics and hence is very useful in the detailed study of thermodynamic processes. For applications to aerodynamics, the concept is even more important since we deal there almost exclusively with gases, and often under conditions where they are nearly perfect.

Measurements of the thermal properties of gases show that for low densities the thermal equation of state approaches the same form for all gases, namely,

$$pv = R(\theta + \theta_0) \tag{1·9}$$

or, in terms of the density $\rho = 1/v$,

$$p = \rho R(\theta + \theta_0)$$

Here θ_0 is a characteristic temperature which turns out to be the *same for all gases*, and R is a characteristic constant for a particular gas.

It is useful to define an *"ideal"* or *"perfect"* gas which satisfies Eq. 1·9 exactly. More precisely, Eq. 1·9 defines a *family* of perfect gases, one for each value of R. Any gas at low enough density approaches a perfect gas, with a particular value of R.

Since θ_0 is found to be the same for all gases, one can define a new, more convenient temperature T,

$$T = \theta + \theta_0$$

and then replace Eq. 1·9 by

$$p = \rho RT \tag{1·10}$$

T is called the *absolute temperature.* It is possible to show that T, which is here defined as a gas temperature, has meaning for all thermodynamic systems. The scale and the zero point of T are determined from the scale and zero point of the thermometer used to measure θ_0. Thus in the Centigrade scale one finds

$$\theta_0 = 273.16°$$

and in the Fahrenheit scale

$$\theta_0 = 459.69°$$

Thus the absolute temperature T can be written

$T = \theta + 273.16$ degrees Centigrade absolute or Kelvin (°K)

$T = \theta + 459.69$ degrees Fahrenheit absolute or Rankine (°R)

R as defined in Eq. 1·10 has the dimensions (velocity)²/temperature. It is related to the velocity of sound, a, in the gas, $R \sim \dfrac{da^2}{dT}$, as will be seen later.

If we rewrite Eq. 1·10 for a given mass M by putting $\rho = M/V$, we have

$$pV = MRT \tag{1·10a}$$

A study of the behavior of different gases led very early to the concept that gases are composed of molecules and that the characteristic parameter of the family of perfect gases defined by Eqs. 1·10 and 1·10*a* is the mass of these molecules. Thus Eq. 1·10*a* can be written in terms of a dimensionless mass ratio $M/m = \mu$ where m denotes the mass of one gas molecule. Written in this reduced or "similarity" form, the family of Eq. 1·10 can be reduced to a single one:

$$\frac{pV}{\mu} = \mathbf{k}T \tag{1·10b}$$

where $\mathbf{k}$ is a *universal* constant, the so-called Boltzmann constant. Instead of m, the mass of a molecule, one often uses the "molecular weight" $\mathbf{m}$ in relative units such that $\mathbf{m}_{\text{oxygen}} = 32$. In terms of $\mathbf{m}$ one has:

$$\frac{pV}{\boldsymbol{\mu}} = \mathbf{R}T \tag{1·10c}$$

where $\boldsymbol{\mu} = M/\mathbf{m}$, and $\mathbf{R}$ is called the *universal gas constant.* One can also use as unit of mass the *mole* and thus make $\boldsymbol{\mu} = 1$ in Eq. 1·10*c*. V then becomes the *mole volume.* We shall not use the mole in the following articles but shall continue to refer to unit mass, and almost always shall use the perfect gas law in the form $p = \rho RT$ (Eq. 1·10).

The internal energy of a perfect gas E is a function of temperature only,

$$E = E(T) \tag{1·11}$$

Equation 1·11 can be taken as the result of experience. We shall see later, however, that it is also a direct consequence of Eq. 1·10.

Often a gas is called "calorically perfect" if Eq. 1·11 simplifies further to

$$E = \text{const.} \cdot T \tag{1·12}$$

Equation 1·12 does not follow directly from Eq. 1·10 by pure thermodynamic reasoning. For certain ranges of temperature it can be justified by experience, and it also follows from statistical mechanics (cf. Chapter 14).†

To judge the degree to which real gases are approximated by Eqs. 1·10 and 1·11, we must say a few words about the equations of state of real gases. (We shall return once more to this subject in Article 1·18.)

Every real gas can be liquefied. The highest temperature at which this is possible is called the *critical temperature* T_c; the corresponding pressure and density are called *critical pressure* p_c and *critical density* ρ_c. These critical variables are characteristic of a gas; they depend upon the intermolecular forces.

An equation of state for real gases must therefore involve at least two parameters besides R, for example T_c and p_c. This is the case in the famous *van der Waals equation* of state which can be used to estimate the approximation in considering real gases, at moderate densities, to be perfect. The *van der Waals* equation is

$$p = \rho RT\left[\frac{1}{1 - \beta\rho} - \frac{\alpha\rho}{RT}\right] \tag{1·13}$$

with

$$\frac{\alpha}{\beta} = \frac{27}{8} RT_c; \quad \frac{\alpha}{\beta^2} = 27 p_c$$

†The zero point for E is undefined in Eq. 1·11; i.e., the energy is defined only up to an additive constant. For a single phase of a single substance this constant is unimportant and may be chosen zero, as in Eq. 1·12. The same is true for other extensive variables of state.

The internal energy of a van der Waals gas is

$$E = E_0(T) - \frac{\alpha}{V} \equiv E(V, T) \tag{1·14}$$

Using Eqs. 1·13, 1·14, and tabulated values of α and β, the degree of approximation of Eqs. 1·10 and 1·11 for any gas can be judged. A more general approach is outlined in Article 1·18.

1·7 The First Law Applied to Reversible Processes. Specific Heats

The differential form of the first law written for unit mass is (Eq. 1·8*b*):

$$de = dq - p\,dv \tag{1·15}$$

If we consider reversible processes, then p and T are the pressure and temperature of the system.

The Specific Heats. A specific heat c is defined by:

$$c = \frac{dq}{dT} \tag{1·16}$$

i.e., c is the heat needed to raise the temperature of unit mass of the system by 1 degree. The value of c depends on the process in which q is added. For a simple system the energy e and the pressure p depend only upon v and T. Hence any reversible process can be plotted as a curve in a v, T diagram (Fig. 1·2). Of course, we could choose as well a p, T or p, v diagram, since $p = p(v, T)$.

Thus, if we know the specific heat for two different processes, we know c for all processes. One usually chooses the specific heat at constant volume c_v and the specific heat at constant pressure c_p. Thus

$$c_v = \left(\frac{dq}{dT}\right)_v \tag{1·17a}$$

$$c_p = \left(\frac{dq}{dT}\right)_p \tag{1·17b}$$

e is a function $e(v, T)$, and hence Eq. 1·15 can be written†

$$de = \frac{\partial e}{\partial v}dv + \frac{\partial e}{\partial T}dT = dq - p\,dv$$

†It is customary in thermodynamics to indicate the variables which are kept constant in forming a partial derivative by a subscript, e.g., $(\partial e/\partial v)_T$ means that T is kept constant. We shall use this notation *only where there could be any doubt about the variable that is considered constant.*

and hence

$$c_v = \frac{\partial e}{\partial T} \tag{1·18}$$

$$c_p = \frac{\partial e}{\partial T} + \left(\frac{\partial e}{\partial v} + p\right)\left(\frac{\partial v}{\partial T}\right)_p \tag{1·19}$$

The natural variable of state for e is v because dv appears explicitly in Eq. 1·15. Keeping v constant leads then to a simple expression like Eq. 1·18, whereas a process at constant p leads to the more complicated expres-

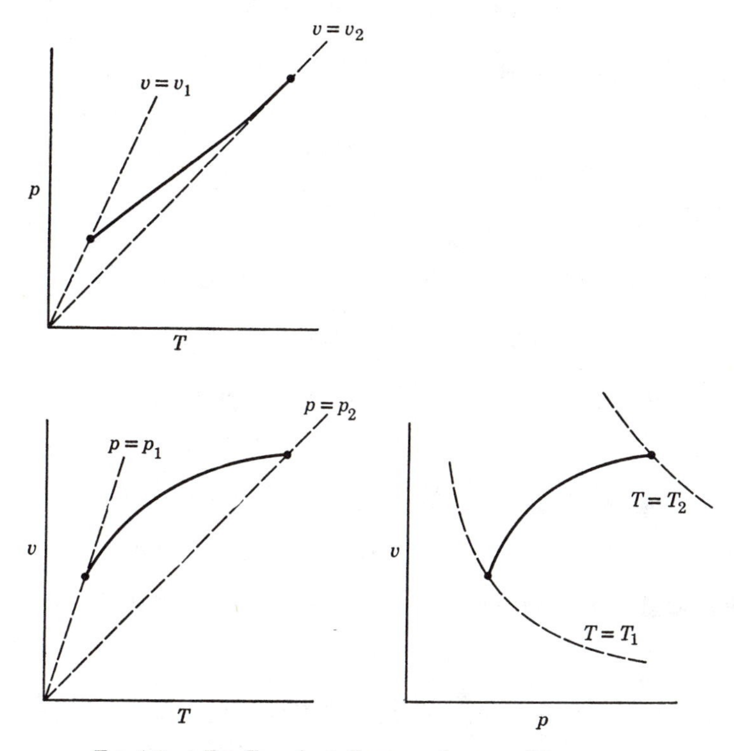

FIG. 1·2 p,T; v,T; and v,p diagrams of a reversible process.

sion Eq. 1·19. It is thus natural to ask whether there exists another variable of state, related to e, for which p is a natural choice of independent variable. This function is the *enthalpy* or *heat function h*, defined by

$$h = e + pv \tag{1·20}$$

or

$$H = E + pV$$

Thus

$$dh = de + p\,dv + v\,dp$$

and the first law can be written as

$$dh = dq + v\,dp \qquad \blacktriangleright(1\cdot21)$$

Again we can form c_v and c_p and find

$$c_v = \frac{\partial h}{\partial T} + \left(\frac{\partial h}{\partial p} - v\right)\left(\frac{\partial p}{\partial T}\right)_v \qquad (1\cdot22)$$

$$c_p = \frac{\partial h}{\partial T} \qquad (1\cdot23)$$

For a perfect gas, we have

$$e = e(T)$$

$$h = e(T) + pv = e(T) + RT = h(T)$$

Consequently, we have, from Eqs. 1·18 and 1·19,

$$c_v = \frac{de}{dT}$$

$$c_p = c_v + p\left(\frac{\partial v}{\partial T}\right)_p = c_v + R$$

and, similarly, from Eqs. 1·22 and 1·23,

$$c_p = \frac{dh}{dT}$$

$$c_v = c_p - v\left(\frac{\partial p}{\partial T}\right)_v = c_p - R$$

Consequently, for a perfect gas we have the important relations:

$$c_p - c_v = R \qquad \blacktriangleright(1\cdot24)$$

$$e(T) = \int c_v\,dT + \text{const.} \qquad \blacktriangleright(1\cdot25)$$

$$h(T) = \int c_p\,dT + \text{const.} \qquad \blacktriangleright(1\cdot26)$$

A gas is sometimes called *calorically perfect* if c_p and c_v are constants, independent of T. In this more specialized case we have

$$e = c_v T + \text{const.} \qquad (1\cdot25a)$$

$$h = c_p T + \text{const.} \qquad (1\cdot26a)$$

Note that Eq. 1·24 holds, regardless of whether c_p and c_v are constants.

The Adiabatic, Reversible Process. Very important in later applications to fluid flow is the adiabatic, reversible process: i.e., a process in which no heat is transferred to or from the system and in which the work is done reversibly. We can then apply (1·15) or (1·21), depending on whether we choose v, T or p, T as the variables. The adiabatic, reversible process is thus given by

$$de = -p\,dv \tag{1·27}$$

or

$$dh = v\,dp \tag{1·28}$$

From Eq. 1·27 it follows that

$$\frac{\partial e}{\partial v}\,dv + \frac{\partial e}{\partial T}\,dT = -p\,dv$$

from Eq. 1·28

$$\frac{\partial h}{\partial p}\,dp + \frac{\partial h}{\partial T}\,dT = v\,dp$$

Consequently for an adiabatic, reversible process we have

$$\frac{dT}{dv} = -\frac{1}{c_v}\left(\frac{\partial e}{\partial v} + p\right) \tag{1·29}$$

$$\frac{dT}{dp} = -\frac{1}{c_p}\left(\frac{\partial h}{\partial p} - v\right) \tag{1·30}$$

$$\frac{dp}{dv} = -\frac{p}{v}\frac{dh}{de} \tag{1·31}$$

Specializing to a perfect gas, we find

$$\frac{v}{T}\frac{dT}{dv} = -\frac{R}{c_v} \tag{1·29a}$$

$$\frac{p}{T}\frac{dT}{dp} = \frac{R}{c_p} \tag{1·30a}$$

$$\frac{v}{p}\frac{dp}{dv} = -\frac{c_p}{c_v} \tag{1·31a}$$

Since $R = c_p - c_v$, the right-hand sides of these relations can be expressed in terms of the ratio $\frac{c_p}{c_v} = \gamma$. γ will depend upon T except if it is assumed that the gas is also calorically perfect. In any case the relations for a perfect gas are integrable. Thus, for example, $\ln v = -\int \frac{dT}{T(\gamma - 1)}$; if γ is constant, the relations become very simple indeed:

$$v = \text{const. } T^{-1/(\gamma-1)} \qquad \blacktriangleright(1{\cdot}29b)$$

$$p = \text{const. } T^{\gamma/(\gamma-1)} \qquad \blacktriangleright(1{\cdot}30b)$$

$$p = \text{const. } v^{-\gamma} \qquad \blacktriangleright(1{\cdot}31b)$$

We shall later see that an adiabatic, reversible process is *isentropic:* i.e., a process for which the entropy S remains constant.

These examples do not exhaust, of course, the applications of the first law to reversible processes. But they outline the general procedure.

1·8 The First Law Applied to Irreversible Processes

The first typical irreversible process that we shall discuss is the *adiabatic expansion of a gas*. The system of interest consists of a vessel with heat-insulated, rigid walls. A diaphragm divides the vessel into two volumes V_1 and V_2 (Fig. 1·3). Both volumes are filled with the same gas at the same

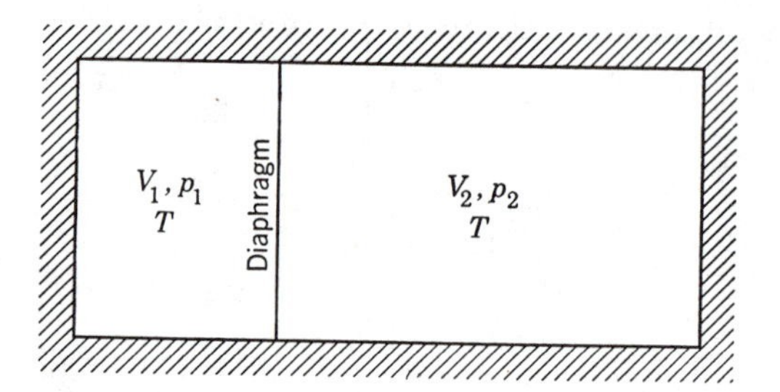

FIG. 1·3 Heat-insulated vessel with dividing diaphragm.

temperature T. The pressures p_1 and p_2, however, are different. The theory of the *flow* under such conditions will occupy us in later chapters; here we are interested only in the thermodynamics of the setup.

Let the diaphragm be ruptured at a time $t = 0$. A violent flow of the gas ensues; a shock wave propagates into the low-pressure side, an expansion wave into the high-pressure side, and by reflection and refraction a complicated wave system is set up. This subsides under the action of viscosity and internal heat conductivity, and eventually the gas is at rest again, in a new state of thermodynamic equilibrium. We now apply the first law to the change from the initial state to the final one.

No heat has left the surroundings, and no work has been performed by the surroundings. Hence from Eq. 1·5 we have

$$E_B - E_A = 0 \qquad (1{\cdot}32)$$

The internal energy in an expansion process is conserved.

If the gas can be approximated by a perfect gas, then $E = E(T)$ and hence Eq. 1·32 yields:

$$T_B = T_A \tag{1·32a}$$

In the adiabatic, irreversible expansion of a perfect gas, initial and final state have the same temperature. Historically Gay Lussac performed a similar experiment; he measured the initial and final temperature and, since he found them equal, reasoned that $E = E(T)$ for a perfect gas.

As a second example, consider a similar setup but now assume the gas to be at the same pressure p in both compartments but at different temperatures T_1 and T_2. The diaphragm is assumed to be heat insulating. If now the partition is removed, the temperature tends to equalize and again currents are set up, but eventually subside. For the initial and final states Eq. 1·32 applies again.

In terms of the specific energy e, using M_1 and M_2 for the masses of the gas in the compartments:

$$\begin{aligned} E_A &= M_1e(T_1, v_1) + M_2e(T_2, v_2) \\ E_B &= (M_1 + M_2)e(T_B, v_B) \end{aligned} \tag{1·33}$$

Specializing to a perfect gas, we thus find

$$M_1e(T_1) + M_2e(T_2) = (M_1 + M_2)e(T_B) \tag{1·33a}$$

If the gas is also calorically perfect, then $e = c_vT$, and T_B is obtained explicitly:

$$T_B = \frac{M_1T_1 + M_2T_2}{M_1 + M_2} \tag{1·33b}$$

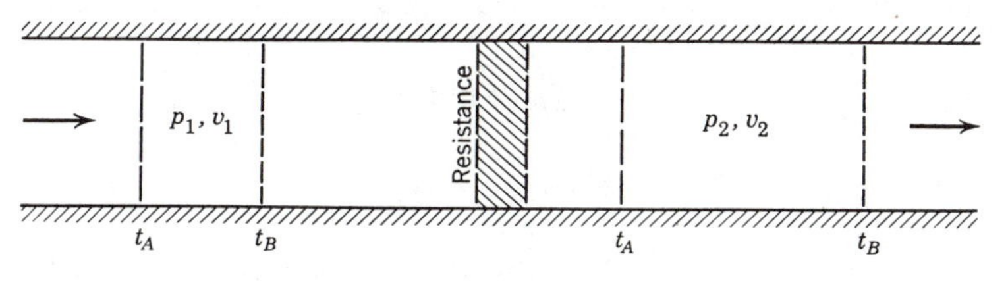

FIG. 1·4 Throttling process.

As a final example, we consider the most important irreversible process for application to fluid mechanics, namely, the *throttling process or Joule-Thomson process*. Here we apply the first law directly to a *moving fluid* under *stationary* conditions, compared to the *static* conditions in the previous examples. We consider a heat-insulated pipe which contains at one cross-section a *resistance*, e.g., a throttling valve, a porous plug, or a screen (Fig. 1·4). The fluid flows slowly from the left to the right through this resistance, so slowly that its kinetic energy per unit mass, $\frac{1}{2}u^2$, is negligible compared to the enthalpy per unit mass. That this condition is the proper criterion to use will become clear later, when we discuss cases where the kinetic energy is included.

As our system we choose the mass of fluid contained between two control surfaces, one to the left, the other to the right of the resistance. The control surfaces move with the fluid and, together with the walls of the pipe, form the enclosure. The rest of the fluid and the exterior of the pipe and the resistance are the surroundings.

We now observe the state of the system at two times t_B and t_A, which we can choose conveniently so that in the time $t_B - t_A$ unit mass has passed through the resistance. In this time the internal energy has changed owing to the passage of unit mass from side (1) to side (2):

$$\Delta E = e_2 - e_1 \tag{1·34}$$

No heat has been transferred from the surroundings but a certain amount of work W has been done, since the control surfaces have been displaced by different amounts against different pressures. (It is evidently possible to replace the control surfaces by pistons without changing the physical picture.)

The net work done W is thus given by

$$W = W_1 + W_2 = \int_{(1)} p\,dv + \int_{(2)} p\,dv$$

Since p_1 and p_2 are constant, we have

$$W = p_1v_1 - p_2v_2 \tag{1·35}$$

because the *change of volume* is equal to the volume per unit mass on either side. The first law then yields:

$$\Delta E = W$$

or

$$e_1 + p_1v_1 = e_2 + p_2v_2$$

that is,

$$h_1 = h_2 \tag{1·36}$$

Thus, *in adiabatic flow through a resistance the enthalpy per unit mass upstream and downstream of the resistance is the same.* This result is of fundamental importance for many applications in fluid flow, and is easily generalized to high-speed flow, as will be seen in Chapter 2. Specializing to a perfect gas, we have

$$T_2 = T_1 \tag{1·36a}$$

Historically the experiment was performed by Joule and Thomson (Lord Kelvin) to establish the form of the energy function for gases, by measuring the temperatures T_1 and T_2.

1·9 The Concept of Entropy. The Second Law

The processes discussed in the previous article, as well as others, such as Joule's paddle wheel experiment, have in common the fact that they proceed in *one direction*. An adiabatic system will equalize the pressure or the temperature, as in the expansion of a gas or the mixing of a gas with initially different temperatures; a gas will flow through a resistance only if the pressure is larger upstream than downstream; and so on. General observations like these form the experimental bases for the second law of thermodynamics. It remains to abstract from those experiences a well-defined law which can be formulated mathematically.

Consider again the case of an expanding gas. Is it possible to decide from measurements performed in two equilibrium states, which one will change spontaneously into the other? Obviously we could say that a pressure discontinuity is unstable and that a state in which one exists would spontaneously change to a state of uniform pressure. In the second example of Article 1·8, we could make a similar statement about the temperature discontinuity. Such a measure of irreversibility is possible but not very useful since it is not general enough. We are trying to find a general criterion such that these simple conditions come out as special cases.

This may be found in a new variable of state which is called S, the entropy. S and T are both concepts that are particular to thermodynamics. In fact, one can introduce S formally as follows: E, the internal energy, has the character of a *potential energy*; for a *reversible change of state* of an *adiabatically enclosed system*, we have (e.g., Eq. 1·27)

$$p = -\frac{\partial E}{\partial V}$$

i.e., the *pressure* is equal to the derivative of the internal energy with respect to the *volume*, or the *force* is equal to the derivative of E with respect to *displacement*. We can ask whether a similar relation can be found for the temperature T. Hence we try to write E as a function of two variables such that the partial derivatives are p and T, respectively; this variable we call S. Thus, let us write tentatively,

$$E = E(S, V)$$

such that

$$-p = \left(\frac{\partial E}{\partial V}\right)_S \qquad T = \left(\frac{\partial E}{\partial S}\right)_V \tag{1·37}$$

and hence

$$dE = \left(\frac{\partial E}{\partial V}\right) dV + \left(\frac{\partial E}{\partial S}\right) dS = -p\,dV + T\,dS$$

The first law in the form (1·8*a*) yields

$$dE = -p\,dV + dQ$$

and hence we identify:

$$T\,dS = dQ \tag{1·37a}$$

Hence $T\,dS$ is equal to the element of heat added in a *reversible process*. Integrating, we have

$$S_B - S_A = \int_A^B \frac{dQ}{T} \tag{1·38}$$

where the integral is to be evaluated for a reversible process leading from a state A to a state B.

S was defined as a variable of state in Eq. 1·37. Thus to relate it to dQ the system has to be in thermodynamic equilibrium, and hence we have to assume a *reversible* process. We now demonstrate that S as defined by Eq. 1·37 or Eq. 1·38 has the properties that we expect. For this purpose, we shall work out Eq. 1·38 in detail for a simple irreversible process, e.g., the mixing of a calorically perfect gas of initially nonuniform temperature (the second example of Article 1·8). We shall consider the process through which we return the system reversibly to the initial condition: state B consists of a gas with mass $(M_1 + M_2)$ at a temperature T_B. We intend to return the system to state A where the mass M_1 has the temperature T_1 and the mass M_2, the temperature T_2. For simplicity, choose $M_1 = M_2 = M$.

We proceed as follows: we partition the containers into two equal parts. Hence there will be equal masses in each portion. We now heat one compartment to T_1; the other we cool to T_2 (assuming $T_1 > T_2$) reversibly, i.e., by a slow addition or removal of heat. We then find that the temperatures are correct for state A but that the pressures are different, say p_1 and p_2. We now move the partition slowly toward the low-pressure side until the pressures are equal. This last step has to be carried out keeping the temperatures constant, i.e., with additional heat transfer to the system.

Thus we can compute the contributions $\int \frac{dQ}{T}$ in two steps:

(*a*)

$$\int \frac{dQ}{T} = \int \frac{dE + p\,dV}{T} = \int_{T_B}^{T_1} \frac{dE}{T} + \int_{T_B}^{T_2} \frac{dE}{T}$$

or

$$\int_{(a)} \frac{dQ}{T} = -Mc_v\left(\ln\frac{T_B}{T_1} + \ln\frac{T_B}{T_2}\right) \tag{1·39a}$$

(*b*)

$$\int \frac{dQ}{T} = \int \frac{dH - V\,dp}{T} = -\frac{1}{T_1}\int_{p_1}^{p} V\,dp - \frac{1}{T_2}\int_{p_2}^{p} V\,dp$$

or

$$-\int_{(b)} \frac{dQ}{T} = MR\left(\ln\frac{p}{p_1} + \ln\frac{p}{p_2}\right) \tag{1·39b}$$

But T_B is related to T_1, T_2 by Eq. 1·33*b*

$$T_B = \frac{T_1 + T_2}{2}$$

and from the equation of state

$$\frac{p}{p_1} = \frac{T_B}{T_1}; \quad \frac{p}{p_2} = \frac{T_B}{T_2}$$

Hence the entropy difference becomes, with $c_p = R + c_v$,

$$S_B - S_A = -\int_{(a)} \frac{dQ}{T} - \int_{(b)} \frac{dQ}{T} = Mc_p \ln\frac{(T_1 + T_2)^2}{4T_1T_2} \tag{1·40}$$

Now, $\dfrac{(T_1 + T_2)^2}{4T_1T_2} > 1$, whenever $T_1 \neq T_2$ and equal to 1 if $T_1 = T_2$. Hence $S_B > S_A$ except in the trivial case where the temperatures are initially equal.

For the first example of Article 1·8, the expansion of a perfect gas, one can show similarly that

$$S_B - S_A = MR\ln\frac{(p_1 + p_2)^2}{4p_1p_2} \tag{1·41}$$

On the other hand, for the adiabatic, reversible process of Article 1·7, $dE + p\,dV = 0$, and hence Eq. 1·37 or Eq. 1·37*a* leads immediately to $S_B - S_A = 0$. Consequently the function S defined by Eqs. 1·37 or 1·38 has, at least for the perfect gas, the expected property. It can be shown that this is equally true for all possible thermodynamic systems. The proof can be given by using the so-called Carnot cycle, or by the more elegant and direct method of Caratheodory. For this proof the reader is referred to the standard textbooks in thermodynamics.

We formulate the second law then as follows:

(*a*) There exists an extensive variable of state S, the entropy, and an intensive variable T, the absolute temperature. The entropy difference between two states A and B is given by

$$S_B - S_A = \int_A^B \frac{dQ}{T}$$

where the integral refers to any reversible process leading from A to B, and T is identical with the temperature defined by the perfect gas law.

(*b*) For a *closed* system, i.e., one that exchanges neither heat nor work with the surroundings, S increases in any spontaneous process. The system has reached equilibrium if S has reached a maximum.

It is possible to show that, for any natural, i.e., irreversible process, $dS > \frac{dQ}{T}$. Indeed, in the irreversible change of a closed system, $dQ = 0$ but $dS > 0$. Thus, if we allow an *arbitrary process*, we can rewrite Eq. 1·38:

$$S_B - S_A \geqq \int_A^B \frac{dQ}{T} \tag{1·42}$$

For a perfect gas, S or the specific entropy s can be given explicitly as a function of V and T, or p and T. Like E and H, S is determined only up to an additive constant:

$$S = \int \frac{dE + p\,dV}{T} = M \int \frac{de + p\,dv}{T} + \text{const.}$$

Hence

$$\frac{S}{M} = s = \int c_v \frac{dT}{T} + R \ln v + \text{const.} \tag{1·43a}$$

From $S = \int \frac{dH - V\,dp}{T}$, we also obtain

$$\frac{S}{M} = s = \int c_p \frac{dT}{T} - R \ln p + \text{const.} \tag{1·43b}$$

The above relations are for a thermally perfect gas. If it is also calorically perfect (constant c_p and c_v), the equations may be written

$$s - s_1 = c_p \ln T/T_1 - R \ln p/p_1 \tag{▶1·43c}$$

$$s - s_1 = c_v \ln T/T_1 + R \ln v/v_1 \tag{▶1·43d}$$

where (p_1, v_1, T_1) are the conditions at some reference state.

1·10 The Canonical Equation of State. Free Energy and Free Enthalpy

For a small reversible change of state, we have $T\,dS = dQ$ and thus:

$$dE = T\,dS - p\,dV \tag{1·44a}$$

or

$$dH = T\,dS + V\,dp \tag{1·44b}$$

Equations 1·44*a* and 1·44*b* involve now *only variables of state*, since dQ has been replaced by $T\,dS$. From these equations it becomes clear that the natural choice of variables for E are S and V, and for H, S and p. Thus

Eq. 1·44 implies:

$$\left(\frac{\partial E}{\partial S}\right)_V = T; \quad \left(\frac{\partial E}{\partial V}\right)_S = -p \tag{1·45a, b}$$

$$\left(\frac{\partial H}{\partial S}\right)_p = T; \quad \left(\frac{\partial H}{\partial p}\right)_S = V \tag{1·46a, b}$$

If $E(S, V)$ is known for a simple system, Eq. 1·45 yields *both* the caloric and thermal equations of state. The same is true if H is known as a function $H(S, p)$. S and T and V and p are called *conjugate variables*. The relations

$$E = E(S, V)$$

$$H = H(S, p)$$

are sometimes called "canonical equations of state"; each one suffices to describe a simple system completely. The second form is used in the so-called *Mollier diagram*, where a process is represented in a plane with H and S as coordinates. The Mollier diagram is very convenient for a graphical representation of a flow process. For instance, fluid flow through a heat-insulated duct, like a wind tunnel, is partially isentropic (S = const.) and partially isenthalpic (H = const.). As an example for the canonical equation of state, consider again the thermally and calorically perfect gas. Here, as follows easily from Eqs. 1·43*c* and 1·26*a*, the canonical equation written for unit mass is

$$h = \text{const. } c_p \exp (s/c_p) p^{R/c_p} \tag{1·47}$$

Equations 1·46*a* and 1·46*b* thus give

$$T = \frac{\partial h}{\partial s} = \text{const. } \exp (s/c_p) p^{R/c_p} = \frac{h}{c_p} \tag{1·48a}$$

$$v = \frac{\partial h}{\partial p} = \text{const. } R \exp (s/c_p) p^{(R/c_p)-1} = \frac{Rh}{c_p p} \tag{1·48b}$$

Hence Eq. 1·48*a* yields the caloric equation of state $h = c_pT +$ const., and Eq. 1·48*b* yields the thermal equation of state $pv = RT$.

It is not always practical to use S and V, or S and p, as independent variables. It is natural to ask whether one can construct functions related to E, S, and H which have V, T and p, T as natural variables. A similar consideration led us already to introduce H in Eq. 1·20.

We introduce F, the so-called *free energy* (sometimes called available energy or work function), and G the *free enthalpy* (sometimes called Gibbs' free energy, and also thermodynamic potential) defined by:

$$F = E - TS \tag{1·49a}$$

$$G = H - TS \tag{1·49b}$$

From Eqs. 1·44*a* and 1·44*b* we thus find:

$$dF = dE - S\,dT - T\,dS = -S\,dT - p\,dV \qquad (1\cdot50a)$$

$$dG = dH - S\,dT - T\,dS = -S\,dT + V\,dp \qquad (1\cdot50b)$$

Hence the natural variables are V, T for F and p, T for G. We have, analogous to Eq. 1·45,

$$\left(\frac{\partial F}{\partial T}\right)_V = -S; \quad \left(\frac{\partial F}{\partial V}\right)_T = -p \qquad (1\cdot51a)$$

$$\left(\frac{\partial G}{\partial T}\right)_p = -S; \quad \left(\frac{\partial G}{\partial p}\right)_T = V \qquad (1\cdot51b)$$

For example, the specific free enthalpy of a perfect gas is:

$$g = \int c_p\,dT - T\int c_p\frac{dT}{T} + RT\ln p - Ts_0 + h_0 \qquad ▶(1\cdot52)$$

as follows immediately from Eqs. 1·26 and 1·43*b*.

1·11 Reciprocity Relations

Often very useful a relation between the caloric and thermal equations of state is obtained easily from the differential form of the second law:

$$T\,ds = dh - v\,dp \qquad (1\cdot53)$$

From $s = s(p, T)$ it follows that

$$ds = \frac{\partial s}{\partial T}dT + \frac{\partial s}{\partial p}dp$$

and similarly, since $h = h(p, T)$,

$$dh = \frac{\partial h}{\partial T}dT + \frac{\partial h}{\partial p}dp$$

hence from Eq. 1·53

$$\frac{\partial s}{\partial T} = \frac{1}{T}\frac{\partial h}{\partial T} \qquad (1\cdot54a)$$

$$\frac{\partial s}{\partial p} = \frac{1}{T}\left(\frac{\partial h}{\partial p} - v\right) \qquad (1\cdot54b)$$

Equations 1·54 are called "reciprocity relations." From Eqs. 1·54 we can eliminate s by differentiating Eq. 1·54*a* with respect to p, Eq. 1·54*b* with respect to T, and subtracting; thus we get:

$$\frac{1}{T}\frac{\partial^2 h}{\partial p\,\partial T} - \frac{\partial}{\partial T}\frac{1}{T}\left(\frac{\partial h}{\partial p} - v\right) = 0$$

or

$$\frac{\partial h}{\partial p} = v - T\frac{\partial v}{\partial T} \tag{1·55}$$

Equation 1·55 relates the caloric and thermal equations of state.

For instance, using the perfect gas as a test case,

$$v = \frac{RT}{p}$$

and hence

$$\frac{\partial h}{\partial p} = \frac{RT}{p} - T\frac{R}{p} = 0$$

consequently

$$h = h(T)$$

Relations similar to Eq. 1·55 can also be written for $e(v, T)$.

1·12 Entropy and Transport Processes

Irreversible changes of state necessarily involve currents in the system. The increase in entropy during an irreversible process must thus be related to these currents.

One may proceed beyond classical thermodynamics and inquire whether it is possible to define the *rate of entropy increase* during the irreversible process. One can say that, during the process, entropy is produced and left in the system. In this way, one can define a continuity equation for the specific entropy, in much the same way that one obtains a continuity equation for the density. Namely, one can formulate a relation for the rate of change of specific entropy in a fixed volume owing to two contributions: the net *flux* of entropy into the volume and the *production* of entropy inside the volume. Special examples of this procedure will be left for later chapters. Here we are mainly interested in the production term. Clearly, the production of entropy must depend on the "currents," since it vanishes if the currents vanish. The currents, on the other hand, depend on the spatial rate of variation of variables of state, such as temperature, mass concentration, etc., and on transport parameters, such as heat conductivity, diffusivity, etc.

The formal relations of entropy production have to satisfy at least the following requirements:

(*a*) The production must vanish if the currents vanish.

(*b*) The production must be positive.

For the simplest case of small one-dimensional currents in an isotropic medium (which includes all fluids of classical fluid mechanics), this leads to the following result. If σ denotes the rate of entropy change per unit time and unit volume, then, for the case of heat conduction,

$$\sigma = \frac{k}{T^2}\left(\frac{dT}{dx}\right)^2$$

where k denotes the coefficient of heat conductivity. The entropy production due to variations of momentum in the direction of flow, as in sound waves or shock waves, is given by

$$\sigma = \frac{\tilde{\mu}}{T}\left(\frac{du}{dx}\right)^2$$

where $\tilde{\mu}$ denotes a coefficient of viscosity.

1·13★ Equilibrium Conditions

A system has reached a state of stable equilibrium if no further spontaneous processes are possible. For a *spontaneous process*, $dS \geq \frac{dQ}{T}$. Hence the system is in a stable equilibrium if for any process:

$$\delta S \leq \frac{\delta Q}{T} \tag{1·56}$$

or, using the first law,

$$T\,\delta S - (\delta E + p\,\delta V) \leq 0 \tag{1·57a}$$

$$T\,\delta S - (\delta H - V\,\delta p) \leq 0 \tag{1·57b}$$

The notation δQ, δS, etc., refers to a so-called "virtual" variation, i.e., a small change of the variables compatible with the constraints of the system. From Eq. 1·57 follow specific conditions for the "thermodynamic potentials" S, E, H, F, and G for various constraints.

1. For a closed system, e.g., the expanding gas in a fixed, heat-insulated vessel, $\delta E = 0$, $\delta V = 0$, and hence

$$\delta S \leq 0 \tag{▶1·58}$$

i.e., the entropy reaches a maximum. (For any virtual process S can only decrease; hence S has a maximum.)

2. If V and T or p and T are kept constant, we introduce $F = E - TS$ and $G = H - TS$. It follows from Eq. 1·57 that in equilibrium

$$\delta F \geq 0 \quad \delta T = 0 \quad \delta V = 0 \tag{▶1·59}$$

$$\delta G \geq 0 \quad \delta T = 0 \quad \delta p = 0 \tag{▶1·60}$$

Free energy F and free enthalpy G have a minimum of equilibrium under the respective constraints.

These two cases are the most important cases for our purposes. Others are, of course, possible.

The study of thermodynamic equilibria was developed by Gibbs, who patterned it after the mechanical equilibrium conditions.

1·14★ Mixtures of Perfect Gases

The working fluid in aerodynamics is almost always a mixture of gases, and the range of variables of state is usually such that the gases are nearly perfect. Hence mixtures of perfect gases represent the most important thermodynamic systems for aeronautical applications.

Such a mixture is a *homogeneous system* if in equilibrium all variables of state are uniform throughout the system. But the mixture is *not a simple system* since the masses x_i of the component gases need to be specified in addition to the usual variables of state of a simple system.

Thus we shall have, for example,

$$E = E(S, V, x_i) \tag{1·61a}$$

$$H = H(S, p, x_i) \tag{1·61b}$$

If the component gases are *inert*, no reactions will occur between them and hence for a given mass of mixture the x_i will be constants, independent of temperature and pressure. In this case, the mixture will behave like a *single perfect gas*. For example, air is a mixture of N_2, O_2, and traces of A, CO_2, etc. Over a fairly wide range of temperatures and pressures, the component gases are inert and air can be treated as a single perfect gas.

On the other hand, any diatomic gas (or, of course, any polyatomic gas) will *dissociate* at high temperatures and form a *reacting mixture*. Thus nitrogen at high temperatures will consist of a mixture of N_2 and N and the masses x_{N_2} and x_N present will depend upon the pressure and temperature. The same is true for chemically reacting gases and also for the process of ionization.

Now Eq. 1·61*a* or Eq. 1·61*b* gives a complete thermodynamic description of the mixture provided we know the x_i as functions of, say, p and T. This relation is furnished by the so-called *law of mass action* which follows from the equilibrium conditions of the previous article, as will be shown presently.

To apply the equilibrium conditions to a mixture, we have first to see how the variables of state of the compound system are obtained from those of the component gases.

A given mass of gas brought into a volume V will fill it completely; the pressure will adjust itself accordingly. This is characteristic for a gas. For a *perfect gas*, this filling of a volume and the adjustment of the pressure are *independent of the presence* of any other perfect gases in V. Hence, in a mixture of perfect gases, one can assign a partial pressure p_i to the ith component gas. p_i will satisfy the perfect gas equation:

$$p_i = \rho_i R_i T = \frac{x_i}{V} R_i T \tag{1·62}$$

and the total pressure p of the mixture is equal to the sum of the partial pressures:

$$p = \sum p_i \tag{1·63}$$

Equations 1·62 and 1·63 are known as *Dalton's law.*

Now the specific energy and enthalpy of the ith gas depend only on the temperature, which is the same for all the gases. Consequently E and H of the mixture are:

$$E = \sum E_i = \sum x_i e_i(T) \tag{1·64a}$$

$$H = \sum H_i = \sum x_i h_i(T) \tag{1·64b}$$

The specific entropies s_i depend upon volume and temperature or pressure and temperature. The entropy S of the mixture becomes

$$S = \sum S_i = \sum x_i s_i(v_i, T) \tag{1·65}$$

or

$$S = \sum S_i = \sum x_i s_i(p_i, T) \tag{1·66}$$

Hence, in forming the entropy of the mixture as the sum of the component entropies, we have to be a little careful as to which variables to choose. V and T are the same for all gases in the mixture; but $v_i = V/x_i$ and p_i are not the same. If we want to choose pressure and temperature as variables, we have to take into account that each gas has its own partial pressure p_i. Equation 1.66 will be the more convenient for us to use.

Since we do know H_i and S_i explicitly from Eqs. 1·26 and 1·43*b*, we have:

$$H = \sum x_i \left[\int (c_p)_i \, dT + (h_0)_i \right] \tag{1·67}$$

$$S = \sum x_i \left[\int (c_p)_i \frac{dT}{T} - R_i \ln p_i + (s_0)_i \right] \tag{1·68}$$

For the derivation of the law of mass action, we shall need the free enthalpy $G = H - TS$, which follows from Eqs. 1·67 and 1·68 directly:

$$G = \sum x_i [\omega_i(T) + R_i T \ln p_i] \tag{1·69}$$

where ω_i is written short for

$$\omega_i \equiv \int (c_p)_i \, dT - T \int (c_p)_i \frac{dT}{T} + (h_0)_i - (s_0)_i T \tag{1·69a}$$

1·15★ The Law of Mass Action

Consider now a mixture of reacting perfect gases. A reaction is described by symbolic relations like:

$$2H_2 + O_2 \leftrightarrows 2H_2O$$

or

$$2N \leftrightarrows N_2 \quad \text{etc.}$$

For example, two molecules of hydrogen plus one molecule of oxygen form two molecules of water vapor. If x_1, x_2, x_3 are the masses of H_2, O_2, and H_2O, respectively, we see that a change dx_1 is not independent of dx_2 and dx_3, since the reaction relation has to be satisfied. Call m_1, m_2, m_3 the masses of the H_2, O_2, and H_2O molecules and introduce a variable λ such that:

$$dx_1 = 2m_1\,d\lambda$$

$$dx_2 = m_2\,d\lambda$$

$$dx_3 = -2m_3\,d\lambda$$

Then any change in λ gives the proper relation between dx_1, dx_2, and dx_3 for the H_2O reaction. In general, we can thus introduce λ by:

$$dx_i = \nu_i m_i\,d\lambda \tag{1·70}$$

where ν_i stands for the integers in the reaction relation.

We are now ready to apply the equilibrium condition. We choose as potential the free enthalpy G. Equilibrium exists if, for a given pressure and temperature, G has a minimum. Now $G = G(p, T, x_i)$ but p and T are fixed. Furthermore the x_i can be expressed in terms of the known constants ν_i and m_i and the variable λ. Hence the equilibrium condition becomes:

$$\frac{dG}{d\lambda} = 0 \tag{1·71}$$

Since G is given by Eq. 1·69,

$$G = \sum x_i[\omega_i(T) + R_iT \ln p_i]$$

we have

$$\frac{dG}{d\lambda} = \sum \left\{\frac{dx_i}{d\lambda}[\omega_i(T) + R_iT \ln p_i] + x_iR_iT\frac{d}{d\lambda}\ln p_i\right\} \tag{1·72}$$

The last term has to be written since the p_i depends on the x_i and hence on λ (Eq. 1·62), but it vanishes because

$$\sum x_iR_iT\frac{d \ln p_i}{d\lambda} = \sum \frac{x_iR_iT}{p_i}\frac{dp_i}{d\lambda}$$

but

$$p_iV = x_iR_iT \quad \text{and} \quad p = \sum p_i$$

Thus

$$\sum \frac{x_iR_iT}{p_i}\frac{dp_i}{d\lambda} = V\sum \frac{dp_i}{d\lambda} = V\frac{dp}{d\lambda} = 0$$

because $p = \text{const.}$

Thus we have finally, from Eqs. 1·72 and 1·71 with Eq. 1·70,

$$\sum [\nu_i m_i\omega_i(T) + \nu_i m_i R_i T \ln p_i] = 0$$

or, since $R_i = \mathbf{k}/m_i$,

$$\sum \ln p_i^{\nu_i} = -\frac{1}{\mathbf{k}T}\sum \nu_i m_i \omega_i(T) \tag{1·73}$$

which is the *law of mass action*. Changing to the exponential, Eq. 1·73 becomes

$$p_1^{\nu_1}p_2^{\nu_2}p_3^{\nu_3}\cdots \equiv \prod p_i^{\nu_i} = K(T) \tag{1·74}$$

$$K(T) = \exp\frac{-\sum \nu_i m_i \omega_i(T)}{\mathbf{k}T}$$

The derivation of Eq. 1·74 is due to Gibbs. Sometimes a so-called "standard state" is introduced; i.e., a reference pressure $\tilde{p}_i$ is chosen for each gas. We can then subtract the expression $\nu_i \ln \tilde{p}_i$ from each side of Eq. 1·73. Thus we find

$$\sum \ln\left(\frac{p_i}{\tilde{p}_i}\right)^{\nu_i} = -\frac{1}{\mathbf{k}T}\sum \nu_i m_i[\omega_i(T) + R_iT\ln \tilde{p}_i] \tag{1·73a}$$

But

$$\omega_i(T) + R_iT\ln\tilde{p}_i = g(T, \tilde{p}_i)$$

which is the free enthalpy evaluated at T and $\tilde{p}_i$, the "standard" free enthalpy. Thus Eq. 1·73*a* is sometimes written in the literature

$$\sum \ln\left(\frac{p_i}{\tilde{p}_i}\right)^{\nu_i} = -\frac{\Delta\tilde{g}}{\mathbf{k}T} \tag{1·73b}$$

where $\Delta\tilde{g}$ is short for $\sum \nu_i m_i g(T, \tilde{p}_i)$. For example, one may choose $\tilde{p}_i$ equal to 1 atmosphere for all i. Then

$$\sum \nu_i \ln p_i = -\frac{\Delta\tilde{g}}{\mathbf{k}T} \tag{1·73c}$$

where the p_i are measured in atmospheres and $\tilde{g}$ refers to the "standard free enthalpy at 1 atmosphere."

1·16★ Dissociation

As a special, very important example, consider the dissociation of a diatomic gas like oxygen. At sufficiently high temperature ($\approx 3000°K$), the gas will consist of both O_2 and O and will be a reacting gas mixture.

Let x_1 and x_2 be the masses of O and O_2 present in the mixture. Consider unit mass of mixture, i.e., $x_1 + x_2 = 1$, and introduce the "degree of dissociation" $\alpha = \dfrac{x_1}{x_1 + x_2}$. Thus:

$$x_1 = \alpha \quad x_2 = 1 - \alpha$$

The enthalpy h of the mixture is thus given in terms of h_1 and h_2, the specific enthalpies of O and O_2, by:

$$h = \alpha h_1 + (1 - \alpha)h_2 = h_2 + \alpha(h_1 - h_2) \tag{1·75}$$

Equation 1·75 is thus the caloric equation of state for the mixture. It gives the important result, $h = h(p, T)$ because $\alpha = \alpha(p, T)$! *This mixture of perfect gases does not behave like a perfect gas,* for which h depends on T only. The thermal equations of state are obtained from

$$\begin{aligned} p_1 &= \frac{x_1}{v} R_1 T = \frac{\alpha}{v} R_1 T \\ p_2 &= \frac{x_2}{v} R_2 T = \frac{1 - \alpha}{v} R_2 T \end{aligned} \tag{1·76}$$

Since the molecular mass of O_2 is twice the molecular mass of O, we have $R_1 = 2R_2$ and hence from Eq. 1·76

$$p = p_1 + p_2 = \frac{R_2 T}{v}(1 + \alpha) \tag{1·77}$$

Hence the thermodynamic behavior of the mixture is determined as soon as $\alpha(p, T)$ is known. α is given by the law of mass action, Eq. 1·74. The reaction equation is

$$2O \leftrightarrows O_2$$

hence $\nu_1 = 2$; $\nu_2 = -1$, and consequently Eq. 1·74 becomes†

$$\frac{{p_1}^2}{p_2} = K(T)$$

and, using Eqs. 1·76 and 1·77, we have

$$\frac{4\alpha^2}{1 - \alpha^2} = \frac{K(T)}{p} \tag{1·78}$$

†A simpler derivation of Eq. 1·78 may be found in Exercise 1·10.

$K(T)$ is given by Eq. 1·74 and hence is a known function of T. For our purposes here it is sufficient to note that K can be tabulated or plotted for any reaction of the type considered. Knowing K, we have in Eqs. 1·75, 1·77, and 1·78 a sufficient number of equations to derive any thermodynamic property of the mixture.

Two more general relations can be easily derived for this specific example:

(*a*) The *heat of dissociation* l_D, i.e., the heat necessary to dissociate unit mass of O_2 at constant p and T is given by:

$$l_D = h_1 - h_2 \tag{1·79}$$

This follows immediately from the first law:

$$dq = dh - v\,dp = \frac{\partial h}{\partial T}\,dT + \left(\frac{\partial h}{\partial p} - v\right)dp + \frac{\partial h}{\partial \alpha}\,d\alpha$$

thus

$$l_D \equiv \left(\frac{dq}{d\alpha}\right)_{p,T} = \frac{\partial h}{\partial \alpha}$$

and from Eq. 1·75 one obtains Eq. 1·79.

(*b*)

$$\frac{d \ln K}{dT} = \frac{l_D}{R_2 T^2} \tag{1·80}$$

To derive Eq. 1·80, we use Eqs. 1·75, 1·77, and the reciprocity relation (Eq. 1·55):

$$\frac{\partial h}{\partial p} = v - T\frac{\partial v}{\partial T}$$

We get first

$$l_D\frac{\partial \alpha}{\partial p} = -\frac{R_2 T^2}{p}\frac{\partial \alpha}{\partial T}$$

and, using Eq. 1·78, we obtain Eq. 1·80 directly.

Both Eqs. 1·79 and 1·80 can be obtained in general and are not restricted to this special problem.

The heat of dissociation l_D is a function of T. However its variation with T is often small enough to be negligible in first approximation. In this case Eq. 1·80 can be integrated and yields:

$$K = \text{const. } e^{-l_D/R_2 T} \tag{1·80a}$$

Figure 1·5 shows a plot of ln K versus $1/T$ for O_2. The curve is very nearly a straight line and hence Eq. 1·80*a* is a good approximation. l_D/R_2 has evidently the dimension of a temperature and hence can be called a *characteristic temperature for dissociation*, θ_D.

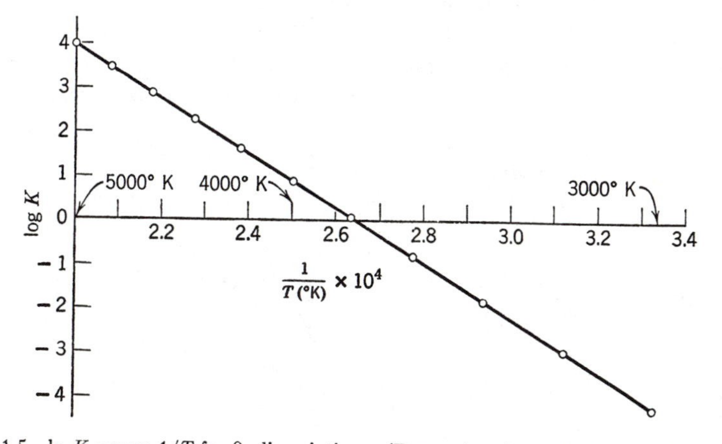

FIG. 1·5 ln K versus $1/T$ for 0_2 dissociation. (Data taken from H. W. Woolley, "Effect of Dissociation on Thermodynamic Properties of Pure Diatomic Gases," *NACA Tech. Note* 3270.)

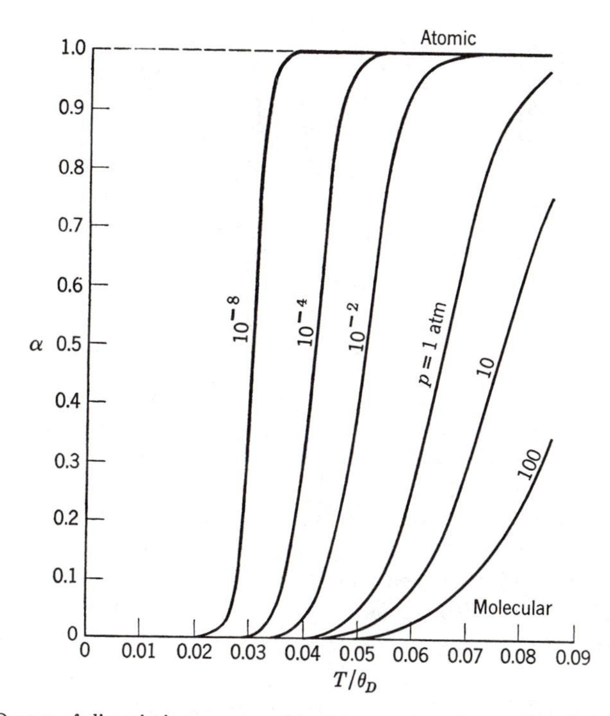

FIG. 1·6 Degree of dissociation versus reduced temperature for a dissociating diatomic gas. (Data taken from *NACA Tech. Note* 3270.)

If the variables of state of a dissociating diatomic gas are plotted as functions of p and T, the plot differs for different gases. If instead of T a reduced temperature θ_D/T is introduced, a single graph can be used approximately for all diatomic gases. Figures 1·6 and 1·7 show typical plots for α and h.

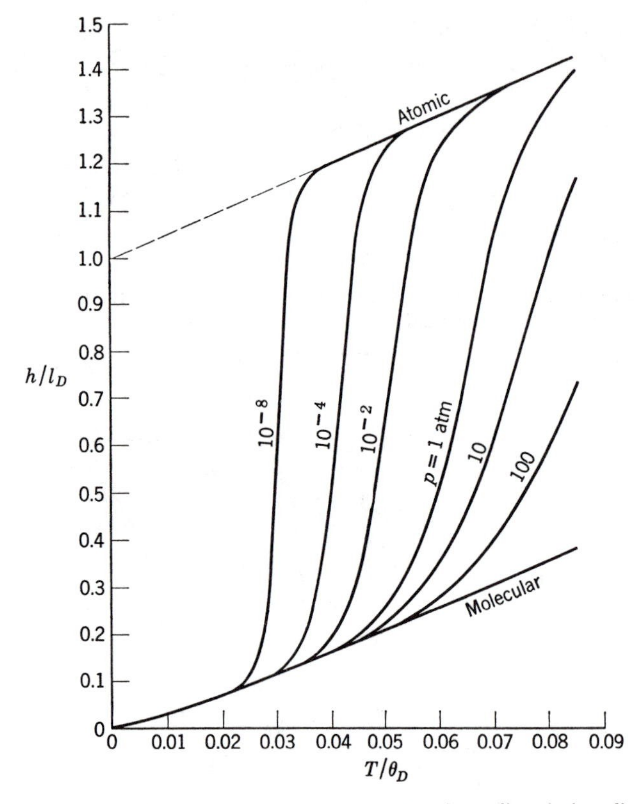

FIG. 1·7 Reduced enthalpy versus reduced temperature for a dissociating diatomic gas. (Data taken from *NACA Tech. Note* 3270.)

Applications of the mass law to dissociation of polyatomic gases or to cases where neutral component gases are also present follow the same lines. Finally the important problem of *ionization* can be—to a good approximation—handled like a *dissociation* problem.

For example, assume that at a temperature still higher than dissociation we deal with atomic oxygen, i.e., with the gas O. If the temperatures become sufficiently high, the O atom may lose an electron ϵ and the gas becomes a mixture of O^+, ϵ, and O, i.e., of *ionized* O, electrons, and neutral O. At these high temperatures all three can be treated as perfect gases, and hence we have a reaction equation:

$$O^+ + \epsilon \leftrightarrows O$$

to which the mass action law, and the same considerations as used above, again apply.

1·17★ Condensation

In some applications, specifically in the design of high-speed wind tunnels, gases at comparatively low temperatures are also important. Hence a *heterogeneous system* composed of a gaseous and a liquid phase of a substance is sometimes of interest in gasdynamics.

The corresponding thermodynamic relations can be obtained from the equilibrium conditions. This conventional derivation is left for the exercises. Here we look at the problem in a different way: In gasdynamics we usually deal with a condensing gas in which the liquid phase is formed in minute droplets carried along with the stream, and in which the vapor is nearly a perfect gas. In such a case *we can consider these droplets like the molecules of a very heavy gas* and again apply the law of mass action to the "reaction" between the gas and the droplets; e.g., for condensing nitrogen, N_2, the "reaction equation" is:

$$nN_2 = D$$

where D stands for "droplet" and n is a very large number, even for the small droplets of interest in gasdynamics.

If we apply the law of mass action (Eq. 1·74) to this reaction, we have

$$\frac{p_1{}^n}{p_2} = K(T) \tag{1·81}$$

or in logarithmic form:

$$n \ln p_1 - \ln p_2 = \ln K(T) = -\frac{1}{\mathbf{k}T}(nm_1\omega_1 - m_2\omega_2)$$

where p_1, m_1, ω_1, refer to the N_2 gas; p_2, m_2, ω_2 to the "droplet" gas. Now $m_2 = nm_1 = nm$ say, and hence

$$n \ln p_1 - \ln p_2 = -\frac{n}{RT}(\omega_1 - \omega_2) \tag{1·82}$$

For large n the second term becomes negligible, and $p = p_1 + p_2 \doteq p_1$. Thus, dividing by n, we find:

$$\ln p = -\frac{\omega_1 - \omega_2}{RT} = \frac{1}{n}\ln K$$

$$\frac{d \ln p}{dT} = \frac{1}{n}\frac{d \ln K}{dT} = \frac{l_v}{RT^2} \tag{▶1·83}$$

Here l_v is the "*heat of vaporization*" introduced exactly like the heat of dissociation in Eqs. 1·79 and 1·80. Equation 1·83 is one form of the Clapeyron-Clausius equation for the pressure in phase equilibrium. The latent heat l_v now is a measure for the energy required to break one of the droplets up into its component molecules. The similarity with the dissociation and ionization process is thus apparent, and this is the reason for the approach given here. In fact, l_v as defined here will include surface tension effects which are usually introduced separately. Enthalpy, entropy, etc., can be obtained exactly as in the dissociation process.

1·18 Real Gases in Gasdynamics

Most explicit results in the theory of compressible fluid flow are worked out for a perfect gas with constant specific heat. This is done because the simplicity of the working fluid makes it possible to bring out the physics of the flow without much cumbersome computational detail. The thermally and calorically perfect gas is, however, an idealization, and real gases will deviate from it to a greater or lesser degree.

We shall devote this last article of Chapter 1 to a brief résumé of these so-called "real gas effects." Fortunately the number of gases of interest in aeronautical application is comparatively small, if we exclude combustion. Usually one has to deal with gases like O_2, N_2, NO, He, etc., seldom with H_2O and CO_2 and very rarely with more complex gases than the last two. Usually the main interest lies with comparatively high temperatures and low pressures. Thus the discussion of "real gas effects" is much simplified since it can be restricted to a few gases and a limited range of pressure and temperature.

The Thermal Equation of State. The equation of state of a real gas can be written formally in terms of the "compressibility factor" Z as:

$$\frac{pv}{RT} = Z(p, T) \tag{1·84}$$

Thus $Z = 1$ gives the perfect gas relation. Deviations of Z from unity are mainly due to two effects: At *low temperatures* and *high pressures*, the intermolecular forces become important. These are the so-called van der Waals forces which account for the possibility of liquefying a gas. At *high temperatures* and *low pressures*, dissociation and ionization processes occur and Z differs from unity because these processes change the number of particles; with this formulation, R is a constant reference value (cf. Eq. 1·77).

The effect of the van der Waals forces can be expressed in first approximation in terms of the so-called "second virial coefficient," and Eq. 1·84 can be written

$$\frac{pv}{RT} = Z = 1 + b(T)\frac{p}{RT} \tag{1·85}$$

The correction term bp/RT differs from gas to gas because b/R is evidently a characteristic function for a particular gas. However, for the limited number of gases with which we are concerned, it is possible to introduce dimensionless variables and rewrite Eq. 1·85 in a universal form.

Let T_c, p_c, v_c denote the critical temperature, pressure, and specific volume. Note that $b(T)$ has the dimension of a volume. Equation 1·85 can be written in dimensionless variables:

$$\frac{pv}{RT} = 1 + \frac{p_c v_c}{RT_c}\frac{p}{p_c}\frac{T_c}{T}\frac{b(T)}{v_c} \tag{1·86}$$

Now $p_c v_c/RT_c = \kappa$ is a constant with very nearly the same value, $\kappa = 0.295$, for the gases of interest (see also table at the end of the book). b/v_c is nearly a universal function of T_c/T for these gases. Consequently we can write Eq. 1·86 in the convenient form

$$\frac{pv}{RT} = 1 + \frac{p}{p_c}\phi\left(\frac{T_c}{T}\right) \tag{1·87}$$

$\phi(T_c/T)$ is the same function for all our gases. For numerical estimates the following values of ϕ are convenient:†

T_c/T	0.1	0.2	0.4	0.6	0.8
ϕ	0.009	0.015	−0.005	−0.067	−0.18

Values for p_c, T_c are given in the tables at the end of the book.

The equation of state of a dissociating, diatomic gas was given in Article 1·16 (Eq. 1·77):

$$\frac{pv}{RT} = 1 + \alpha \tag{1·88}$$

Hence Z is here simply related to the *degree of dissociation* α which in turn is a function of both p and T. The characteristic temperature for dissociation was defined in terms of the heat of reaction l_D by $\theta_D = l_D/R$. (The same definition could have been used for the effects due to van der Waals' forces in terms of the heat of vaporization l_v; it is, however, customary to use the critical data in this case.)

Now l_D is related to the force with which *the atoms are held within* the molecule; l_v, to the force *between molecules*. Consequently $l_D \gg l_v$, and

†The values for b/v_c are taken from p. 140, E. A. Guggenheim, *Thermodynamics*, North Holland Publishing Co., Amsterdam, 1950.

hence the temperature range in which dissociation effects occur is usually far removed from the range where the van der Waals forces are important. Thus in a given problem in gasdynamics one has rarely to worry about possible simultaneous effects of both.

The same reasoning applies to ionization, where the forces that bind an electron in an atom and the corresponding ionization energy l_i are involved.

The Caloric Equation of State. For a perfect gas the caloric equation of state is

$$h = \int c_p \, dT + h_0 \tag{1·89}$$

and c_p and c_v are related by

$$c_p - c_v = R \tag{1·90}$$

Equations 1·89 and 1·90 are direct consequences of $pv = RT$. The variation of c_p with T cannot be given by thermodynamics but has to come from experiments or statistical mechanics. Classical statistical mechanics leads to a simple expression for c_p in terms of n_f, the "number of degrees of freedom" of the appropriate molecular model (cf. Chapter 14).

$$c_p = \frac{n_f + 2}{2} R \tag{1·91}$$

For a smooth sphere or a mass point, $n_f = 3$. This is a good model for monatomic gases such as He, A, etc., and indeed experiments show that for such gases

$$c_p = \tfrac{5}{2}R$$

over a very wide range of temperatures, from near condensation up to ionization. Thus monatomic gases are very nearly calorically perfect.

The simplest model for a diatomic molecule is a rigid dumbbell, with $n_f = 5$. Diatomic gases (and also "linear" triatomic gases like CO_2) near room temperature also have specific heats corresponding to Eq. 1·91 with $n_f = 5$, i.e.,

$$c_p = \tfrac{7}{2}R$$

However at high temperatures c_p increases, with T, above $\frac{7}{2}R$,† because the atoms in a diatomic molecule are not rigidly bound but can vibrate. Classically, this vibration would add 2 degrees of freedom, and we would expect to find

$$c_p = \tfrac{9}{2}R$$

†At low temperatures, c_p can *decrease* below $\frac{7}{2}R$ because the rotational degrees of freedom "freeze." However, for N_2, O_2, etc., the characteristic temperature for rotation is in the neighborhood of 2° K and hence out of the range of interest here. Rotational effects are important only for H_2.

The temperature at which the specific heat changes from $\frac{7}{2}$ to $\frac{9}{2}R$, as well as the function $c_p(T)$ in the transition region, is beyond classical statistical mechanics, but can be given by quantum statistical mechanics. The specific heat for a diatomic gas in the range of temperatures between dissociation and the region of van der Waals' effects is closely represented by:

$$\frac{c_p}{R} = \frac{7}{2} + \left[\frac{\theta_v/2T}{\sinh\,(\theta_v/2T)}\right]^2 \tag{1·92}$$

where θ_v denotes a characteristic temperature for the vibrational energy. (Values of θ_v for different diatomic gases are given in Table I at the back of the book.) Equation 1·92 shows that for $T \ll \theta_v$, $c_p/R \rightarrow 7/2$, and for $T \gg \theta_v$, $c_p/R \rightarrow 9/2$ as expected. The statistical origin of Eq. 1·92 is outlined in Article 14·7.

Triatomic gases like H_2O show a similar behavior; however they start with 6 degrees of freedom, and hence the constant term is $\frac{8}{2} = 4$, and more than one mode of vibration are usually important.

Finally, any deviation in the thermal equation of state from the perfect gas relation affects the caloric equation of state, since both are related by the relation Eq. 1·55

$$\frac{\partial h}{\partial p} = v - T\left(\frac{\partial v}{\partial T}\right)_p$$

Using the equation of state in the form of Eq. 1·84, we have:

$$\frac{\partial h}{\partial p} = -\frac{RT^2}{p}\left(\frac{\partial Z}{\partial T}\right)_p \tag{1·93}$$

Furthermore the difference $c_p - c_v \neq R$, because from Eq. 1·22

$$c_p - c_v = \left(v - \frac{\partial h}{\partial p}\right)\left(\frac{\partial p}{\partial T}\right)_v$$

and thus

$$c_p - c_v = R\,\frac{\left(Z + T\dfrac{\partial Z}{\partial T}\right)^2}{Z - p\dfrac{\partial Z}{\partial p}} \tag{1·94}$$

This brief discussion indicates the limits of the applicability of the perfect gas approximation.

Summary. (1) At low temperatures and high pressures, a gas becomes thermally and calorically imperfect due to the intermolecular forces. Then $Z \neq 1$ and $h = h(p, T)$.

(2) At high temperatures and low pressures, $Z = 1$ and $h = h(T)$, but

$c_p = c_p(T)$, because the contribution to c_p from the vibrational modes depends on the temperature.

(3) At still higher temperature, $Z \neq 1$ and $h = h(p, T)$ because of dissociation and ionization, i.e., because of processes that change the number of particles.

(4) A gas is both thermally and calorically perfect if $T_c \ll T \ll \theta_v$ and $p \ll p_c$.

(5) For monatomic gases the effects due to vibrational modes and dissociation are absent.

CHAPTER 2

One-Dimensional Gasdynamics

2·1 Introduction

We shall begin our study of the motion of compressible fluids with the case of *one-dimensional flow*. This definition applies to flow in a channel or tube, such as that illustrated in Fig. 2·1, which may be described by specifying the variation of the cross-sectional area along its axis, $A = A(x)$, and in which the flow properties are uniform over each cross-section, that is, $p = p(x)$, $\rho = \rho(x)$, etc. Similarly, the velocity u, which is normal to the cross-section, should be uniform over each section, $u = u(x)$. These quantities may also be functions of time t, if the flow is *nonstationary*, or *nonsteady*.

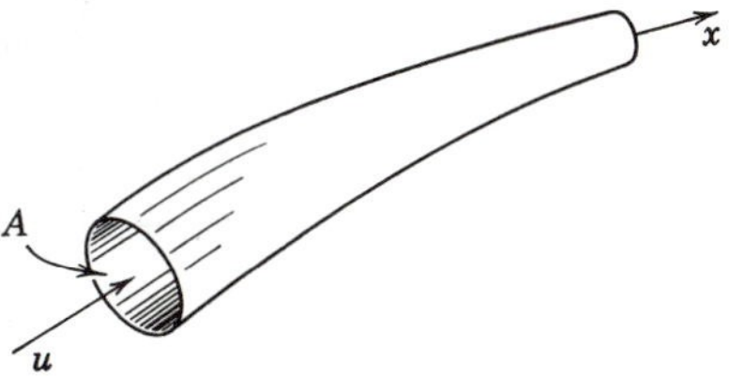

FIG. 2·1 One-dimensional flow in a stream tube.

These conditions are not as restrictive as they may appear. For instance, if there are sections over which the flow conditions are not uniform it is still possible to apply the results between sections where they are uniform, that is, one-dimensional. Even at nonuniform sections the results may often be applied to suitable mean values.

Furthermore, the one-dimensional results are applicable to the individual stream tubes of a general, three-dimensional flow, x being the coordinate along the stream tube.† We shall see in Chapter 7 what additional relations are needed for this application.

For an incompressible fluid, practically all the information about a one-dimensional flow is contained in the kinematic relation, "u is inversely proportional to A"; the pressure is obtained from the (independent) Bernoulli equation. In compressible flow, on the other hand, the variation of the density makes the continuity and momentum equations interdependent, and the relation between velocity and area is then not so simple.

†In Chapter 7 we use s for the streamline coordinate, but x appears to be more convenient here.

2·2 The Continuity Equation

If the flow in the tube of Fig. 2·1 is *steady*, the mass of fluid which passes a given section must eventually pass all the other sections farther downstream. This is simply a statement of the law of conservation of mass. At any two sections where conditions are uniform, the mass flows are equal, that is

$$\rho_1 u_1 A_1 = \rho_2 u_2 A_2 \tag{2·1}$$

This form of the continuity equation is very general, since it holds even if conditions *between* the sections are not uniform. If the flow *is* uniform at every section, the equation may be written

$$\rho u A = \text{const.} = m \tag{2·2}$$

and applied continuously along the tube. It may be differentiated, to give the differential form of the steady continuity equation,

$$\frac{d}{dx}(\rho u A) = 0 \tag{2·3}$$

If the flow is nonsteady, the continuity equation is obtained as follows. In Fig. 2·2 the mass contained between sections 1 and 2 is $\rho A \,\Delta x$, and in-

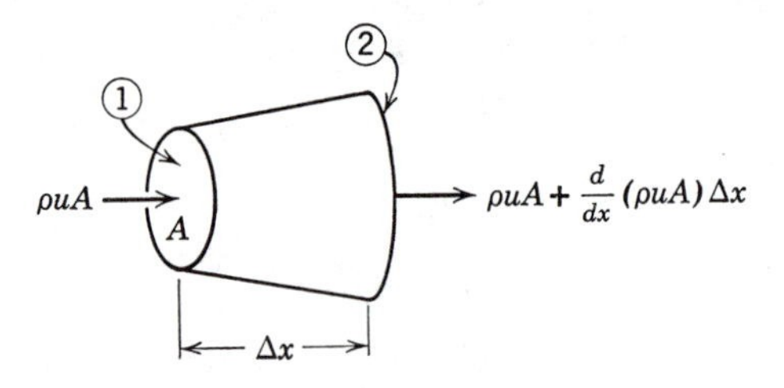

FIG. 2·2 Flow through a tube segment.

creases at the rate $\frac{\partial}{\partial t}(\rho A \,\Delta x)$. This must be equal to the flow through 1 minus the flow through 2, that is, the net inflow,†

$$-\frac{\partial}{\partial x}(\rho u A)\,\Delta x = \frac{\partial}{\partial t}(\rho A \,\Delta x)$$

Δx does not depend on time and may be divided through, to give

$$\frac{\partial}{\partial x}(\rho u A) + \frac{\partial}{\partial t}(\rho A) = 0 \tag{2·4}$$

†We are excluding cases in which the excess may be due to a *source* between the sections.

2·3 The Energy Equation

In Article 1·8, the first law of thermodynamics was used to relate the equilibrium conditions in a throttling process. This is a *flow process*, as compared to cases like the irreversible expansion experiment, also described in Article 1·8, in which the equilibrium states are static. In the latter case,

$$e_1 = e_2$$

whereas in the throttling process

$$h_1 = h_2$$

The difference between them is due to the "flow work" in the latter case. Thus *for a flowing fluid the basic thermodynamic quantity is the enthalpy*, rather than the internal energy.

In the throttling process of Article 1·8, the flow was assumed to be so slow that its kinetic energy was negligible. The extension to arbitrary flow speeds may be made by simply including the kinetic energy in the total energy of the fluid. The enthalpy, h, of the fluid is defined as the enthalpy that would be measured by an "observer" moving with the fluid. The condition of equilibrium is that relative to this observer there should be no currents of energy, momentum, etc. (cf. Article 1·5). It is also easy to include external heat addition in the equation and to allow area changes in the flow, and thus to obtain the energy equation for one-dimensional flow. The details, which may be compared with the Joule-Thomson process of Article 1·8, are as follows.

We select for the "system" a definite portion of fluid (Fig. 2·3), between sections 1 and 2. In the lower part of the figure, this system is shown bounded by pistons at 1 and 2, instead of fluid. These are equivalent to the fluid they replace, and they help to clarify the work done on the system, which consists of a definite mass of fluid.

During the small time interval in which the fluid is displaced, to a region bounded by sections 1′ and 2′, a quantity of heat, q, is added. According to the energy law,

$$q + \text{work done} = \text{increase of energy} \tag{2·5}$$

To compute the work, assume that the volume displaced at 1 is the specific volume v_1, corresponding to a *unit mass*. Then, for steady conditions, the displacement at 2 is also unit mass, that is, a volume v_2. The work done on the system by the pistons, during this displacement, is $p_1v_1 - p_2v_2$. (There might be additional external work, on a machine between 1 and 2, but this will be omitted.) We have then,

$$\text{Work done} = p_1v_1 - p_2v_2 \tag{2·6}$$

Finally, we have to compute the change in the energy of the system. The flowing fluid possesses not only *internal energy* e but also *kinetic energy* $\frac{1}{2}u^2$ (per unit mass). Thus its local energy is $e + \frac{1}{2}u^2$, per unit mass. Comparing the energy of the system after the displacement with that before, it

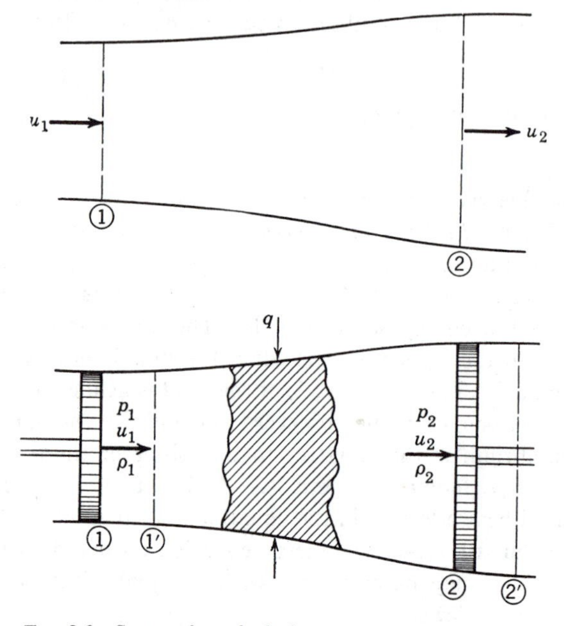

FIG. 2·3 System for calculating energy relations in flow.

will be seen that there has been an increase of energy, $e_2 + \frac{1}{2}u_2^2$, corresponding to the displacement from 2 to 2′, and a decrease, $e_1 + \frac{1}{2}u_1^2$, at 1. The net result is

$$\text{Increase of energy} = (e_2 + \tfrac{1}{2}u_2^2) - (e_1 + \tfrac{1}{2}u_1^2) \tag{2·7}$$

The energy equation for *steady flow*† is then

$$q + p_1v_1 - p_2v_2 = (e_2 + \tfrac{1}{2}u_2^2) - (e_1 + \tfrac{1}{2}u_1^2) \tag{2·8}$$

The notation may be simplified by introducing the enthalpy,

$$h = e + pv$$

Then

$$q = h_2 - h_1 + \tfrac{1}{2}u_2^2 - \tfrac{1}{2}u_1^2 \qquad \blacktriangleright(2·8a)$$

As shown in the figure, q is "external heat," which is added from outside the walls. (Heat of condensation, chemical reaction, etc., is included in h.)

†The equation for nonsteady flow is developed in Chapter 7 (see Eq. 7·25).

For $q = 0$, we have the *adiabatic energy equation*

$$h_2 + \tfrac{1}{2}u_2^2 = h_1 + \tfrac{1}{2}u_1^2 \qquad (2\cdot 9)$$

Equations 2·8 and 2·9 relate conditions at two *equilibrium states* (1) and (2) of the flow. *They are valid even if there are viscous stresses, heat transfer, or other nonequilibrium conditions between sections* 1 *and* 2, *so long as* (1) *and* (2) *themselves are equilibrium states.*

If equilibrium exists all along the flow, the equilibrium equation is valid continuously and may be written

$$h + \tfrac{1}{2}u^2 = \text{const.} \qquad (2\cdot 10)$$

which applies to every section. It is then possible to write the differential form

$$dh + u\,du = 0 \qquad (2\cdot 10a)$$

If the gas is *thermally perfect*,† h depends only on T, and the equation may be written

$$c_p\,dT + u\,du = 0 \qquad (2\cdot 10b)$$

If it is also *calorically perfect*, c_p is constant, and so

$$c_pT + \tfrac{1}{2}u^2 = \text{const.} \qquad (2\cdot 10c)$$

(A gas which is both thermally and calorically perfect will sometimes be simply called perfect, except in cases where the distinction must be made explicitly.)

2·4 Reservoir Conditions

The constant in Eq. 2·10 may be conveniently evaluated at a place where $u = 0$ and the fluid is in equilibrium. Thus

$$h + \tfrac{1}{2}u^2 = h_0 \qquad (2\cdot 11)$$

The quantity h_0 is called the *stagnation* or *reservoir enthalpy*, for it must be the enthalpy in a large reservoir, like that illustrated in Fig. 2·4, where the velocity is practically zero.

Furthermore, if there is no heat addition to the flow between the two reservoirs in Fig. 2·4, then h_0 is also the value of the enthalpy in the second reservoir. That is, in *adiabatic* flow,

$$h'_0 = h_0 \qquad (2\cdot 13)$$

This is really no more than the result obtained for the throttling process of Article 1·8. The "resistance" or "throttle" is now the connecting tube between the reservoirs, which we shall presently consider in more detail.

†I.e., satisfies the perfect gas law, Eq. 1·10.

For a *perfect gas*, $h = c_p T$, and the energy equation is

$$c_p T + \tfrac{1}{2}u^2 = c_p T_0 \tag{2·12}$$

where T_0 is the stagnation or reservoir temperature. For a perfect gas, the temperatures in the two reservoirs are the same, no matter what the respective pressures may be, that is,

$$T'_0 = T_0 \tag{2·13a}$$

If the gas is not perfect, Eq. 2·13*a* does not necessarily hold, but Eq. 2·13 does.

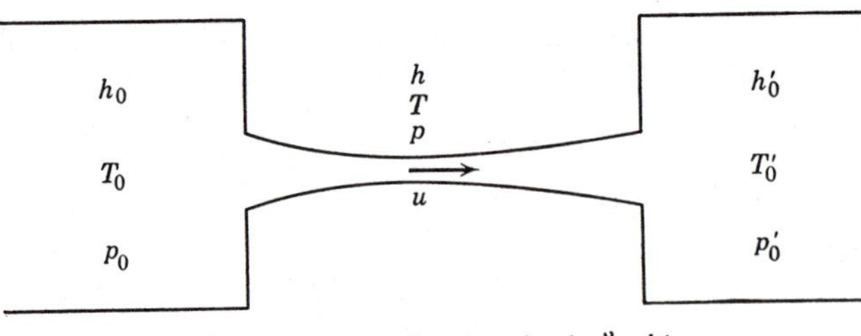

FIG. 2·4 Flow between two reservoirs.

According to the second law of thermodynamics, the entropy in the second reservoir cannot be smaller than in the first one, that is,

$$s'_0 - s_0 \geq 0 \tag{2·14}$$

To check the consequences for a perfect gas, we have from Eq. 1·43*c*,

$$s'_0 - s_0 = R \ln p_0/p'_0 + c_p \ln T'_0/T_0 \tag{2·15}$$

The last term is zero since $T'_0 = T_0$, and thus it is necessary that

$$p_0/p'_0 \geq 1 \tag{2·14a}$$

This agrees with the intuitive notion that the pressure in the downstream reservoir cannot be greater than that in the upstream one. This result is actually true for any gas, since it follows from the definition of entropy† that

$$\left(\frac{\partial s}{\partial p}\right)_h = -\frac{1}{\rho T} < 0$$

so that an increase in entropy, at constant stagnation enthalpy, must result in a decrease of stagnation pressure.

†$ds = \dfrac{1}{T}\left(dh - \dfrac{1}{\rho}\,dp\right)$

The irreversible increase of entropy indicated in Eq. 2·14, and the corresponding decrease of stagnation pressure, is due to the *production of entropy* in the flow between the reservoirs. Only if there are no dissipative processes, that is only if the flow is in equilibrium throughout, will there be no production of entropy. The equality signs, $s'_0 = s_0$ and $p'_0 = p_0$, apply only to such an *isentropic flow.*

The conditions that we have been calling reservoir or stagnation conditions are also called *total* conditions, e.g., total pressure. These terms are also applied in a broader sense, to define conditions at any point in the flow, not necessarily at the reservoirs. *The local total conditions at any point in the flow are the conditions that would be attained if the flow there were brought to rest isentropically.*

For example, in an adiabatic flow of a perfect gas the local stagnation temperature is everywhere T_0, but the local stagnation pressure is smaller than, or at most equal to, p_0; the value depends on the amount of dissipation which the fluid has undergone up to the point in question.

Since the imaginary local stagnation process is an isentropic one, the local stagnation entropy is by definition equal to the local static entropy, that is, $s'_0 = s'$, and thus it is not necessary to use the subscript. For a perfect gas then the local entropy is related to the total pressure by Eq. 2·15, that is,

$$s' - s = -R \ln p'_0/p_0 \tag{2·15a}$$

since $T'_0 = T_0$. Thus a measurement of the local total pressure furnishes a measure of the entropy of the flow. Under suitable conditions the measurement may be made by means of a simple pitot probe (see Chapter 5).

For stagnation conditions to exist it is not enough that the velocity be zero; it is also necessary that equilibrium conditions exist. For example, a thermometer immersed in a flow will not measure the local total temperature, even though it brings the flow to rest on its surface, for the fluid at the surface is not in a state of equilibrium; the large viscous stresses and heat transfer which are usually present correspond to large "currents" of energy and momentum. Surface pressures and temperatures, in the presence of viscous shear and heat conduction, will be discussed in Chapter 13.

2·5 Euler's Equation

In this section we shall apply Newton's law to a flowing fluid. It states that

$$\text{Force} = \text{mass} \times \text{acceleration}$$

We take the Eulerian viewpoint, that is, we observe the acceleration of the fluid particle as it encounters the varying conditions in the tube through which it flows.

The acceleration is the time rate of change of velocity, and is due to two effects. First, since conditions vary *along* the tube there is a velocity gradient $\partial u/\partial x$ in the direction of flow. The rate of change of velocity is proportional to this gradient and to the speed with which the particle moves through it. Thus the acceleration due to "convection" through the velocity gradient is

$$u \frac{\partial u}{\partial x}$$

Second, conditions at a *given section* may be changing, if the flow is nonsteady, or nonstationary. This will give a *nonstationary* term,

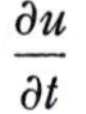

$$\frac{\partial u}{\partial t}$$

The acceleration of the particle is, then, in general,

$$a_x = \frac{\partial u}{\partial t} + u \frac{\partial u}{\partial x} \tag{2·16}$$

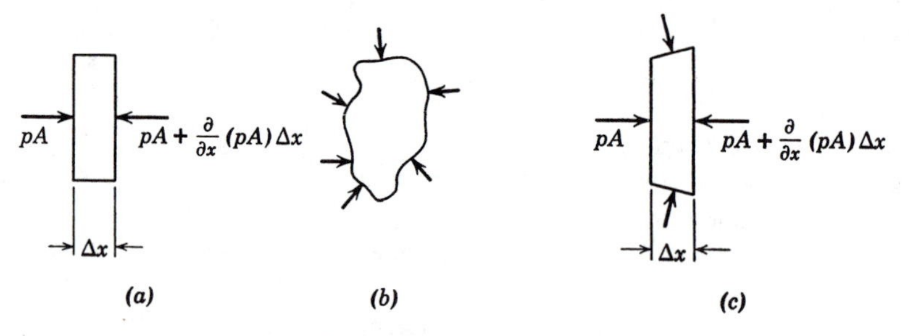

FIG. 2·5 Pressure on a fluid particle.

Next we have to compute the force on the particle. Consider the particle shown in Fig. 2·5*a*. For this simple shape the force in the x-direction is easily computed to be $-\dfrac{\partial p}{\partial x}(\Delta x A)$. Dividing this by $A\,\Delta x$, the volume of the particle, we obtain the force per unit volume, $-\partial p/\partial x$. Finally, dividing through by the density gives the *force per unit mass*,

$$f_x = -\frac{1}{\rho}\frac{\partial p}{\partial x} \tag{2·17}$$

This result is valid for a particle of any shape, for example, the one in Fig. 2·5*b*, as may be proved by an application of Gauss's law (Chapter 7). Thus the result is valid for the particle in Fig. 2·5*c*, and thus for a diverging or converging stream tube. The last result is easily proved directly, by taking into account the pressures on the inclined faces.

This computation of the force does not include any viscous terms, either shearing stresses on the sides or normal viscous stresses on the end sections, and Eq. 2·17 applies only if these are negligible, that is, the flow is nonviscous.

The expressions for the force f_x and acceleration a_x may now be put in Newton's law, which, written for unit mass, is simply

$$a_x = f_x$$

or

$$\frac{\partial u}{\partial t} + u \frac{\partial u}{\partial x} = -\frac{1}{\rho} \frac{\partial p}{\partial x} \tag{2·18}$$

This is called *Euler's equation.*

For steady flow, the first term is zero. The others are then total derivatives and may be written

$$u\,du + \frac{dp}{\rho} = 0 \tag{2·18a}$$

or, in integral form,

$$\frac{u^2}{2} + \int \frac{dp}{\rho} = \text{const.} \tag{2·18b}$$

which is Bernoulli's equation for compressible flow. The integral will be evaluated later. For the moment, it may be noted that for incompressible flow, $\rho = \rho_0$, and Bernoulli's equation has the well-known form

$$\tfrac{1}{2}\rho_0{}^2 + p = \text{const.} \tag{2·18c}$$

2·6 The Momentum Equation

It is often convenient to observe the flow through a space defined by certain fixed surfaces and sections, as we did when deriving the continuity equation (Fig. 2·2). For this point of view, one needs an equation which describes the changes of momentum of the fluid within the "control space." This may be obtained by combining Euler's equation with the continuity equation, as follows.

Multiplying Euler's equation (2·18) by ρA and the continuity equation (2·4) by u gives, respectively,

$$\rho A \frac{\partial u}{\partial t} + \rho u A \frac{\partial u}{\partial x} = -A \frac{\partial p}{\partial x}$$

$$u \frac{\partial}{\partial t} (\rho A) + u \frac{\partial}{\partial x} (\rho u A) = 0$$

Adding these, and combining appropriate terms, yields the one-dimensional momentum equation,

$$\frac{\partial}{\partial t}(\rho u A) + \frac{\partial}{\partial x}(\rho u^2 A) = -A\frac{\partial p}{\partial x} = -\frac{\partial}{\partial x}(pA) + p\frac{\partial A}{\partial x}$$

Then, integrating this, with respect to x between any two sections, gives

$$\frac{\partial}{\partial t}\int_1^2 (\rho u A)\,dx + (\rho_2 u_2^2 A_2 - \rho_1 u_1^2 A_1) = (p_1 A_1 - p_2 A_2) + \int_1^2 p\,dA$$

The first integral is the momentum of the fluid enclosed between 1 and

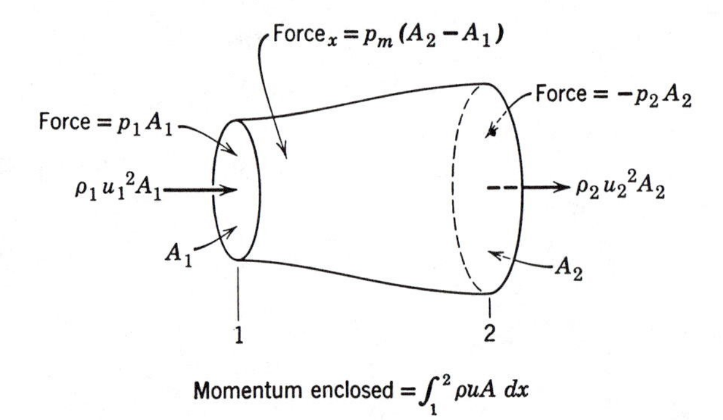

FIG. 2·6 Forces and flux of momentum at a control space.

2 (Fig. 2·6), and the last integral may be evaluated by defining a mean pressure, p_m. Thus

$$\frac{\partial}{\partial t}\int_1^2 (\rho u A)\,dx + (\rho_2 u_2^2 A_2 - \rho_1 u_1^2 A_1) = (p_1 A_1 - p_2 A_2) + p_m(A_2 - A_1) \tag{2·19}$$

The left-hand side of this equation is the rate of change of momentum in the space between 1 and 2; it consists of two terms, the contribution from *nonstationary* changes inside the space, and the contribution from transport or flux of momentum into the space, through the end sections. The right-hand side is the force in the x-direction, due to the pressures on the end sections and on the walls.

For steady flow the first term in Eq. 2·19 is zero.

The integral form of the momentum equation is actually more general than has been indicated so far, since it is valid even when there are frictional forces and regions of dissipation *within* the control space, provided that these are absent at the reference sections, 1 and 2. This generality occurs for the following reason. The integration of the differential momentum equation corresponds, physically, to a summation of the forces on adjacent fluid elements, and of the flows into them, as illustrated in Fig. 2·7. The forces on adjacent *internal* faces are equal and opposite, and so they cancel in the summation. Similarly the inflows and outflows through adjacent faces cancel in the summation. One is left with the force and the flux at the boundaries of the control space. If there is a nonequilibrium region *inside* this space, it does not affect the integral result.

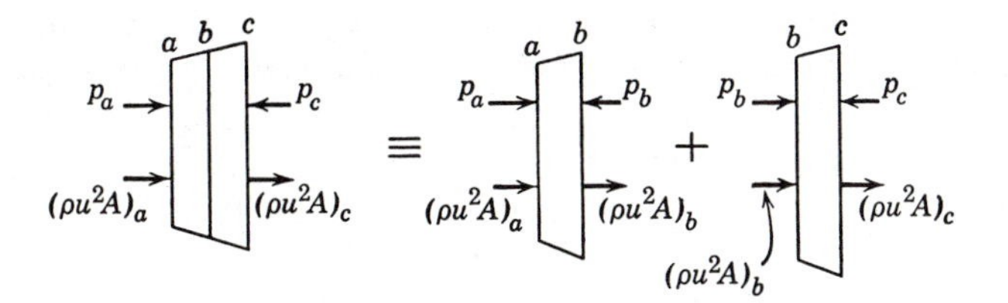

FIG. 2·7 Illustrating the summation of volume elements in the control space.

One sometimes takes the integral momentum equation, instead of Euler's equation, to be axiomatic, as we shall do in Chapter 7.

For *steady* flow in a duct of *constant area*, the momentum equation becomes especially simple:

$$\rho_2 u_2^2 - \rho_1 u_1^2 = p_1 - p_2 \tag{2·20}$$

2·7 Isentropic Conditions

It was stated in Article 2·4 that a flow which is adiabatic and in equilibrium is *isentropic*. This may be verified from the energy and momentum equations which have been derived. For adiabatic, nonconducting flow, the energy equation

$$dh + u\,du = 0$$

applies all along the flow, and similarly in the absence of friction forces Euler's equation

$$u\,du + dp/\rho = 0$$

is applicable. Elimination of u from these two equations gives the relation between the thermodynamic variables,

$$dh - dp/\rho = 0$$

But the entropy of the flow is related to these variables by Eq. 1·53, from which

$$ds = \frac{1}{T}(dh - dp/\rho) = 0$$

or

$$s = \text{const.} \tag{2·21}$$

along the flow. Thus an *adiabatic, nonviscous, nonconducting flow is isentropic.* In this case, then, either the momentum or the energy equation may be replaced by Eq. 2·21. For a perfect gas this condition may also be written

$$p/p_0 = (\rho/\rho_0)^\gamma = (T/T_0)^{\gamma/(\gamma-1)} \tag{2·21a}$$

The conditions of equilibrium cannot be strictly attained in a real, nonuniform flow, since a fluid particle must adjust itself continuously to the new conditions that it encounters. The rate at which the adjustments must be made depends on the gradients in the flow, and is a measure of the degree to which the fluid is out of equilibrium. This furnishes, in fact, an exact measure of the rate of production of entropy. In one-dimensional flow, the entropy production terms are (see Chapter 13)

$$\frac{\tilde{\mu}}{T}\left(\frac{\partial u}{\partial x}\right)^2 \quad \text{and} \quad \frac{k}{T^2}\left(\frac{\partial T}{\partial x}\right)^2$$

They depend on the squares of the velocity and temperature gradients, respectively, and so are always positive. The coefficients $\tilde{\mu}$ and k are the coefficients of viscosity and conductivity.

In a real fluid these entropy production terms are never strictly zero, since gradients are always present and since $\tilde{\mu}$ and k are finite. However, the idealized flows, which are obtained by neglecting them, form a large and useful part of the subject matter of aerodynamics, and of fluid mechanics in general. The usual way of expressing this idealization is to describe the fluid as nonviscous† and nonconducting ($\tilde{\mu} = 0$ and $k = 0$).

This idealization is better, the smaller the actual values of $\tilde{\mu}$ and k. However, this is a relative matter, for even in a fluid of extremely low viscosity and conductivity the nonequilibrium terms will not be negligible in regions where the gradients are extremely high. Such regions must occur in all real flows. They appear as boundary layers, wakes, vortex cores, and, in supersonic flow, as shock waves. The modern progress in the understanding of fluid flow is in large part due to the discovery of how to analyze these narrow, nonequilibrium regions separately and then fit them into the nonviscous flow field.

2·8 Speed of Sound; Mach Number

A parameter that is fundamental in compressible flow theory is the *speed of sound, a.* As will be shown in the next chapter, it is the speed at which small disturbances (waves) are propagated through a compressible fluid. Its relation to the compressibility of the fluid is given by

$$a^2 = \left(\frac{\partial p}{\partial \rho}\right)_s \tag{2·22}$$

†When the term "nonviscous" is used alone, it usually also implies "nonconducting."

The disturbances produced in a fluid by a sound wave (more precisely, the temperature and velocity gradients produced) are so small that each fluid particle undergoes a nearly isentropic process.† For purposes of computing the wave speed the process is assumed to be strictly isentropic (see Articles 3·2 and 3·6). Accordingly, the derivative in Eq. 2·22 is evaluated with the isentropic relation. In a perfect gas this is

$$p = \text{const. } \rho^{\gamma}$$

giving

$$a^2 = \frac{\gamma p}{\rho} = \gamma RT \qquad \blacktriangleright(2\cdot23)$$

In a *flowing* fluid, the speed of sound is a significant measure of the effects of compressibility only when it is compared to the speed of the flow. This introduces a dimensionless parameter which is called the *Mach number*,

$$M = u/a \qquad \blacktriangleright(2\cdot24)$$

It is the most important parameter in compressible flow theory.

It is evident that M will vary from point to point in a given flow, not only because u changes but also because a depends on the local conditions, according to Eq. 2·23. The local value of a is related to the local value of u, according to a relation that will be obtained later. For the present, it need only be noted that in adiabatic flow an increase of u always corresponds to an increase of M.

If the flow speed exceeds the local speed of sound, the Mach number is greater than 1 and the flow is called *supersonic*. For Mach numbers below 1, it is *subsonic*.

2·9 The Area-Velocity Relation

Some of the effects of compressibility are simply demonstrated by considering steady adiabatic flow in a stream tube of varying area (Fig. 2·8). From the continuity equation (2·2) we have

$$\frac{d\rho}{\rho} + \frac{du}{u} + \frac{dA}{A} = 0 \qquad (2\cdot25)$$

For incompressible flow, $d\rho = 0$, this reduces to the simple result that increase of velocity is proportional to decrease of area. The way in which this is modified by compressibility may be found by using Euler's equation

†The reason usually given is that the changes in a sound wave are so *fast* that a fluid particle cannot lose nor gain heat. Such explanations not only miss the point, but are misleading. In fact, the process is isentropic because the changes that a particle undergoes are *slow enough* to keep the velocity and temperature gradients small. For example, if the frequency of a constant amplitude wave is increased sufficiently, these gradients become so large that the processes cannot be considered isentropic.

(2·18a) to find the relation between density and velocity changes. We may rewrite it in the form,†

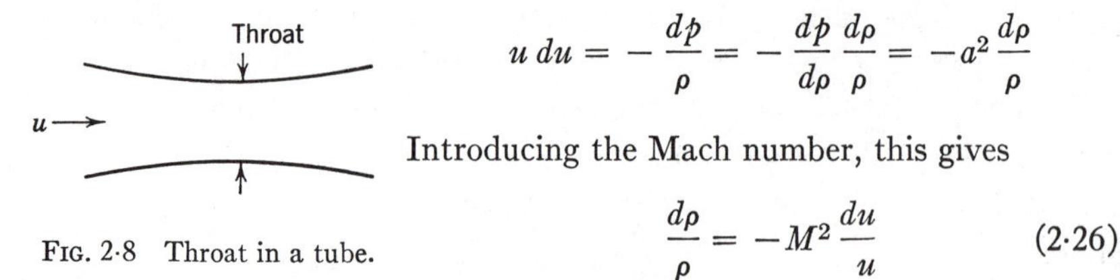

$$u\,du = -\frac{dp}{\rho} = -\frac{dp}{d\rho}\frac{d\rho}{\rho} = -a^2\frac{d\rho}{\rho}$$

Introducing the Mach number, this gives

$$\frac{d\rho}{\rho} = -M^2\frac{du}{u} \tag{2·26}$$

Fig. 2·8 Throat in a tube.

The role of M as a measure of compressibility is well illustrated here. At very low Mach numbers the density changes are so small, compared to the velocity changes, that they may be neglected in computing the flow field; that is, it may be considered that $\rho =$ const. Equivalent definitions of "incompressible flow" are $a = \infty$ or $M = 0$.

Substitution of this relation in Eq. 2·25 gives the area-velocity relation

$$\frac{du}{u} = \frac{-dA/A}{1 - M^2} \tag{▶2·27}$$

which shows the following Mach number effects:

(1) At $M = 0$, a decrease in area gives a *proportional* increase in velocity.

(2) For values of M between 0 and 1, that is, at subsonic speeds, the relation is qualitatively the same as for incompressible flow, a decrease of area giving an increase in velocity, but the effect on the velocity is relatively greater, since the denominator is less than 1.

(3) At supersonic speeds, the denominator becomes negative and an increase of speed is produced by an *increase* of area. For one accustomed to thinking in terms of incompressible flow, this behavior may appear quite remarkable. It is due to the fact that at supersonic speeds the "density decreases faster than the velocity increases," so that the area must increase to maintain continuity of mass. This may be seen from Eq. 2·26, which shows that for $M > 1$ the decrease in density is greater than the increase in velocity.

There remains the question of what happens at *sonic* speed, $M = 1$. Consider a tube in which the velocity increases continuously, from zero, and eventually becomes supersonic. The above discussion shows that the tube must converge in the subsonic and diverge in the supersonic portion. Just at $M = 1$ there must be a *throat* (Fig. 2·8). This is clear also from Eq. 2·27, which shows that, at $M = 1$, du/u can be finite only if $dA/A = 0$. The same argument applies to the case where the velocity decreases continuously from supersonic to subsonic. The important conclusion is that

†Adiabatic, nonviscous flow is isentropic; hence $dp/d\rho \equiv (\partial p/\partial \rho)_s$. Cf. Art. 3·6.

$M = 1$ *can be attained only at a throat* of the tube. (The inverse does not hold; that is, M is not necessarily 1 at a throat. But, if it is not, then Eq. 2·27 shows that a throat corresponds to $du = 0$; that is, the velocity attains a maximum or a minimum there, depending on whether the flow is subsonic or supersonic.)

Near $M = 1$, the flow is very sensitive to changes in the area, since the denominator in Eq. 2·27 is then small.

2·10 Results from the Energy Equation

It was shown in Article 2·3 that in adiabatic flow the energy equation *for a perfect gas*† is

$$\tfrac{1}{2}u^2 + c_p T = c_p T_0 \tag{2·28}$$

With the expression $a^2 = \gamma RT$ for the speed of sound, this gives

$$\frac{u^2}{2} + \frac{a^2}{\gamma - 1} = \frac{a_0{}^2}{\gamma - 1} \qquad \blacktriangleright(2·29)$$

Then, multiplying the last equation by $(\gamma - 1)/a^2$ gives

$$\frac{a_0{}^2}{a^2} = \frac{T_0}{T} = 1 + \frac{\gamma - 1}{2} M^2 \qquad \blacktriangleright(2·30)$$

The *isentropic relations* (Eq. 2·21*a*) may then be used to obtain

$$\frac{p_0}{p} = \left(1 + \frac{\gamma - 1}{2} M^2\right)^{\gamma/(\gamma-1)} \qquad \blacktriangleright(2·31)$$

$$\frac{\rho_0}{\rho} = \left(1 + \frac{\gamma - 1}{2} M^2\right)^{1/(\gamma-1)} \qquad \blacktriangleright(2·32)$$

In Eqs. 2·28, 2·29, and 2·30 the values of T_0 and a_0 are constant throughout the flow, so that they may be taken as those in the actual reservoir. In Eqs. 2·31 and 2·32 the values of p_0 and ρ_0 are the *local* "reservoir values." They are constant throughout only if the flow is isentropic.

These relations between the thermodynamic variables and the Mach number are tabulated in Tables II and III at the back of the book for air ($\gamma = 1.40$).

Instead of the reservoir, we may use any other point in the flow for evaluating the constant in the energy equation. A particularly useful one is the point where $M = 1$, that is, a throat. The flow variables there are called "sonic," and are denoted by the superscript *. Thus the flow speed and sound speed are u^* and a^*, respectively. But, since the Mach number

†If the gas is not calorically perfect, it is not possible to write $c_p T$ instead of h.

is 1, these are equal, that is, $u^* = a^*$. The energy equation then gives

$$\frac{u^2}{2} + \frac{a^2}{\gamma - 1} = \frac{u^{*2}}{2} + \frac{a^{*2}}{\gamma - 1} = \frac{1}{2}\frac{\gamma + 1}{\gamma - 1} a^{*2} \tag{2·33}$$

Comparing with Eq. 2·29, the relation between the speed of sound in the throat and the reservoir is

$$\frac{a^{*2}}{a_0^2} = \frac{2}{\gamma + 1} = \frac{T^*}{T_0} \tag{2·34}$$

Thus for a given fluid the sonic and the reservoir temperatures are in a fixed ratio, so that T^* is constant throughout an adiabatic flow. For air the numerical values are

$$T^*/T_0 = 0.833, \quad a^*/a_0 = 0.913 \tag{2·35}$$

The sonic pressure and density ratios may also be obtained, by using the isentropic relations with Eq. 2·34 or by setting $M = 1$ in Eqs. 2·31 and 2·32. They are

$$\begin{aligned} \frac{p^*}{p_0} &= \left(\frac{2}{\gamma + 1}\right)^{\gamma/(\gamma-1)} = 0.528 \\ \frac{\rho^*}{\rho_0} &= \left(\frac{2}{\gamma + 1}\right)^{1/(\gamma-1)} = 0.634 \end{aligned} \tag{2·35a}$$

Of course, it is not necessary that a throat actually exist in the flow for the sonic values to be used as reference.

In some problems, especially transonic ones, the speed ratio u/a^* is convenient. This will sometimes be denoted by

$$M^* = u/a^* \tag{2·36}$$

(This is not strictly in agreement with our convention for starred quantities, because according to it, $M^* = 1$. However, this license is convenient and should cause no confusion.) The relation between M^* and M may be obtained by dividing both sides of Eq. 2·33 by u^2. The resulting expression may be solved for M^{*2} or M^2, giving

$$\left(\frac{u}{a^*}\right)^2 \equiv M^{*2} = \frac{\frac{\gamma + 1}{2} M^2}{1 + \frac{\gamma - 1}{2} M^2} = \frac{\gamma + 1}{\frac{2}{M^2} + \gamma - 1} \tag{2·37a}$$

$$M^2 = \frac{M^{*2}}{\frac{\gamma + 1}{2} - \frac{\gamma - 1}{2} M^{*2}} = \frac{2}{\frac{\gamma + 1}{M^{*2}} - (\gamma - 1)} \tag{2·37b}$$

From these it may be seen that $M^* < 1$ for $M < 1$, and $M^* > 1$ for $M > 1$.

2·11 Bernoulli Equation; Dynamic Pressure

The energy equation (2·28) may be rewritten by using the gas law $p = R\rho T$ to eliminate T. This gives

$$\frac{1}{2}u^2 + \frac{\gamma}{\gamma - 1}\frac{p}{\rho} = \frac{\gamma}{\gamma - 1}\frac{p_0}{\rho_0}$$

This equation is valid for adiabatic flow. For isentropic conditions, the relation $p/\rho^\gamma = p_0/\rho_0{}^\gamma$ may be used to eliminate ρ, resulting in

$$\frac{1}{2}u^2 + \frac{\gamma}{\gamma - 1}\frac{p_0}{\rho_0}\left(\frac{p}{p_0}\right)^{(\gamma-1)/\gamma} = \frac{\gamma}{\gamma - 1}\frac{p_0}{\rho_0} \tag{2·38}$$

Equation 2·38 is the integrated Bernoulli equation (2·18*b*).

In compressible flow, the *dynamic pressure* $\frac{1}{2}\rho u^2$ is no longer simply the difference between stagnation and static pressures, as in the incompressible case. It depends now on Mach number as well as on the static pressure. For a perfect gas, the relation is

$$\frac{1}{2}\rho u^2 = \frac{1}{2}\rho a^2 M^2 = \frac{1}{2}\rho\left(\frac{\gamma p}{\rho}\right)M^2 = \frac{1}{2}\gamma p M^2 \tag{2·39}$$

The dynamic pressure is used for normalizing pressure and force coefficients. For example, the pressure coefficient is

$$C_p = \frac{p - p_1}{\frac{1}{2}\rho_1 U^2} = \frac{p - p_1}{\frac{1}{2}\gamma p_1 M_1{}^2} = \frac{2}{\gamma M_1{}^2}\left(\frac{p}{p_1} - 1\right) \tag{2·40}$$

where U and M_1 are the reference velocity and the Mach number.

For isentropic flow, Eq. 2·31 may be used to rewrite this in terms of the local Mach number:

$$C_p = \frac{2}{\gamma M_1{}^2}\left\{\left[\frac{2 + (\gamma - 1)M_1{}^2}{2 + (\gamma - 1)M^2}\right]^{\gamma/(\gamma-1)} - 1\right\} \tag{2·40a}$$

Finally, introducing $M_1{}^2 = U^2/a_1{}^2$, $M^2 = u^2/a^2$ and using the energy equation

$$\frac{u^2}{2} + \frac{a^2}{\gamma - 1} = \frac{U^2}{2} + \frac{a_1{}^2}{\gamma - 1}$$

to eliminate a^2, the pressure coefficient may be obtained in the form

$$C_p = \frac{2}{\gamma M_1{}^2}\left\{\left[1 + \frac{\gamma - 1}{2}M_1{}^2\left(1 - \frac{u^2}{U^2}\right)\right]^{\gamma/(\gamma-1)} - 1\right\} \tag{2·40b}$$

2·12 Flow at Constant Area

Consider adiabatic, constant-area flow (Fig. 2·9*a*) through a nonequilibrium region (shown shaded). If sections 1 and 2 are outside this region, then the equations of continuity, momentum, and energy are

$$\rho_1 u_1 = \rho_2 u_2 \tag{2·41a}$$

$$p_1 + \rho_1 u_1^2 = p_2 + \rho_2 u_2^2 \tag{2·41b}$$

$$h_1 + \tfrac{1}{2}u_1^2 = h_2 + \tfrac{1}{2}u_2^2 \tag{2·41c}$$

The solution of these gives the relations that must exist between the flow parameters at the two sections; it will be worked out presently.

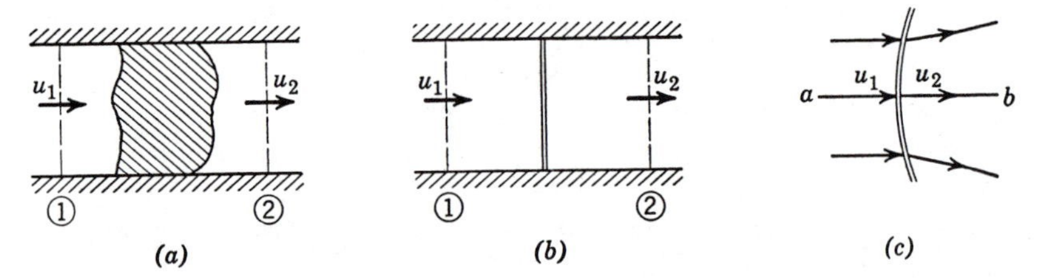

FIG. 2·9 Illustrating a change of equilibrium conditions in constant area flow. (*a*) Uniform conditions on either side of a region of nonuniformity or dissipation; (*b*) normal shock wave; (*c*) shock wave normal to flow on streamline *a–b*.

There is no restriction on the size or details of the dissipation region so long as the reference sections are outside it. In particular, it may be idealized by the vanishingly thin region, shown in Fig. 2·9*b*, across which the flow parameters are said to "jump." The control sections 1 and 2 may then be brought arbitrarily close to it. Such a discontinuity is called a *shock wave.*† Of course, a real fluid cannot have an actual discontinuity, and this is only an idealization of the very high gradients that actually occur in a shock wave, in the transition from state 1 to 2. These severe gradients produce viscous stress and heat transfer, i.e., nonequilibrium conditions, inside the shock.

The mechanism of shock-wave formation, as well as some details of conditions inside the dissipation region, will be discussed later. For application to most aerodynamic problems, it is sufficient to calculate the jumps in the equilibrium values, and to represent the shock as a discontinuity. Since the reference sections may be brought arbitrarily near to the shock, the device of a constant area duct is no longer needed, that is, the results always

†In this book a shock wave will always be represented by a double line, as in Fig. 2·9*b* and *c*. No implication about its structure is intended. On shadowgraphs (e.g., Fig. 2·10*b*) a shock wave appears as a dark line followed by a bright line, for reasons explained in Article 6·13.

apply *locally* to conditions on either side of a shock, provided it is normal to the streamline (Fig. 2·9*c*).

Of course, the shock relations may be applied to equilibrium sections of real constant-area ducts, such as the one shown in Fig. 2·9*a*, but it is necessary that the friction forces on the walls be negligible, since there are no friction terms in the momentum equation. An example is the constant-area supersonic diffuser, in which an adverse pressure gradient reduces the wall friction to negligible values. The diffusion occurs through a complicated, three-dimensional process involving interactions between shock waves and boundary layer. For equilibrium to be attained the diffuser must be *long*, in curious contrast to the normal shock, for which equilibrium is reached in a very short distance.

Figure 2·10*a* shows an example of compression in a constant-area duct; Fig. 2·10*b* shows, in contrast, an example of a normal shock wave.

2·13 The Normal Shock Relations for a Perfect Gas

Equations 2·41 are the general equations for a normal shock wave. It will usually be necessary to solve them numerically (see Exercise 3·6). However, for a gas that is thermally and calorically perfect it is possible to obtain explicit solutions in terms of the Mach number M_1 ahead of the shock.

Dividing the two sides of the momentum equation (2·41*b*), respectively, by $\rho_1 u_1$ and $\rho_2 u_2$, which are equal from the continuity equation, gives

$$u_1 - u_2 = \frac{p_2}{\rho_2 u_2} - \frac{p_1}{\rho_1 u_1} = \frac{a_2^2}{\gamma u_2} - \frac{a_1^2}{\gamma u_1}$$

Here the perfect gas relation $a^2 = \gamma p/\rho$ has been used. Then a_1^2 and a_2^2 may be replaced by using the energy equation for a perfect gas,

$$\frac{u_1^2}{2} + \frac{a_1^2}{\gamma - 1} = \frac{u_2^2}{2} + \frac{a_2^2}{\gamma - 1} = \frac{1}{2}\frac{\gamma + 1}{\gamma - 1}a^{*2}$$

After some rearrangement there is obtained the simple relation

$$u_1 u_2 = a^{*2} \tag{2·42}$$

This is known as the Prandtl or Meyer relation.

In terms of the speed ratio $M^* = u/a^*$, this equation is

$$M^*_2 = 1/M^*_1 \tag{2·43}$$

Now $M^* \gtrless 1$ corresponds to $M \gtrless 1$, and thus the Prandtl relation shows that the velocity change across a normal shock must be from supersonic to subsonic, or vice versa. It will be shown later that only the former is

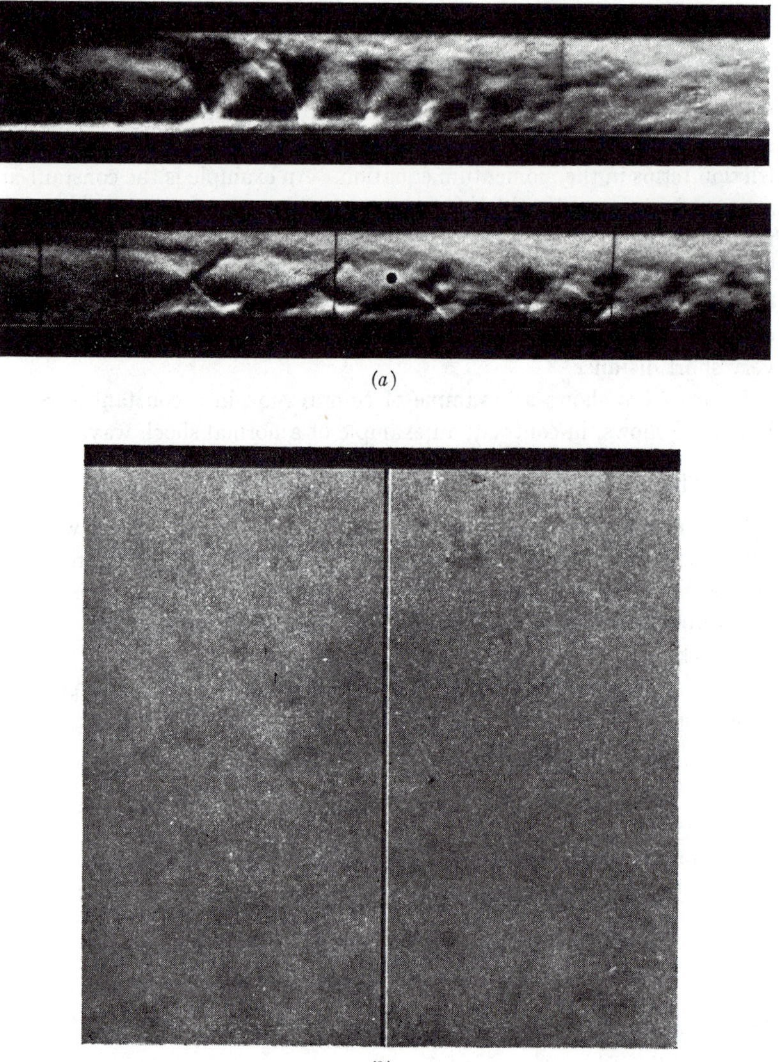

(*a*)

(*b*)

FIG. 2·10 Compression in constant-area flow. (*a*) Example of extended compression regions in a square duct. The flow is compressed from supersonic to subsonic through complex patterns of shocks interacting with the boundary layer. (J. Lukasiewicz, *J. Aeronaut. Sci.*, *20* (1953), p. 618.)

(*b*) Example of a very narrow compression region. The flow is compressed from supersonic to subsonic through a single normal shock wave. Shadowgraph of shock wave moving from right to left in a shock tube. (Walker Bleakney, D. K. Weimer, and C. H. Fletcher, *Rev. Sci. Instr.*, *20* (1949), p. 807.)

possible, as might be anticipated since the effect of dissipation can hardly be such as to increase the velocity.

The correspondence between M^* and M was obtained in Article 2·10. It is

$$M^{*2} = \frac{(\gamma + 1)M^2}{(\gamma - 1)M^2 + 2} \tag{2·44}$$

When used to replace M^*_1 and M^*_2 in the Prandtl equation, it gives the relation between the Mach numbers

$$M_2{}^2 = \frac{1 + \frac{\gamma - 1}{2} M_1{}^2}{\gamma M_1{}^2 - \frac{\gamma - 1}{2}} \tag{▶2·45}$$ †

The ratio of the velocities may also be written quite simply:

$$\frac{u_1}{u_2} = \frac{u_1{}^2}{u_1 u_2} = \frac{u_1{}^2}{a^{*2}} = M^*{}_1{}^2 \tag{2·46}$$

This, together with Eq. 2·44, is useful in the derivation of other expressions. For instance, using the continuity equation, the ratio of the densities is

$$\frac{\rho_2}{\rho_1} = \frac{u_1}{u_2} = \frac{(\gamma + 1)M_1{}^2}{(\gamma - 1)M_1{}^2 + 2} \tag{▶2·47}$$

To obtain the pressure relation, we have, from the momentum equation,

$$p_2 - p_1 = \rho_1 u_1{}^2 - \rho_2 u_2{}^2 = \rho_1 u_1 (u_1 - u_2)$$

the last step being due to the continuity equation. In dimensionless form, this is

$$\frac{p_2 - p_1}{p_1} = \frac{\rho_1 u_1{}^2}{p_1}\left(1 - \frac{u_2}{u_1}\right)$$

Finally, with $a_1{}^2 = \gamma p_1/\rho_1$ and Eq. 2·47 for u_2/u_1, the pressure jump is obtained in the form

$$\frac{p_2 - p_1}{p_1} = \frac{\Delta p_1}{p_1} = \frac{2\gamma}{\gamma + 1}(M_1{}^2 - 1) \tag{2·48}$$

The ratio $\Delta p_1/p_1$ is often used to define the *shock strength.* One may also use the ratio

$$\frac{p_2}{p_1} = 1 + \frac{2\gamma}{\gamma + 1}(M_1{}^2 - 1) \tag{▶2·48a}$$

The temperature ratio may be obtained from Eqs. 2·47 and 2·48*a*, using

†This and other shock-wave relations are tabulated in Table IV at the end of the book.

$T_2/T_1 = (p_2/p_1)(\rho_1/\rho_2)$. Alternatively, starting directly from the energy equation, and using Eq. 2·46, we find, after some rearrangement,

$$\frac{a_2^2}{a_1^2} = \frac{T_2}{T_1} = 1 + \frac{2(\gamma - 1)}{(\gamma + 1)^2} \frac{\gamma M_1^2 + 1}{M_1^2} (M_1^2 - 1) \tag{2·49}$$

Finally we may compute the change of entropy. From Eq. 1·43*c* it may be rewritten in the form

$$\frac{s_2 - s_1}{R} = \ln \left[\left(\frac{p_2}{p_1} \right)^{1/(\gamma-1)} \left(\frac{\rho_2}{\rho_1} \right)^{-\gamma/(\gamma-1)} \right]$$

By using Eqs. 2·47 and 2·48*a* for ρ_2/ρ_1 and p_2/p_1,

$$\frac{s_2 - s_1}{R} = \ln \left[1 + \frac{2\gamma}{\gamma + 1} (M_1^2 - 1) \right]^{1/(\gamma-1)} \left[\frac{(\gamma + 1)M_1^2}{(\gamma - 1)M_1^2 + 2} \right]^{-\gamma/(\gamma-1)} \tag{2·50}$$

A more convenient form may be obtained by letting $M_1^2 - 1 = m$. Then

$$\frac{s_2 - s_1}{R} = \ln \left\{ \left(1 + \frac{2\gamma}{\gamma + 1} m \right)^{1/(\gamma-1)} (1 + m)^{-\gamma/(\gamma-1)} \left(\frac{\gamma - 1}{\gamma + 1} m + 1 \right)^{\gamma/(\gamma-1)} \right\}$$

This expression is still exact. For values near $M_1 = 1$, m is small, and the expression may be simplified, since each of the terms in parentheses is like $1 + \epsilon$, with $\epsilon \ll 1$. Each of the three terms obtained by the logarithmic product rule is like $\ln(1 + \epsilon)$, which has the series expansion $\epsilon - \epsilon^2/2 + \epsilon^3/3 + \cdots$. Collecting terms, it is found that the coefficients of m and m^2 are zero, leaving

$$\frac{s_2 - s_1}{R} = \frac{2\gamma}{(\gamma + 1)^2} \frac{m^3}{3} + \text{higher-order terms}$$

that is,

$$\frac{s_2 - s_1}{R} \doteq \frac{2\gamma}{(\gamma + 1)^2} \frac{(M_1^2 - 1)^3}{3} \tag{2·51}$$

Since the entropy cannot decrease in adiabatic flow, Eq. 2·51 shows that $M_1 \geq 1$. Thus of the two possibilities obtained earlier, only the jump from supersonic to subsonic conditions is possible. Examination of Eqs. 2·47, 2·48, and 2·49 shows that the corresponding jumps in density, pressure, and temperature are from lower to higher values. The shock is said to *compress* the flow.

An important result is that the increase of entropy is *third order* in $(M_1^2 - 1)$. This may be written in terms of shock strength by using Eq. 2·48 in Eq. 2·51, whence

$$\frac{s_2 - s_1}{R} \doteq \frac{\gamma + 1}{12\gamma^2} \left(\frac{\Delta p_1}{p_1} \right)^3 \tag{2·52}$$

Thus, a small but finite change of pressure, for which there are corresponding first-order changes of velocity, density, and temperature, gives only a third-order change of entropy. A weak shock produces a nearly isentropic change of state.

Finally we can find the change of stagnation or total pressure. Since $T_{02} = T_{01}$, the change of entropy is related to the total pressures, as shown in Article 2·4, by

$$\frac{s_2 - s_1}{R} = \ln \frac{p_{01}}{p_{02}} \tag{2·53}$$

We can write $p_{02} = p_{01} + \Delta p_{01}$, and $\Delta s/R = -\ln\left(1 + \frac{\Delta p_{01}}{p_{01}}\right)$. For weak shocks, $\Delta p_{01}/p_{01} \ll 1$, and then

$$\frac{s_2 - s_1}{R} \doteq -\frac{\Delta p_{01}}{p_{01}} = \frac{2\gamma}{(\gamma + 1)^2} \frac{(M_1^2 - 1)^3}{3} \tag{2·53a}$$

Thus, a small entropy change is directly proportional to the change of total pressure. It follows that the change of total pressure is also third order in the shock strength.

The exact expression for the ratio of total pressures may be obtained from Eqs. 2·53 and 2·50. It is

$$\frac{p_{02}}{p_{01}} = \left[1 + \frac{2\gamma}{\gamma + 1}(M_1^2 - 1)\right]^{-1/(\gamma-1)} \left[\frac{(\gamma + 1)M_1^2}{(\gamma - 1)M_1^2 + 2}\right]^{\gamma/(\gamma-1)} \tag{2·54}$$

CHAPTER 3

One-Dimensional Wave Motion

3·1 Introduction

Disturbances created in a fluid by a moving body are propagated or communicated to other parts of the fluid. The motion of the disturbances relative to the fluid is called *wave motion*, and the speed of propagation is called the *wave speed*. This is the mechanism by which the various parts of the body interact with the fluid and with each other, and by which the forces on the body are established. Thus, wave motion underlies all problems of fluid motion, though it is not always convenient nor necessary to describe it explicitly.

In this chapter we shall study only the case of one-dimensional motion, in a tube of constant area, that is, the case of "plane waves." Since some of the examples may be generated by the motion of a piston in the tube, they are sometimes called "piston problems."

The results of this chapter are not essential to the following ones, in which we shall again return to the case of steady flow, from the conventional aerodynamic viewpoint. But a study of wave motion, even in the simple one-dimensional case, is helpful in understanding some of the underlying mechanisms in two- and three-dimensional steady flow. The motion in a tube is also of considerable practical importance in connection with shock tubes, starting processes in wind tunnels, etc.

3·2 The Propagating Shock Wave

In Article 2·13 we studied the flow conditions across a shock wave, assuming it to be stationary, as illustrated again in Fig. 3·1*a*. The fluid flows through the shock with speed u_1. Alternatively we may say that the shock is propagating *through the fluid* with speed u_1. This is illustrated more explicitly in Fig. 3·1*b*, which is obtained from Fig. 3·1*a* by a uniform velocity transformation. The fluid ahead of the shock wave is at rest, the wave is propagating into it with the wave speed

$$c_s = u_1 \tag{3·1a}$$

and the fluid behind the wave is "following" with the speed

$$u_p = u_1 - u_2 \tag{3·1b}$$

The static densities, pressures, and temperatures on either side of the shock are not affected by the transformation, and are still related by Eqs. 2·47, 2·48, and 2·49. Thus, a shock wave propagating into a stationary fluid sets it into motion and raises its pressure, density, and temperature.

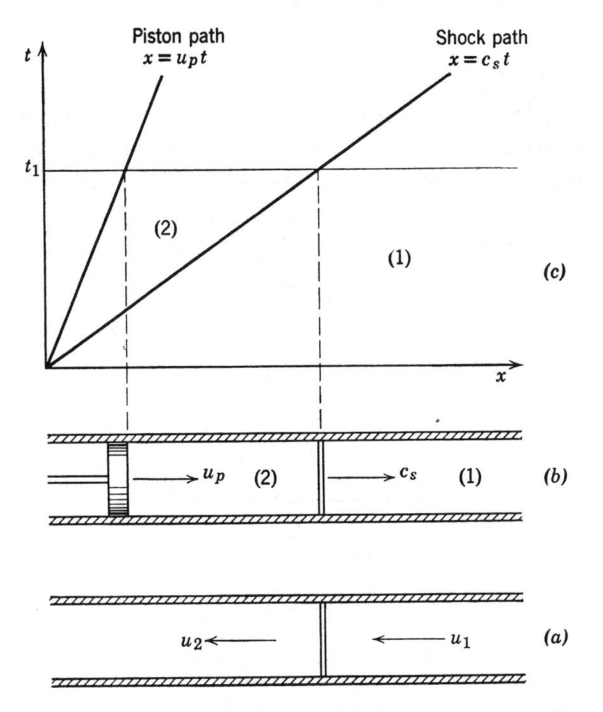

FIG. 3·1 Stationary and propagating shocks. (*a*) Shock stationary; (*b*) fluid at (1) stationary (conditions at time t_1); (*c*) *x-t* diagram of the flow in (*b*).

It may be assumed that the fluid behind the shock wave is followed by a driving piston, moving at the speed u_p, as illustrated in Fig. 3·1*b*. In fact, all the conditions that we have calculated are satisfied by the motion illustrated on the *x-t* diagram (Fig. 3·1*c*), in which the piston is started impulsively at time $t = 0$ with speed u_p. It establishes a shock wave which runs ahead at the speed c_s. The pressure on the piston is p_2. The region of compressed fluid, between the shock wave and piston, increases in length at the rate $(c_s - u_p)$.

Even if the piston attains the speed u_p gradually, rather than impulsively, a uniform state of motion is eventually attained, since nonuniformities in the compressed region catch up to the shock front. This effect will be discussed in Article 3·10.

This example of a moving piston illustrates the general effect described in the introduction; that is, a moving body establishes a certain wave motion and a corresponding "piston pressure" on its surface.

The "jump" relations across the shock, which were obtained in Article 2·13, may be rewritten in terms of c_s and u_p by using the transformation equations (3·1). For instance, the Mach number of the shock is

$$M_1 = c_s/a_1$$

where $a_1^2 = (dp/d\rho)_1$.

For application to practical problems it is often more convenient to use the pressure ratio p_2/p_1 as the basic independent parameter. All other quantities may then be computed in terms of this ratio and conditions in the undisturbed fluid. For instance, by using Eq. 2·48*a* to relate p_2/p_1 to the Mach number, the shock velocity, for a perfect gas, is

$$c_s = M_1 a_1 = a_1 \left(\frac{\gamma - 1}{2\gamma} + \frac{\gamma + 1}{2\gamma} \frac{p_2}{p_1} \right)^{1/2} \tag{3·2}$$

The density ratio and temperature ratio are given by the Rankine-Hugoniot relations (Exercise 2·5)

$$\frac{\rho_2}{\rho_1} = \frac{1 + \dfrac{\gamma + 1}{\gamma - 1} \dfrac{p_2}{p_1}}{\dfrac{\gamma + 1}{\gamma - 1} + \dfrac{p_2}{p_1}} = \frac{u_1}{u_2} \tag{3·3}$$

$$\frac{T_2}{T_1} = \frac{p_2}{p_1} \frac{\dfrac{\gamma + 1}{\gamma - 1} + \dfrac{p_2}{p_1}}{1 + \dfrac{\gamma + 1}{\gamma - 1} \dfrac{p_2}{p_1}} \tag{3·4}$$

The fluid velocity behind the shock is

$$u_p = u_1 - u_2 = c_s(1 - u_2/u_1)$$

If Eqs. 3·2 and 3·3 are substituted, this becomes

$$u_p = \frac{a_1}{\gamma} \left(\frac{p_2}{p_1} - 1 \right) \left\{ \frac{\dfrac{2\gamma}{\gamma + 1}}{\dfrac{p_2}{p_1} + \dfrac{\gamma - 1}{\gamma + 1}} \right\}^{1/2} \tag{3·5}$$

A *weak shock* is defined as one for which the normalized pressure jump is very small, that is,

$$\frac{\Delta p}{p_1} = \frac{p_2 - p_1}{p_1} \ll 1$$

The other disturbances are then correspondingly small, as may be seen by expanding the above equations in series and retaining only the first terms in $\Delta p/p_1$. This gives

$$\frac{\Delta\rho}{\rho_1} \doteq \frac{1}{\gamma}\frac{\Delta p}{p_1} \doteq \frac{u_p}{a_1} \tag{3·6a}$$

$$\frac{\Delta T}{T_1} \doteq \frac{\gamma - 1}{\gamma}\frac{\Delta p}{p_1} \tag{3·6b}$$

$$c_s \doteq a_1\left(1 + \frac{\gamma + 1}{4\gamma}\frac{\Delta p}{p_1}\right) \tag{3·6c}$$

The last equation shows that the speed of very weak shocks is nearly equal to a_1.

A *very strong shock* is defined as one for which the pressure ratio p_2/p_1 is very large. In this case we have

$$\frac{\rho_2}{\rho_1} \to \frac{\gamma + 1}{\gamma - 1} \tag{3·6aa}$$

$$\frac{T_2}{T_1} \to \frac{\gamma - 1}{\gamma + 1}\frac{p_2}{p_1} \tag{3·6bb}$$

$$c_s \to a_1\left(\frac{\gamma + 1}{2\gamma}\frac{p_2}{p_1}\right)^{1/2} \tag{3·6cc}$$

$$u_p \to a_1\sqrt{\frac{2}{\gamma(\gamma + 1)}\frac{p_2}{p_1}} \tag{3·6dd}$$

3·3 One-Dimensional Isentropic Equations

The propagating shock wave illustrates the following general problem: Given the form of the "disturbance" at some time t, how does it change subsequently and what are the relations between the various quantities involved? In general the problem cannot be reduced to a corresponding stationary one, as in the preceding example, and the solution must be obtained from the nonstationary equations of motion. We shall first consider the case of adiabatic, nonviscous motion.

The differential equations of continuity (Eq. 2·4) and momentum (Eq. 2·18) were obtained in Chapter 2. For a tube of constant cross-sectional area, A may be factored out of the continuity equation, giving

$$\frac{\partial\rho}{\partial t} + \rho\frac{\partial u}{\partial x} + u\frac{\partial\rho}{\partial x} = 0 \tag{▶3.7}$$

Euler's form of the momentum equation is

$$\frac{\partial u}{\partial t} + u\frac{\partial u}{\partial x} + \frac{1}{\rho}\frac{\partial p}{\partial x} = 0 \tag{3·8}$$

This equation, without friction terms, is valid only if the velocity gradients are small enough to ensure that friction is negligible. It implies further (there being no external heat addition) that isentropic conditions exist. The isentropic relation between p and ρ may be left, for the present, in the functional form

$$p = p(\rho)$$

Since the pressure thus depends explicitly on the density, it is not an independent variable of the problem, and may be replaced in the momentum equation by means of the relation

$$\frac{\partial p}{\partial x} = \frac{dp}{d\rho}\frac{\partial \rho}{\partial x} = a^2\frac{\partial \rho}{\partial x} \tag{3·9}$$

where the pressure-density relation, $dp/d\rho$, is represented by a^2.

We define the disturbances or "perturbations" to be relative to the portion of the fluid that is *at rest*, with $u = 0$, $\rho = \rho_1$. Values of u different from zero, and values of ρ different from ρ_1, are called perturbations. They are not necessarily small.

It is convenient to introduce the definition

$$\rho = \rho_1(1 + \tilde{s}) \tag{3·10a}$$

The dimensionless quantity $\tilde{s} = (\rho - \rho_1)/\rho_1$ is called the *condensation*.†

With these definitions, the equations of motion may be written

$$\rho_1\frac{\partial \tilde{s}}{\partial t} + \rho_1\left(\frac{\partial u}{\partial x} + \tilde{s}\frac{\partial u}{\partial x}\right) + \rho_1 u\frac{\partial \tilde{s}}{\partial x} = 0 \tag{3·11a}$$

$$\frac{\partial u}{\partial t} + u\frac{\partial u}{\partial x} + \frac{a^2}{1 + \tilde{s}}\frac{\partial \tilde{s}}{\partial x} = 0 \tag{3·11b}$$

These two equations determine the relation between the fundamental disturbance quantities, that is, the *particle velocity* u and *the condensation* $\tilde{s}$. The pressure is related to $\tilde{s}$ by the isentropic relation; for a *perfect gas*, this is

$$\frac{p}{p_1} = \left(\frac{\rho}{\rho_1}\right)^\gamma = (1 + \tilde{s})^\gamma \tag{3·10b}$$

Similarly, the temperature is obtained from

$$\frac{T}{T_1} = (1 + \tilde{s})^{\gamma-1} \tag{3·10c}$$

†The usual notation is s, but we shall use $\tilde{s}$ to distinguish it from the entropy.

3·4 The Acoustic Equations

The equations obtained in the last article are exact insofar as the motion may be considered frictionless and nonconducting. However, they are not readily integrable, the principal difficulty being due to the *nonlinear* terms, such as $u(\partial u/\partial x)$ and $\tilde{s}(\partial u/\partial x)$, in which the *dependent* variables appear as coefficients of their derivatives. The equations can be *linearized*, and a considerable simplification realized, by making the assumption of *small disturbances.*

For instance, assuming $\tilde{s} \ll 1$, it is possible in Eqs. 3·11 to neglect $\tilde{s}(\partial u/\partial x)$ compared to $\partial u/\partial x$. The terms $u\, \partial\tilde{s}/\partial x$ and $u\, \partial u/\partial x$ are of the same order and may also be neglected. (This will be made more precise later when the relation between u and $\tilde{s}$ is worked out.) The a^2, appearing in Eq. 3·11*b*, may be expanded in Taylor series about the value $a_1{}^2$ in the undisturbed fluid, that is,

$$a^2 = \frac{dp}{d\rho} = \left(\frac{dp}{d\rho}\right)_1 + \left(\frac{d^2p}{d\rho^2}\right)_1 (\rho - \rho_1) = a_1{}^2 + \rho_1\tilde{s}\left(\frac{d^2p}{d\rho^2}\right)_1$$

Thus a^2 also differs from its value in the undisturbed fluid by a small quantity, and so the last term in Eq. 3·11*b* may be approximated by $a_1{}^2(\partial\tilde{s}/\partial x)$.

For small disturbances, then, the exact equations of motion may be approximated by a set of equations that contains no non-linear terms,

$$\frac{\partial \tilde{s}}{\partial t} + \frac{\partial u}{\partial x} = 0 \qquad \blacktriangleright(3{\cdot}12a)$$

$$\frac{\partial u}{\partial t} + a_1{}^2 \frac{\partial \tilde{s}}{\partial x} = 0 \qquad \blacktriangleright(3{\cdot}12b)$$

These are called the *acoustic equations* by virtue of the fact that the disturbances due to a sound wave are, by definition, very small.

The corresponding approximations in the isentropic relations (Eqs. 3·10*b* and 3·10*c*) for a perfect gas give

$$\frac{p}{p_1} = 1 + \gamma\tilde{s} \qquad (3{\cdot}13a)$$

$$\frac{T}{T_1} = 1 + (\gamma - 1)\tilde{s} \qquad (3{\cdot}13b)$$

Either of the dependent variables may be eliminated from Eqs. 3·12. Since $\dfrac{\partial^2 u}{\partial x\, \partial t} = \dfrac{\partial^2 u}{\partial t\, \partial x}$, cross differentiation gives

$$\frac{\partial^2 \tilde{s}}{\partial t^2} - a_1{}^2 \frac{\partial^2 \tilde{s}}{\partial x^2} = 0 \qquad \blacktriangleright(3{\cdot}14a)$$

and similarly

$$\frac{\partial^2 u}{\partial t^2} - a_1{}^2 \frac{\partial^2 u}{\partial x^2} = 0 \tag{3·14b}$$

This equation, which governs both $\tilde{s}$ and u, is called the *wave equation.* It is typical for phenomena in which a "disturbance" is propagated with a definite *signal velocity* or *wave velocity.* The signal velocity here, as we shall see, is a_1.

The solution of the wave equation may be written in very general form. It is

$$\tilde{s} = F(x - a_1 t) + G(x + a_1 t) \tag{3·15a}$$

where F and G are *arbitrary* functions of their arguments. This may be checked by direct substitution in the wave equation: Let $\xi = x - a_1 t$ and $\eta = x + a_1 t$. Then

$$\frac{\partial \tilde{s}}{\partial t} = \frac{dF}{d\xi}\frac{\partial \xi}{\partial t} + \frac{dG}{d\eta}\frac{\partial \eta}{\partial t} = a_1(-F' + G')$$

where the prime denotes differentiation with respect to the argument. Continuing in this way, it may be shown that $\partial^2 \tilde{s}/\partial t^2 = a_1{}^2(F'' + G'')$ and $\partial^2 \tilde{s}/\partial x^2 = F'' + G''$, so that Eq. 3·14*a* is satisfied.

Similarly the solution for u may be written in terms of two arbitrary functions,

$$u = f(x - a_1 t) + g(x + a_1 t) \tag{3·15b}$$

Of course, f and g are related to F and G, for u must be related to $\tilde{s}$ by the original equations (3·12). These are satisfied if

$$f = a_1 F \tag{3·15c}$$

$$g = -a_1 G \tag{3·15d}$$

as may again be checked by direct substitution.

3·5 Propagation of Acoustic Waves

The character of the solution (Eq. 3·15*a*) may be illustrated by first taking $G = 0$, so that the density distribution at time t is given by

$$\tilde{s} = F(x - a_1 t)$$

This represents a disturbance, or *wave*, which at time $t = 0$ had the (arbitrary) shape

$$\tilde{s} = F(x)$$

and which now, at time t, has exactly the *same shape*, but with corresponding points displaced a distance $a_1 t$ to the right (Fig. 3·2). That is, the velocity of each point in the wave, and thus of the wave itself, is a_1.

A wave such as this, in which the propagation velocity is in one direction, is called a *simple wave.* If it is assumed that $F = 0$ and that the wave is described by

$$\tilde{s} = G(x + a_1 t)$$

one obtains a simple wave propagating to the *left* with speed a_1. The general solution (Eq. 3·15) is a superposition of the two kinds of *simple* wave, one

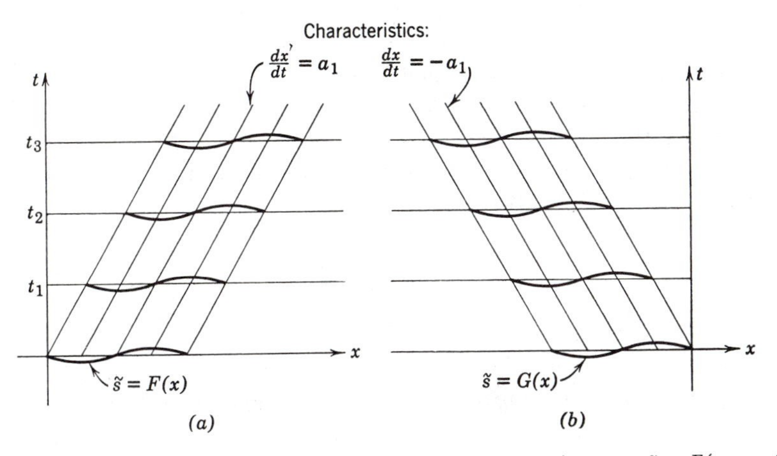

FIG. 3·2 Simple waves in the *x-t* plane. (*a*) Rightward propagating wave, $\tilde{s} = F(x - a_1 t)$; (*b*) leftward propagating wave, $\tilde{s} = G(x - a_1 t)$.

of which propagates to the left and the other to the right. Any acoustic wave may be resolved into two simple waves. The method is illustrated in the example of Article 3·8.

The lines in the *x-t* plane (Fig. 3·2) which trace the progress of the waves, that is, the lines of slope $dx/dt = \pm a_1$, are called the *characteristics* of the wave equation.

3·6 The Speed of Sound

The quantity $a = \sqrt{dp/d\rho}$ is called the *speed of sound*, or acoustic speed, since it is the speed with which disturbances propagate through the fluid. In the equations from which this result was obtained, in the preceding articles, it was assumed that friction is negligible, and so the result is applicable only to disturbances which are small enough for the assumption to be valid. Sound waves are, by definition, "small enough"; the criterion is that the velocity gradients due to the disturbances be so small that friction forces are negligible, and that $u/a_1 \ll 1$.

It follows that the motion in a sound wave is isentropic, since the entropy production, which depends on the squares of the velocity gradients (and temperature gradients), is also negligible. Thus the pressure-density relation in a sound wave must be the *isentropic* one, and a^2 should correctly be written

$$a^2 = \left(\frac{\partial p}{\partial \rho}\right)_s \tag{3·16}$$

The amplitude of ordinary audible sound is small enough so that the local production of entropy is, in fact, negligible, and Eq. 3·16 is very accurate for computing the speed of propagation.† An estimate of the error for sound of finite amplitude may be obtained from Eq. 3·6*c*.

The parameter a^2 plays a central role in compressible flow theory because it is, in fact, the pressure-density relation for the fluid. It is used to eliminate the pressure from the momentum equation (e.g., Articles 2·9 and 7·12) by means of the substitution

$$\frac{\partial p}{\partial x} = \frac{dp}{d\rho}\frac{\partial \rho}{\partial x} = a^2 \frac{\partial \rho}{\partial x}$$

The use of the total derivative implies that p is explicitly a function of ρ, that is, $p = p(\rho)$, while the introduction of a^2 implies that this pressure-density relation is the isentropic one. Thus the above substitution is valid only for isentropic flows. In cases with friction, heat addition, or any non-isentropic process, the pressure depends also on entropy, $p = p(\rho, s)$ and then

$$\frac{\partial p}{\partial x} = a^2 \frac{\partial \rho}{\partial x} + \left(\frac{\partial p}{\partial s}\right)_\rho \frac{\partial s}{\partial x}$$

in which the entropy term containing $\partial s/\partial x$ appears explicitly.

Finally it may be noted that a^2 may always be evaluated from the equation of state, using Eq. 3·16. For a perfect gas (Eq. 3·10*b*) this gives

$$a^2 = \gamma p/\rho = \gamma RT$$

a^2 is often used as replacement for the temperature, for instance, in the energy equation, but this substitution is valid only for a perfect gas. When evaluated for air ($\gamma = 1.4$, $R = 1715$ ft-lb/slug° F $= 2.87 \times 10^6$ erg/g° C), this relation gives for the speed of sound the value

$$a = 1095 \text{ ft/sec} \quad \text{at} \quad T = 500° \text{ R } (41° \text{ F})$$

$$a = 348 \text{ m/sec} \quad \text{at} \quad T = 300° \text{ K } (27° \text{ C})$$

The values at other temperatures may be conveniently obtained from these

†Friction (and local entropy production) is negligible for computing the speed of ordinary sound, but the cumulative effect on the amplitude is *not* negligible.

by using the relation

$$\frac{a_2}{a_1} = \sqrt{\frac{T_2}{T_1}}$$

3·7 Pressure and Particle Velocity in a Sound Wave

The pressure disturbance which accompanies the density wave may be obtained from Eq. 3·13*a*, for a perfect gas. It is

$$\frac{p - p_1}{p_1} = \frac{\Delta p}{p_1} = \gamma \bar{s} \tag{3·17}$$

Thus the pressure wave has the same shape as the density wave, differing only by the constant factor γ.

As the wave progresses through the fluid, the pressure disturbance sets the fluid in motion, giving it a velocity u, which is called the *particle velocity*. This is not to be confused with the *wave speed* a_1, which is ordinarily much higher. The corresponding velocities in the shock problem (Article 3·2) are u_p and c_s.

A simple wave, $\bar{s} = F(x - a_1 t)$, propagating to the right, produces a velocity disturbance which from Eqs. 3·15 is

$$u = a_1 F(x - a_1 t) = a_1 \bar{s} \tag{3·18a}$$

In a leftward propagating wave,

$$u = -a_1 G(x + a_1 t) = -a_1 \bar{s} \tag{3·18b}$$

These relations between $\bar{s}$ and u, in the two kinds of simple wave, are shown in Fig. 3·3.

The various parts of the wave are called *condensations*† or *rarefactions*, depending on whether the density is higher or lower than the undisturbed density ρ_1.

The effect that the wave produces on the fluid depends on the *gradient* of this density (and pressure) distribution and on the *direction* of motion of the wave. Thus the portion of a wave that increases the density, as it passes, is called a *compression*, and that which decreases the density is an *expansion*.

The corresponding distributions of particle velocity, shown in Fig. 3·3, are

$$u = \pm a_1 \bar{s}‡ \tag{3·19}$$

for compressions and expansions, respectively. It may be seen that a com-

†This is the origin of the name for the quantity $\bar{s}$ defined in Eq. 3·10*a*.

‡From this result it may be seen that $u \frac{\partial \bar{s}}{\partial x} = \bar{s} \frac{\partial u}{\partial x}$, which makes it possible to make a precise comparison of all the terms in Eqs. 3·11. It justifies the neglect of the nonlinear terms in obtaining Eqs. 3·12.

pression accelerates the fluid in the direction of wave motion, whereas an expansion decelerates it. In the general case of a nonsimple wave, which is a superposition of two simple waves (Eqs. 3·18), the relation between particle velocity and density is

$$\frac{u}{a_1\tilde{s}} = \frac{F - G}{F + G}$$

and is variable in space and time.

For later reference it may be noted that in the limit of vanishingly small disturbances the perturbation quantities may be written in differential

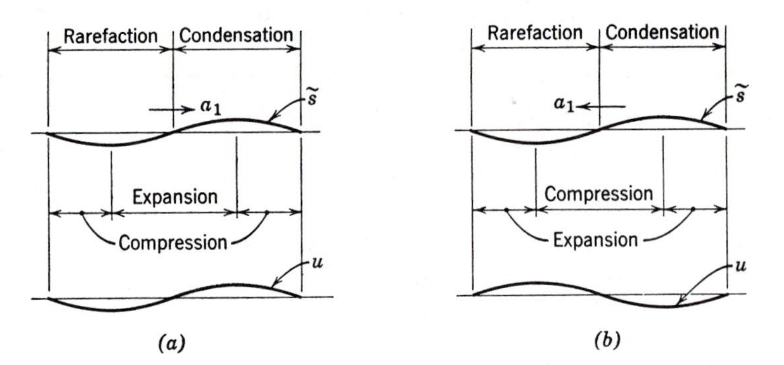

FIG. 3·3 Illustrating regions in simple acoustic waves. (*a*) Rightward propagating wave: $\tilde{s} = F(x - a_1t)$, $u = a_1F(x - a_1t)$; (*b*) leftward propagating wave: $\tilde{s} = G(x + a_1t)$, $u = -a_1G(x + a_1t)$.

form. u and $\tilde{s}$ are replaced by the differential quantities du and $d\rho/\rho_1$ and Eqs. 3·19 and 3·17 are then written

$$du = \pm a_1 \frac{d\rho}{\rho_1} \qquad (3\cdot20a)$$

$$\frac{dp}{p_1} = \gamma \frac{d\rho}{\rho_1} \qquad (3\cdot20b)$$

from which we may also obtain

$$dp = \frac{\gamma p_1}{\rho_1} d\rho = \pm \rho_1 a_1 \, du \qquad (3\cdot20c)$$

3·8 "Linearized" Shock Tube

To illustrate the application of the acoustic equations to a specific problem, consider the shock tube (Fig. 3·4). This is simply a tube which is divided by a membrane or diaphragm into two chambers in which the pressures are different. When the membrane is suddenly removed (broken), a wave motion is set up. If the pressure difference is so small that the motion may be approximately described by the acoustic equations, the shock tube might be called "acoustic" or "linearized." The exact equations for a shock tube are obtained in Article 3·12.

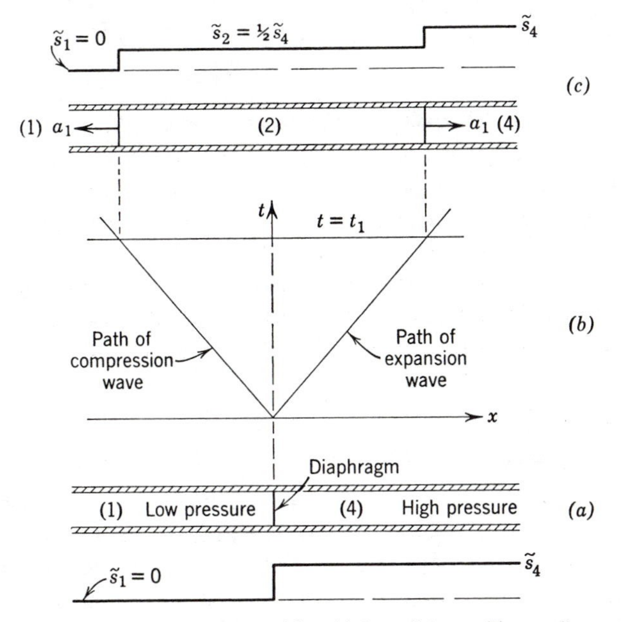

FIG. 3·4 Acoustic model of shock tube. (*a*) Initial conditions; (*b*) *x-t* diagram; (*c*) conditions at time t_1.

The low-pressure chamber (1) may be taken as reference ($\tilde{s}_1 = 0$). At the initial time $t = 0$, immediately after the membrane is removed, the wave has the shape shown in the figure, that is, a "step" distribution of density. The particle velocity at this first instant is everywhere zero. Thus the description of the wave at $t = 0$ is given by

$$\tilde{s}(x, 0) = F(x) + G(x) = \tilde{s}_0(x) = \begin{cases} \tilde{s}_4 & x > 0\dagger \\ 0 & x < 0 \end{cases}$$

$$u(x, 0) = a_1 F(x) - a_1 G(x) = 0$$

The simultaneous solution of these equations gives

$$F(x) = G(x) = \tfrac{1}{2}\tilde{s}_0(x) = \begin{cases} \tfrac{1}{2}\tilde{s}_4 & x > 0 \\ 0 & x < 0 \end{cases}$$

†The subscripts are chosen to correspond to the shock-tube flow discussed in Article 3·12.

Thus the description of the motion at any subsequent time is given by

$$\tilde{s}(x, t) = \tfrac{1}{2}\tilde{s}_0(x - a_1 t) + \tfrac{1}{2}\tilde{s}_0(x + a_1 t) = \begin{cases} \tilde{s}_4 & x > a_1 t \\ \tfrac{1}{2}\tilde{s}_4 & -a_1 t < x < a_1 t \\ 0 & x < -a_1 t \end{cases}$$

$$u(x, t) = \tfrac{1}{2}a_1\tilde{s}_0(x - a_1 t) - \tfrac{1}{2}a_1\tilde{s}_0(x + a_1 t) = \begin{cases} 0 & x > a_1 t \\ -\tfrac{1}{2}a_1\tilde{s}_4 & -a_1 t < x < a_1 t \\ 0 & x < -a_1 t \end{cases}$$

The distribution of density, at time t_1, is shown in Fig. 3·4. A compression wave is propagated into the low-pressure side (1), and an expansion wave, of equal strength, is propagated into the high-pressure side (4). (Actually a discontinuous expansion of finite strength cannot exist, as will be shown in Article 3·10, but in the acoustic theory this distinction between the nature of compression and expansion waves does not appear.) The particle velocity in the region (2) between shock and expansion is uniform. It will be seen from the above solution that it has the value

$$u = u_2 = \tfrac{1}{2}a_1\tilde{s}_4$$

3·9 Isentropic Waves of Finite Amplitude

The simplifying properties of acoustic waves, such as constant wave velocity and permanent shape of a simple wave, are due to the linearization of the equations, which depends on the assumption of infinitesimal amplitudes and gradients. If such an assumption cannot be made, the conditions at a given point in the wave may not be approximated by those in the undisturbed fluid. The wave velocity then differs from point to point, and a simple wave becomes distorted as it propagates. To describe such finite disturbances it is necessary to solve the complete nonlinear equations (3·7) and (3·8). The solution was obtained by Riemann and Earnshaw about a century ago. Here, instead of describing the formal mathematical solution,† we shall take a more physical approach. It may be summarized in the statement that from the point of view of an observer moving with the local particle velocity the acoustic theory applies *locally*.

The situation is illustrated in Fig. 3·5, which shows a wave of finite amplitude, on which is superimposed a "wavelet" at $x = x_n$. Relative to an observer moving with the local fluid velocity, u_n, the wavelet propagates with the local acoustic speed $a_n = (dp/d\rho)_n^{1/2}$, whereas, relative to the fixed frame of reference in the undisturbed fluid, it propagates with the speed $c_n = a_n + u_n$. We may now dispense with the wavelet. It is clear that, at *any* point x_n, the local wave speed is given by this expression, or, in

†Cf. Exercise 3·4.

general,

$$c = a \pm u \tag{3·21}$$

The negative sign is to be taken if the wave is travelling to the left. The sign convention is that u is a velocity (same sign as x) while c and a are speeds (always positive).

The wave speed is no longer constant, as in the linearized theory, for not only is the acoustic speed a variable but also the particle velocity u may no longer be neglected. To evaluate these in terms of the density,

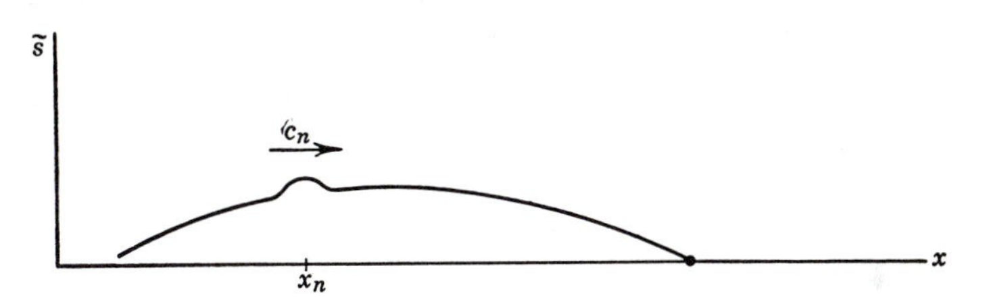

FIG. 3·5 "Wavelet" on a finite amplitude wave.

the acoustic theory is applied locally. First, the local acoustic speed may be evaluated in terms of the density. Using the isentropic relation to eliminate p from $a^2 = \gamma p/\rho$, we obtain, for a perfect gas,

$$a = a_1 \left(\frac{\rho}{\rho_1}\right)^{(\gamma-1)/2} \tag{3·22}$$

Next, the particle velocity is evaluated in terms of the density by applying Eq. 3·20*a* *locally*, that is,

$$du = \pm a \left(\frac{d\rho}{\rho}\right)$$

This becomes integrable when a is replaced by Eq. 3·22, as follows:

$$u = \pm \int_{\rho_1}^{\rho} a \frac{d\rho}{\rho} = \pm \frac{2a_1}{\gamma - 1}\left[\left(\frac{\rho}{\rho_1}\right)^{(\gamma-1)/2} - 1\right] = \pm \frac{2}{\gamma - 1}(a - a_1) \tag{3·23}$$

$$a = a_1 \pm \frac{\gamma - 1}{2} u \tag{3·23a}$$

When this expression for the local acoustic speed is substituted in Eq. 3·21, the equation for wave speed becomes

$$c = a_1 \pm \frac{\gamma + 1}{2} u \tag{3·21a}$$

or

$$c = a_1 \left\{ 1 \pm \frac{\gamma + 1}{\gamma - 1} \left[\left(\frac{\rho}{\rho_1} \right)^{(\gamma-1)/2} - 1 \right] \right\} \quad \blacktriangleright (3 \cdot 21b)$$

where a_1 is the speed of sound in the undisturbed fluid.

3·10 Propagation of Finite Waves

The consequences of this "nonlinear" wave speed may be illustrated by considering again the propagation of a simple wave, now allowing it to have finite amplitude. For comparison with Fig. 3·2, we assume that the

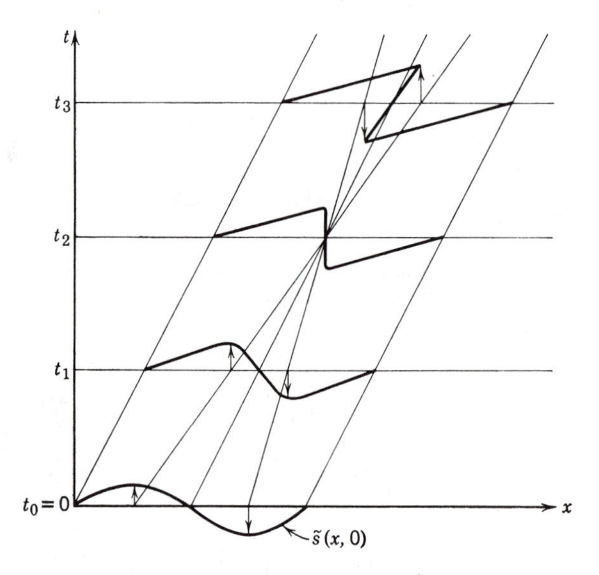

FIG. 3·6 Propagation of a wave of finite amplitude. (Arrows help to identify points in the wave.)

initial density distribution is that shown in Fig. 3·6 at $t = t_0 = 0$. For the rightward-propagating wave, the signs in the equations of the previous article are to be taken positive. According to Eq. 3·21*b*, the wave speed is higher than a_1 in regions of condensation ($\rho > \rho_1$) and lower than a_1 in regions of rarefaction. This means that the wave distorts as it propagates, the regions of higher condensation tending to overtake those of lower condensation, as shown in Fig. 3·6. In regions of higher condensation, the *characteristic lines* (Article 3·5) are inclined more, since the slope is inversely proportional to the wave speed.

In terms of the compression and expansion regions defined in Fig. 3·3, the net effect is to steepen compression regions and to flatten expansion regions, in which the characteristic lines converge and diverge, respectively. In a compression region, the characteristics would eventually cross, leading to the situation shown for $t = t_3$. But this would be physically impossible, for it implies three values of density at a given point x. Actually, before this happens, or even before the situation of $t = t_2$ is reached, the velocity and temperature gradients in the compression regions become so great that *friction and heat transfer effects* become important. These have a diffusive action which counteracts the steepening tendency. The two opposing effects achieve a balance, and the compression portion of the wave becomes "stationary," in the sense that it propagates without further distortion. It is then a shock wave.

In compression regions, the isentropic relations are valid only up to the time that friction and heat transfer first become important. On the other hand, once a stationary balance between diffusive and steepening terms has been reached, the conditions *across* the wave front are given by the shock wave relations of Article 3·2; the conditions *inside* the shock are discussed in Article 13·12. The intermediate, nonstationary, nonisentropic interval can be treated only with the full nonstationary equations, including friction and heat transfer terms, and in this sense, is the most complicated one.

For finite amplitudes, then, there is an important difference in the behavior of compression and expansion waves. A compression wave tends to steepen and reach a "stationary" state in which it is no longer isentropic. An expansion wave, on the other hand, always remains isentropic, for it tends to flatten and so further reduce the velocity and temperature gradients. It never achieves a "stationary" condition, corresponding, in shock wave theory, to the fact that there are no "expansion shocks."

The mechanisms at work here are basically similar to those in the two-dimensional, steady supersonic flows to be discussed in Chapter 4, and the resulting effects are analogous. For example, in the expansion regions of steady, two-dimensional supersonic flow, the characteristics or Mach lines diverge, whereas in the compression regions they converge to form a shock wave.

In the shock wave motion discussed in Article 3·2 it is assumed that the piston is set into motion impulsively, with the velocity u_p, and thus the shock wave starts out instantaneously, with the speed c_s. It may be seen now, from the preceding discussion, that, even if the velocity u_p is attained gradually, so that the compression is isentropic at first, the steepening effect will eventually establish the same shock wave as in the impulsive case. In fact, this effect will always tend to steepen and smooth a wave in which the pressure increases. For example, the complicated motion in the initial instants of an explosion or in the bursting of a shock-tube diaphragm develops into a wave motion with a steep shock front.

3·11 Centered Expansion Wave

If, instead of moving the piston *into* the fluid as in Fig. 3·1, the piston is *withdrawn* as in Fig. 3·7, an expansion wave is produced. If the piston starts impulsively, with speed $|u_p|$, the distribution of particle velocity in the first instant is a "step." But in an expansion wave this step cannot be

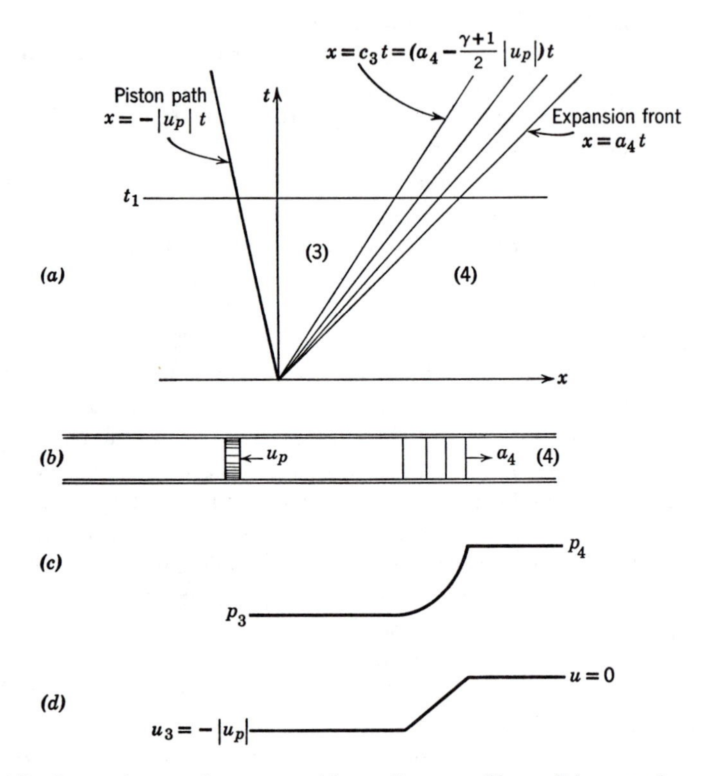

FIG. 3·7 Centered expansion wave. (*a*) *x-t* diagram; (*b*) conditions at time t_1; (*c*) pressure distribution at t_1; (*d*) fluid velocity distribution at t_1.

maintained; as soon as the wave starts propagating it begins to "flatten." At some later time t_1 the particle velocity has a linear distribution, as shown in Fig. 3·7*d*, and the pressure has a corresponding distribution, shown in Fig. 3·7*c*.

The front of the wave propagates with speed a_4 into the undisturbed fluid (4)†, that is, in the direction *opposite* to the piston and fluid motion. (The front of any *isentropic* wave propagates at the speed of sound of the undisturbed fluid.) The wave speed in the portions of the wave behind the

†See the footnote on page 73.

front is given by Eq. 3·21*a*, that is,

$$c = a_4 + \frac{\gamma + 1}{2} u$$

c decreases continuously through the wave, since $u < 0$. The fan of straight lines shown in Fig. 3·7 are lines of constant c, and thus of constant u and ρ. These lines are the characteristics. With increasing time the fan becomes wider, that is, the wave becomes "flatter," and the gradients of velocity, density, etc., become smaller. Thus *the wave remains isentropic.*

The terminating characteristic is given by

$$x/t = c_3 = a_4 - \frac{\gamma + 1}{2} |u_p|$$

and slopes to right or left, depending on whether $a_4 \gtrless \frac{1}{2}(\gamma + 1)|u_p|$. Between the terminating characteristic and the piston, the fluid properties have the uniform values ρ_3, p_3, a_3, etc. For a perfect gas, they are given in terms of the conditions in (4) by the isentropic relations of Article 3·9. For example, from Eq. 3·23,

$$\frac{\rho_3}{\rho_4} = \left(1 - \frac{\gamma - 1}{2} \frac{|u_p|}{a_4}\right)^{2/(\gamma-1)} \tag{3·24a}$$

$$\frac{p_3}{p_4} = \left(1 - \frac{\gamma - 1}{2} \frac{|u_p|}{a_4}\right)^{2\gamma/(\gamma-1)} \tag{3·24b}$$

The pressure ratio p_3/p_4 defines the strength of the expansion wave.

The maximum expansion that can be obtained corresponds to $\rho_3 = 0$, and is obtained when $|u_p| = 2a_4/(\gamma - 1)$, according to Eq. 3·24*a*. In this case, $p_3 = T_3 = 0$, that is, all the fluid energy is converted into kinetic energy of flow. If the piston velocity is higher than this limiting value, it produces no further effect on the flow.

The wave that we have described here, produced by an impulsive withdrawal of the piston, is called a *centered* expansion wave, after the fan-like set of characteristic lines in the x-t plane. With the centered expansion wave and the shock wave at our disposal, it is possible to obtain the solutions to some other finite amplitude wave problems. It must be remembered, however, that shock waves, and finite amplitude waves in general, cannot be simply superimposed. Some interactions may be treated by the method of characteristics, which is described in Chapter 12 for the case of steady supersonic flow, for which the equations are analogous.

3·12 The Shock Tube

The solutions worked out in preceding articles, for shock waves and expansion waves, may be used to analyze flow conditions in the *shock tube.*

Its basic principle was described in Article 3·8, where an "acoustic," or linearized, solution was obtained for the case of very small pressure ratio.

The flow regions and nomenclature are illustrated in Fig. 3·8. A diaphragm at $x = 0$ separates the low-pressure (expansion) chamber (1)

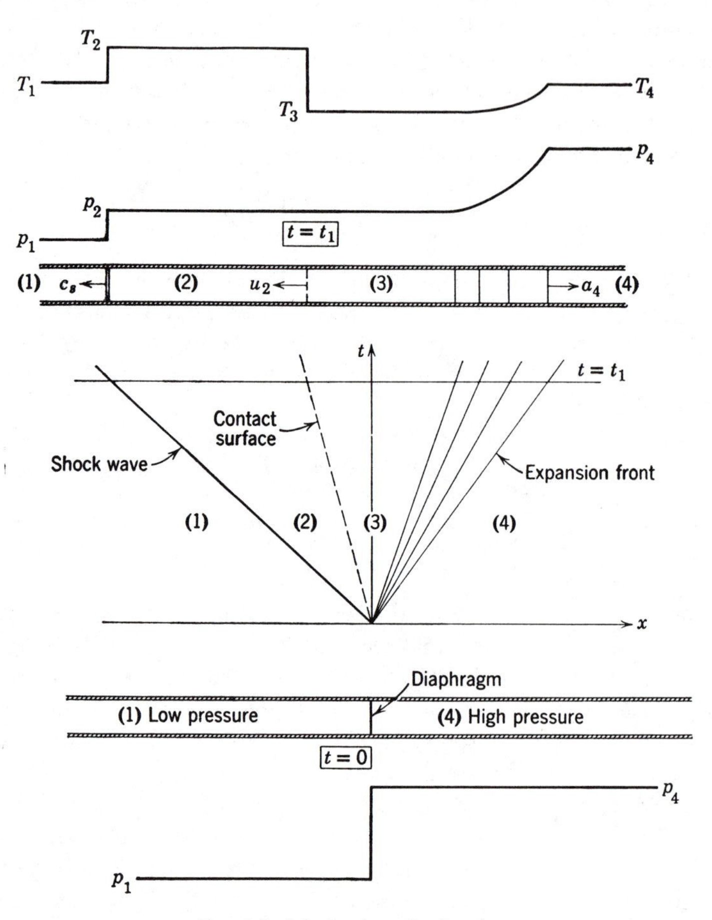

FIG. 3·8 Motion in a shock tube.

from the high-pressure (compression) chamber (4). The basic parameter of the shock tube is the *diaphragm pressure ratio*, p_4/p_1. The two chambers may be at different temperatures, T_1 and T_4, and may contain different gases with gas constants R_1 and R_4.

At the initial instant, when the diaphragm is burst, the pressure distribution is ideally a "step," as shown for $t = 0$. This "splits," as in the linearized example, into a shock wave, which propagates into the expansion chamber with speed c_s, and an expansion wave, which propagates into the compression chamber with the speed a_4 at its front. The condition of the fluid which is traversed by the shock is denoted by (2), and that of the fluid traversed by the expansion wave is denoted by (3). The interface between regions 2 and 3 is called the *contact surface.* It marks the boundary between the fluids which were initially on either side of the diaphragm. Neglecting diffusion, they do not mix, but are permanently separated by the contact surface, which is like the front of a piston, driving into the low-pressure chamber.

On either side of the contact surface, the temperatures, T_2 and T_3, and the densities, ρ_2 and ρ_3, may be different, but it is necessary that the pressure and fluid velocity be the same, that is,

$$p_2 = p_3$$

$$u_2 = u_3$$

(The latter must thus be the velocity of the contact surface.) These two conditions are sufficient to determine the *shock strength*, p_2/p_1, and the expansion strength, p_3/p_4, in terms of the diaphragm pressure ratio, p_4/p_1, as follows. The values of u_2 and u_3 may be calculated from Eqs. 3·5 and 3·24, which are for shock and expansion waves, respectively. Slightly rearranged, and with subscripts to correspond to the present case, they give

$$u_2 = a_1\left(\frac{p_2}{p_1} - 1\right)\sqrt{\frac{2/\gamma_1}{(\gamma_1 + 1)p_2/p_1 + (\gamma_1 - 1)}} \tag{3·25a}$$

$$u_3 = \frac{2a_4}{\gamma_4 - 1}\left[1 - \left(\frac{p_3}{p_4}\right)^{(\gamma_4-1)/2\gamma_4}\right] \tag{3·25b}$$

Elimination of $u_2 = u_3$ from these equations, and substitution of $p_3 = p_2$, results in the basic shock-tube equation,

$$\frac{p_4}{p_1} = \frac{p_2}{p_1}\left[1 - \frac{(\gamma_4 - 1)(a_1/a_4)(p_2/p_1 - 1)}{\sqrt{2\gamma_1}\sqrt{2\gamma_1 + (\gamma_1 + 1)(p_2/p_1 - 1)}}\right]^{-2\gamma_4/(\gamma_4-1)} \tag{3·26}$$

This gives the shock strength p_2/p_1 implicitly as a function of the diaphragm pressure ratio p_4/p_1. The expansion strength is then obtained from

$$\frac{p_3}{p_4} = \frac{p_3}{p_1}\frac{p_1}{p_4} = \frac{p_2/p_1}{p_4/p_1} \tag{3·27}$$

Once the shock strength is known, all other flow quantities are easily determined from the normal shock relations (Articles 2·13 and 3·2).

Although the values of u and p calculated across the shock and expansion must be identical, this is not necessary for ρ and T, and in fact they usually are different. The region behind the shock is different from the one behind the expansion; they are separated by the contact surface. On the two sides of the contact surface, the pressures and the fluid velocities are the same but the densities and temperatures are different. The temperature T_3 behind the expansion wave is given by the isentropic relation,

$$\frac{T_3}{T_4} = \left(\frac{p_3}{p_4}\right)^{(\gamma_4-1)/\gamma_4} = \left(\frac{p_2/p_1}{p_4/p_1}\right)^{(\gamma_4-1)/\gamma_4} \tag{3·28}$$

The temperature T_2 behind the shock is given by the Rankine-Hugoniot relation (Eq. 3·4),

$$\frac{T_2}{T_1} = \frac{1 + \dfrac{\gamma_1 - 1}{\gamma_1 + 1}\dfrac{p_2}{p_1}}{1 + \dfrac{\gamma_1 - 1}{\gamma_1 + 1}\dfrac{p_1}{p_2}} \tag{3·29}$$

The corresponding ratios for the speed of sound and the density are then easily written. The velocity of the contact surface may be obtained from either Eq. 3·25*a* or Eq. 3·25*b*. Other shock-tube equations are given in the Exercises.

Experimentally it is not possible to start the flow in the ideal way, since the bursting or shattering of the diaphragm is a complicated, three-dimensional phenomenon. Nevertheless a plane shock wave is developed within a few diameters, by the steepening effect associated with compression waves. It is found that at low pressure ratios the speed and strength of the shock agrees well with the theoretical predictions (Fig. 3·9), but that discrepancies begin to appear at higher pressure ratios. These are due to several effects: (1) three dimensional and finite time effects during rupture of the diaphragm; (2) viscous effects related to formation of a boundary layer on the shock tube wall; (3) changes in gas properties at the high temperatures behind the shock.

Less information is available about the other flow regions, but it appears that the above effects have an important influence on them, especially at high pressure ratios.

The duration of flow is limited by the lengths of the expansion and compression chambers, since the shock wave and expansion wave reflect from the ends of the chambers and eventually interact with each other.

Many adaptations and applications of the shock tube have been discovered. For instance, the uniform flow behind the shock wave may be used as a short-duration wind tunnel. In this role the shock tube is similar to an intermittent or blow-down tunnel, with the difference that the duration of flow is much shorter, usually of the order of a millisecond. On the other hand, the operating conditions (particularly the high stagnation

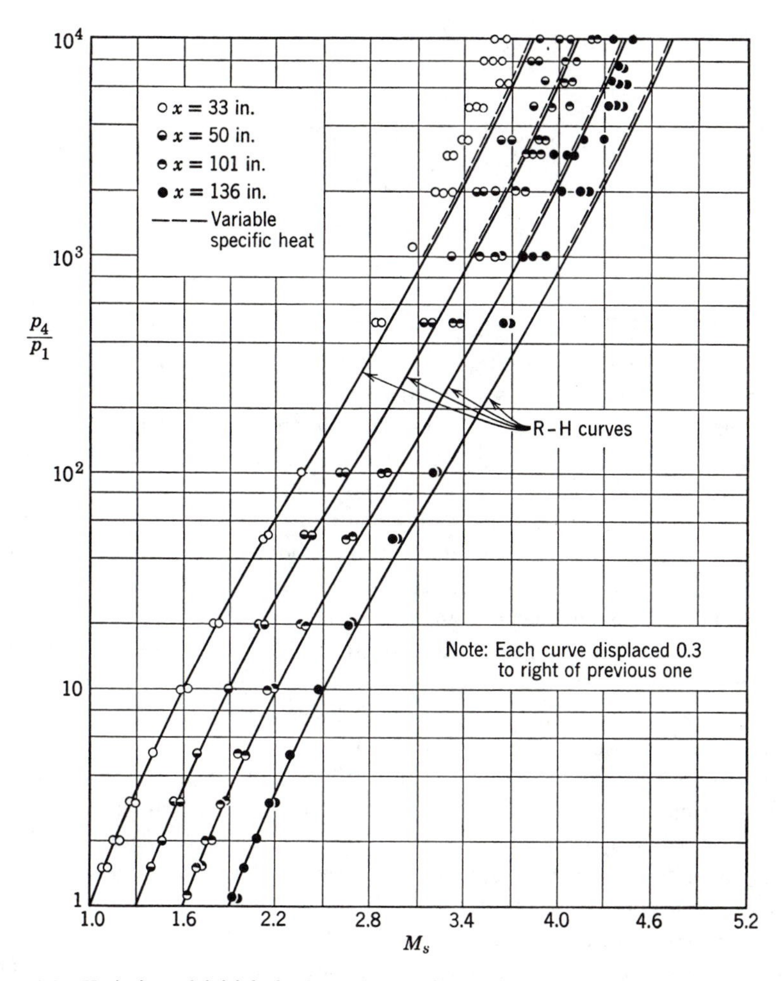

FIG. 3·9 Variation of initial shock wave velocity with diaphragm pressure ratio and distance (x). (Air in both chambers.) (Solid lines are Rankine-Hugoniot values for a perfect gas. Dashed lines are corrections for the variation of specific heat with temperature.) I. I. Glass and G. N. Patterson, *J. Aeronaut. Sc.*, *22* (1955), p. 73.

enthalpies) which are possible cannot be easily obtained with other types of facility.

The abrupt changes of flow condition at the shock front may be utilized for studying transient aerodynamic effects, and for studies of dynamic and thermal response.

In the field of molecular physics, the shock tube provides a simple means for producing fast changes in the state of a fluid, in order to observe relaxation effects, reaction rates, etc. With the high enthalpies that are attainable, it is possible to study dissociation, ionization, etc.

C H A P T E R 4

Waves in Supersonic Flow

4·1 Introduction

The one-dimensional example of a piston moving in a fluid, discussed in the preceding chapter, shows that there is a certain relation between the wave motion in the fluid and the pressure on the piston. This example is significant, for it is the simplest case of what might be described as generalized "piston motion." That is, the motion of any arbitrary body in a fluid at rest may be interpreted from this point of view. The disturbances in those parts of the fluid that are adjacent to the body are transmitted to other parts of the body by the propagation of waves. The wave motion that is established is the one that is compatible with the motion of the body, and this in turn determines the pressures on the body. Of course, in the three-dimensional case, it is necessary to use the general, three-dimensional equations of wave propagation, but in principle it is mainly a matter of greater geometrical complexity (see Article 9·19).

This point of view for studying the motion of a body in a fluid is not always a convenient one. In the case of *steady* motion, particularly, it is usually much simpler to view the motion from a reference system in which the body is stationary and the fluid flows over it. One usually does not think then in terms of wave motion, especially if the flow is subsonic. However, if the relative speed is *supersonic*, the waves cannot propagate ahead of the immediate vicinity of the body, and the *wave system travels with the body*. Thus, in the reference system in which the body is stationary the wave system is also stationary, and the correspondence between the wave system and the flow field is then quite direct.

In this chapter we shall mainly consider problems of *steady, two-dimensional (plane) supersonic flow*. Using the fact that in this case there is a steady wave system, we shall find solutions by an indirect approach. That is, we shall first study the conditions under which simple stationary waves may exist in the flow, and then find the flow boundaries to which they correspond or which may be fitted to them. In this procedure the *limited upstream influence* in a supersonic field is very helpful, for it allows flows to be analyzed or constructed step by step, which is a method that is not possible in the subsonic case.

4·2 Oblique Shock Waves

We shall first investigate the conditions for a stationary shock wave to be *oblique* to the flow direction (Fig. 4·1). This could be treated in the same way as the normal shock (Article 2·13) directly from the equations of motion, taking into account the additional component of velocity. However, some calculation can be avoided by making use of the normal shock results in the following way.

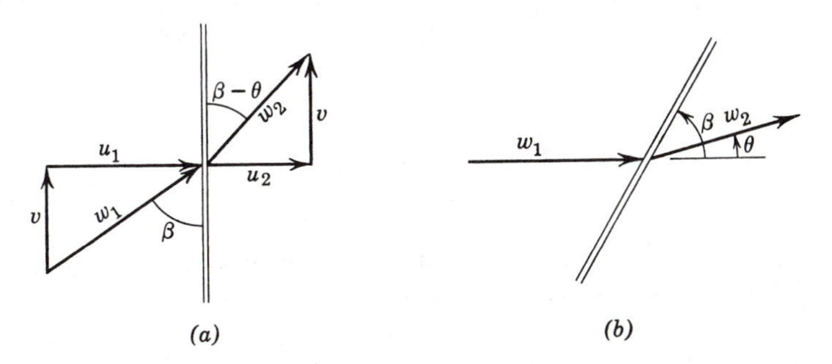

FIG. 4·1 Flow through an oblique shock wave. (*a*) Resolution of velocity components; (*b*) conventional nomenclature.

If a uniform velocity v is superimposed on the flow field of the normal shock, the resultant velocity ahead of the shock may be adjusted to *any* direction simply by adjusting the magnitude and direction of v. If v is taken parallel to the shock wave, as shown in Fig. 4·1*a*, the resultant velocity ahead of the shock is $w_1 = \sqrt{u_1^2 + v^2}$ and its inclination to the shock is given by $\beta = \tan^{-1} (u_1/v)$. Now, since u_2 is not the same as u_1, the inclination of the flow after the shock is different from that ahead, that is, the flow *turns* abruptly at the shock. Since u_2 is always less than u_1, the turn is always towards the shock, and the angle of deflection θ, defined in Fig. 4·1, is positive. In part (*b*) of the figure, the flow of (*a*) has been rotated so that the initial flow direction is more conventionally aligned.

The relations for the conditions before and after the shock may now be determined easily, since the superposition of a uniform velocity v does not affect the static pressures and other static parameters, which were defined for the normal shock (Article 2·13). The only modification is that the initial Mach number is now $M_1 = w_1/a_1$ and that $u_1 = w_1 \sin \beta$, or

$$\frac{u_1}{a_1} = M_1 \sin \beta \tag{4·1}$$

Thus wherever u_1/a_1 occurs in Eqs. 2·47, 2·48, 2·49, and 2·50 it is to be replaced by $M_1 \sin\beta$. This gives the corresponding relations for the oblique shock:

$$\frac{\rho_2}{\rho_1} = \frac{(\gamma+1)M_1^2\sin^2\beta}{(\gamma-1)M_1^2\sin^2\beta+2} \tag{4·2}$$

$$\frac{p_2-p_1}{p_1} = \frac{2\gamma}{\gamma+1}(M_1^2\sin^2\beta-1) \tag{4·3}$$

$$\frac{T_2}{T_1} = \frac{a_2^2}{a_1^2} = 1 + \frac{2(\gamma-1)}{(\gamma+1)^2}\frac{M_1^2\sin^2\beta-1}{M_1^2\sin^2\beta}(\gamma M_1^2\sin^2\beta+1) \tag{4·4}$$

$$\frac{s_2-s_1}{R} = \ln\left[1+\frac{2\gamma}{\gamma+1}(M_1^2\sin^2\beta-1)\right]^{1/(\gamma-1)} \times\left[\frac{(\gamma+1)M_1^2\sin^2\beta}{(\gamma-1)M^2\sin^2\beta+2}\right]^{-\gamma/(\gamma-1)} = \ln\frac{p_{01}}{p_{02}} \tag{4·5}$$

In other words, the ratios of the static thermodynamic variables depend only on the *normal* component of the velocity. From the normal shock analysis, this component must be supersonic, that is, $M_1\sin\beta \geq 1$. This sets a *minimum wave inclination* for a given Mach number. The *maximum* wave inclination, of course, is that for a normal shock, $\beta = \pi/2$. Thus we have, for a given initial Mach number, a range of possible wave angles,

$$\sin^{-1}\frac{1}{M} \leq \beta \leq \frac{\pi}{2} \tag{4·6}$$

For each wave angle β (at given M_1) there is a corresponding deflection angle θ. The relation between them is obtained in the following article.

The Mach number M_2 after the shock may be obtained by noting that $M_2 = w_2/a_2$ and (from Fig. 4·1a) that $u_2/a_2 = M_2\sin(\beta-\theta)$. This is to be substituted for M_2 in Eq. 2·45, giving

$$M_2^2\sin^2(\beta-\theta) = \frac{1+\frac{\gamma-1}{2}M_1^2\sin^2\beta}{\gamma M_1^2\sin^2\beta-\frac{\gamma-1}{2}} \tag{4·7}$$

Numerical values of the oblique shock relations for a perfect gas, with $\gamma = 1.40$, may be obtained from the charts at the end of the book.

4·3 Relation between β and θ

From Fig. 4·1, the following two relations may be obtained

$$\tan\beta = \frac{u_1}{v} \tag{4·8a}$$

$$\tan(\beta - \theta) = \frac{u_2}{v} \tag{4·8b}$$

Eliminating v, and using continuity and Eq. 4·2, gives

$$\frac{\tan(\beta - \theta)}{\tan \beta} = \frac{u_2}{u_1} = \frac{\rho_1}{\rho_2} = \frac{(\gamma - 1)M_1^2 \sin^2 \beta + 2}{(\gamma + 1)M_1^2 \sin^2 \beta} \tag{4·9}$$

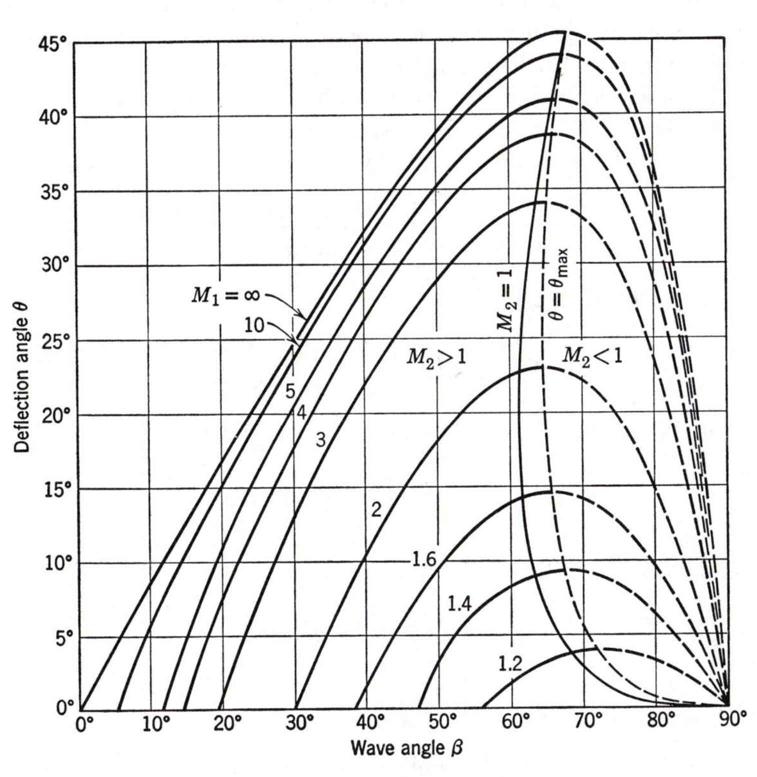

FIG. 4·2 Oblique shock solutions.

For given M_1, this is an implicit relation between θ and β. With a little trigonometric manipulation, it may be rewritten to show the dependence of θ explicitly,

$$\tan \theta = 2 \cot \beta \frac{M_1^2 \sin^2 \beta - 1}{M_1^2(\gamma + \cos 2\beta) + 2} \tag{4·10}$$

This expression becomes zero at $\beta = \pi/2$ and at $\beta = \sin^{-1}(1/M_1)$, which are the limits of the range defined in Eq. 4·6. Within this range θ is positive, and must therefore have a maximum value. This is shown in Fig. 4·2, in which the relation between θ and β is plotted for various values of M_1. For each value of M_1 there is a maximum value of θ.

If $\theta < \theta_{max}$, then for each value of θ and M there are two possible solutions, having different values of β. The larger value of β gives the stronger shock. On Fig. 4·2, the strong shock solutions are indicated by dashed lines.

Also shown on the figure is the locus of solutions for which $M_2 = 1$. *In the solution with strong shock the flow becomes subsonic.* In the solution with weak shock the flow remains supersonic, except for a small range of values of θ slightly smaller than θ_{max}.

The relation between θ and β may be obtained in another useful form by rearranging Eq. 4·9 in the following way. Divide the numerator and denominator of the right-hand side by $(1/2)M_1^2 \sin^2 \beta$, and then solve for

$$\frac{1}{M_1^2 \sin^2 \beta} = \frac{\gamma + 1}{2} \frac{\tan (\beta - \theta)}{\tan \beta} - \frac{\gamma - 1}{2}$$

Further reduction then gives

$$M_1^2 \sin^2 \beta - 1 = \frac{\gamma + 1}{2} M_1^2 \frac{\sin \beta \sin \theta}{\cos (\beta - \theta)} \qquad \blacktriangleright(4\cdot11)$$

For small deflection angles θ, this may be approximated by

$$M_1^2 \sin^2 \beta - 1 \doteq \left(\frac{\gamma + 1}{2} M_1^2 \tan \beta\right) \cdot \theta \qquad (4\cdot11a)$$

If M_1 is very large, then $\beta \ll 1$ but $M_1\beta \gg 1$, and Eq. 4·11a reduces to

$$\beta = \frac{\gamma + 1}{2} \theta \qquad (4\cdot11b)$$

4·4 Supersonic Flow over a Wedge

In nonviscous flow, any streamline can be replaced by a solid boundary. Thus the oblique shock flow, described in the last article, provides the solution to *supersonic flow in a corner*, as illustrated in Fig. 4·3a. For given values of M_1 and θ, the values of β and M_2 are determined (Charts 1 and 2). For the present, we shall consider only cases in which $M_2 > 1$. This restricts the oblique shock to the weaker of the two possible ones and also requires that θ be smaller than θ_{max}. The other (stronger) solution, as well as the cases $\theta > \theta_{max}$, will be discussed later.

By symmetry, the flow over a wedge of nose angle 2θ is also obtained (Fig. 4·3b). But it is not essential that the wedge be symmetric. In Fig. 4·3c, the flow on each side of the wedge is determined only by the inclination of the surface on that side. This is a general result: *if the shock waves are attached to the nose, the upper and lower surfaces are independent,* for there is no influence on the flow upstream of the waves.

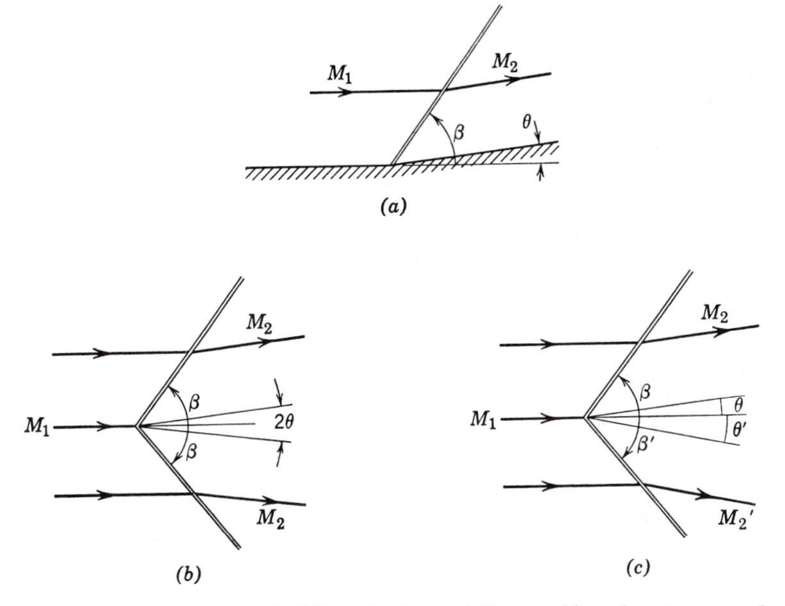

FIG. 4·3 Supersonic flows with oblique shocks. (*a*) Corner; (*b*) wedge at zero angle of attack; (*c*) wedge at angle of attack.

4·5 Mach Lines

The portion of Fig. 4·2 that we are at present considering ($M_2 > 1$) shows that a decrease of the wedge angle θ corresponds to a decrease of the wave angle β. When θ decreases to zero, β decreases to the limiting value μ (Fig. 4·4*b*) which is given, from Eq. 4·11, by the expression

$$M_1^2 \sin^2 \mu - 1 = 0 \tag{4·12}$$

It will be seen that the strength of the "wave," as given by any of the jump quantities (Eqs. 4·2–4·5), also becomes zero. There is, in fact, no disturbance in the flow. In Fig. 4·4*b* there is no longer anything unique about the point P; it might be any point in the flow. The angle μ is simply a characteristic angle associated with the Mach number M by the relation

$$\mu = \sin^{-1} \frac{1}{M} \tag{4·13}$$

It is called the *Mach angle.*

The lines of inclination μ which may be drawn at any point in the flow field are called *Mach lines*, and, sometimes, *Mach waves*. The latter name, however, is misleading for it is often used, ambiguously, for the weak but finite waves that are produced by small disturbances.

If the flow is not uniform, then μ varies, with M, and the Mach lines are curved.

At any point P in the field (Fig. 4·4c), there are always two lines which intersect the streamline at the angle μ. (In three-dimensional flow, the

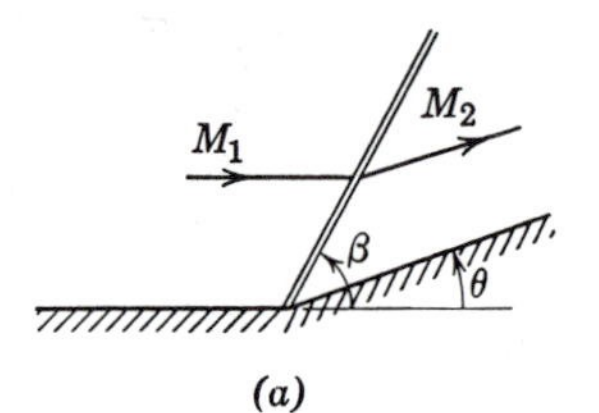

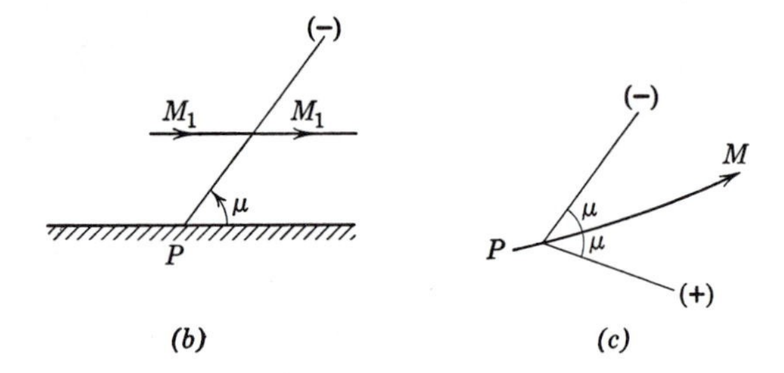

FIG. 4·4 Comparison of oblique shocks. (a) Oblique shock due to deflection θ; (b) degeneration to Mach line as $\theta \to 0$; (c) left- and right-running Mach lines at an arbitrary point in the flow.

Mach lines define a conical *surface*, with vertex at P.) Thus a two-dimensional supersonic flow is always associated with two families of Mach lines. These may be conveniently distinguished by the labels $(+)$ and $(-)$, as shown in Fig. 4·4c. Those in the $(+)$ set run to the right of the streamline, and those in the $(-)$ set run to the left. They are also called *characteristics*, a name that derives from the mathematical theory of the hyperbolic differential equations that describe the flow. They are, in fact, analogous to the two families of characteristics that trace the propagation of one-dimensional waves in the x-t plane (Article 3·5). In Chapter 12 it will be shown how the characteristics system provides the basis of a method of computation.

Like the characteristics in the x-t plane, Mach lines have a distinguished direction, that is, the direction of flow, which is the direction of "increasing time." This is related in an obvious way to the fact that there is no *upstream* influence in supersonic flow.

4·6 Piston Analogy

The analogy between characteristics in the acoustic problem and Mach lines in two-dimensional supersonic flow applies also to other aspects of wave propagation, as shown in Fig. 4·5. In Fig. 4·5*a* the x-t plane of Chapter 3 has been rotated so that the t-axis is horizontal. At $t = 0$ the piston is started impulsively with velocity u_p, producing a shock wave which runs ahead at the speed c_s. The traces of the piston and shock wave in the x-t plane correspond to the wedge surface and oblique shock in the x_1-x_2 plane (Fig. 4·5*b*). In this analogy the x_1-axis is "*time-like.*"

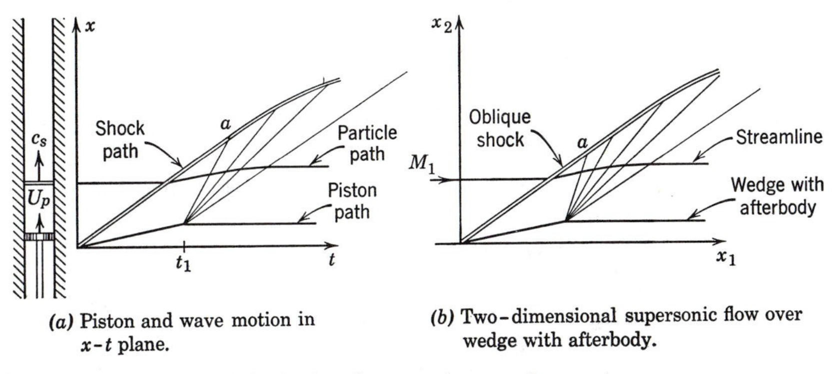

(*a*) Piston and wave motion in $x-t$ plane.

(*b*) Two-dimensional supersonic flow over wedge with afterbody.

FIG. 4·5 Analogy between the x-t and x_1-x_2 planes.

If the piston is *stopped* impulsively at $t = t_1$, then its subsequent history is a line at constant x. On the wedge, this corresponds to a shoulder followed by an afterbody with surface parallel to the free stream. In the piston problem, this produces a centered expansion wave, starting from t_1. In the wedge problem, there is an analogous expansion wave, centered at the shoulder. Its position is indicated by the straight Mach lines, inclined at the local Mach angles to the streamline. The flow relations for this expansion "fan" will be given in Article 4·10.

In the x-t plane the front of the expansion wave catches up to the shock wave at the point a. In the subsequent interaction, the shock wave is continually attenuated and its velocity decreases (approaching the acoustic speed a_1), until very far from the initial point its strength is negligible. (Not shown are very weak waves that are reflected back toward the piston.) Similarly, in the flow over the wedge, the effect of the shoulder is not felt by the shock ahead of the point a, where the leading Mach wave of the expansion fan "overtakes" the shock. This limited upstream influence in supersonic flow is a most important feature, for it allows a step-by-step construction of flow fields.

It should be remarked that the analogy outlined here is a formal one and that the actual geometries in the two planes cannot be literally related to each other. For instance, it is possible for the leading Mach lines from the shoulder of the wedge to be inclined *upstream*, but in the piston problem the characteristics cannot possibly be inclined toward negative t. Nevertheless the analogy is useful, especially for visualizing the general nature of interactions in the x-t plane.

4·7 Weak Oblique Shocks

For *small* deflection angles θ, the oblique shock equations reduce to very simple expressions. The approximate relation, from which others may be derived, was already given in Eq. 4·11a,

$$M_1^2 \sin^2 \beta - 1 \doteq \left(\frac{\gamma + 1}{2} M_1^2 \tan \beta\right) \cdot \theta$$

For small θ, the value of β is close to either $\pi/2$ or μ, depending on whether $M_2 < 1$ or $M_2 > 1$ (see Fig. 4·2). For the present, we are considering only the latter case ($M_2 > 1$), for which we may use the approximation

$$\tan \beta \doteq \tan \mu = \frac{1}{\sqrt{M_1^2 - 1}}$$

The preceding equation then reduces to the form

$$M_1^2 \sin^2 \beta - 1 \doteq \frac{\gamma + 1}{2} \frac{M_1^2}{\sqrt{M_1^2 - 1}} \theta \tag{4·14}$$

This is the basic relation for obtaining all other approximate expressions, since all the oblique shock relations depend on the *normal* component, $M_1 \sin \beta$. Thus, from Eq. 4·3 the pressure change is easily obtained,

$$\frac{p_2 - p_1}{p_1} = \frac{\Delta p}{p} \doteq \frac{\gamma M_1^2}{\sqrt{M_1^2 - 1}} \theta \tag{4·15}$$

This shows that *the strength of the wave is proportional to the deflection angle.*

The changes in the other flow quantities, except the entropy, are also proportional to θ. The change of entropy, on the other hand, is proportional to the *third* power of the shock strength (Article 2·13) and hence to the third power of the deflection angle

$$\Delta s \sim \theta^3 \tag{4·16}$$

To find explicitly the deviation of the wave angle β from the Mach angle μ, we may put

$$\beta = \mu + \epsilon$$

where $\epsilon \ll \mu$. In the expansion of $\sin (\mu + \epsilon)$ the approximations $\sin \epsilon \doteq \epsilon$,

$\cos \epsilon \doteq 1$ may then be used to write

$$\sin \beta \doteq \sin \mu + \epsilon \cos \mu$$

Since, by definition, $\sin \mu = 1/M_1$, and $\cot \mu = \sqrt{M_1^2 - 1}$, this gives

$$M_1 \sin \beta \doteq 1 + \epsilon\sqrt{M_1^2 - 1} \tag{4·17a}$$

or

$$M_1^2 \sin^2 \beta \doteq 1 + 2\epsilon\sqrt{M_1^2 - 1} \tag{4·17b}$$

Comparing with Eq. 4·14, the relation between ϵ and θ is

$$\epsilon = \frac{\gamma + 1}{4} \frac{M_1^2}{M_1^2 - 1} \theta \tag{4·18}$$

For a finite deflection angle θ, the direction of the wave differs from the Mach direction by an amount ϵ, which is of the same order as θ.

We shall also need the change of flow speed across the wave. This may be found from the ratio (see Fig. 4·1)

$$\frac{w_2^2}{w_1^2} = \frac{u_2^2 + v^2}{u_1^2 + v^2} = \frac{(u_2/v)^2 + 1}{(u_1/v)^2 + 1} = \frac{\tan^2 (\beta - \theta) + 1}{\tan^2 \beta + 1} = \frac{\cos^2 \beta}{\cos^2 (\beta - \theta)}$$

where Eqs. 4·8 have been used to replace u_2/v and u_1/v. In the last term we have from Eq. 4·17b

$$\cos^2 \beta = 1 - \sin^2 \beta = \frac{M_1^2 - 1}{M_1^2}\left(1 - \frac{2\epsilon}{\sqrt{M_1^2 - 1}}\right)$$

A similar expression for $\cos^2 (\beta - \theta)$ is obtained from this by replacing ϵ by $\epsilon - \theta$. The final result, after dropping all terms of order θ^2 and higher, is

$$\frac{w_2}{w_1} \doteq 1 - \frac{\theta}{\sqrt{M_1^2 - 1}} \tag{4·19a}$$

$$\frac{\Delta w}{w_1} \doteq - \frac{\theta}{\sqrt{M_1^2 - 1}} \tag{4·19b}$$

4·8 Supersonic Compression by Turning

A shock wave increases the pressure and density of the fluid passing through it, that is, it *compresses* the flow. A simple method for compressing a supersonic flow is to *turn* it through an oblique shock, by deflecting the wall through an angle θ, shown in Fig. 4·6*a*.

The turn may be subdivided into several segments, which make smaller corners of angle $\Delta\theta$, as illustrated in Fig. 4·6*b*. Compression then occurs through successive oblique shocks. These shocks divide the field near the wall into segments of uniform flow. Further out they must intersect each other, since they are convergent, but for the present we shall consider only the flow near the wall. In this region, each segment of flow is independent of the following one; that is, the flow may be constructed step by step, proceeding downstream. This property of *limited upstream influence* exists

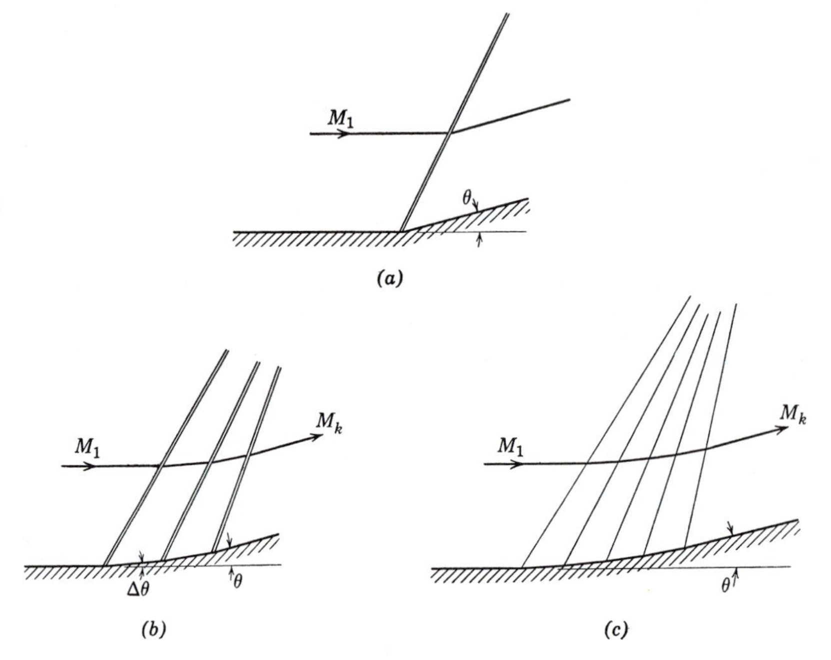

FIG. 4·6 Compression of a supersonic flow by turning an angle θ. (*a*) Single shock of strength θ; (*b*) several weaker shocks, each of strength $\Delta\theta$; (*c*) smooth, continuous compression.

as long as the deflection does not become so great that the flow becomes subsonic.

To compare the compression in the two cases, (*a*) and (*b*) of Fig. 4·6, we may use the approximate expressions for weak shocks, obtained in the previous article. For each wave in (*b*), we have

$$\Delta p \sim \Delta\theta$$

$$\Delta s \sim (\Delta\theta)^3$$

If there are n segments in the complete turn,

$$\theta = n\,\Delta\theta$$

then the overall pressure and entropy changes are

$$p_k - p_1 \sim n\,\Delta\theta \sim \theta$$

$$s_k - s_1 \sim n(\Delta\theta)^3 \sim n\,\Delta\theta(\Delta\theta)^2 \sim \theta(\Delta\theta)^2$$

Thus, if the compression is obtained by using a large number of weak waves, the entropy increase can be reduced very much, compared to a single shock giving the same net deflection. It decreases as $1/n^2$.

By continuing the process of subdivision, the segments can be made vanishingly small, $\Delta\theta \to 0$, and, in the limit, the *smooth turn* of Fig. 4·6*c* is obtained. The entropy increase then becomes vanishingly small, that is, the *compression is isentropic.*

This limiting procedure also gives the following results. (1) The shocks become vanishingly weak, their limiting positions being straight *Mach lines* of which a few are shown in Fig. 4·6*c*. (2) Each segment of uniform flow becomes vanishingly narrow and finally coincides with a Mach line. Thus, on each Mach line, the flow inclination and Mach number are constant. (3) The limited upstream influence is preserved; the flow upstream of a given Mach line is not affected by downstream changes in the wall.† (4) The approximate expression for the change of speed across a weak shock

$$\frac{\Delta w}{w} = -\frac{\Delta\theta}{\sqrt{M_1^2 - 1}}$$

becomes a *differential* expression,

$$\frac{dw}{w} = -\frac{d\theta}{\sqrt{M_1^2 - 1}} \tag{4·20}$$

Similarly, the other approximate relations of Article 4·7 may be written in differential form.

Equation 4·20 applies continuously through the isentropic turn; when integrated (Article 4·10), it gives a relation between θ and M, which for the present we shall simply write in the form

$$\theta = fn(M) \tag{4·20a}$$

We may now return to the question of what happens farther out in the stream, where the shock waves of Fig. 4·6*b* and the Mach lines of Fig. 4·6*c* converge. The intersection of shocks will be considered in Article 4·12; here we are interested only in the continuous compression, which is illustrated again in Fig. 4·7*a*.

Due to the convergence of the Mach lines, the change from M_1 to M_2 on streamline *b* occurs in a shorter distance than on streamline *a*, and thus the gradients of velocity and temperature on *b* are higher than those on *a*. An intersection of Mach lines would imply an infinitely high gradient, for there would be two values of M at one point. However, this cannot occur, for, in the region where the Mach lines converge, and before they cross, the

†Thus, in the limit of smooth flow, the velocities and flow inclinations must be *continuous*, but their derivatives may still be discontinuous.

gradients become high enough so that conditions are no longer isentropic. It is not possible to discuss the details here; qualitatively, a shock wave is developed in the manner illustrated.

Figure 4·7*b* shows a small-scale view of the shock formation. Far from the corner, we must have the simple, oblique shock discussed in Article 4·2,

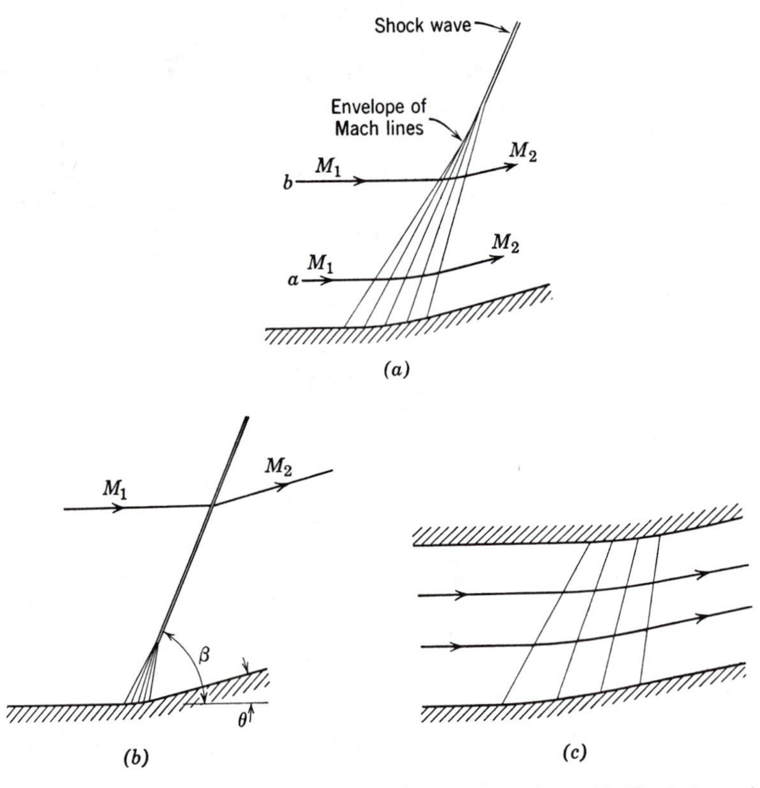

Fig. 4·7 Illustrating Mach line convergence in a compression. (*a*) Shock formation; (*b*) shock formation shown on a smaller scale; (*c*) channel conforming to streamlines of a smooth compression.

for M_1 and θ. In fact, if the straight walls on either side of the corner extend to infinity, then scale is a relative matter, and from "very far away" the turn looks like a sharp corner.

The convergence of Mach lines in a compression is a typical nonlinear effect: decreasing Mach number and increasing flow inclination both tend to make successive Mach lines steeper. The analogous nonlinear effect leading to the formation of shock waves in nonstationary flow was discussed in Article 3·10.

If a wall is placed along one of the streamlines, say *b* in Fig. 4·7 where the gradients are still small enough for the flow to be isentropic, then an isentropic compression in a curved channel is obtained, as shown in Fig. 4·7*c*. It may be noted that, since this flow is isentropic, it may be reversed, without violating the second law of thermodynamics. The reverse flow is an expansion flow.

4·9 Supersonic Expansion by Turning

Up to now we have been considering only turns that are concave, that is, turns in which the wall is deflected "into" the flow. What happens in a convex turn, in which the wall is deflected "away" from the oncoming stream? In particular, what happens in supersonic flow over a convex corner, like that illustrated in Fig. 4·8*a*?

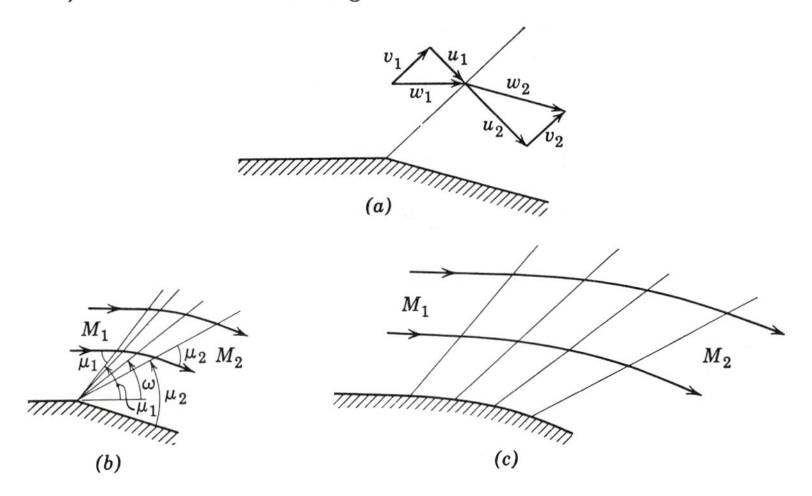

FIG. 4·8 Supersonic expansion. (*a*) Not possible on thermodynamic grounds; (*b*) centered expansion wave; (*c*) simple expansion.

A turn through a single oblique wave, like that illustrated in Fig. 4·8*a*, is not possible. It would require, as shown by a simple vector consideration, that the normal component of velocity, u_2, after the shock, be greater than the normal component ahead, since the tangential components, v_1 and v_2, must be equal. Although this would satisfy the equations of motion, it would lead to a decrease of entropy, as shown in Article 2·13, and thus is not physically possible.

What actually happens is the following. The nonlinear mechanism which tends to steepen a compression (Article 4·8) produces the opposite effect in an expansion. Instead of being convergent, Mach lines are *divergent*, as shown in Fig. 4·8*b* and *c*, and consequently there is a tendency to decrease gradients. Thus *an expansion is isentropic* throughout.

The expansion at a corner (Fig. 4·8*b*) occurs through a *centered wave*, defined by a "fan" of straight Mach lines. This may be concluded from one of several arguments, as follows.

(1) The flow up to the corner is uniform, at Mach number M_1, and thus the leading Mach wave must be straight, at the Mach angle μ_1. The same argument, together with the limited upstream influence, may be applied to each succeeding portion of flow. The terminating Mach line stands at the angle μ_2 to the downstream wall.

(2) Since there is no characteristic length to define a scale in this configuration, any variation of flow parameters can only be with respect to angular position, measured from the corner. That is, flow parameters must be constant along "rays" from the corner. This is the "conical flow" argument which is referred to in Article 4·21.

(3) A centered wave is indicated in the analogy with the piston problem (Fig. 4·5).

This centered wave, more often called a *Prandtl-Meyer expansion fan*, is the counterpart, for a convex corner, of the oblique shock at a concave corner.

Figure 4·8*c* shows a typical expansion over a continuous, convex turn. Since the flow is isentropic, it is reversible. For example, in the channel formed by any two streamlines, the forward flow is expansive and the reverse flow is compressive.

The relation between flow inclination and Mach number in these isentropic turns is given, in functional form, in Eq. 4·20*a*,

$$\theta = fn(M)$$

We have yet to evaluate the function.

4·10 The Prandtl-Meyer Function

Equation 4·20, which gives the differential relation between θ and M in an isentropic compression or expansion by turning, may be written

$$-d\theta = \sqrt{M^2 - 1}\,\frac{dw}{w}$$

or

$$-\theta + \text{const.} = \int \sqrt{M^2 - 1}\,\frac{dw}{w} = \nu(M) \tag{4·21a}$$

To evaluate the integral and thus find the explicit form of the function ν, we may rewrite w in terms of M by using the relations

$$w = aM$$

and

$$\frac{a_0^2}{a^2} = 1 + \frac{\gamma - 1}{2} M^2$$

from which

$$\frac{dw}{w} = \frac{dM}{M} + \frac{da}{a} = \frac{dM}{M}\left(\frac{1}{1 + \frac{\gamma - 1}{2} M^2}\right)$$

The function $\nu(M)$ is then

$$\nu(M) = \int \frac{\sqrt{M^2 - 1}}{1 + \frac{\gamma - 1}{2} M^2} \frac{dM}{M}$$

$$= \sqrt{\frac{\gamma + 1}{\gamma - 1}} \tan^{-1} \sqrt{\frac{\gamma - 1}{\gamma + 1}(M^2 - 1)} - \tan^{-1} \sqrt{M^2 - 1} \qquad (4{\cdot}21b)†$$

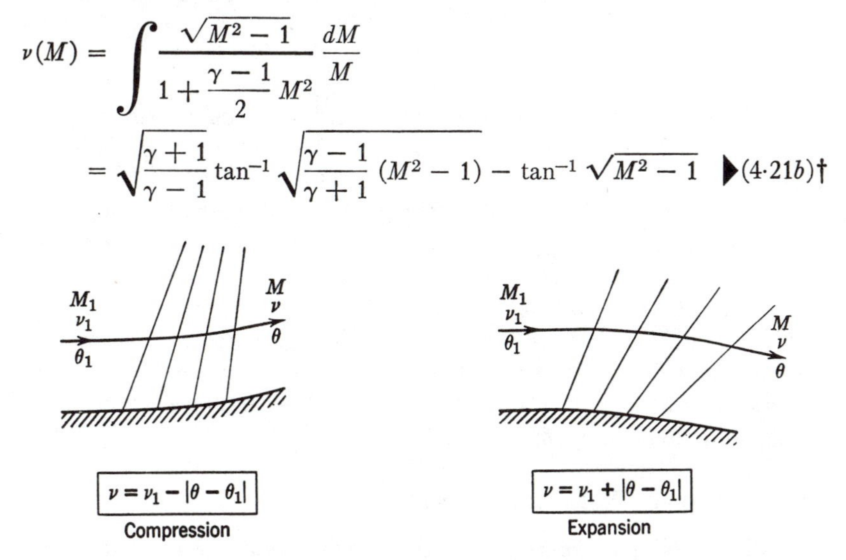

FIG. 4·9 Relation of ν and θ in simple isentropic turns.

It is called the *Prandtl-Meyer function.* The constant of integration has been arbitrarily chosen so that $\nu = 0$ *corresponds to* $M = 1$. The dimensionless (radian) form for ν which results from Eq. 4·21*b* is usually converted into degrees for convenience in calculating flow inclinations. Tabulated values of M, for integer values of ν (in degrees), are given in Table V at the end of the book. Corresponding values of pressure ratio, etc., are obtained from the isentropic equations (Article 2·10). Some of these are also shown in Table V. Values of ν are also tabulated versus M in Table III.

Thus, a supersonic Mach number M is always associated with a definite value of the function ν. As M varies from 1 to ∞, ν increases monotonically from 0 to ν_{max}, where

$$\nu_{max} = \frac{\pi}{2}\left(\sqrt{\frac{\gamma + 1}{\gamma - 1}} - 1\right) \qquad (4{\cdot}22)$$

The relation of the Prandtl-Meyer function ν, and thus of Mach number

†The manipulations in this integration are somewhat tedious; a less direct but simpler method of obtaining this result is given in Exercise 4·7.

M, to the flow inclination θ is illustrated in Fig. 4·9, for both compression and expansion turns. In the preceding articles, θ was considered positive in the direction of compressive deflection, but here the question of algebraic signs will be avoided by using absolute values of flow deflection. Thus for compression and expansion turns, respectively,

$$\nu = \nu_1 - |\theta - \theta_1| \qquad \text{(compression)} \qquad (4\cdot23a)$$

$$\nu = \nu_2 + |\theta - \theta_2| \qquad \text{(expansion)} \qquad (4\cdot23b)$$

In specific cases, a sign convention for θ can easily be adopted, if needed.

In a compression turn, ν decreases, whereas, in an expansion turn, it increases, in each case by an amount equal to the flow deflection. The initial value $\nu_1 = \nu(M_1)$ may be found in Table V; the calculated value of ν, for any value of θ, then gives the corresponding value of M. Usually it is convenient to set $\theta_1 = 0$, since only the deflection matters.

As an example, a flow at Mach number $M_1 = 2$ corresponds to $\nu_1 =$ 26.38° (Table V). If this flow is turned through a *compression* of 10°, the final value of ν is 16.38° and the corresponding Mach number is 1.652 (interpolation in Table V); if it is turned through an *expansion* of 10°, then $\nu =$ 36.38° and $M =$ 2.386.

4·11 Simple and Nonsimple Regions

The isentropic compression and expansion waves discussed in the preceding several articles are called *simple waves*. A simple wave is distinguished by the straight Mach lines, with constant conditions on each one, and by the simple relation (Eq. 4·23) between flow deflection and Prandtl-Meyer function.

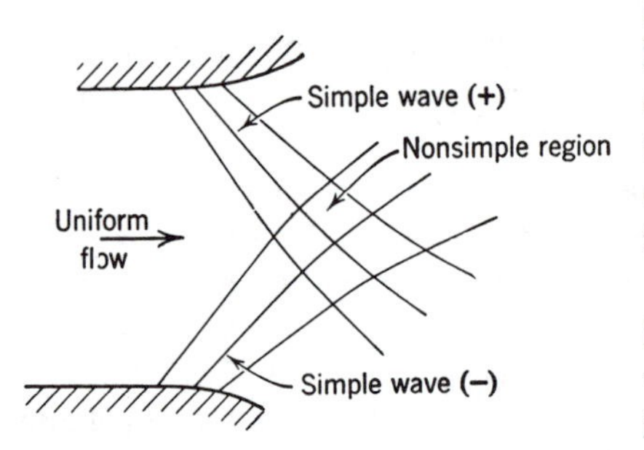

FIG. 4·10 Regions in isentropic supersonic flow.

A wave belongs to one of two families (+ or –), depending on whether the wall that produces it is to the left or right of the flow, respectively, as shown in Fig. 4·10 (cf. also Fig. 4·4).

In the region where two simple waves, of opposite family, interact with each other, the flow is *nonsimple*, that is, the relation between ν and θ is not the simple one given in Eqs. 4·23. Such a region may be treated by the method of characteristics, described in Chapter 12, where some additional discussion of simple and nonsimple regions may be found.

4·12 Reflection and Intersection of Oblique Shocks

If an oblique shock is intercepted by a wall it is "reflected," as shown in Fig. 4·11*a*. The incident shock deflects the flow through an angle θ toward the wall. A second reflected shock, of the opposite family, is required to turn it back again an amount θ and thus satisfy the constraint of the wall.

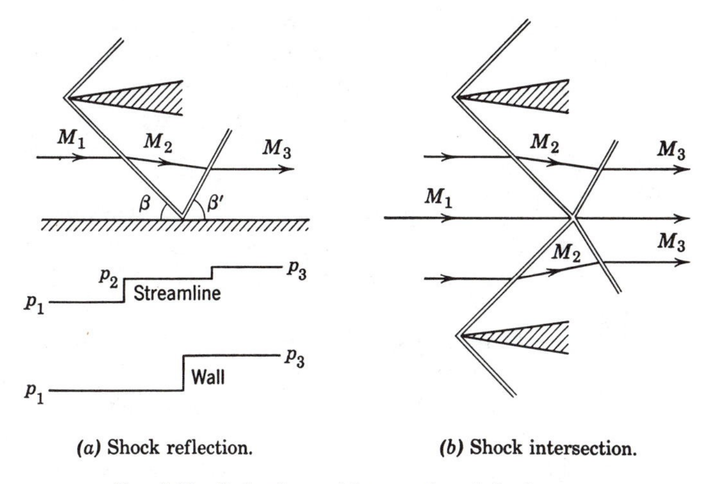

(*a*) Shock reflection.

(*b*) Shock intersection.

Fig. 4·11 Reflection and intersection of shock waves.

Although the deflections produced by the two shocks are equal in magnitude, the pressure ratios are not, since $M_2 < M_1$. Typical distributions of pressure along a streamline and along the wall are shown in the figure. The strength of the reflection may be defined by the overall pressure ratio,

$$\frac{p_3}{p_1} = \frac{p_3}{p_2}\frac{p_2}{p_1}$$

It is the product of the individual shock strengths.

The reflection is, in general, not specular; that is, the inclination β' of the reflected shock is not the same as the inclination β of the incident one. There are two contributing effects: both Mach number and flow inclination ahead of the second shock are smaller than those ahead of the first shock. The two effects are opposite, and the net result depends on the particular values of M_1 and θ. It cannot be written explicitly in general form but may easily be found, for particular cases, from the shock charts. For high Mach numbers $\beta' < \beta$, whereas for low Mach numbers $\beta' > \beta$, with the reversal Mach number depending on θ.

The wall streamline in Fig. 4·11*a* may also be identified with the central streamline of the symmetric flow shown in Fig. 4·11*b*. This gives the *intersection* of two shocks of equal strength but of opposite families. The shocks "pass through" each other, but are slightly "bent" in the process. The flow downstream of the shock system is parallel to the initial flow.

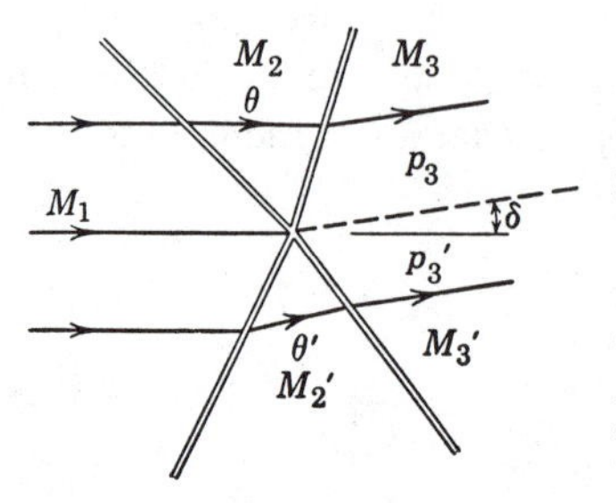

Fig. 4·12 Intersection of shocks of different strengths.

If the intersecting shocks are of *unequal* strengths (Fig. 4·12), a new feature appears. The streamline through the intersection point divides the flow into two portions, which experience different changes in traversing the shock wave system. The overall result must be such that the two portions have the *same pressure* and the *same flow direction*. The latter is no longer necessarily that of the free stream. These two requirements determine the final direction δ and the final pressure p_3.

All other parameters are then also determined, but they do not have the same values on the two sides of the dividing streamline (shown broken). This streamline is, in fact, a *slipstream*, or shear layer, since the magnitudes of the velocity on either side of it are different. It is also called a *contact* surface, because of the fact that the temperature and density on either side are different. Basically, these differences are related to the fact that the net entropy changes experienced by the fluid are different on the two sides of the intersection. Note the analogous effect in the shock tube (Article 3·12).

4·13 Intersection of Shocks of the Same Family

If the two shocks are of the *same* family, produced, for example, by successive corners in the same wall, then the configuration is like that of Fig. 4·13*a*. Here the two shocks cannot "pass through" each other but must coalesce to form a single stronger branch. Here again, the flow on either side of the intersection point, *o*, experiences different entropy changes and a slipstream *od* is produced. An additional wave *oe*, of the opposite family, is needed to equalize the pressures on the two sides of the slipstream. This may be either a compression or expansion wave, depending on the particular configuration and Mach number, but in any event it is very much weaker than the primary waves.

If the second shock *bo* is much weaker than the first one *ao*, then *oe* is always a compression. This interaction may be conveniently described as follows: the second shock is partly "transmitted," along *oc*, thus augmenting the first one, and partly "reflected," along *oe*.

Similarly, in the interaction of an expansion wave with a shock wave of the same family (Fig. 4·13*c*), the main effect is an attenuation of the shock, but there is also a partial reflection of the expansion, along Mach lines of the opposite family. These reflected waves are always *very much weaker* than

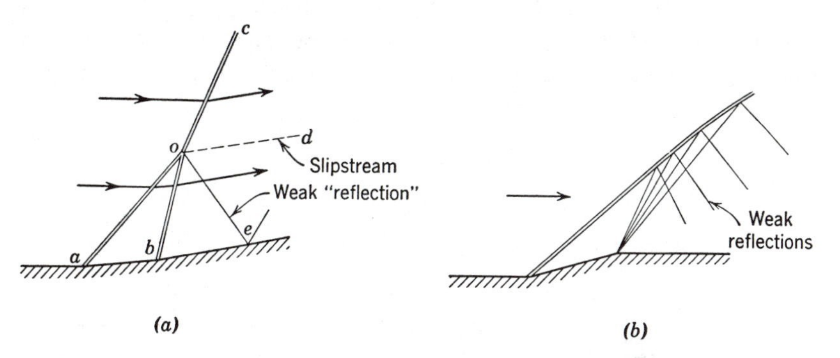

FIG. 4·13 Intersection of waves of the same family. (*a*) Formation of stronger shock from two shocks; (*b*) attenuation of shock by expansion wave.

the primary ones, and may be neglected in all but the strongest interactions. Instead of the single slipstream *od*, there is a whole *region* of vorticity, that is, an entropy field, downstream of the interaction.

4·14 Detached Shocks

So far we have restricted ourselves to cases in which the flow after the shock remains supersonic (cf. Fig. 4·2). We have now to return to the questions that were postponed in Article 4·4, concerning the possible oblique shock solutions: What flow configurations does the other (stronger) solution represent? Second, what occurs when the wall deflection, or wedge angle θ, is greater than θ_{max}? We shall consider the second question first.

There is in fact no rigorous analytical treatment for problems in which the deflection angles are greater than θ_{max}. Experimentally, it is observed that the flow configurations are like those sketched in Fig. 4·14.

The flow is compressed through a curved shock wave which is detached from the wedge and stands at some distance ahead of it. The shape of the shock and its detachment distance *depend on the geometry of the body* and on the Mach number M_1.

On the central streamline, where the shock is normal, as well as on the nearby ones, where it is nearly normal, the flow is compressed to subsonic conditions. Farther out, as the shock becomes weaker, it becomes less steep, approaching asymptotically to the Mach angle. Thus conditions along the detached shock wave run the whole range of the oblique shock solution for the given Mach number (cf. Fig. 4·14*b*, which shows one curve from Fig. 4·2). In such configurations one does find shock inclinations corresponding to the other (strong) solution. Shock segments of the strong solution may also appear as parts of other complex configurations.

The complexity of this and similar examples is related to the appearance of *subsonic* regions. When the flow behind the shock is subsonic, the shock is no longer independent of the far-downstream conditions. A change of geometry or pressure in the subsonic portion affects the entire flow up to the shock, which must then adjust itself to the new conditions.

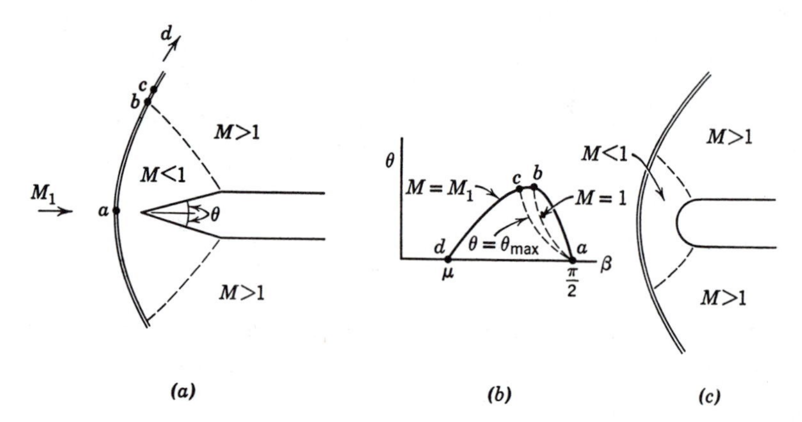

FIG. 4·14 Detached shock waves. (*a*) Detached shock on wedge with afterbody (at lettered points on shock, flow deflection is given by corresponding points in *b*); (*b*) flow deflections vs. wave angle for fixed M_1; (*c*) detached shock on plate with rounded nose.

In the case of a blunt-nosed body (Fig. 4·14*c*), the shock wave is detached at all Mach numbers, and is similar to the one described above. Conversely, a wedge of half-angle $\theta > \theta_{max}$ is a "blunt-nosed" body so far as the oncoming flow is concerned.

Examples of flow over a wedge with afterbody are shown in Figs. 4·5*b*, 6·15, and 11·1. For given wedge angle θ, the sequence of events with decreasing Mach number is as follows. (1) When M_1 is high enough the shock wave is attached to the nose; the straight portion is independent of the shoulder and afterbody (cf. Fig. 4·5*b*). As M_1 decreases, the shock angle increases. (2) With further decrease of Mach number, a value is reached for which conditions after the shock are subsonic. The shoulder now has an effect on the whole shock, which may become curved, even though still attached. These conditions correspond to the region between the lines $M_2 = 1$ and $\theta = \theta_{max}$, on Fig. 4·2. (3) At the Mach number corresponding to θ_{max}, the shock wave starts to detach. This is called the detachment Mach number. (4) With further decrease of M_1, the detached shock moves upstream of the nose.

A similar sequence of events occurs in flow over a cone with cylindrical afterbody. The detachment Mach numbers, which are lower than for wedges, may be found in Fig. 4·27*a*.

The shock shape and detachment distance cannot, at present, be theoretically predicted. Figure 4·15 shows some measurements of shock wave

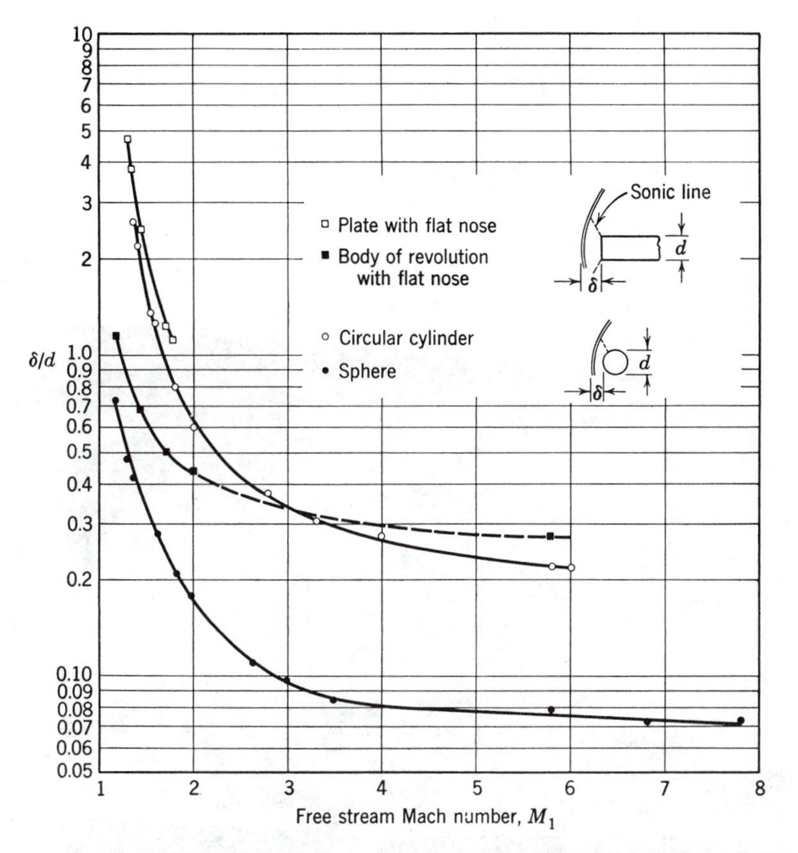

FIG. 4·15 Shock wave detachment distance for plane and axially symmetric flow. Sources of experimental data: G. E. Solomon, *N.A.C.A. Tech. Note* 3213; W. Griffith, *J. Aeronaut. Sci., 19* (1952); Heberle, Wood, and Gooderum, *N.A.C.A. Tech. Note* 2000; C. S. Kim, *J. Phys. Soc. Japan, 11* (1956); U. S. Naval Ordnance Lab., unpublished data; California Institute of Technology, various unpublished data (Alperin, Hartwig, Kubota, Oliver, Puckett). Data are for air, with constant specific heat.

detachment distance on two-dimensional and axially symmetric bodies with flat noses and circular (or spherical) noses. The curves for the flat-nosed plate and for the flat-nosed body of revolution also define the *limiting* positions for detached shock waves on wedges and cones, respectively. That is, a detached shock wave on a wedge or cone, with afterbody, has practically the same position as if the forebody (wedge or cone) were not there.† The reason is that the shock position is controlled mainly by the position of the sonic line, which for these cases is fixed at the shoulder.

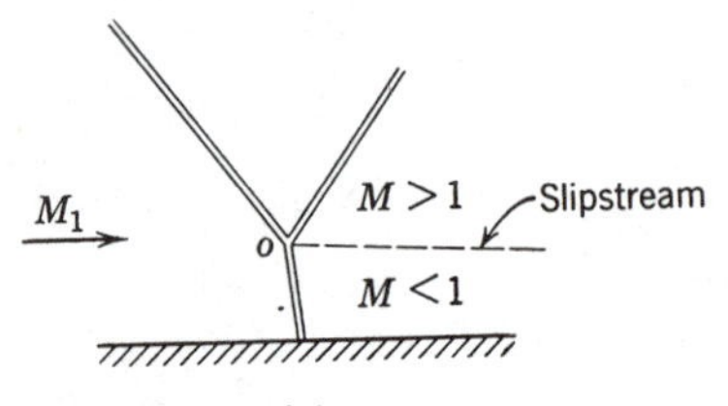

(a)

(a)

Fig. 4·16 (a) Mach reflection. (b) Schlieren photograph of a Mach reflection. $M_1 = 1.38$; wedge angle $2\theta = 10°$. A faint slipstream is visible downstream of the triple point. A pitot probe is embedded in the boundary layer downstream of the reflection point.

4·15 Mach Reflection

The complications due to the appearance of subsonic regions in the flow are again encountered in shock *reflections*, when they are too strong to give

†Wayland Griffith, "Shock Tube Studies of Transonic Flow over Wedge Profiles," *J. Aeronaut. Sci.*, *19* (1952), 249.

the simple reflections of Article 4·13. That is, if M_2 after the incident shock is lower than the detachment Mach number for θ, then no solution with simple oblique waves is possible. Here the configurations are similar to those shown in Fig. 4·16. They are called *Mach reflections*. A normal, or nearly normal, shock that appears near the wall forms with the incident and "reflected" shocks a triple intersection point at o. Due to the difference in entropy on streamlines above and below the triple point, the streamline which extends downstream from the triple point is a slipstream. It is visible in the photograph (Fig. 4·16*b*).

The subsonic region, behind the nearly normal leg, makes it impossible to give a purely local description of the configuration. Attempts at a local description of the triple point would appear to be fruitless, since the only fixed parameter, in a given problem, is the incident shock. The strengths of the reflected shock, the nearly normal leg, and the slipstream are all arbitrary. They may be arranged in an infinite number of ways, all compatible locally at o. The triple-point solution which actually occurs in a particular problem, as well as the position of the triple point, are determined by the downstream conditions, which influence the subsonic part of the flow.

4·16 Shock-Expansion Theory

The oblique shock wave and the simple isentropic wave furnish the building blocks for constructing (or analyzing) many problems in two-dimensional, supersonic flow by simply "patching" together appropriate combinations of the two solutions. Some examples of flow over two-dimensional airfoil sections are shown in Fig. 4·17, together with the corresponding pressure distributions.

Consider the diamond-section airfoil in Fig. 4·17*a*. The nose shock compresses the flow to the pressure p_2, the centered expansion at the shoulder expands it to the pressure p_3, and the trailing edge shock recompresses it to (nearly) the free-stream value, p_4. There is a drag on the airfoil, due to the overpressure on the forward face and underpressure on the rearward face. For unit span, it is

$$D = (p_2 - p_3)t \qquad (4\cdot24)$$

where t is the thickness of the section at the shoulder. The values of p_2 and p_3 are easily found from shock charts and tables of the Prandtl-Meyer function ν.

This introduces the phenomenon of *supersonic wave drag*. In supersonic flow, drag exists even in the idealized, nonviscous fluid. It is fundametally different from the frictional drag and separation drag that are associated with boundary layers in a viscous fluid. (Of course, the wave drag is ultimately "dissipated" through the action of viscosity, within the shock waves.

But it does not *depend* on the actual value of the viscosity coefficient. This is another statement of the fact that the entropy change across a shock is independent of the detailed nonequilibrium process inside the shock.)

The second example in Fig. 4·17*b* is a curved airfoil section, which has *continuous* expansion along the surface. For the shocks to be attached, it is necessary that nose and tail be wedge-shaped, with half-angle less than θ_{max}.

The third example in Fig. 4·17*c* shows a flat plate at angle of attack α_0. The streamline ahead of the leading edge is straight, since there is no upstream influence, and the portions of flow over the upper and lower surfaces are independent. Thus the flow on the upper side is turned through an

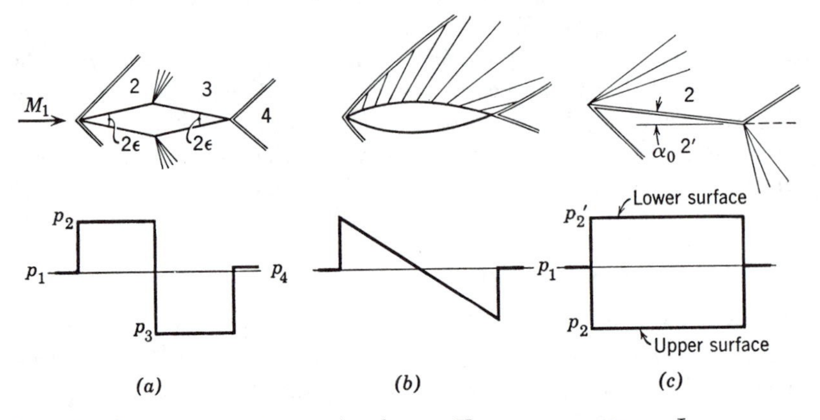

FIG. 4·17 Illustrating shock-expansion theory. Upper: wave patterns. Lower: pressure distributions. (*a*) Symmetrical diamond section; (*b*) lens section; (*c*) lifting flat plate.

expansion angle α_0 by means of a centered wave at the nose, whereas that on the lower side is turned through a compression angle α_0 by means of an oblique shock. From the uniform pressures on the two sides, the lift and drag are computed very simply. They are

$$\begin{aligned} L &= (p'_2 - p_2)c \cos \alpha_0 \\ D &= (p'_2 - p_2)c \sin \alpha_0 \end{aligned} \tag{4·25}$$

where c is the chord. The increase in entropy for flow along the upper surface is not the same as for flow along the lower surface because the shock waves occur at different Mach numbers on the two sides. Consequently, the streamline from the trailing edge is a slipstream, inclined at a small angle relative to the free stream.

So far no mention has been made, in these examples, of the interaction between the shocks and expansion waves. For this it is necessary to examine a larger portion of the flow field, as in Fig. 4·18, which shows smaller-scale versions of two of the previous examples. The expansion fans attenuate the oblique shocks, making them weak and curved. At large distances they approach asymptotically the free-stream Mach lines.

The reflected waves were not shown in the previous figure (4·17), for in shock-expansion theory they are neglected. Their effect is small, but in an exact analysis it would have to be considered. For a diamond airfoil and

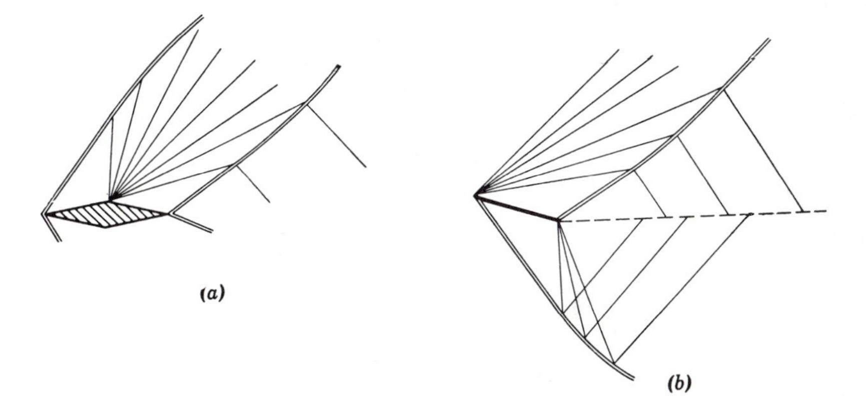

FIG. 4·18 Attenuation of waves by interaction. (*a*) Wave system for diamond airfoil; (*b*) wave system for lifting flat plate. (Reflected waves are very weak.)

a lifting flat plate, however, the reflected waves do not intercept the airfoil at all, and hence do not affect the shock-expansion result for the pressure distribution.

The wave system extends to very large distances from the body, and is such as ultimately to reduce all disturbances to vanishing strength.

4·17 Thin Airfoil Theory

The shock-expansion theory of the preceding article gives a simple and general method for computing lift and drag, applicable as long as the shocks are attached. However, the results cannot be expressed in concise, analytical form, and the theory is mainly used for obtaining numerical solutions. If the airfoil is thin and at a small angle of attack, that is, if all flow inclinations are small, then the shock-expansion theory may be approximated by using the approximate relations for weak shocks and expansions. This leads to simple analytical expressions for the lift and drag.

The basic approximate expression (Eq. 4·15) for calculating pressure changes is

$$\frac{\Delta p}{p} \doteq \frac{\gamma M^2}{\sqrt{M^2 - 1}} \Delta\theta$$

Since, with the weak wave approximation, the pressure p will never be greatly different from p_1, nor M greatly different from M_1, we may make the further approximation, correct to first order,

$$\frac{\Delta p}{p_1} \doteq \frac{\gamma M_1^2}{\sqrt{M_1^2 - 1}} \Delta\theta$$

Finally, if we refer all pressure changes to the free-stream pressure (p_1), and all direction changes to the free-stream direction (zero), we have

$$\frac{p - p_1}{p_1} = \frac{\gamma M_1^2}{\sqrt{M_1^2 - 1}} \theta$$

where θ is the inclination relative to the free-stream direction. The sign of θ will be clear in each application.

The pressure coefficient, which is defined by Eq. 2·40,

$$C_p = \frac{p - p_1}{q_1} = \frac{2}{\gamma M_1^2} \frac{p - p_1}{p_1}$$

then becomes

$$C_p = \frac{2\theta}{\sqrt{M_1^2 - 1}} \tag{4·26}$$

This is the basic relation for thin airfoil theory. It states that *the pressure coefficient is proportional to the local flow inclination.*

With the above result, it is a simple matter to obtain the lift and drag coefficients for various airfoil sections. Thus for the *flat plate* (Fig. 4·17c) at a small angle of attack α_0, the pressure coefficients on the upper and lower surfaces are

$$C_p = \mp \frac{2\alpha_0}{\sqrt{M_1^2 - 1}} \tag{4·27}$$

The lift and drag coefficients are

$$C_L = \frac{(p_L - p_U)c \cos \alpha_0}{q_1 c} = (C_{p_L} - C_{p_U}) \cos \alpha_0$$

$$C_D = \frac{(p_L - p_U)c \sin \alpha_0}{q_1 c} = (C_{p_L} - C_{p_U}) \sin \alpha_0$$

Since α_0 is small, we may use $\cos \alpha_0 \doteq 1$, $\sin \alpha_0 \doteq \alpha_0$. Thus with Eq. 4·27, we have for the lift and drag coefficients of a lifting flat plate,

$$C_L = \frac{4\alpha_0}{\sqrt{M_1^2 - 1}}$$

$$C_D = \frac{4\alpha_0^2}{\sqrt{M_1^2 - 1}} \tag{4·28}$$

The aerodynamic center is at midchord. The ratio $D/L^2 = \frac{1}{4}\sqrt{M_1^2 - 1}$ is independent of α_0.

As a second example, consider the *diamond section* airfoil of nose angle 2ϵ, at zero angle of attack, as shown in Fig. 4·17*a*. The pressure coefficients on the front and rear faces are

$$C_p = \pm \frac{2\epsilon}{\sqrt{M_1^2 - 1}}$$

and so the pressure difference is

$$p_2 - p_3 = \frac{4\epsilon}{\sqrt{M_1^2 - 1}} q_1$$

giving a drag

$$D = (p_2 - p_3)t = (p_2 - p_3)\epsilon c = \frac{4\epsilon^2}{\sqrt{M_1^2 - 1}} q_1 c$$

Thus

$$C_D = \frac{4\epsilon^2}{\sqrt{M_1^2 - 1}} = \frac{4}{\sqrt{M_1^2 - 1}} \left(\frac{t}{c}\right)^2 \qquad (4\cdot29)$$

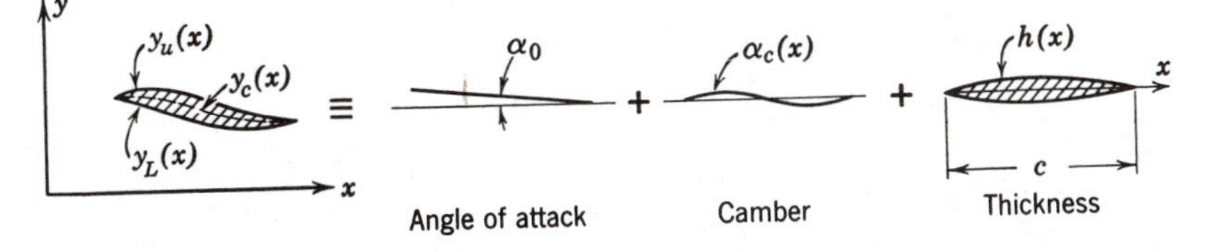

FIG. 4·19 Linear resolution of arbitrary airfoil into lift, camber, and drag.

The form of the lift and drag coefficients in these two examples is typical. A general result, applicable to an arbitrary thin airfoil, may be obtained as follows. Figure 4·19 shows an airfoil that has thickness, camber, and angle of attack. The pressure coefficients on the upper and lower surfaces are

$$C_{p_U} = \frac{2}{\sqrt{M_1^2 - 1}} \frac{dy_U}{dx}$$
$$C_{p_L} = \frac{2}{\sqrt{M_1^2 - 1}} \left(- \frac{dy_L}{dx}\right) \qquad (4\cdot30)$$

where y_U and y_L are the respective profiles. The profile may be resolved into a symmetrical thickness distribution $h(x)$ and a camber line of zero thickness $y_c(x)$. Thus

$$\begin{aligned} \frac{dy_U}{dx} &= \frac{dy_c}{dx} + \frac{dh}{dx} = -\alpha(x) + \frac{dh}{dx} \\ \frac{dy_L}{dx} &= \frac{dy_c}{dx} - \frac{dh}{dx} = -\alpha(x) - \frac{dh}{dx} \end{aligned} \tag{4·31}$$

where $\alpha(x) = \alpha_0 + \alpha_c(x)$ is the local angle of attack of the camber line. The lift and drag are given by

$$L = q_1 \int_0^c (C_{p_L} - C_{p_U})\, dx$$

$$D = q_1 \int_0^c \left[C_{p_L}\left(-\frac{dy_L}{dx}\right) + C_{p_U}\left(\frac{dy_U}{dx}\right)\right] dx$$

Putting Eqs. 4·30 and 4·31 into these expressions, we obtain

$$L = \frac{2q_1}{\sqrt{M_1^2 - 1}} \int_0^c \left(-2\frac{dy_c}{dx}\right) dx = \frac{4q_1}{\sqrt{M_1^2 - 1}} \int_0^c \alpha(x)\, dx$$

$$\begin{aligned} D &= \frac{2q_1}{\sqrt{M_1^2 - 1}} \int_0^c \left[\left(\frac{dy_L}{dx}\right)^2 + \left(\frac{dy_U}{dx}\right)^2\right] dx \\ &= \frac{4q_1}{\sqrt{M_1^2 - 1}} \int_0^c \left[\alpha(x)^2 + \left(\frac{dh}{dx}\right)^2\right] dx \end{aligned}$$

The integrals may be replaced by average values, denoted by a bar, for example,

$$\bar{\alpha} = \frac{1}{c} \int_0^c \alpha(x)\, dx$$

Since by definition $\bar{\alpha}_c = 0$, we have

$$\bar{\alpha} = \overline{(\alpha_0 + \alpha_c)} = \overline{\alpha_0} + \overline{\alpha_c} = \alpha_0$$

Similarly,

$$\overline{\alpha^2} = \overline{(\alpha_0 + \alpha_c)^2} = \overline{\alpha_0{}^2} + 2\overline{\alpha_0 \alpha_c} + \overline{\alpha_c{}^2} = \alpha_0{}^2 + \overline{\alpha_c{}^2}$$

The lift and drag coefficients, L/q_1c and D/q_1c, may then be written

$$\begin{aligned} C_L &= \frac{4\bar{\alpha}}{\sqrt{M_1^2 - 1}} = \frac{4\alpha_0}{\sqrt{M_1^2 - 1}} \\ C_D &= \frac{4}{\sqrt{M_1^2 - 1}}\left[\overline{\left(\frac{dh}{dx}\right)^2} + \overline{\alpha^2(x)}\right] \\ &= \frac{4}{\sqrt{M_1^2 - 1}}\left[\overline{\left(\frac{dh}{dx}\right)^2} + \alpha_0{}^2 + \overline{\alpha_c{}^2(x)}\right] \end{aligned} \tag{4·32}$$

These are the general expressions for the lift and drag coefficients of a thin airfoil in supersonic flow.

In thin-airfoil theory the drag splits into three parts: a "drag due to thickness," a "drag due to lift," and a "drag due to camber." The lift coefficient depends only on the mean angle of attack.

This splitting into lift and drag problems is typical of "small perturbation" problems in supersonic flow; it is an expression of the fact that the governing differential equations are *linear*, as will be seen in Chapters 8 and 9. However, the condition of small perturbations, e.g., small flow deflections, is not sufficient to ensure linearity, as we shall see in the case of transonic flow (Chapter 12), and hypersonic flow (Article 8·2). In those cases, even the small perturbation equations are nonlinear, and the effects of thickness and angle of attack cannot be obtained by simple superposition.

Finally, one may note that in computing the lift we did not need to impose the *Kutta condition* explicitly, as is necessary in subsonic flow. The Kutta condition distinguishes between leading and trailing edges, and thus establishes a definite direction for increasing time. In this sense it is satisfied also in the supersonic case, the direction of increasing time being established by the choice of a direction (downstream) for the waves (cf. Article 4·5).

4·18★ Flat Lifting Wings

It will not be possible in this book to go into the extensive subject of supersonic wings. It need only be remarked that most of the theory is based on a linearized equation, on the assumption of small flow inclinations, so that a problem may always be split into a thickness case and a lifting case.

Due to the limited regions of influence in supersonic flow, the two-dimensional results may often be applied directly. For a single example, consider the flat, rectangular wing of Fig. 4·20*a*, at an angle of attack α_0. The fact that the span is finite is felt only within the tip region, inside the Mach cone from each tip. The inner part, between the tips, behaves as if it were part of an infinite wing, that is, the flow there is two-dimensional, and the pressure coefficient has the two-dimensional value C_{p0}, given in Eq. 4·27. Thus the inner part of the wing carries two-dimensional lift.

In the tip region, the lift falls off, to zero at the side edge, where the pressures must equalize. Two typical spanwise pressure distributions, along the trailing edge and the midchord, respectively, are shown in Fig. 4·20*b*. Other spanwise distributions are similar, for, in fact, the pressures are constant along "rays" from the tip of the leading edge (cf. comment in Article 4·21). In the tip region, the *average* pressure coefficient is $\frac{1}{2}C_{p0}$, as may be seen from the symmetry of the pressure distribution there. Thus the lift coefficient of the whole wing is less than the two-dimensional value

C_{L0}, by an amount

$$\Delta C_L = \frac{\frac{1}{2}C_{L0}(\frac{1}{2}cd + \frac{1}{2}cd)}{cb} = \frac{\frac{1}{2}C_{L0}}{b/d} = \frac{\frac{1}{2}C_{L0}}{(b/c)(c/d)}$$

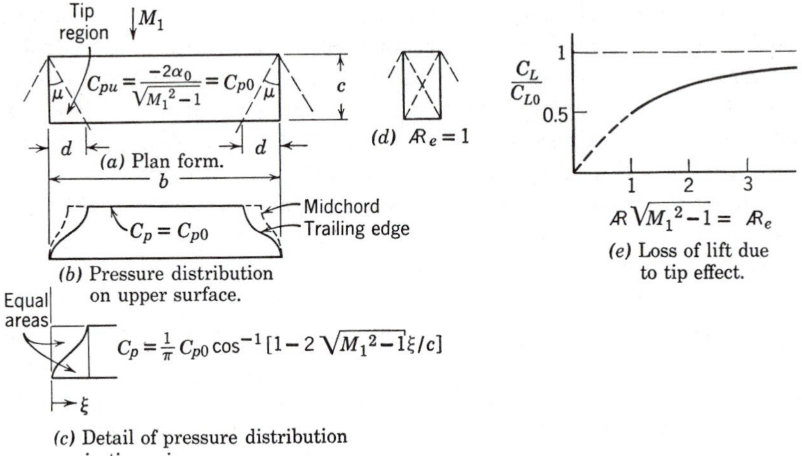

FIG. 4·20 Lift of flat rectangular wing.

The terms in the denominator are

$$b/c = AR = \text{aspect ratio}$$

$$c/d = \cot\mu = \sqrt{M_1^2 - 1}$$

Substituting these in the above expression for the lift deficiency gives the lift coefficient of the rectangular wing, in terms of the two-dimensional value

$$\frac{C_L}{C_{L0}} = 1 - \frac{1}{2AR\sqrt{M_1^2 - 1}} \tag{4·33}$$

This is plotted in Fig. 4·20*e*. The expression is valid for values of $AR\sqrt{M_1^2 - 1}$ as small as 1, corresponding to the case shown in Fig. 4·20*d*, in which the tip Mach lines are just beginning to intersect the opposite side edges. For lower values the expression is different, but tends to zero as shown by the broken line.

Thus the effect of tips, that is, finite aspect ratio, is to reduce the lift, just as in subsonic flow, but with a rather different effect on the pressure distribution. It may be seen from Eq. 4·33 that there is also an effect of the Mach number. The effects of aspect ratio and Mach number may be combined in terms of an *effective aspect ratio*

$$\mathcal{R}_e = \mathcal{R}\sqrt{M_1^2 - 1}$$

This may also be shown from general similarity considerations (Chapter 10). For a wing of fixed dimensions the effective aspect ratio increases with increasing Mach number. This, of course, is due to the fact that the tip Mach cones become narrower.

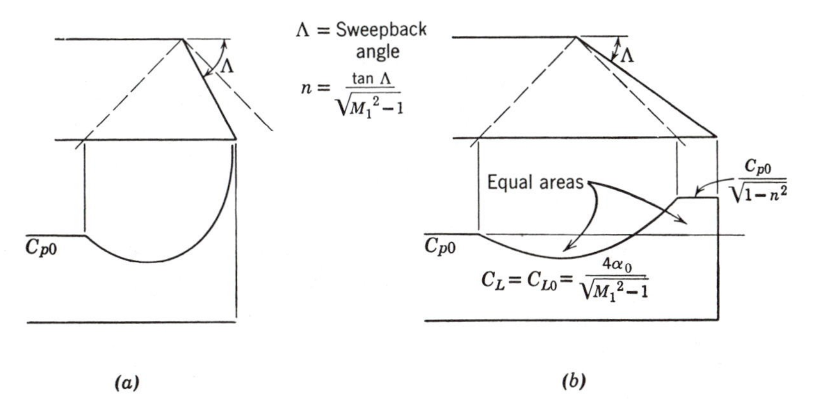

FIG. 4·21 Pressure distributions on nonrectangular wing tips. (*a*) Subsonic edge; (*b*) supersonic edge.

Two more examples of flat lifting wings are given in Fig. 4·21, only the tip regions being shown. The side edge in Fig. 4·21*a* (and in Fig. 4·20*a*), which is downstream of the tip Mach cone, is called a *subsonic edge*, whereas the edge in Fig. 4·21*b*, ahead of the Mach cone, is called a supersonic edge. At a supersonic edge the pressure need not be equalized since the upper and lower surfaces are independent of each other and there is no loss of lift due to "leakage." Although there is a redistribution of pressure, as shown, the lift coefficient of the wing with supersonic edges is the two-dimensional value C_{L0}.† The subsonic edge, on the other hand, has the typical leading edge singularity of subsonic thin-airfoil theory.

4·19★ Drag Reduction

Article 4·17 introduced the phenomenon of *wave drag* in supersonic flow. For two-dimensional airfoils wave drag may be separated into drag due to thickness, lift, and camber, and can be reduced only by reducing each one separately. However, on three-dimensional wings and on combinations such as biplanes and wing-body combinations, it is possible to arrange the

†P. A. Lagerstrom and M. D. Van Dyke, "General Considerations about Planar and Non-Planar Lifting Systems," *Rep. S.M.* 13432 (1949), Douglas Aircraft Co.

elements in such a way that there is beneficial *interference* between them, resulting in reduction of drag.

A famous example is the *Busemann biplane* which depends on mutual *cancellation* of waves between the two planes. The principle of wave cancellation is illustrated in Fig. 4·22*a*. The upper part shows the ordinary

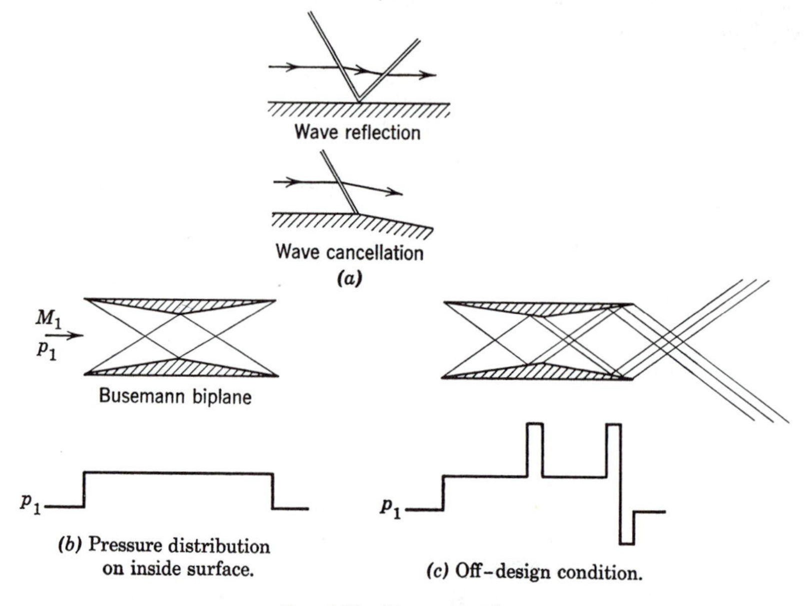

FIG. 4·22 Drag reduction.

reflection of a shock, discussed in Article 4·12; the lower part shows how the reflected shock may be "canceled," by turning the wall parallel to the flow direction behind the incident shock. One might say that the reflected shock is canceled by an expansion at the corner.

The Busemann biplane, at zero lift, is illustrated in Fig. 4·22*b*. Part of the expansion wave at the shoulder cancels the compression wave from the leading edge of the opposite plane and produces the *symmetrical* pressure distribution shown in the figure. The wave drag is zero.

At Mach numbers off the design value, there is partial cancellation of drag, as illustrated in Fig. 4·22*c*.

The disappearance of drag corresponds to the fact that there are no waves outside the system, the contribution from those inside being negligible in the first-order theory. If the Mach number is different from the design value, waves "escape" from the system and the drag is no longer zero. However, there will also be complete cancellation at certain other Mach numbers, namely, those for which the leading edge compression wave inter-

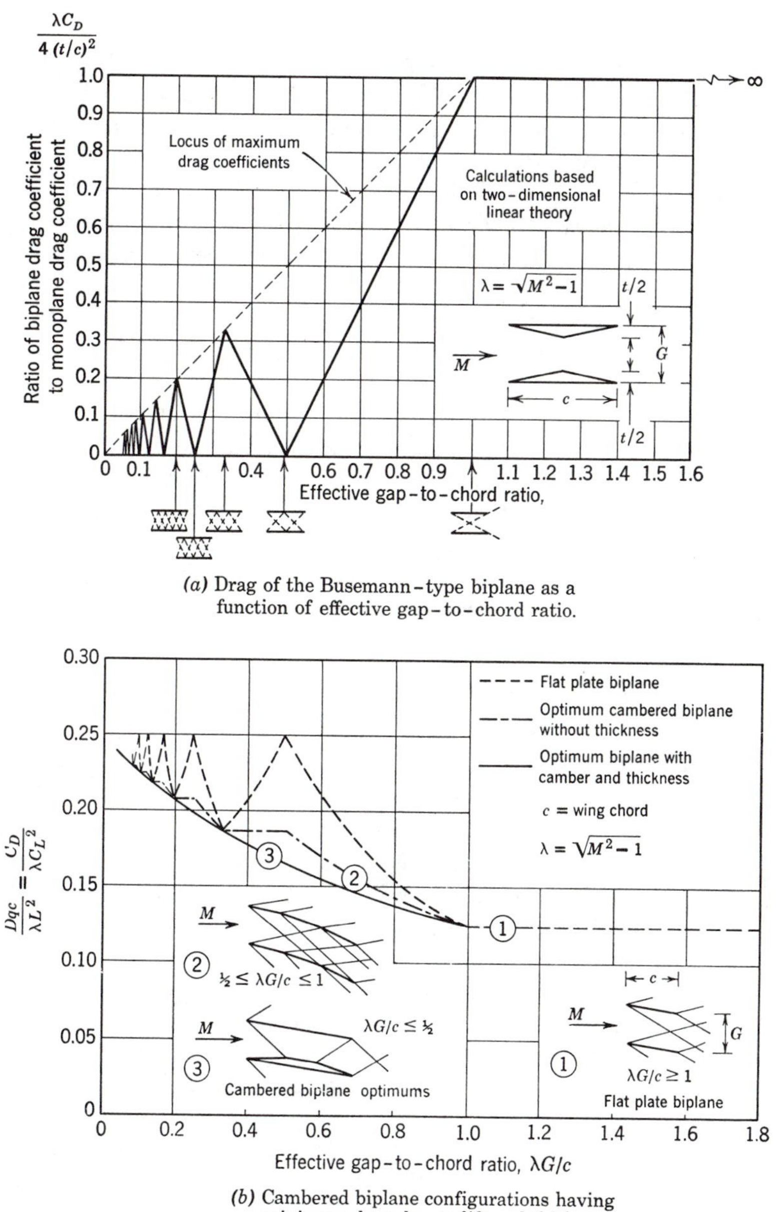

(*a*) Drag of the Busemann-type biplane as a function of effective gap-to-chord ratio.

(*b*) Cambered biplane configurations having minimum drag due to lift and thickness.

FIG. 4·23 Supersonic biplanes. Courtesy of (*a*) B. J. Beane, "Notes on the Variation of Drag with Mach Number of a Busemann Biplane," *Rep. S.M.* 18737, Douglas Aircraft Co.; (*b*) R. M. Licher, "Optimum Two-Dimensional Multiplanes in Supersonic Flow," *Rep. S.M.* 18688, Douglas Aircraft Co.

cepts one of the shoulders. This is illustrated in Fig. 4·23*a*, which shows zero drag at cancellation Mach numbers, and finite values at other intermediate Mach numbers. For values of $\sqrt{M_1^2 - 1}\, G/c > 1$, the leading edge wave completely "misses" the opposite airfoil, there is no interference, and the full "monoplane" drag is obtained.

If the Busemann biplane is lifting, drag due to lift is produced. Lift, thickness, and camber effects are no longer separable as in the monoplane (or "planar") system. They may, however, be combined to produce beneficial interference.

Figure 4·23*b* gives some examples for which the ratio D/L^2 may be optimized. In this case the best values occur when the effective gap is large enough so that there is *no* interference. For smaller gaps, the value of D/L^2 increases, but can be improved by using camber or thickness, as shown.

On three-dimensional planar wings, it is also possible to obtain beneficial interference by suitable distribution of thickness, camber, and twist, and to find optimum configurations for given fixed parameters, e.g., constant lift. Wing-body combinations may also be optimized. Inclusion of the effects of boundary layer friction and separation extends the scope of the problem even further. The whole problem of drag reduction is a very important one.

4·20★ The Hodograph Plane

A coordinate system which is often very useful in fluid mechanics problems is that of the *hodograph plane*, in which the *velocity components* are the coordinates, or independent variables. This plane is useful, in the first place, simply for the presentation of data or solutions, and for graphical (vector) analysis. But a much deeper reason for its importance is that certain problems that are *nonlinear* in the physical plane become *linear* when reformulated in the hodograph plane, that is, with the velocity components as independent variables. The latter application is briefly discussed in Chapter 11, in connection with the transonic equation.

The representation of the flow through an oblique shock wave is shown in Fig. 4·24. The physical plane is shown in (*a*) and the hodograph plane in (*b*). A point in the physical plane is located in the hodograph plane by plotting its velocity components (u, v). Thus a vector in the hodograph plane, drawn from the origin to the point in question, is the corresponding velocity vector. In this example, the *whole flow field* upstream of the shock plots into the single point A, whereas the flow field downstream plots into B. They are the only two velocities in this example. The mapping from the physical plane to the hodograph plane, or vice versa, is sometimes most singular! Furthermore, the mapping is not unique. In the physical plane, many different flows may be obtained from the oblique shock solution by making various choices for the boundaries, compatible with the streamlines. In the hodograph plane, these will all be represented by the two points, A and B.

For a given upstream velocity U_1, or Mach number M_1, there is a whole family of solutions, with θ as a parameter. That is, as w_2 changes, with θ, the point B traces out a certain locus, which is called a *shock polar*. This locus is shown in Fig. 4·24c, where the velocities are plotted in terms of the velocity ratios $(u/a^*, v/a^*)$. It will be noted that for a given deflection

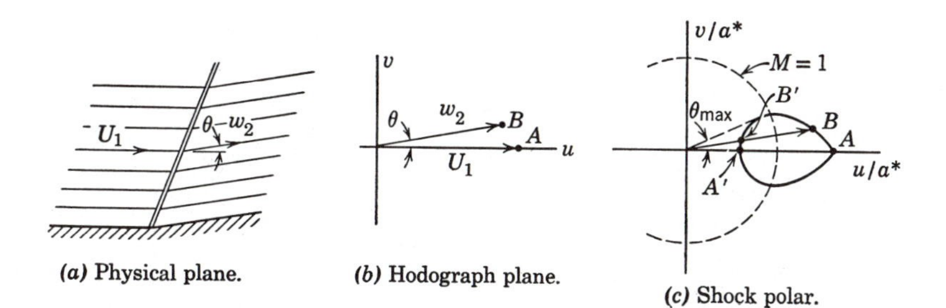

(*a*) Physical plane. (*b*) Hodograph plane. (*c*) Shock polar.

FIG. 4·24 Representation of oblique shock solutions in the hodograph plane.

angle θ, there are two solutions, shown by B and B'. For the latter, the downstream flow is subsonic, as may be seen from the position of the *sonic circle*. This other strong "solution" has been discussed in Article 4·14. Also shown is the occurrence of a maximum value of θ, beyond which there is no oblique shock solution.

Flow through a normal shock is represented by the two points A and A'.

For each value of the initial Mach number, point A, there is a new shock polar. They all have the same general shape, and each one encloses the ones for lower values of M_1.

Next, we consider the simple isentropic expansions and compressions of Article 4·10, which may also be plotted in the hodograph plane. Consider the Prandtl-Meyer expansion shown in Fig. 4·25a. In the hodograph plane (b) it is represented by the *curve* AB, on which each point represents one of the velocities on a given streamline. *All* the streamlines map into AB. This is again a singular mapping—AB may represent a whole family of flows in the physical plane. The reverse flow BA is the map of the compression shown in (c).

The Prandtl-Meyer function (Article 4·10) for all supersonic Mach numbers may be represented as shown in Fig. 4·25d by the full lines, which represent expansions from $M = 1$ to $M = \infty$. The lower curve is for θ increasing, and the upper one for θ decreasing. The equation of this curve, an *epicycloid*, may be obtained directly from the Prandtl-Meyer function (Eqs. 4·21 and 4·23). It is

$$\pm\theta + \theta_1 = \sqrt{\frac{\gamma+1}{\gamma-1}} \tan^{-1} \sqrt{\frac{\gamma-1}{\gamma+1}(M^2-1)} - \tan^{-1}\sqrt{M^2-1}$$

where M_1 may be rewritten in terms of w_1/a^*. For different initial directions, θ_1, the curve is simply displaced through the angle θ_1; this gives a whole family of (identical) epicycloids. One of these, shown by the broken line, includes the segment AB of the flow in (*c*).

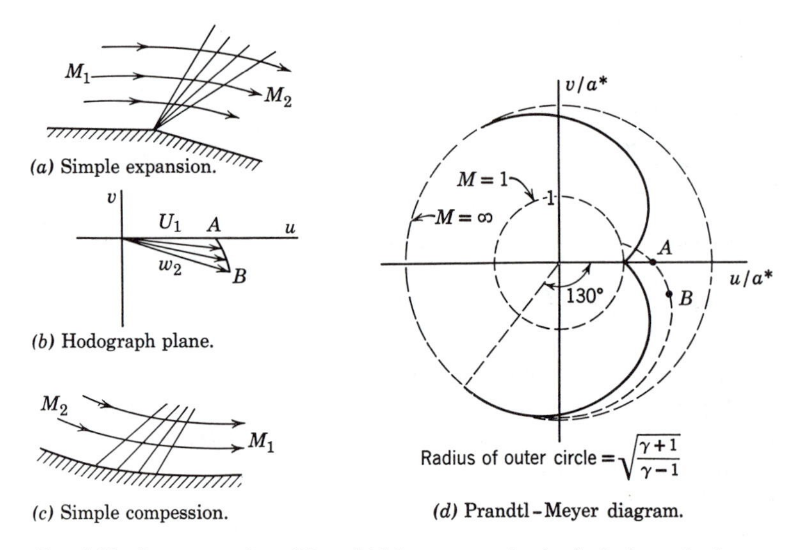

Fig. 4·25 Representation of Prandtl-Meyer expansion in the hodograph plane.

Both the examples illustrated in this article, that is, flow through oblique shocks and simple isentropic flows, have singular mappings onto the hodograph plane. In general, the flow in the physical plane will map onto a whole *region* in the hodograph plane, but singular points and curves, like those in the above examples, may appear. The problem of fitting given, physical boundary conditions to a hodograph solution may be a most difficult one.

4·21 Cone in Supersonic Flow

The flow field for a cone is not so simple as for a wedge. In the latter case one can fit the oblique shock solution directly, since the flow in the region between the shock and the wedge is *uniform*. In the three-dimensional case of the cone, however, a uniform flow downstream of the shock is not possible, for it would not satisfy the equation of continuity.

However, the flow possesses another property that is very helpful in the analysis. Because of the limited upstream influence, it is sufficient to assume that the cone is semi-infinite (Fig. 4·26). Since there is then no characteristic length in the problem, in relation to which others may be compared, the only possible variation of flow properties is with respect to the angle ω.

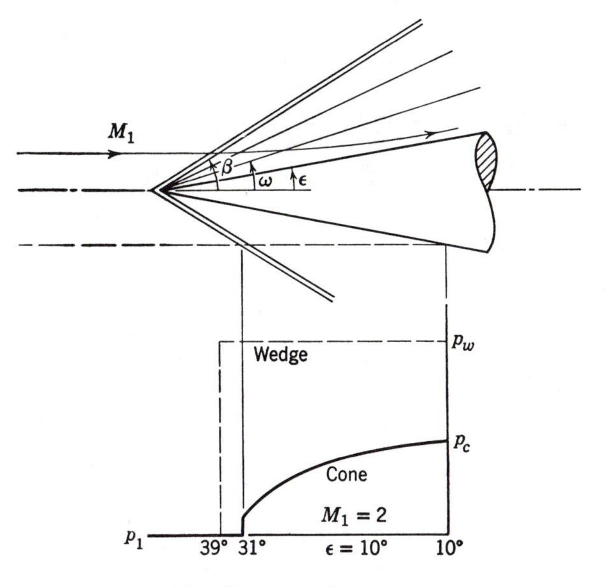

FIG. 4·26 Supersonic flow over a cone.

That is, *conditions are constant on each "ray" from the vertex.* Such a flow field is called conical.† Due to the axial symmetry, we need consider only the rays in any meridian plane, as shown in Fig. 4·26. Between the shock and the cone, the conditions vary from one ray to another, and thus along the streamlines which cross them.

The solution, first given by Busemann and in a different form by Taylor and Maccoll, consists of fitting an isentropic conical flow to a conical shock. For the isentropic part of the flow, the three-dimensional equations (Chapter 7) are rewritten in terms of the single conical variable ω, resulting in an ordinary, nonlinear differential equation, which has to be solved numerically.‡ At the shock, the conditions are given directly by the "jump" relations for the simple oblique shock (Articles 4·2 and 4·3), since this solution always applies *locally* to any shock surface. The flow behind the

†The above argument cannot be applied to a cone in *subsonic* flow, for in that case the boundary conditions at infinity cannot be satisfied by a conical field.

‡Z. Kopal, *Tables of Supersonic Flow about Cones*, MIT, 1947.

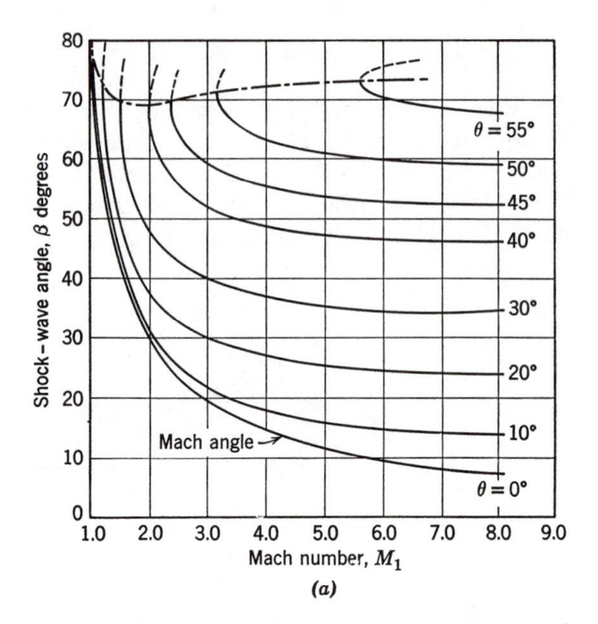

FIG. 4·27(*a*) Cone in supersonic flow. Shock angle versus Mach number for various cone angles. Courtesy of J. W. Maccoll, "The Conical Shock Wave Formed by a Cone Moving at High Speed," *Proc. Roy. Soc. A*, *159* (1937), p. 459.

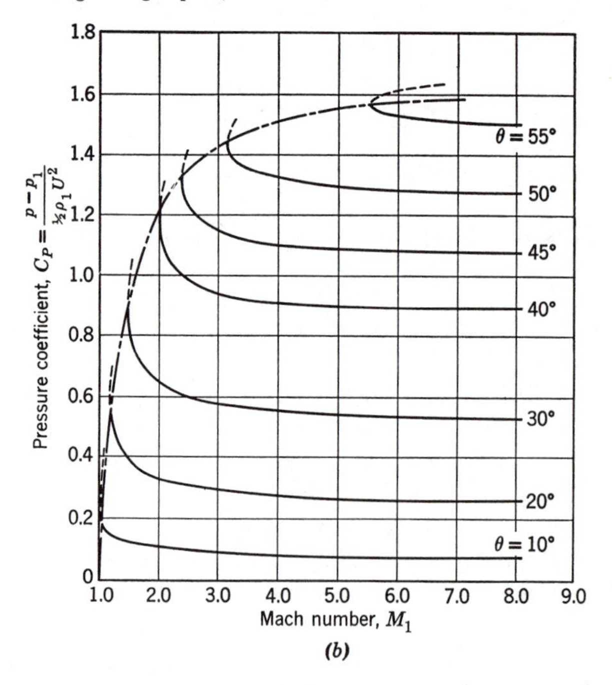

FIG. 4·27(*b*) Pressure coefficient versus Mach number for various cone angles. Courtesy of J. W. Maccoll, "The Conical Shock Wave Formed by a Cone Moving at High Speed," *Proc. Roy. Soc. A*, *159* (1937), p. 459.

shock has to match the isentropic, conical field, and this condition determines the solution.

A typical result, for $M_1 = 2$ and cone of half-angle 10°, is shown in Fig. 4·26. The pressure distribution shown is along a line parallel to the cone axis. Part of the compression occurs through the shock, but there is additional, isentropic compression up to the surface pressure p_c. The streamlines are curved. Also shown for comparison is the pressure distribution for a wedge of the same angle. Due to the three-dimensional effect, the compression by a cone is much weaker than by a wedge, as shown by the lower surface pressure and smaller shock wave angle. Another difference is that detachment occurs at a much lower Mach number.

Plots of wave angle and surface pressure, as functions of Mach number and cone angle, are given in Fig. 4·27.

CHAPTER 5

Flow in Ducts and Wind Tunnels

5·1 Introduction

In this chapter the one-dimensional flow theory of the preceding chapters, together with the oblique shock results, is used to describe some practical examples of compressible flow. Specific examples of flow in nozzles and wind tunnels serve to illustrate some of the basic flow relations, and the methods of applying them. These are quite general and may easily be extended to other cases, some of which are included in the Exercises.

5·2 Flow in Channel of Varying Area

The converging-diverging channel is the basic aerodynamic element in most methods of obtaining prescribed flows. We shall see later how it is incorporated in wind tunnels and other aerodynamic systems. For the

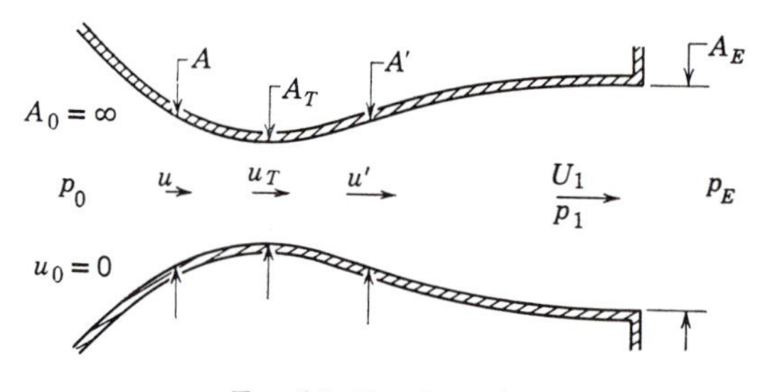

FIG. 5·1 Laval nozzle.

present it is sufficient to study the simplest configuration, shown in Fig. 5·1, in which the channel is supplied with fluid at high pressure p_0 at the inlet and exhausts into a lower pressure p_E at the outlet. This is called a *Laval nozzle.* The nozzle may be attached, at the inlet, to a high-pressure reservoir and allowed to discharge into the atmosphere; alternately, it may be attached, at the outlet, to a vacuum tank and allowed to draw its supply from the atmosphere. To obtain very high pressure ratios it may be convenient to use high pressure at the inlet as well as suction at the outlet.

To investigate the variation of flow parameters along the nozzle, we shall make the approximation that the flow is *one dimensional*; that is, conditions across each section are uniform. We are interested in the relation of these conditions to the section area. The degree to which the flow is *actually* one dimensional depends on the flow inclinations. Thus, in a long, "slender" nozzle the conditions are very nearly one dimensional (neglecting viscous effects). In any case, the results will always apply exactly to sections where the flow actually is uniform (for example, the test section) and to "average" conditions at other sections. The methods for exact analysis of the non-uniform sections will be given later, in Chapter 12.

The qualitative features of compressible flow in the channel of Fig. 5·1, already discussed in Article 2·9, may be reviewed briefly. If the flow is *subsonic* throughout, the maximum velocity occurs at the section of minimum area, called the *throat*. In a subsonic wind tunnel this is the *test section*. It further follows that at two equal sections, A and A', upstream and downstream of the throat, the velocities are equal. The flow is "symmetrical" with respect to the throat. On the other hand, if the flow becomes supersonic, it must be "unsymmetrical," with subsonic conditions upstream and supersonic conditions downstream of the throat. The throat itself must then be *sonic*.

These simple facts have some remarkable consequences. At subsonic speeds a decrease of the downstream pressure (p_E) increases the velocity at the throat (and correspondingly at other sections), but, once the speed at the throat becomes sonic, it can increase no further, for *sonic conditions can exist only at the throat*. A further decrease of p_E cannot change the speed at the throat nor in the upstream portion of the nozzle. This fixes a maximum value on the amount of fluid that can be forced through the nozzle by decreasing the exit pressure.

What, then, actually occurs if p_E is decreased below the value needed to attain sonic flow at the throat? To answer this question it is necessary first to investigate the possible pressure distributions along the nozzle.

5·3 Area Relations

To relate conditions at any two sections in steady flow we may start with the one-dimensional continuity equation:

$$\rho_1 u_1 A_1 = \rho_2 u_2 A_2$$

It is convenient to use *sonic conditions* as reference; thus,

$$\rho u A = \rho^* u^* A^* \tag{5·1}$$

If the flow is subsonic throughout, then A^* is a fictitious throat area that does not actually occur in the flow. On the other hand, if sonic and supersonic conditions are attained, then $A^* = A_t$, where A_t is the area of the actual throat.

If we remember that $u^* = a^*$, Eq. 5·1 may be rewritten

$$\frac{A}{A^*} = \frac{\rho^*}{\rho} \cdot \frac{a^*}{u} = \frac{\rho^*}{\rho_0} \frac{\rho_0}{\rho} \frac{a^*}{u}$$

The ratios on the right-hand side are given in Chapter 2, in Eqs. 2·35*a*, 2·32, and 2·37*a*, as functions of the Mach number. After some reduction, the area-Mach number relation is found to be

$$\left(\frac{A}{A^*}\right)^2 = \frac{1}{M^2}\left[\frac{2}{\gamma+1}\left(1 + \frac{\gamma-1}{2}M^2\right)\right]^{(\gamma+1)/(\gamma-1)} \tag{5·2}$$

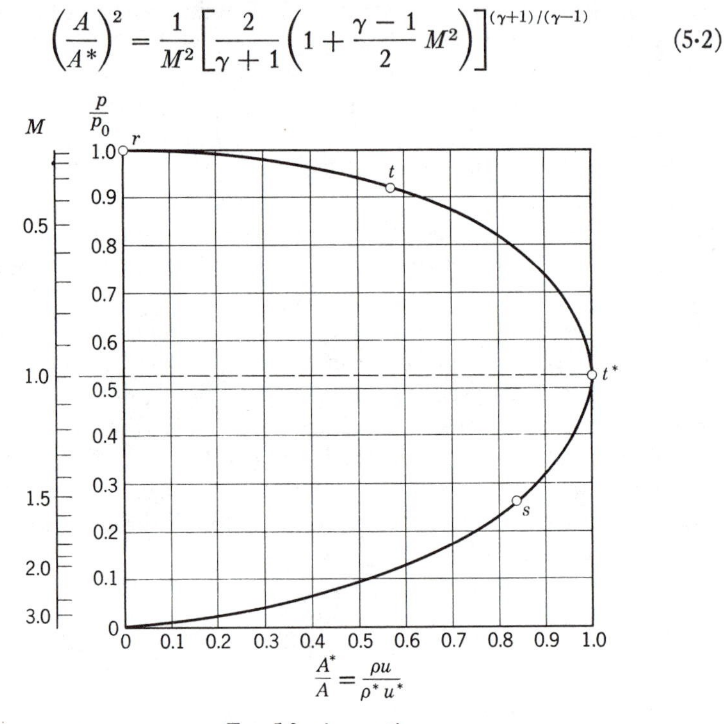

FIG. 5·2 Area ratio.

This result is valid only for *isentropic* flow, since the relation used for ρ_0/ρ is the isentropic one. The relation between area and any other flow parameter is now easily obtained through the Mach number. For example, by using Eq. 2·31, the relation between area and pressure may be rewritten in the form

$$\frac{A^*}{A} = \frac{\rho u}{\rho^* u^*} = \frac{\left[1 - \left(\frac{p}{p_0}\right)^{(\gamma-1)/\gamma}\right]^{1/2}\left(\frac{p}{p_0}\right)^{1/\gamma}}{\left(\frac{\gamma-1}{2}\right)^{1/2}\left(\frac{2}{\gamma+1}\right)^{(1/2)(\gamma+1)/(\gamma-1)}} \tag{5·3}$$

Equations 5·2 and 5·3 are plotted in Fig. 5·2.

5·4 Nozzle Flow

Using the area relations, we may now plot the distributions of Mach number and pressure along a given nozzle. The curve in Fig. 5·2 is double valued, with a subsonic and a supersonic branch. The flow starts out along

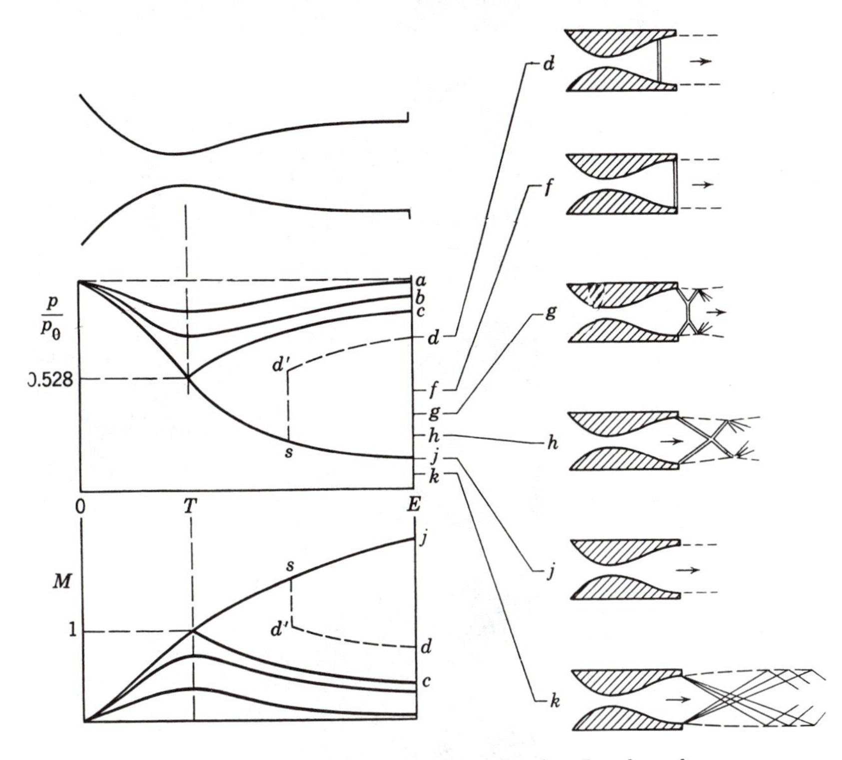

FIG. 5·3 Effect of pressure ratio on flow in a Laval nozzle.

the subsonic branch from r. If the flow is subsonic throughout, then the throat area A_t is less than A^*; the flow reaches only the point t and then returns along the same branch. On the other hand, if the throat becomes sonic, then $A_t = A^*$, the flow in the throat is at the branch point t^*, and the flow in the downstream section lies on the supersonic branch, up to some point s.

Typical cases are given in Fig. 5·3, which shows the form of the theoretical pressure distributions and the wave configurations for several exit pressures. For the upper curves, a, b, c, which lie entirely on the subsonic branch, the exit pressure controls the flow throughout the channel. On curve c the throat has just become sonic, and so the pressure at the throat, and upstream

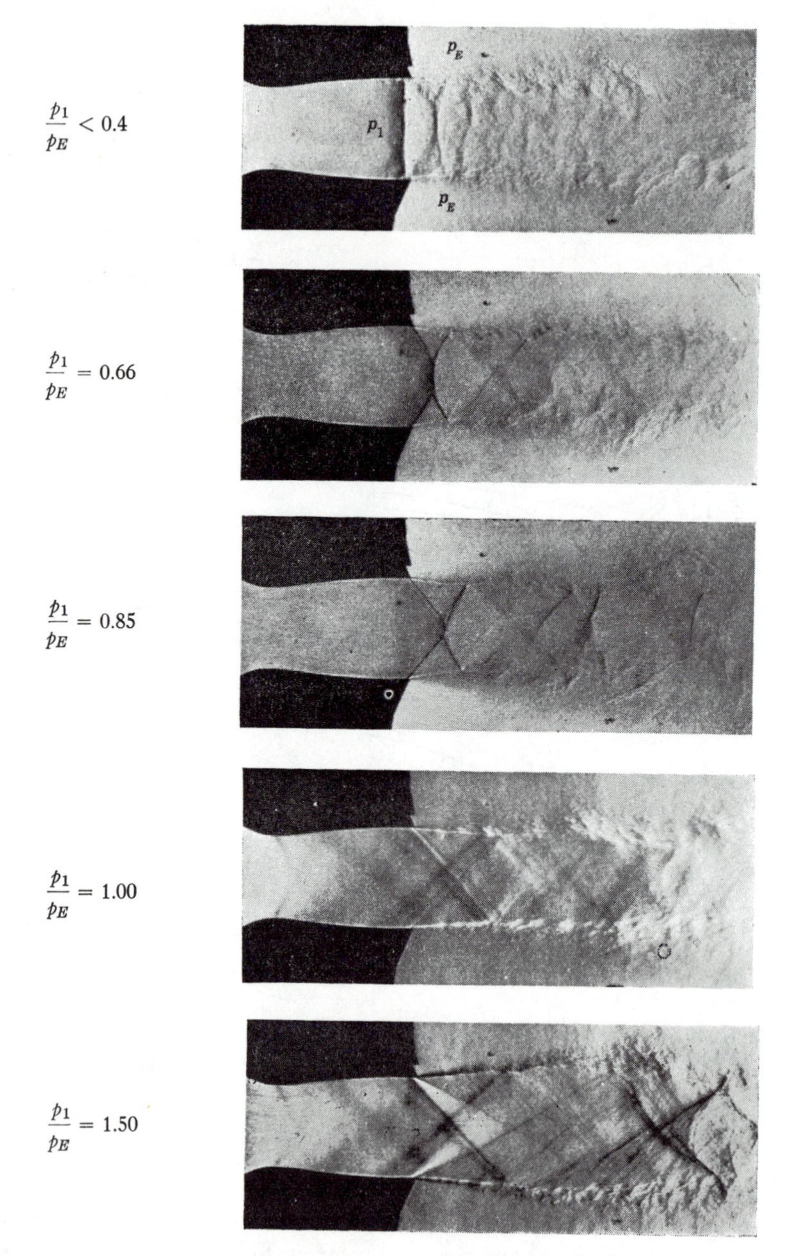

FIG. 5·4 Schlieren photographs of flow from a supersonic nozzle at different back pressures. The photographs, from top to bottom, may be compared with Fig. 5·3, sketches *d, g, h, j, k*, respectively. Reproduced from: L. Howarth (ed.), *Modern Developments in Fluid Dynamics, High Speed Flow*, Oxford, 1953.

of it, can decrease no further. There is now another possible isentropic solution, on the supersonic branch, which terminates at j. But, for this flow to exist, the exit pressure must be p_j, considerably lower than p_e. What of the pressures that lie between these values? For them, *there are no isentropic solutions.*

For example, consider an exit pressure p_d, somewhat lower than p_e. It could be reached by a subsonic flow, shown by the broken line $d'd$, along which the pressure ratio p/p_{0d} satisfies the area relation (Eq. 5·3), provided that p_{0d} *be less than* p_0. It requires that a *nonisentropic* process occur somewhere along the flow. There are many possibilities—the simplest assumption, without introducing viscous effects, is that the increase of entropy occurs at a single, normal shock. The location of the shock, at s, must be at the Mach number that will give just the right p_{0d} to reach point d. For each position s of the shock along the supersonic branch there is a corresponding exit pressure that may be reached.

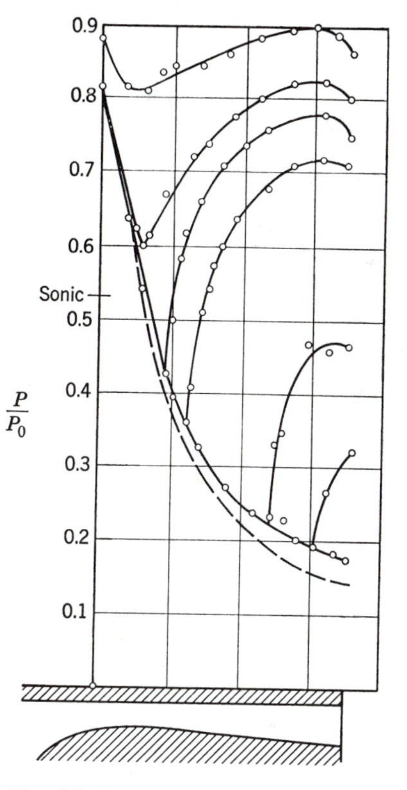

FIG. 5·5 Observed pressures on the wall of an expanding nozzle, for various exit pressures. From Ackeret's article in the *Handbuch der Physik*, Vol. VII.

As the exit pressure is decreased the shock moves downstream, finally reaching the exit when the pressure there reaches a value p_f. With further decrease of exit pressure the flow in the nozzle is no longer affected, the pressure adjustment being made through systems of oblique waves, as shown by the sketches in Fig. 5·3 and the photographs of Fig. 5·4.

Figure 5·5 gives an example of observed pressure distributions along a nozzle wall for cases in which the shock is inside the nozzle, that is for $p_E < p_f$. The positions of pressure rise mark the shock locations, for various values of p_E. Because of the interaction between the shock wave and the boundary layer on the nozzle walls, the observed pressure rise is not a "step," as in the ideal theory (see Article 13·16).

For exit pressures lower than p_f, the flow up to the exit is completely supersonic. Thus a Laval nozzle may be used as a supersonic wind tunnel, provided that

$$p_E \leq p_f$$

This, in fact, is the principle of the open-circuit type of supersonic wind tunnel, operating from a high-pressure reservoir or into a vacuum receiver, or both. If enough power is available, continuous flow may be obtained. Otherwise, this arrangement is used as an *intermittent* or *blow-down* wind tunnel.

5·5 Normal Shock Recovery

If the nozzle discharges directly into the receiver, the minimum pressure ratio for full supersonic flow in the test section is

$$\left(\frac{p_0}{p_E}\right)_{\min} = \frac{p_0}{p_f}$$

where p_f is the value of p_E at which the normal shock will stand at the nozzle exit (Fig. 5·3). However, if a *diffuser* is attached to the exit, as shown

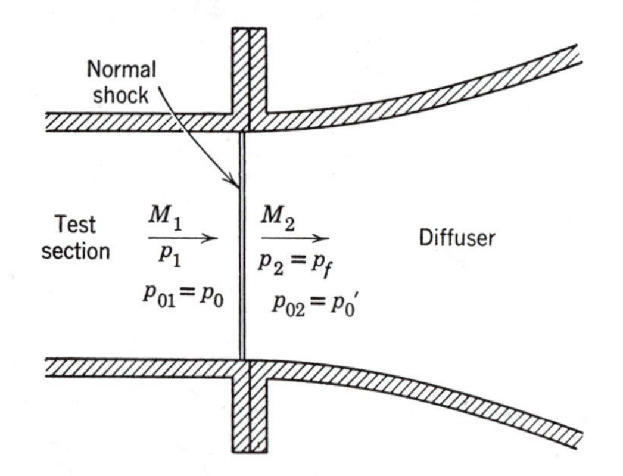

FIG. 5·6 Diffusion through a normal shock followed by a subsonic diffuser.

in Fig. 5·6, operation at a lower pressure ratio is possible, since the subsonic flow downstream of the shock may be decelerated isentropically, in principle, to the stagnation pressure p'_0. The pressure ratio required then is the ratio of stagnation pressures across a normal shock wave at the test section Mach number M_1, that is,

$$\lambda_s = \frac{p_0}{p'_0} = \left[1 + \frac{2\gamma}{\gamma + 1}(M_1^2 - 1)\right]^{1/(\gamma-1)} \left[\frac{(\gamma - 1)M_1^2 + 2}{(\gamma + 1)M_1^2}\right]^{\gamma/(\gamma-1)} \tag{5·4}$$

This ratio was given previously in Eq. 2·54.

In practice, the diffuser of Fig. 5·6 does not give the expected recovery; the interaction of shock wave and boundary layer produces a flow that is different from the above model, and one that usually results in lower recovery.

There is another configuration that more nearly realizes the normal shock recovery, namely, one with a long constant-area duct ahead of the subsonic diffuser. Such a duct, provided it is long enough, gives nearly the same recompression as a normal shock, even though the mechanism is quite different (Article 2·12). The compression occurs through a system of shocks interacting with the thickened boundary layer. It is remarkable that, *in supersonic flow*, these dissipative processes actually result in fairly efficient recompression, whereas in subsonic flow the effects of friction can only reduce the recoverable pressure. Admittedly, recovery through such a dissipative system is not the most efficient recovery, but it is often the most practical. Its virtue is its stability with respect to variations of inlet conditions. The effects of friction can in any event not be avoided, and, though it is possible to design more efficient ducts for *specific conditions*, these may perform quite badly at off-design points.

The "normal shock pressure recovery" assuming full subsonic recovery, given by Eq. 5·4, is a convenient reference, or standard, for comparing the performance of actual supersonic diffusers and wind tunnels.

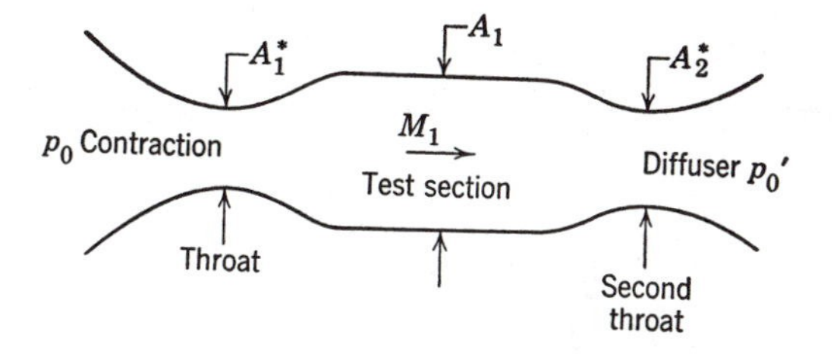

FIG. 5·7 Laval nozzle discharging into a second throat.

5·6 Effects of Second Throat

Ideally, it is possible to operate at even lower pressure ratios than the one for normal shock recovery. Consider the arrangement shown in Fig. 5·7. If the supersonic flow at the test section could be isentropically compressed to sonic conditions at the *second throat*, it could then be decelerated subsonically in the diffuser. In this idealized case, no shock waves would occur and the pressure would be completely recovered, giving $p'_0 = p_0$. For these conditions the second throat area would have to be the same as the first, $A^*_2 = A^*_1$. Since there are no losses in this ideal, shock-free, frictionless flow, it requires no pressure difference, that is, no power, to maintain it.

However, it would still be necessary to establish a pressure difference initially, in order to *start* it. The starting processes are not completely understood, and may be different in different installations, but the following one-dimensional model agrees fairly well with experimental results, at least at the lower supersonic Mach numbers (Article 5·7). The flow which is first established in the test section must be subsonic; it is followed by supersonic flow. Thus, if the conditions are considered quasi-stationary, they must be similar to those corresponding to *d* in Fig. 5·3, with a normal shock in the test section.† There must then be sufficient pressure ratio to make up for the shock losses, and the channel geometry must be such as to accommodate the flow. Specifically, the second throat area must be large enough to accommodate the mass flow when there is a normal shock ahead of it in the test section. The *minimum* permissible area corresponds to sonic conditions, and may be obtained from the continuity equation,

$$\rho^*_2 a^*_2 A^*_2 = \rho^*_1 a^*_1 A^*_1$$

In adiabatic flow, $a^*_2 = a^*_1$ since $T^*_2 = T^*_1$, and $\rho^*_1/\rho^*_2 = p^*_1/p^*_2 = p_{01}/p_{02}$. Therefore the minimum starting area for the second throat is given by the ratio

$$\frac{A^*_2}{A^*_1} = \frac{p_{01}}{p_{02}} \left(\equiv \frac{p_0}{p'_0} \right) = \lambda_s \tag{5·5}$$

Instead of this ratio, it is often convenient to deal with the ratio of test-section area to diffuser-throat area, that is, A_1/A^*_2. This is called the *diffuser contraction ratio*, ψ. Thus, the maximum permissible contraction ratio for starting is

$$\psi_{\max} = \frac{A_1}{A^*_2} = \frac{A_1}{A^*_1}\frac{A^*_1}{A^*_2} = \frac{A_1}{A^*_1}\frac{1}{\lambda_s} = f(M_1) \tag{5·5a}$$

λ_s is given by Eq. 5·4, and A_1/A^*_1 is the area-Mach number relation given in Eq. 5·2. With these, the maximum starting contraction ratio is obtained as a function of test section Mach number M_1; it is plotted in Fig. 5·8*a*.

With this minimum area for the second throat, and with larger ones, the shock wave is able to "jump" from the test section to the downstream side of the diffuser throat. This is picturesquely called "swallowing" the shock. The test section is then fully supersonic, but unfortunately so is the second throat and part of the diffuser. We have apparently only shifted the scene of the loss to a shock in the diffuser. What can be done now, in principle, is to reduce the area of the second throat, *after the flow has started*. As A^*_2 is reduced the shock in the diffuser moves upstream toward the throat, becoming weaker, until, when $A^*_2 = A^*_1$, it just reaches the throat, at

†These are assumptions. A more realistic model might have to take into account nonstationary effects, the possibility of oblique shocks, and the role of boundary layer development.

vanishing strength. The flow is then the ideal one, with supersonic test section and isentropic diffuser.

In practice, it is not possible to reduce A^*_2 to this ideal value. However, some contraction after starting is possible, up to some value at which the boundary layer effects prevent sufficient mass flow for maintaining a supersonic test section, and the flow "breaks down."

5·7 Actual Performance of Wind Tunnel Diffusers

The performance of a real diffuser is dominated by the behavior of the boundary layer, which was entirely neglected in the preceding discussion. In the strong *adverse pressure gradient*, which necessarily exists in a diffuser, the boundary layer tends to thicken and separate, thus changing the effective geometry of the channel. This invariably occurs in a way which impairs the recompression, for the adjustment is such as to relieve the adverse pressure gradient. The effect, already present in subsonic flow, is even more complicated in the supersonic case, where there is always strong interaction between the shock waves and boundary layer (Article 13·16), leading usually to local separation and complex patterns of shock reflection. It is not possible, with present knowledge, to treat these effects theoretically and thus to design the optimum (real) diffuser.

Some experimental results collected by Lukasiewicz† are given in Fig. 5·8. Observed values of the minimum second-throat area for starting are given in (*a*). They agree well with the "theoretical" result of Eq. 5·5*a*, except at the higher Mach numbers, where smaller starting areas are found to be possible. Figure 5·8*b* gives the values of pressure ratio λ corresponding to the minimum starting area. Thus, this data also represents the optimum operating pressure ratio for diffusers with *fixed* throat, for which the area must be large enough to allow starting.

If the diffuser throat is adjustable, so that the second throat area can be decreased *after starting*, improved recovery is possible. The minimum area is that which causes breakdown by not permitting sufficient mass flow. Optimum recovery usually occurs at slightly larger area than the breakdown value. Some experimental values of optimum recovery are given in Fig. 5·8*c*. The line marked "Normal Shock at ψmax" corresponds to recovery through a normal shock in a throat contracted to the theoretical starting ratio.

A discussion of diffusers may be found in the article by Lukasiewicz.

5·8 Wind Tunnel Pressure Ratio

The diffuser pressure ratio given in Fig. 5·8 is for all practical purposes the wind tunnel pressure ratio, which must be provided by the compressor

†J. Lukasiewicz, *J. Aeronaut. Sci.*, *20* (1953), 617.

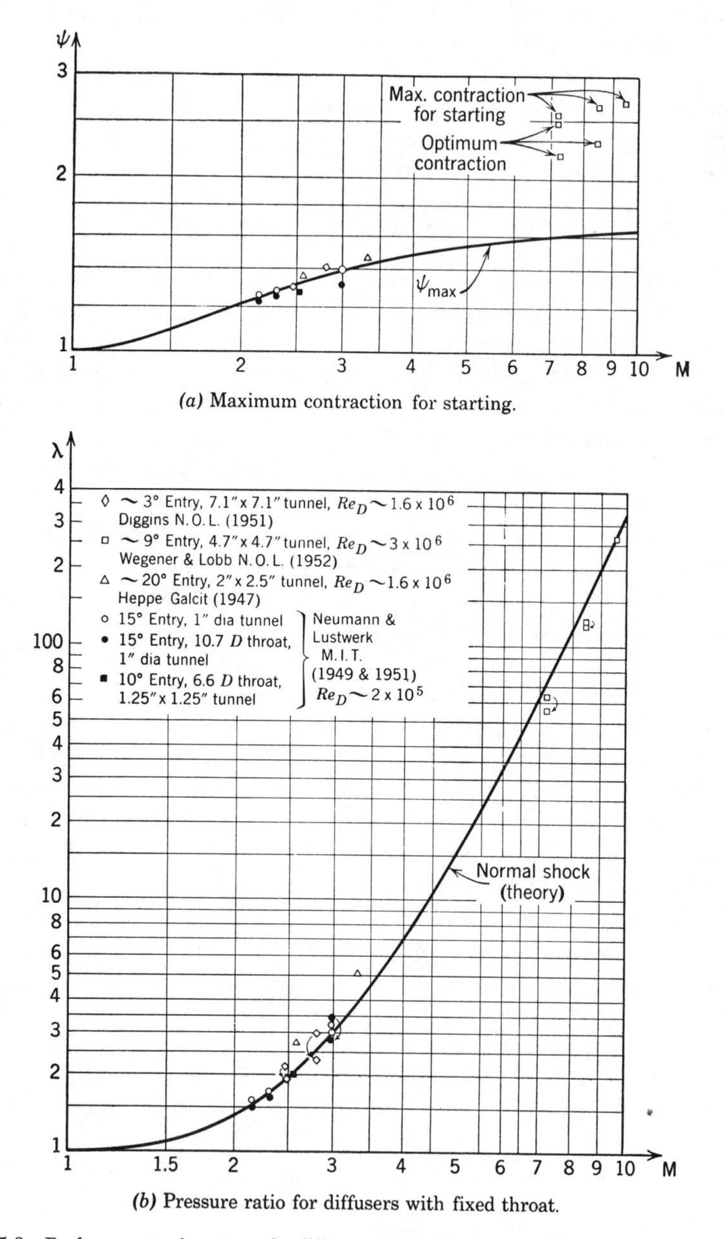

(a) Maximum contraction for starting.

(b) Pressure ratio for diffusers with fixed throat.

FIG. 5·8 Performance of supersonic diffusers. *(a)* Maximum contraction for starting. *(b)* Pressure ratio for diffusers with fixed throat. *(c)* Optimum pressure ratio for diffusers with adjustable throat. From J. Lukasiewicz, *J. Aeronaut. Sci.*, *20*, 617 (1953).

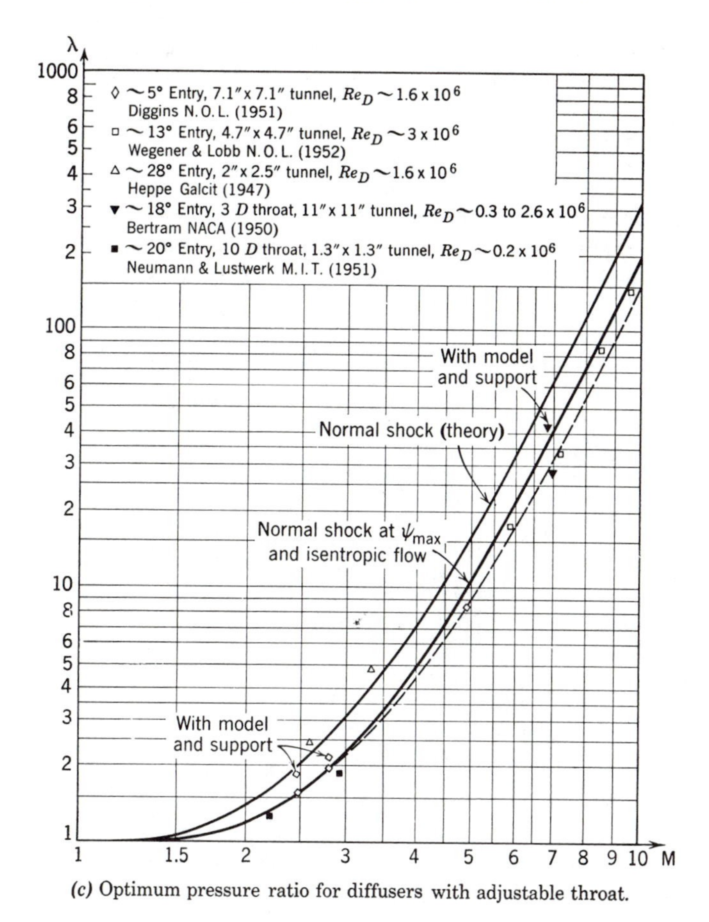

(c) Optimum pressure ratio for diffusers with adjustable throat.

plant. The pressure drop in the ducting, corners, coolers, etc., may be included in the compressor characteristics (Article 5·11).

Probably the most serious unknown factor is the *interference effect* of the model. It may have a strong influence on the performance of the diffuser, through the effects of its wake on the sidewall boundary layers and on the effective diffuser geometry. It will not necessarily be a helpful interference, and may seriously reduce the overall recovery. Some examples are shown in Fig. 5·8. Since the model configurations may be quite varied, and since their effects are not easily predicted, conservative estimates should be used.

If the diffuser has an adjustable throat, operation is possible at the optimum pressure ratios of Fig. 5·8*c*. Nevertheless, the compressor must be able to provide the pressure ratios of Fig. 5·8*b*, which are the minimum required for *starting*. After starting, the diffuser throat may be closed down to the optimum *running* area; the corresponding decrease in pressure ratio may be an important factor in the power economy of large installations.

5·9 Supersonic Wind Tunnels

Figure 5·9 shows a typical arrangement of a closed-circuit, continuously operating, supersonic wind tunnel. The basic elements are the following.

(1) The *supply section* is the section of largest area in the circuit. The speed is so low that stagnation conditions (p_0, ρ_0, T_0) exist there.

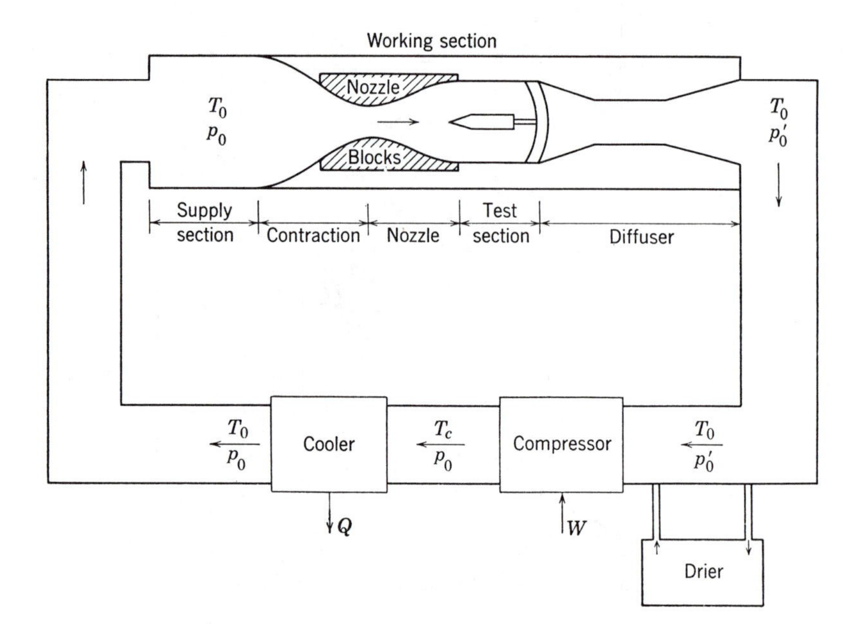

FIG. 5·9 Typical continuous, closed-return supersonic wind tunnel.

(2) The *contraction*, from the supply section to the throat, must be of such a shape that the air is uniformly accelerated. This requires gentle wall curvature, to ensure monotonically decreasing pressure. There must be no adverse pressure gradient at any point, otherwise separation may occur. The *length* of the contraction has little effect on the boundary layer thickness at the throat.

(3) The *supersonic nozzle* expands the flow from sonic speed, at the throat, to the test-section speed. For a given Mach number, the area ratio between test section and throat must have the value given by Eq. 5·2, but in addition *the nozzle must have the correct shape to give uniform, shockfree flow* in the test section. Methods for designing nozzle shapes are discussed in Chapter 12. The nozzle shape, and thus the Mach number, may be changed by changing

the nozzle blocks (Fig. 5·9). Some tunnels use a "flexible nozzle," which allows continuous and rapid adjustment of Mach number. Basically, flexible nozzle walls are steel plates which may be elastically deformed to the required wall contours.

(4) The *test section*, of essentially constant area (some divergence is needed to allow for boundary layer growth), houses the model support, calibration probes, etc. It is usually provided with high-quality glass windows for optical observation of the model and flow.

(5) The *diffuser*, discussed already in preceding articles, decelerates the flow and discharges it into the return circuit.

(6) The *compressor* recompresses the flow to the supply pressure p_0, raising its temperature to a value T_c.

(7) The *cooler* reduces the temperature to the supply value T_0, leaving the pressure practically unchanged, at p_0.

(8) The *drier* by-passes a small part of the air in the circuit and dries it. This is necessary to prevent condensation of water vapor at the low temperatures of the test section. Most driers use beds of anhydrous material, such as alumina or silica gel, over which the by-passed air is circulated. Once the circuit is filled with dry air, the drier has only to remove the moisture introduced by leakage. In a nonreturn tunnel, on the other hand, it must have sufficient capacity to dry the *entire* mass flow.

In a hypersonic wind tunnel it may be necessary to heat the air in the supply section, in order to keep T_0 sufficiently high to prevent condensation of the air components (principally nitrogen) in the test section.

5·10 Wind Tunnel Characteristics

The flow in the closed-circuit wind tunnel of Fig. 5·9 comprises a steady-state system. *The energy of the system is constant*, the work performed by the compressor being just equal to the heat removed by the cooler,

$$W = Q$$

Thus the *compressor-cooler combination adds no net energy to the circuit.* Its function is to *remove the entropy* that is being continually produced by the dissipative processes in the flow.

Since the velocities in the return circuit are low, the air entering the compressor-cooler unit is practically at the condition (p'_0, T_0) of the diffuser outlet, and the air leaving is at the supply condition (p_0, T_0). Thus the entropy that is removed, per unit mass, is given by Eq. 2·15*a*,

$$\Delta S = R \log p_0/p'_0 = R \log \lambda \tag{5·6}$$

where λ is the *wind tunnel pressure ratio* (Article 5·8). The *rate* of entropy removal is

$$m\,\Delta S = mR \log \lambda \tag{5·6a}$$

where m is the mass flow, that is, the mass passing a given section per unit time.

Equation 5·6*a* contains the two fundamental quantities in the wind tunnel flow system, namely, the *overall pressure ratio* λ and the *mass flow* m (or, alternatively, the *volumetric flow*, $V = m/\rho$).

The above relations may be used to calculate an *ideal value of the power* required for given pressure ratio and mass flow. In the wind tunnel circuit of Article 5·9, the compressor and cooler are arranged as shown in Fig. 5·10

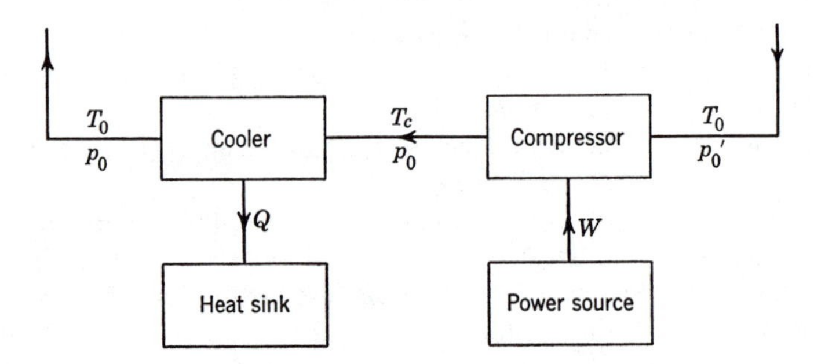

FIG. 5·10 Schematic diagram of the compressor-cooler unit.

Assuming that the heat transfer at the cooler occurs reversibly, at constant pressure, the relation between the heat per unit mass and the temperatures is

$$q = c_p(T_c - T_0) = w$$

The temperature ratio T_c/T_0 across the cooler is the same as that across the compressor. Assuming the compression to be reversible, T_c/T_0 is related to the pressure ratio by the isentropic relation,

$$\frac{T_c}{T_0} = \left(\frac{p_0}{p'_0}\right)^{(\gamma-1)/\gamma} = \lambda^{(\gamma-1)/\gamma}$$

Thus the heat that must be removed, and the work that must be supplied, *per unit mass* of fluid, is

$$w = c_p T_0(\lambda^{(\gamma-1)/\gamma} - 1) = \frac{\gamma}{\gamma - 1} RT_0(\lambda^{(\gamma-1)/\gamma} - 1)$$

The *rate* of work, that is, the ideal wind tunnel *power*, is then given by

$$P = mw = m\frac{\gamma}{\gamma - 1} RT_0(\lambda^{(\gamma-1)/\gamma} - 1) \tag{5·7}$$

This may be rewritten in terms of more convenient wind tunnel parameters by writing out the mass flow m. Using the throat for the reference section,

$$m = \rho^* a^* A^* = \left(\frac{2}{\gamma + 1}\right)^{(1/2)(\gamma+1)/(\gamma-1)} \rho_0 a_0 A^* \qquad (5\cdot8)$$

The second equation here was obtained by using the values of ρ^*/ρ_0 and a^*/a_0 from Article 2·10. For air, the relation becomes

$$m = 0.579 \rho_0 a_0 A^* \qquad (5\cdot8a)$$

This may also be rewritten in terms of the test section area and Mach number, by using the area-Mach number relation, Eq. 5·2.

Equation 5·8 together with the equation of state for a perfect gas may now be used to rewrite Eq. 5·7 for the power in the form

$$P = \frac{\gamma}{\gamma - 1}\left(\frac{2}{\gamma + 1}\right)^{(1/2)(\gamma+1)/(\gamma-1)} p_0 a_0 A^* (\lambda^{(\gamma-1)/\gamma} - 1) \qquad (5\cdot7a)$$

This shows that the power is proportional to the stagnation pressure, the stagnation speed of sound ($a_0 = \sqrt{\gamma R T_0}$), the throat area, and a function of the overall pressure ratio. For stagnation conditions corresponding to the standard atmosphere, $p_0 = 2116$ lb/ft^2, $a_0 = 1117$ ft/sec, the *horsepower per square foot of throat area is*

$$P' = 8700(\lambda^{2/7} - 1) \qquad (5\cdot9)$$

Since the work of the compressor is not actually done in the ideal, reversible manner, the actual power required is higher than this ideal value. In addition to the mechanical inefficiency, there is always thermodynamic inefficiency due to the fact that a given piece of equipment cannot be ideally matched to the flow requirements under all operating conditions. The matching problem is discussed in the following article.

Multiplying Eq. 5·9 by A^*/A, one obtains the ideal horsepower per square foot of *test section area*,

$$P_1 = 8700(\lambda^{2/7} - 1)A^*/A \qquad (5\cdot9a)$$

5·11★ Compressor Matching

The design of a wind tunnel has either of two objectives: (1) to *select a compressor* for specified test section size, Mach number, and pressure level; or (2) to determine the best utilization of an already available compressor. In the former case the wind tunnel characteristics govern the selection, whereas in the latter case the compressor characteristics must be utilized. In either case the characteristics to be matched are the overall pressure ratio and the mass flow.

Since compressor characteristics are usually given in terms of the *volumetric flow V* rather than the mass flow, it is convenient to give the wind tunnel characteristics also in terms of V. It is related to the mass flow by

$$V = m/\rho$$

Since density varies through the circuit, the volumetric flow also varies, for constant m. For the compressor one specifies the *intake flow*

$$V'_0 = m/\rho'_0 \tag{5·10}$$

which is essentially the same as at the end of the diffuser.

On the other hand, the flow in the supply section is

$$V_0 = m/\rho_0 \tag{5·11}$$

Introducing Eq. 5·8 for m, this becomes

$$V_0 = \left(\frac{2}{\gamma+1}\right)^{(1/2)(\gamma+1)/(\gamma-1)} a_0 A^* = \left(\frac{2}{\gamma+1}\right)^{(1/2)(\gamma+1)/(\gamma-1)} \sqrt{\gamma R T_0}\,\frac{A^*}{A}\,A$$

$$= \text{const.}\sqrt{T_0}\,A\,(A^*/A) \tag{5·11a}$$

Thus, for a given gas, V_0 depends on the stagnation temperature, the test section area, and the test section Mach number.

The relation between the compressor intake flow and the supply section flow, easily obtained from Eqs. 5·10 and 5·11, is

$$\frac{V'_0}{V_0} = \frac{\rho_0}{\rho'_0} = \frac{p_0}{p'_0}\frac{T'_0}{T_0} = \Lambda \tag{5·12}$$

Since $T'_0 = T_0$, Λ is the pressure ratio at which the wind tunnel is actually operating,

$$\Lambda = p_0/p'_0$$

It cannot be smaller than the minimum pressure ratio λ, which will permit operation at the desired Mach number (Fig. 5·8). That is, it is necessary that

$$\Lambda \geq \lambda$$

The relation between the operating pressure ratio and the compressor intake volume may now be rewritten, from Eq. 5·12, in the form

$$\Lambda = \left(\frac{1}{V_0}\right) V'_0$$

On the Λ-V'_0 plot† this is a straight line, through the origin, with slope $1/V_0$, as shown in Fig. 5·11*a*. This is the wind tunnel characteristic for fixed values of T_0, A, and M_1. Operation is possible only above the point o, which corresponds to the minimum pressure ratio λ.

Also shown in Fig. 5·11*a* is a typical compressor characteristic. The

†This method of plotting was communicated to the authors by Dr. P. Wegener.

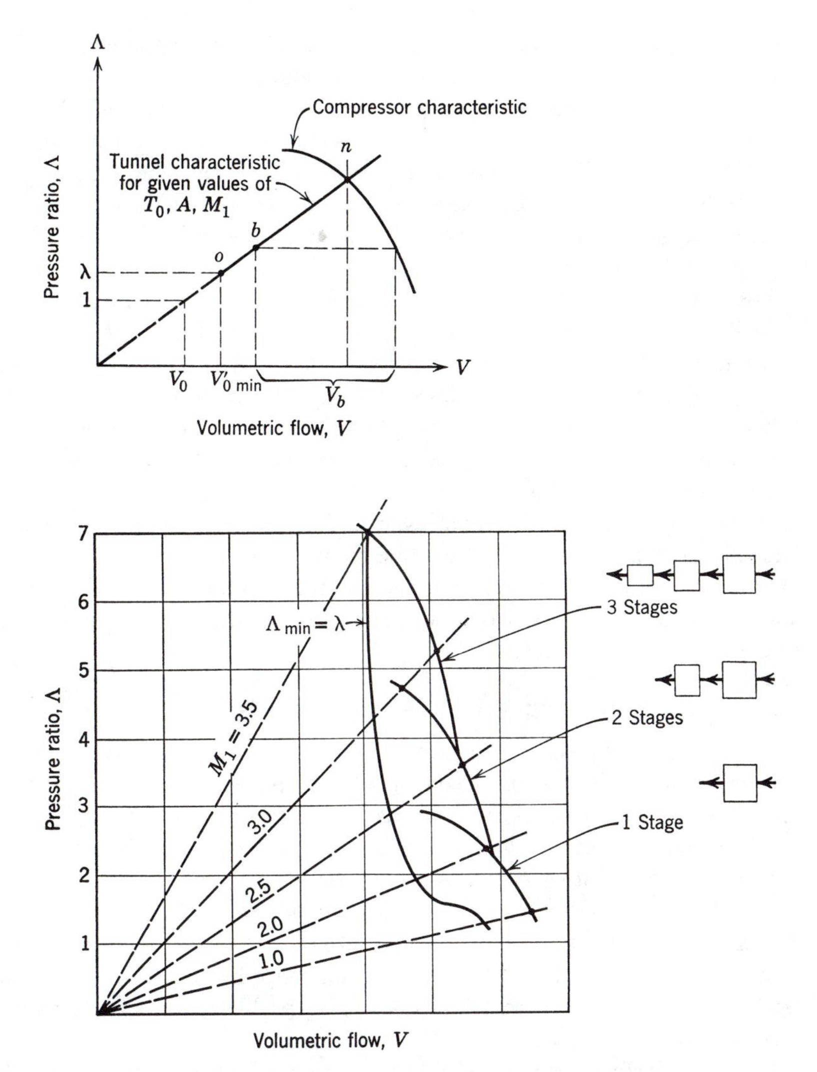

FIG. 5·11 Wind tunnel and compressor characteristics. (*a*) Matching of wind tunnel and compressor characteristics (one test section condition): *n*, match point; *b*, match point with by-pass; *o*, match point at minimum operating pressure ratio. (*b*) Operation over a range of Mach numbers, using multistage compressors.

tunnel will operate at the match point *n*. Evidently, at this Mach number the operation is at a higher pressure ratio than the minimum. The flow will adjust itself to this higher pressure ratio, for example, by a downstream shift of the shock waves in the diffuser. This loss of efficiency is unavoidable if it is desired to operate over a range of Mach number or pressure, since the compressor characteristics usually do not match the minimum wind tunnel requirements over a range of Mach numbers and pressures.

If it is desired to operate at lower pressure ratio but the same Mach number, say at the point b in Fig. 5·11a, then some of the flow (a quantity V_b) may be short-circuited by means of a by-pass with metering valve. The volume flow through the compressor is then increased but the compression ratio is decreased. Whether this results in an increase or decrease of power depends on the pressure level.

For increasing Mach number the slopes of the wind tunnel characteristics, as well as the values of $\Lambda_{min} = \lambda$, increase as shown in Fig. 5·11b, if T_0 and A are fixed. Also shown is the compressor characteristic for a plant which consists of several compressors, staged in series. Each successive compressor boosts the pressure higher, and so must be able to withstand higher pressures. On the other hand, it may be smaller in size than the preceding one, since the volume becomes smaller. The example has been simplified by assuming that each unit consists of a compressor and aftercooler, so that the intake and exhaust temperatures are the same throughout.

In the example of Fig. 5·11b the maximum Mach number which may be obtained is $M = 3.5$, fixed by the intersection of the compressor characteristic with the Λ_{min} curve.

5·12 Other Wind Tunnels and Testing Methods

The closed-circuit, continuous wind tunnel was described in some detail in the preceding articles, for, on the one hand, it is the most useful arrangement for most purposes, and, on the other hand, it contains most of the features that are found in other wind tunnels. We shall now describe the more important of these other types, but only briefly.

(1) *Open-circuit tunnels* do not have a return circuit. The air is simply discharged into the atmosphere. The choice of such an arrangement might be dictated by the nature of the power available or by lack of space for the return circuit. A disadvantage is the drying problem, for a fresh supply must be dried continuously.

(2) An *induction wind tunnel* may be considered when there is available a relatively small supply of air at very high pressure. In the induction method of drive, high-velocity jets injected into the tunnel, just downstream of the diffuser, entrain the air and establish a flow. An advantage is that the compressor does not work directly on the flow and need not be closely matched.

(3) *Intermittent* or *blow-down pressure tunnels* utilize an air supply from a storage tank. Flow is started by means of a quick-opening valve and lasts only until the pressure in the tank decreases to the value that gives the minimum operating pressure ratio. The compelling advantage of this arrangement is its economy, for the power required is small, compared to

that for continuous tunnels. The power is governed by the time allowed for filling of the storage tank, between runs, and by the length of run. These two factors, in fact, sum up the disadvantages of an intermittent tunnel. Short running times may require the use of special instrumentation for obtaining and recording the data, whereas intervals between runs reduce the useful working time of the tunnel, unless this storing time can be utilized for model changes, etc. Nevertheless, because of their economy, intermittent tunnels are widely used, and with satisfactory results.

(4) *Intermittent vacuum tunnels* use a vacuum sphere on the diffuser end, instead of a pressure tank on the supply end. The supply in this tunnel is simply the atmosphere. An advantage over the pressure type is that high pressure ratio can be obtained without subjecting the tank to more than an atmosphere of pressure. Also, the supply stagnation conditions are constant through a run, being atmospheric, although this may also be arranged in the pressure type by using a constant-pressure valve. The test section Reynolds numbers are lower.

(5) *Open-jet tunnels*, of either the continuous or intermittent type, have an open test section, between the end of the nozzle and the diffuser. This permits easy access to the test section but otherwise has no advantages, and requires higher pressure ratios.

Although the wind tunnel is still the basic aerodynamic facility for testing and research, other methods are available. Some of these are becoming increasingly important, owing to the extreme conditions that must often be simulated. They include the following.

The *shock tube* may be used as a short-duration wind tunnel, utilizing the region of uniform flow behind the advancing shock, or behind the contact surface. There is a limit to the flow Mach number behind the shock in a conventional tube, but this may be increased by allowing the flow to expand into a supersonic nozzle. The main attractiveness of this method is that the stagnation temperatures that may be achieved are very much higher than in conventional supersonic and hypersonic tunnels, and the test conditions may be more nearly those of free flight. The problems of instrumentation are similar to those for intermittent tunnels, but they are much more severe because of the shorter durations.

Free-flight techniques provide another method for obtaining data. Models may be launched by means of some propulsion device such as a rocket, or simply dropped from an aircraft. Data are obtained by tracking from the ground, by recording on a recorder inside the model, or by telemetering from a transmitter inside the model.

A free-flight technique under more controlled conditions is the *ballistic range* technique. The flight of the model, which is fired from a gun, is measured by means of photographic stations along the range, or recorded on "cards" which it pierces during its flight.

CHAPTER 6

Methods of Measurement

6·1 Introduction

This chapter is an outline of methods for measuring the variables of compressible flow. It is by no means possible to include all possible measurements; only the better-developed ones are discussed, and these only briefly. The discussion is mainly from the point of view of an observer working with a wind tunnel, but many of the methods are, of course, applicable, or adaptable, to the problem of measurement in flight.

6·2 Static Pressure

The pressure on an aerodynamic surface, such as a wind tunnel wall or airfoil, may be measured by means of a small orifice drilled normal (or nearly so) to the surface, and connected to a manometer or other pressure

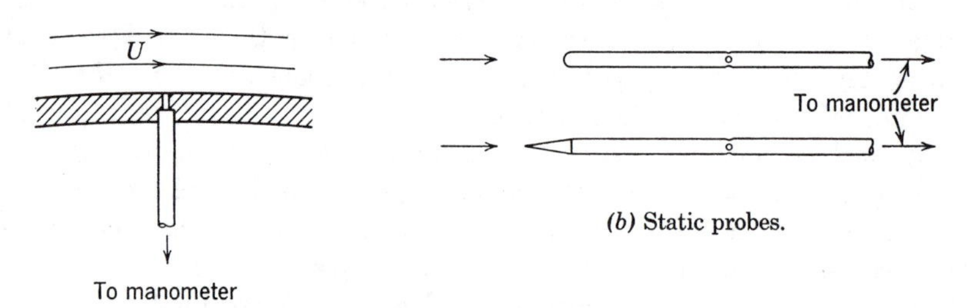

(a) Surface pressure orifice.

(b) Static probes.

FIG. 6·1 Measurement of static pressure.

measuring device (Fig. 6·1*a*). The orifice must be small, and free of imperfections (such as burrs), in order not to disturb the flow locally. A safe rule is that the diameter should be small (say ⅕) compared to the boundary-layer thickness, unless governed by some other factor, such as surface curvature or resolution of pressure gradient. Diameters usually range from 0.01 in., on small models, to 0.1 in. on large installations.

This principle is also used for measuring static pressure at an interior point of the flow, by introducing into the flow a *static probe* (Fig. 6·1*b*) having an orifice at the point of measurement. To disturb the flow as little as possible, the probe must be slender, and aligned parallel to the local flow direction. The nose region will in any case be disturbed, and so the orifice must be downstream of its "region of influence" (10 to 15 diameters). The point of measurement is at the orifice, not at the nose. The sensitivity to yaw may be decreased by using several holes around the circumference, opening to a common manometer lead, so that an average pressure is

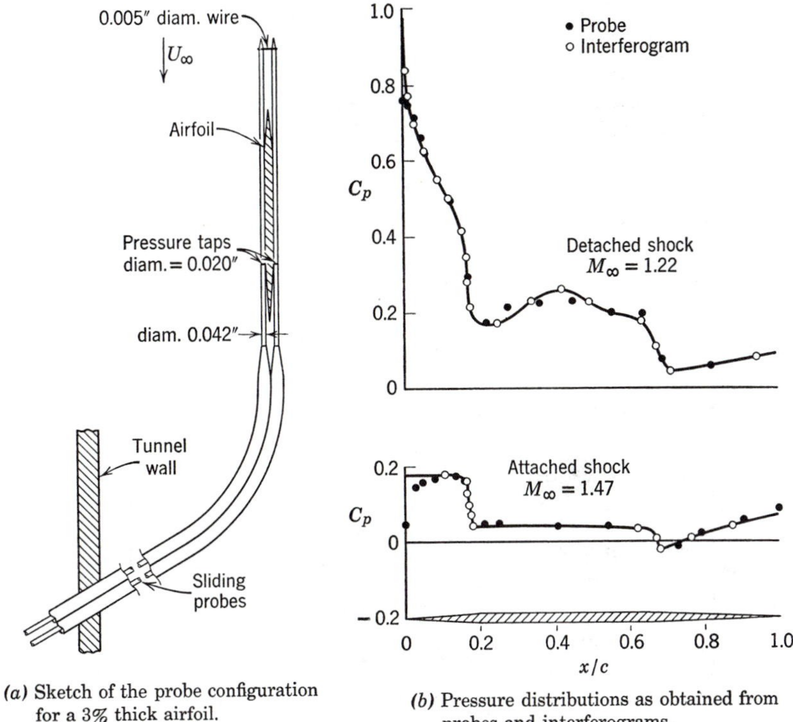

(*a*) Sketch of the probe configuration for a 3% thick airfoil.

(*b*) Pressure distributions as obtained from probes and interferograms.

Fig. 6·2 Measurement of surface pressure, with a static probe. From W. W. Willmarth, *J. Aeronaut. Sci.*, *20* (1953), p. 438.

measured. Even with this precaution it is usually not permissible for the yaw angle to exceed 5°, if 1% accuracy is required.

The round-nosed probe shown in Fig. 6·1*b* may be used for either subsonic or supersonic work, although the one with conical nose is usually preferred for the supersonic case.

Long, slender static probes may also be used for the measurement of pressure on surfaces in which it is not feasible to install orifices, for example, very thin airfoils, as illustrated in Fig. 6·2.

Supersonic static probes are subject to the type of interference from shock waves that is illustrated in Fig. 6·3. This shows a pressure survey

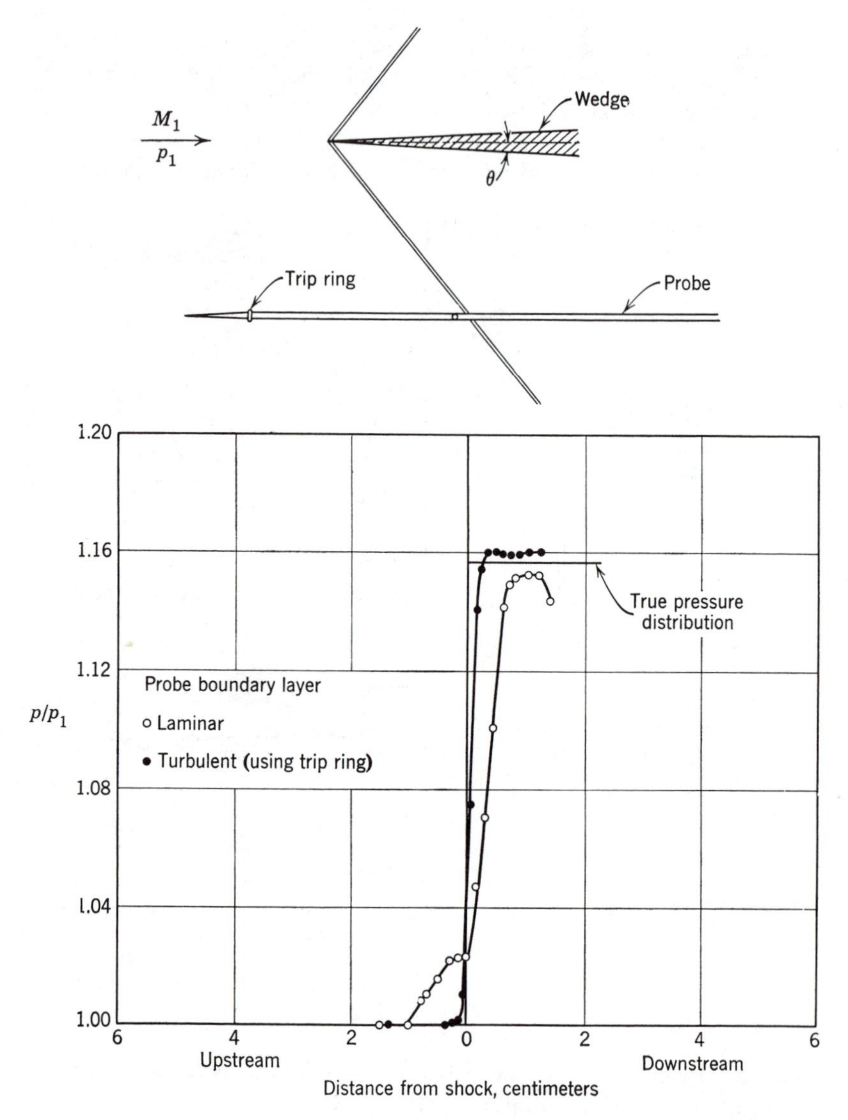

FIG. 6·3 Effect of probe boundary layer on static-pressure measurements through a shock wave. $\theta = 3°$; $M_1 = 1.354$. From Liepmann, Roshko, and Dhawan, "On Reflection of Shock Waves from Boundary Layers," *NACA Rep.* 1100 (1952).

made with a static probe through a weak, oblique shock. The true pressure distribution is a "step," the shock thickness being much too small to show up on this scale, but the probe cannot measure this, since its boundary layer cannot support a large pressure gradient. It interacts with the shock in such a way as to relieve the pressure gradient which it feels, the result being an upstream and downstream "influence" of the pressure rise. The effect on a laminar boundary layer is much greater than on a turbulent one, as indicated by the improvement obtained with a turbulent boundary layer, tripped by means of a small ring on the nose of the probe.

6·3 Total Pressure

The total or stagnation pressure is defined in Article 2·4. It is a measure of the entropy of the fluid. If the fluid at a given point of the flow has experienced only *isentropic* changes in its past history, its total pressure is equal to the reservoir pressure p_0, and may simply be measured in the settling chamber. If, on the other hand, entropy-changing conditions have

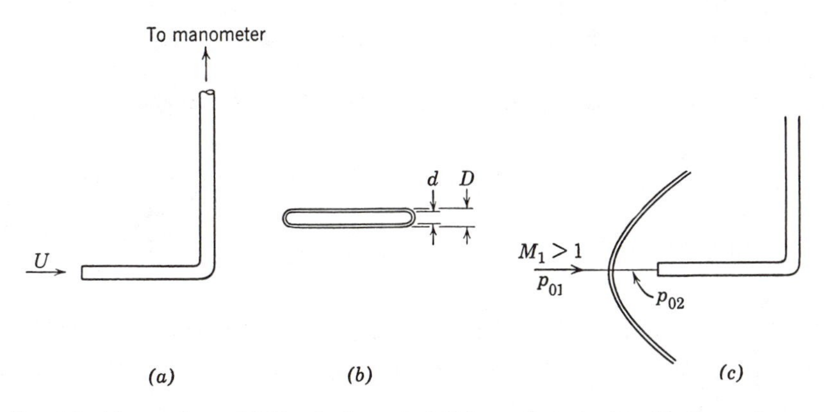

FIG. 6·4 Pitot probes. (*a*) Simple pitot tube; (*b*) front view of tube with flattened opening for boundary-layer work; (*c*) pitot probe in supersonic flow.

been encountered, the *local* pitot pressure p'_0 is different from p_0, and must be obtained from a local measurement. This occurs, for instance, if the fluid has traversed shock waves, or entered boundary layers or wakes, or if external heat has been added.

A pitot tube, for measuring the total pressure, consists of nothing more than an open-ended tube (Fig. 6·4) aligned parallel to the flow, with the open end "facing" the stream. The other end is connected to a manometer. The fluid in the tube is at rest. Except at very low Reynolds numbers (based on tube diameter), the deceleration to rest is isentropic, and the pressure in the tube is the stagnation pressure of the flow at the position defined by the mouth of the tube.

The limits on the size of the opening are similar to those for static pressure orifices.

The pitot probe is less sensitive than the static probe to flow alignment.

A simple open-ended probe, like that shown in Fig. 6·4, gives 1% accuracy up to yaw angles of 20°. Tubes with rounded nose and small orifice (compared to external diameter) are more sensitive.

Pitot tubes are often used for measuring the distribution of stagnation pressure in a boundary layer. The gradient normal to the surface is large, but good accuracy is obtainable with a flattened pitot tube like that shown in Fig. 6·4*b*. The small opening height d gives good resolution, and the large width improves response time. It is quite practicable (with a little patience) to build tubes having dimensions as small as $D = 0.003$ in. and $d = 0.001$ in. The tube is traversed normal to the surface by a suitable micrometer drive, and its position is measured by a micrometer head or a traversing microscope.

In supersonic flow a pitot tube does not indicate the local total pressure, since a detached shock wave stands ahead of the tube (Fig. 6·4*c*). On the stagnation streamline the shock is normal and the ratio of true total pressure to measured pitot pressure is given by Eq. 2·54. Slightly rewritten, this is

$$\frac{p_{01}}{p_{02}} = \left(\frac{2\gamma}{\gamma+1} M_1^2 - \frac{\gamma-1}{\gamma+1}\right)^{1/(\gamma-1)} \left(\frac{1 + \frac{\gamma-1}{2} M_1^2}{\frac{\gamma+1}{2} M_1^2}\right)^{\gamma/(\gamma-1)} \tag{6·1}$$

If M_1 is known, then the true stagnation pressure may be easily computed from this relation. Otherwise an additional measurement is needed.

6·4 Mach Number from Pressure Measurements

The Mach number is one of the most important parameters in compressible flow. It may be obtained from any one of the many flow relationships in which it appears, if the other quantities in the relationship may be measured. Among the most useful of these are the ones involving pressures.

If the fluid at the point of measurement has undergone only isentropic changes, its stagnation pressure may be assumed to be the reservoir pressure, p_0. A measurement of the *static pressure* p then determines the Mach number from the isentropic relation,

$$\frac{p_0}{p} = \left(1 + \frac{\gamma-1}{2} M^2\right)^{\gamma/(\gamma-1)} \tag{6·2}$$

This relation, valid for subsonic and supersonic flow, is used for the determination of Mach number distribution along an aerodynamic surface, using surface static holes, and for flow-field Mach number, using a static probe.

In isentropic *supersonic* flow, the local Mach number may be measured by a pitot tube, using Eq. 6·1, with p_{01} equal to the reservoir pressure. This method is quite as sensitive as the preceding one based on static pressure. It must be certain that no condensation has occurred in the flow ahead of the point of measurement, for this affects the stagnation pressure. Alternately, this measurement, together with a supplementary measurement of Mach number, may be used for *detecting* condensation.

In measurements from aircraft, or in nonisentropic flow, the "reservoir" conditions are not known, and it is necessary to obtain *both* the static pressure and the pitot pressure. If the flow is *subsonic*, the pitot pressure is the true total pressure, and the Mach number is obtained from Eq. 6·2. But if the flow is *supersonic*, the indicated pitot pressure is p_{02}, the stagnation pressure behind the normal shock. The Mach number may be obtained by dividing Eq. 6·1 by Eq. 6·2 in order to eliminate p_0 ($\equiv p_{01}$). This gives

$$\frac{p}{p_{02}} = \frac{\left(\dfrac{2\gamma}{\gamma+1} M_1^2 - \dfrac{\gamma-1}{\gamma+1}\right)^{1/(\gamma-1)}}{\left(\dfrac{\gamma+1}{2} M_1^2\right)^{\gamma/(\gamma-1)}} \tag{6·3}$$

which is called the *Rayleigh supersonic pitot formula.*

The *dynamic pressure* is usually obtained from the static pressure and Mach number, using Eq. 2·39,

$$\frac{1}{2}\rho u^2 = \frac{\gamma}{2} p M^2$$

6·5 Wedge and Cone Measurements

In supersonic flow, it is sometimes convenient to use a wedge or cone instead of a static probe. The pressure on the wedge surface may be related to the flow conditions by the simple oblique shock relations (Chart 2), whereas for the cone the corresponding conical shock theory is used (Fig. 4·27). Cones have the advantage that the detachment Mach number is lower.

For instance, in the case of a wedge symmetrically aligned with the flow, the ratio between the pressure p_a, on the wedge surface, and the stream pressure p_1, may be obtained from the oblique shock charts. When multiplied by p_1/p_0, from Eq. 6·2, a relation is obtained in the form

$$\frac{p_a}{p_0} = f(M_1, \theta)$$

where θ is the half-angle of the wedge and p_0 is the local stagnation pressure.

The wedge may also be used to find the flow inclination by measuring the difference of pressure, $p_b - p_a$, on the two sides. Using the shock charts, it is possible to plot the ratios

$$\frac{p_b + p_a}{2p_0} \quad \text{and} \quad \frac{p_b - p_a}{2p_0}$$

as functions of the Mach number for a given wedge of angle 2θ, and for various flow inclinations α. The measurement of p_a and p_b then gives both Mach number and flow inclination.

For wedges of *small* nose angle, at small inclination, the approximate oblique shock relations (Article 4·7) may be used to obtain these functions in simple, closed form.

The Mach number and flow inclination may also be obtained from a measurement of the *shock wave angles.* This is not usually as accurate or as convenient as the pressure measurements.

6·6 Velocity

The Mach number may be converted into other dimensionless speed ratios by means of the relations developed in Chapter 2. Thus Eq. 2·37*a* gives

$$\left(\frac{u}{a^*}\right)^2 = M^{*2} = \frac{\gamma + 1}{(2/M^2) + (\gamma - 1)} \tag{6·4a}$$

Since

$$\left(\frac{a^*}{a_0}\right)^2 = \frac{2}{\gamma + 1} \tag{6·5}$$

it follows that

$$\left(\frac{u}{a_0}\right)^2 = \frac{2}{(2/M^2) + (\gamma - 1)} \tag{6·4b}$$

In Eqs. 6·4*a* and 6·4*b* a_0 and a^* must be the *local* values. If the flow is adiabatic up to the point of measurement, these local values are equal to the reservoir values, and may be measured in the supply section. They may be obtained from a measurement of the reservoir temperature T_0, from the perfect gas relation

$$a_0^2 = \gamma R T_0$$

If there is heat addition ahead of the point of measurement, or if the point is in a nonequilibrium region such as a boundary layer, or if the measurement is being made from an aircraft, it is necessary to make a *local* measurement of T_0. This problem is discussed in the following article.

Using pressures, the determination of velocity requires the measurement of three local quantities, p, p_0, and T_0. In special cases one or more of these measurements may be avoided, as already noted, by using the settling chamber values. For velocity profiles through a boundary layer, it is usually possible to assume that p is constant throughout; it may then be simply measured at a surface orifice.

Methods for *direct* measurement of velocity are based mainly on tracer techniques, using, for example, ions or illuminated particles for tracers.

They are seldom used, owing largely to the technical inconveniences, and also because they do not provide a *local* measurement, since the flight must be timed over a finite distance.

Other methods use velocity-sensitive elements which are *calibrated*. One of these, the hot-wire anemometer, is discussed in Article 6·19.

6·7 Temperature and Heat Transfer Measurements

There is no method for the direct measurement of *static* temperature T in a moving fluid. The temperature indicated by any thermometer immersed in the fluid is higher than T, for there is an increase of temperature through the boundary layer, from the static temperature T at the edge to the *recovery* temperature T_r at the surface. T_r will in general be different on different parts of the surface, depending on geometry, Reynolds number, etc., and the thermometer will indicate a *mean* recovery temperature.†

Indirect determinations of T may be obtained from measurements of pressure and density, using the equation of state; if conditions are isentropic, one of these is sufficient. An *independent*, direct determination may be obtained from a measurement of the acoustic speed, a, using the relation

$$a^2 = \gamma RT \tag{6·6}$$

The value of a is obtained by producing weak pressure pulses (sound waves), of known frequency, in the flow, photographing them by one of the optical techniques described later in this chapter, and measuring the spacing (wavelength) between pulses. In the measurement of wavelength, the velocity of the fluid relative to which the pulses are propagating must be taken into account. This difficulty, together with the fact that an averaging distance is involved, has so far prevented any general adoption of the method.

The measurement of the stagnation or *total* temperature T_0 is in principle simple. The temperature inside a pitot tube, where the flow is brought to rest at equilibrium, should be the stagnation temperature, in either subsonic or supersonic flow, and could be measured by a thermometer placed inside the tube. The difficulty is that equilibrium does not actually exist, since heat is conducted (and radiated) away by the thermometer and the probe walls.‡ This reduces the temperature of the stagnant fluid to a value T_r lower than T_0. It depends on the probe configuration, conductivity of the walls, flow conditions on the outer walls, etc.

Figure 6·5 shows an example of a total temperature probe which uses a thermocouple for a sensing element. The shield and support are designed

†In principle, a direct static temperature measurement could be obtained by allowing the thermometer to move *with* the fluid.

‡This departure from thermal equilibrium affects only the temperature and density, not the pressure.

to keep the rate of heat loss by conduction and radiation to a minimum. To replenish some of the lost energy a small flow through the probe is permitted, by means of a vent hole. With such designs, it has been possible to obtain values of the *recovery factor*,

$$r = \frac{T_r - T_1}{T_0 - T_1}$$

very nearly equal to 1, and, what is more important, to keep r constant over a fairly wide range of conditions, for the probe is a *calibrated* instrument.

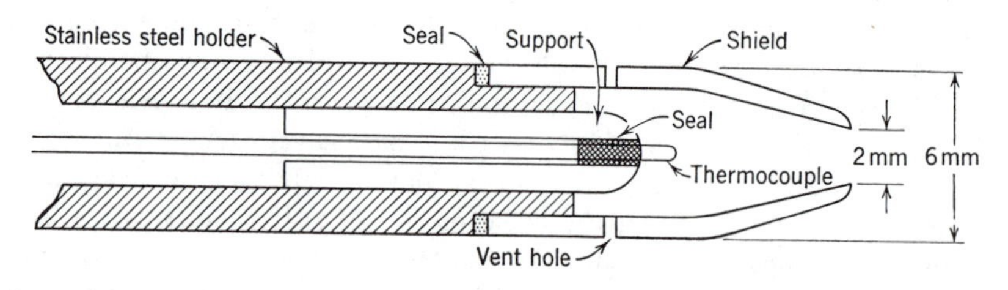

FIG. 6·5 Design of total temperature probe. From Eva M. Winkler, *J. Appl. Phys.*, *25* (1954), p. 231.

For boundary layer profiles a probe with flattened opening, like the pitot probe described in Article 6·3, may be employed.

Inasmuch as the total temperature probe is a *calibrated* instrument, it is possible to use a much simpler calibrated probe, namely, a fine resistance wire aligned normal to the flow. The wire temperature may be determined from a measurement of its resistance (see Article 6·19).

Measurements of *wall temperature*, for use in heat transfer or boundary layer studies, may be made with thermometer elements, such as thermocouples or resistance wires, embedded in the surface. The principal requirement is that the element should not change the conditions of heat transfer that would exist if it were absent. By embedding several thermocouples *in* the wall, at various distances from the surface, data for a temperature gradient may be obtained† and the local heat transfer to the surface may be determined.

Another technique for measurement of heat transfer between the wall and the flow is the *transient* method, in which the rate of temperature change of a portion of the wall provides a measure of the heat transfer through the surface. The heat capacity of the wall material must be known, and it is necessary to minimize heat transfer through sections other than the surface, or to estimate its effect (see Exercise 6·5).

†Lobb, Winkler, and Persh, *J. Aeronaut. Sci.*, *22* (1955), p. 1.

6·8 Density Measurements

If, in addition to the pressure p, there is a measurement of the density ρ, considerable information about the state of the flow becomes available. For instance, a local measurement of p, p_0, and ρ defines all the other flow variables, as follows. The speed of sound is calculated from

$$a = \sqrt{\frac{\gamma p}{\rho}}$$

and the velocity from

$$u = aM$$

where M is obtained from the pressure relation (Eq. 6·2). The local values of T and T_0 are also determined from

$$p = R\rho T$$

$$\frac{T_0}{T} = 1 + \frac{\gamma - 1}{2} M^2$$

In this example, the measurement of density supplements the pressure measurements, but sometimes the measurement of density is the *only* one required. For instance, in isentropic flow (p_0 and T_0 known throughout) a measurement of density gives the pressure and Mach number from

$$\frac{p}{p_0} = \left(\frac{\rho}{\rho_0}\right)^{\gamma}$$

$$\frac{\rho_0}{\rho} = \left(1 + \frac{\gamma - 1}{2} M^2\right)^{1/(\gamma-1)} \tag{6·7}$$

The methods for measuring, or visualizing, the density field depend almost invariably on the effects of the fluid density on some form of electromagnetic radiation. The methods may be broadly classified as those which depend on the *refractive index* (optical methods), on *absorption*, and on *emission*, respectively. Of these, the optical methods are by far the best developed and the ones most widely used. The next articles are devoted to them; the others are briefly discussed in Article 6·17.

6·9 Index of Refraction

The three principal optical methods, schlieren, shadowgraph, and interferometer, depend on the fact that the *speed of light* varies with the density of the medium through which it is passing. The speed c, in any medium, is related to the speed c_0, in vacuum, by the *index of refraction*,

$$n = c_0/c \tag{6·8}$$

For a given substance, and for given wavelengths of light, the index of refraction is a function of density, $n = n(\rho)$. In gases, the speed of light is only slightly less than in vacuum, the difference being *directly proportional to the density*, to very good accuracy. That is,

$$n = 1 + \beta \frac{\rho}{\rho_s} \tag{6·9}$$

where ρ_s, the reference density, is taken at standard conditions.† Equation 6·9 may be regarded as the first-order approximation in a series expansion. The higher-order terms are negligible, except possibly in very dense gases. The values of β for several gases are given in Table 6·1, for light of wavelength $\lambda = 5893$ A (the D-line).

Equation 6·9 shows that in an inhomogeneous medium, such as a flow field of varying density, the index of refraction is also variable. Light passing through one part of the field is retarded differently from that passing through another part. This has two effects: (1) a turning of the wave fronts, that is, a *refraction* of the rays, and (2) a relative phase shift on different rays. The first effect is used in the schlieren and shadow methods, and the second is the basis of the interferometer principle.

TABLE 6·1†

ρ_s at 0° C, 760 mm Hg
$\lambda = \lambda_D = 5893$ A‡

Gas	β
Air	0.000292
Nitrogen	0.000297
Oxygen	0.000271
Carbon dioxide	0.000451
Helium	0.000036
Water vapor	0.000254

†Smithsonian Physical Tables, Washington, 1954.
‡The variation of β with wavelength is small. For example, for air $\beta = 0.000291$ for red (A) and $\beta = 0.000297$ for blue (G).

We shall consider first the refractive effect. Figure 6·6*a* shows light traversing a field in which the density varies in the y-direction, as indicated by the lines of constant density. The lines marked w are the *wave fronts* and the lines orthogonal to them and marked r are the *rays*. Figure 6·6*b* shows the progress of a wave front, from position w_1 to w_2, in a small time interval

$$\tau = d\xi/c$$

†β is dimensionless; one sometimes uses the Gladstone-Dale constant κ, defined by $n = 1 + \kappa\rho$.

If the density is increasing in the y-direction, the speed of the wave is greater along r_a than along r_b, by an amount dc, say. This results in a *turning* of the wave front, of amount

$$d\phi = \frac{\tau|dc|}{d\eta}$$

The ray is turned through the same angle. Thus its *curvature* is

$$\frac{1}{R} = \frac{d\phi}{d\xi} = \frac{1}{c}\left|\frac{dc}{d\eta}\right| = \frac{1}{n}\left|\frac{dn}{d\eta}\right| \tag{6·10}$$

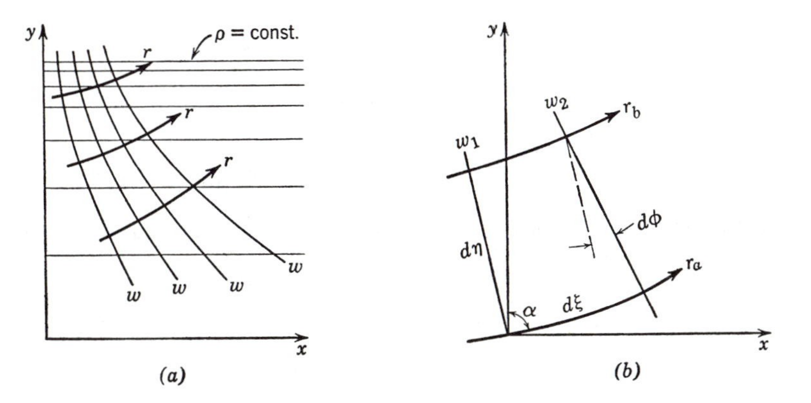

FIG. 6·6 Refraction in a density field. (*a*) Rays (*r*) and wave fronts (*w*); (*b*) orthogonal mesh element formed by rays and wave fronts.

For the density field in Fig. 6·6, this may be written

$$\frac{1}{R} = \frac{\sin\alpha}{n}\frac{dn}{dy}$$

In a general, three-dimensional density field, the above result generalizes to

$$\frac{1}{R} = \frac{\sin\alpha}{n}|\text{grad } n| \tag{6·10a}$$

where grad n is the vector gradient of the n-field and α is the angle between this vector and the ray. *The curvature is in the direction of increasing density.*

This result is often obtained in optics books in another way, from Fermat's principle, which states that "the path taken by a light ray between two points is that one which requires the shortest time."

In the application to a wind tunnel, the ray usually enters at right angles to a side wall, as shown in Fig. 6·7, which is a cross-sectional view of the test section. The curvature shown corresponds to density increasing toward positive y. There may also be curvature in the z-direction, due to gradients along the flow, but it will be sufficient to discuss only the component shown.

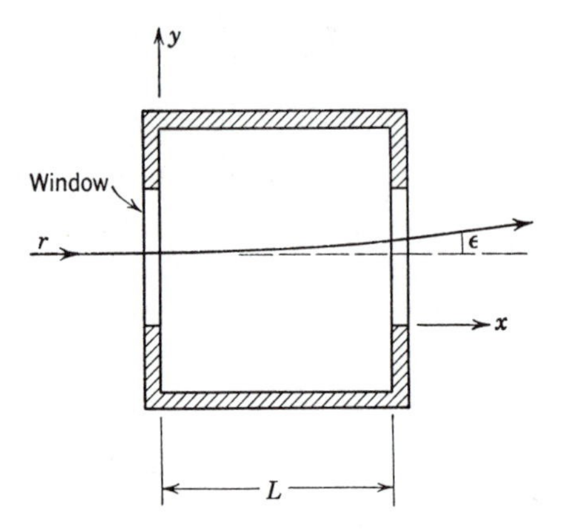

FIG. 6·7 Refraction of a ray passing through a wind tunnel. Curvature shown corresponds to $\partial\rho/\partial y > 0$.

The angular deflection of the ray after it traverses the flow is

$$\epsilon = \int d\phi$$

where the integral is to be taken along the (curved) ray. The ray deflection is usually so small† that the density along the curved ray is nearly the same as along the nearby path, $y = y_1$. Thus

$$\epsilon = \int_0^L \frac{1}{R}\,dx = \int_0^L \left(\frac{1}{n}\frac{dn}{dy}\right)_{y_1} dx \qquad (6\cdot11)$$

where L is the test section width. This is the fundamental relation for fields of small density gradient, such as those which occur in isentropic, or nearly isentropic, portions of the flow.

If the flow is *plane*, so that conditions are the same in every x-plane, the integral gives, with $n_1 \doteq 1$,

$$\epsilon = \frac{L}{R_1} = \frac{L}{n_1}\left(\frac{dn}{dy}\right)_1 = \frac{L\beta}{\rho_s}\left(\frac{d\rho}{dy}\right)_1 \qquad (6\cdot12)$$

This relation shows that for plane flow the deflection of the emerging ray is *proportional to the density gradient.*

†This may not be true at a shock wave or in a boundary layer.

For three-dimensional flow the final deflection will be an integrated effect dependent on all the density gradients encountered.

6·10 Schlieren System

The basic idea of the schlieren system is that *part of the deflected light is intercepted* before it reaches the viewing screen or photographic plate, so that the parts of the field which it has traversed appear darker.† The details may best be explained by reference to the basic system shown in Fig. 6·8.

A beam of parallel, monochromatic light is obtained by passing the light from a source through a lens L_1. (The source is shown in edge view. It usually consists of a rectangular cut-out‡ placed at the focus of a lamp-lens combination as shown in the inset.) The portion of flow field that appears in the final picture is that which is traversed by the beam.

After passing through the test section, the beam is focused by a second lens L_2. If a screen is placed in the focal plane of L_2, an *image of the source* is obtained there. Otherwise the light passes through this plane to an objective lens L_3, which directs it to a screen or photographic plate, located at the *image plane of the test section.*

To understand the focusing, it must be remembered that there are *two* focal planes, one for the source and one for the test section. Considerable confusion may be avoided by considering *pencils* of light, rather than individual rays.

For instance, a point a in the source emits a pencil abc which focuses on a' at the source-image plane. Other points are focused similarly, to form the image of the source. It will be noted that *each* of these pencils completely fills the test section. Thus each point in the source image receives light from *every* portion of the test section.

Now consider the second aspect. The light reaching a point g in the test section is contained within the pencil adg. It is transmitted within the pencil $gd'a'$, which is focused at g' on the viewing screen. This pencil passes through the source image and *completely fills* it. Other pencils from the test section, such as $hd'a'$, are focused at their image points on the screen, and the image of the test section is formed there. (Only one plane of the test section can be focused precisely, but if there is sufficient focal depth the other sections are also sufficiently sharp.)

It will be noted that the individual *rays* passing through the test section are not strictly parallel, owing to the finite size of the source, but that the *pencils* are parallel.

†Another arrangement is to intercept the *undeflected* light and allow the deflected light to pass through. The illumination of the image is then reversed.

‡Typical dimensions are 1 mm. x 1 cm.

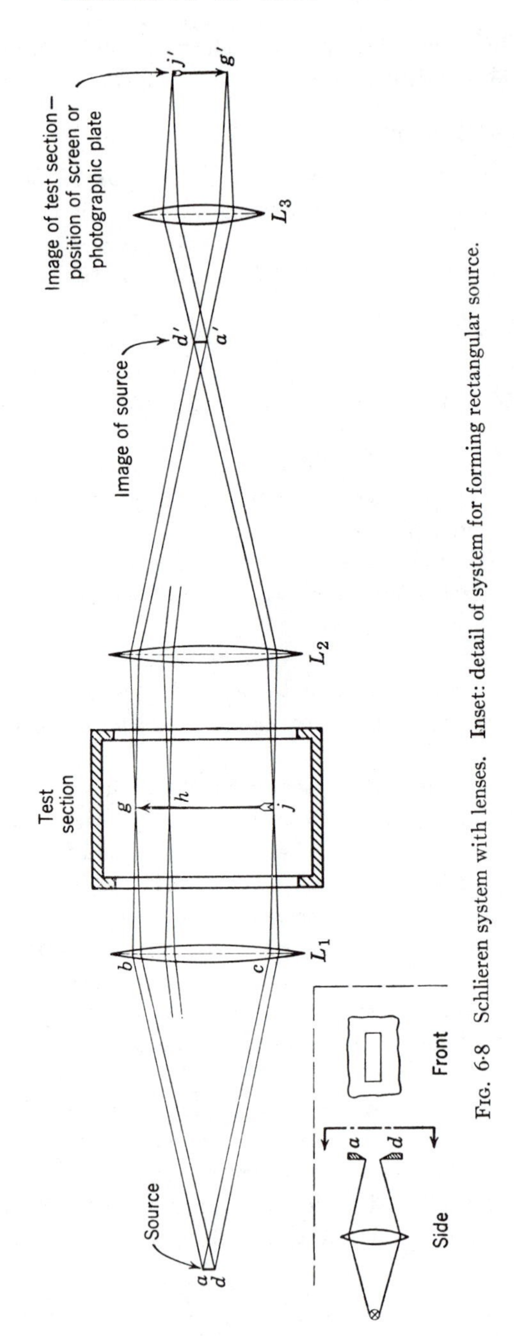

FIG. 6·8 Schlieren system with lenses. Inset: detail of system for forming rectangular source.

6·11 The Knife Edge

It will be noted in Fig. 6·8 that all the pencils of light overlap in only one plane—that of the source image. If part of the light here is intercepted, the illumination at the final screen is decreased. All parts are darkened equally since all the pencils are equally affected. However, if one of the

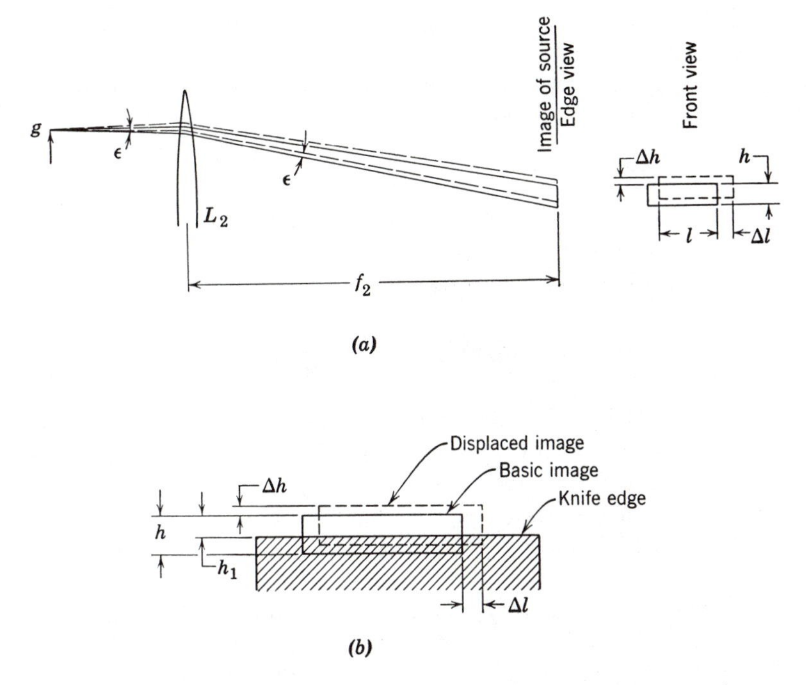

Fig. 6·9 The schlieren knife edge. (*a*) Displacement of part of the light at the source image due to refraction of a pencil of light; (*b*) interception of source image by an opaque cutoff.

pencils were deflected by an angle ϵ at the test section (Fig. 6·9), at the focal plane it would *not* overlap the other pencils. Its image point at the screen would be darkened a different amount, and the point would appear darker or brighter than the rest of the field, depending on how the light is intercepted.

The displacement of a pencil at the source image, due to an angular displacement ϵ at the lens L_2, is

$$\Delta h = f_2 \epsilon \dagger \tag{6·13}$$

where f_2 is the focal length of L_2. There may also be a displacement Δl

†It is assumed that the distance along the pencil is f_2; this is quite accurate for small apertures.

due to deflection in the other direction. A pencil which is deflected arrives at the same point on the viewing screen as the undeflected rays, and thus the image on the viewing screen remains sharp. Only the illumination is affected.

The opaque "cutoff" which is used for intercepting part of the source image ordinarily has a straight edge, called the *knife edge* (Fig. 6·9*b*). It is set parallel to the long side of the rectangular source image, leaving a portion of it, h_1, uncovered. Thus the viewing screen has a general illumination E which is proportional to h_1. A point on the screen that is illuminated by a *deflected* pencil has an additional illumination ΔE which is proportional to Δh. The *contrast* is defined by

$$c = \frac{\Delta E}{E} = \frac{\Delta h}{h_1}$$

This may be written in terms of other parameters of the system by substituting Eqs. 6·13 and 6·12 (the latter is for *plane* flow). Thus

$$c = \frac{f_2 \epsilon}{h_1} = \frac{f_2 L \beta}{h_1 \rho_s} \left(\frac{d\rho}{dy} \right)_1 \tag{6·14}$$

This equation summarizes the effects that determine the contrast on the viewing screen. They are (1) focal length of the second lens, (2) width of test section, (3) refractivity of the fluid, (4) density gradient, and (5) uncovered width of the basic image. Thus, for *plane flow*, the increase or decrease of illumination at the screen is proportional to the *density gradient* in the flow.

The component of density gradient that is displayed depends on the orientation of the source (and knife edge). As may be seen from Fig. 6·9*b*, only the displacement Δh affects the illumination; the light corresponding to displacement Δl is not intercepted. Thus the schlieren method gives the *density gradient normal to the knife edge.* The gradient in any direction in the flow may be obtained by setting the knife edge (and source) normal to it; the settings ordinarily used are *parallel* and *normal* to the general flow direction. Some examples of schlieren photographs are shown in Fig. 5·4, in which the knife edge is vertical, and in Figs. 13·12 and 13·13, in which the knife edge is horizontal. In Fig. 5·4, normal shock waves are prominent while in Figs. 13·12 and 13·13 the boundary layers stand out clearly.

Another parameter for evaluating the performance of the system is the *sensitivity* s, defined as the fractional deflection obtained at the knife edge for unit angular deflection of the ray at the test section,

$$s = \frac{\Delta h / h_1}{\epsilon} = \frac{c}{\epsilon} = \frac{f_2}{h_1} \tag{6·15}$$

Thus the sensitivity depends only on the optical parameters. The contrast may then be written

$$c = s \frac{L\beta}{\rho_s} \left(\frac{d\rho}{dy}\right)_1 \tag{6·14a}$$

6·12 Some Practical Considerations

In addition to these basic performance parameters, some other practical matters must be considered and balanced against each other. For instance, Eq. 6·15 indicates that maximum sensitivity is obtained with minimum h_1, that is, by leaving very little of the source image uncovered. On the other hand, h_1 must be large enough to furnish sufficient illumination at the screen. Thus its minimum value is limited by the brightness of the source. There is also an upper limit on h_1 (or on the size of the source), determined by the maximum illumination desired. The requirements may be different for photographic work and for visual observation.

In general, high sensitivity is desirable, but even here there is an upper limit, determined by the fact that spurious density gradients are encountered by the light; the system should not be so sensitive as to make these visible. Some such "noise" is always present, in the form of density fluctuations in the surroundings of the test section, as well as in turbulent boundary layer fluctuations on the test-section sidewalls. The maximum tolerable sensitivity s_m (cf. Eq. 6·15) is defined by

$$\frac{\Delta c}{c} = s_m \frac{\Delta \epsilon}{\epsilon} \tag{6·16}$$

where Δc is the minimum discernible contrast and $\Delta\epsilon$ is the mean (rms) deflection corresponding to the spurious density fluctuations.

If the method is to be used for quantitative measurements, there is another limitation on sensitivity. This is related to the fact that Δh must not be so large that the secondary image is deflected completely off (or completely onto) the knife edge, for then any additional deflection would not produce corresponding changes of illumination at the screen. For qualitative work this nonlinearity is not so objectionable, and may even be desirable.

Using the above relations, one can in principle obtain quantitative measurements of density gradient, which can be integrated to give the density field. The information may be obtained by scanning the screen with a suitable light meter, or the photograph with a densitometer. In practice, it is best to calibrate any such survey against a known gradient, preferably on the same picture. (For example, if a known gradient in the flow is not available, its equivalent might be introduced with the help of a prism.) Actually, schlieren pictures are seldom used for quantitative evaluation of density. They are, however, indispensable for obtaining qualitative understanding of flows.

Finally, it may be remarked that in practice schlieren systems usually employ mirrors, rather than lenses. For the large apertures required, the mirrors are less expensive; in addition, there is less light lost in reflection at mirrors than in transmission through lenses. A conventional mirror

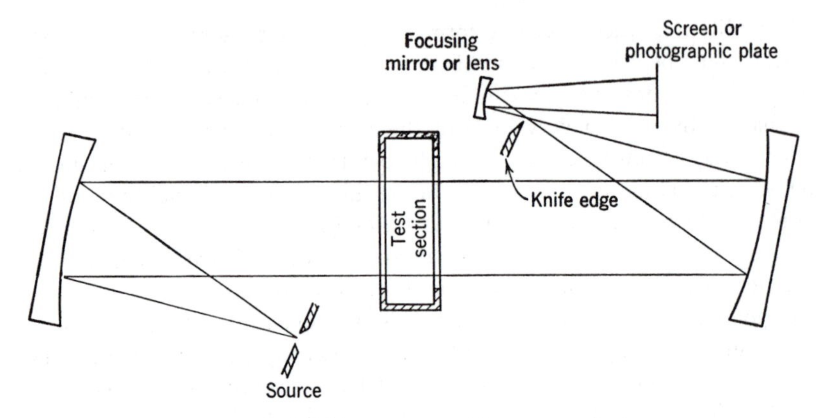

FIG. 6·10 Schlieren system with concave mirrors.

arrangement is shown in Fig. 6·10. For practical reasons the source and its image must usually be off-axis, and this introduces some astigmatism. For instance, the position of the source image is different for "horizontal" and "vertical" knife-edge arrangements, and the knife edge must accordingly be moved from one position to the other.

6·13 The Shadow Method

It was noted that the *positions* of the image points on the viewing screen of the schlieren system are not affected by deflections in the test section, since the deflected rays are also brought to focus in the focal plane, and that this screen is uniformly illuminated when the knife edge is not inserted into the beam. If, on the other hand, the screen is placed at some position other than the focal plane of the test section, the effects of ray deflection will be visible. The best position is close to the test section.

This effect, known as the *shadow* effect, is illustrated in Fig. 6·11, which shows parallel light entering the test section and being intercepted on the other side. On the screen there are bright regions where the pencils crowd together and dark regions where they diverge. At places where the spacing is unchanged the illumination is normal, even though there has been refraction. Thus the shadow effect depends not on the absolute deflection but on the relative deflection of the pencils, that is, on the rate at which they converge on emerging from the test section. This convergence is measured by $d\epsilon/dy$, where ϵ is the deflection, and is proportional to the density gradient (Eq. 6·12). Thus, in plane flow, the shadow effect depends on the *second derivative* of the density. Deflections may also occur in the other direction; the change of illumination depends on the net effect,

$$\Delta E = \frac{\partial^2 \rho}{\partial y^2} + \frac{\partial^2 \rho}{\partial z^2} \tag{6·17}$$

Evidently there is no need for a slit source; a point source is usually preferable. Also the light need not necessarily be parallel when it enters the test

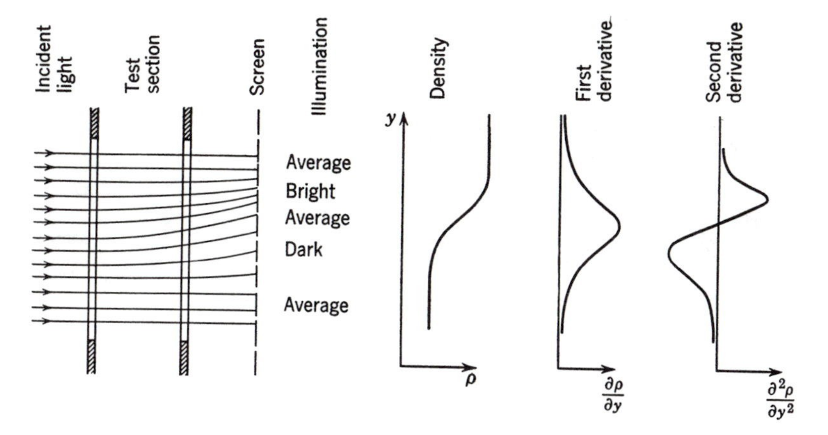

FIG. 6·11 The shadow effect.

section; the first lens may be dispensed with, and the divergent light from the source may be used directly.

This simplicity makes the shadow technique rather less expensive than the schlieren method, and it may often be used to advantage, particularly where the finer details of a density field are not required, or need to be suppressed. An example of a shadowgraph is given in Fig. 13·14. A shock wave always appears as a dark line followed by a light line, as may be seen by referring to Fig. 6·11, which shows the general shape of the density profile through a shock wave. A useful analogy is to imagine the effect on the light to be the same as that obtained by passing it through a lens having the same shape as the profile. It will be seen that the dark and bright lines correspond to maximum and minimum values of $d^2\rho/dy^2$, near the front and the back of the shock wave, respectively. In addition, there is some displacement of these lines relative to the true position of the shock, an effect that increases as the screen is moved farther from the test section.

6·14 Interference Method

In the schlieren and shadow techniques, described in the previous articles, the fields displayed on the final image correspond to the first and second derivatives, respectively, of the density. From them the density field itself can be obtained, in principle, by integration, but in practice the accuracy is not very good. Fortunately, there is available an optical method which gives the density field directly. It is based on the *principle of interference.*

The principle is illustrated in Fig. 6·12, which shows two rays arriving at point p on the screen by two different paths. Interference between them

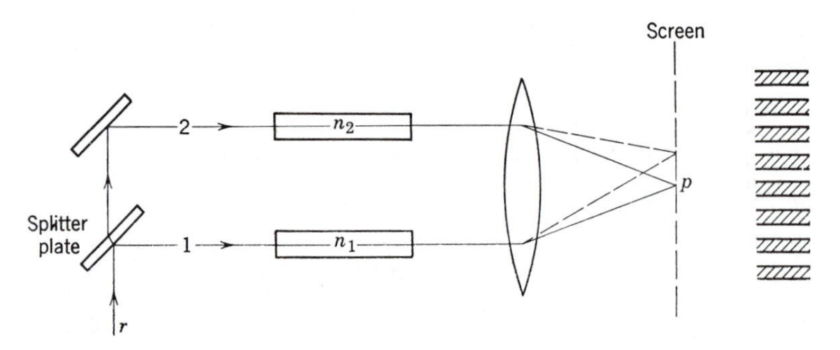

FIG. 6·12 Formation of interference fringes.

is possible if they are *coherent.* To be coherent they must have had the *same origin* in the source, so that there is a definite phase relation between them. Whether they reinforce or annul each other at the screen depends on their relative phase there, which in turn depends on the difference in the lengths of their paths. If the rays pass through identical media, the only path differences are due to the geometrical disposition, which may be arranged to give a set of fringes on the screen, as shown in Fig. 6·12. We may single out for attention one point, p, on this primary fringe pattern. If now a different medium (the test section) is placed in the path of ray 2, leaving the other ray undisturbed, the effective path, or *optical path,* of ray 2 is changed, and with it the order of interference at p. If the optical path is changed by an amount $N\lambda$, where λ is the wavelength of the light and N is an integer, then the order of interference at p is changed by the amount N; that is, a shift of N fringes is observed there.

For instance, by taking pictures with and without flow, the fringe shift at a point p may be determined, from which the equivalent change of optical path may be calculated and related to the change of density introduced into the path. The relations are easily obtained. If the index of refraction in the test section is increased from n_1 to n_2, the light speed there is decreased from c_0/n_1 to c_0/n_2 (Eqs. 6·18) and the additional time needed to traverse the test section is

$$\Delta t = \frac{L}{c_2} - \frac{L}{c_1} = \frac{L}{c_0}(n_2 - n_1)$$

The corresponding change in optical path is

$$\Delta L = c_0\,\Delta t = L(n_2 - n_1) \tag{6·18a}$$

and the *fringe shift* is

$$N = \frac{\Delta L}{\lambda} = \frac{L}{\lambda}(n_2 - n_1) = \beta\frac{L}{\lambda}\left(\frac{\rho_2 - \rho_1}{\rho_s}\right) \tag{6·18b}$$

where Eq. 6·9 has been used to relate the index of refraction to the density of the fluid in the test section.

If the density is variable in the test section, the net change of optical path is the integrated effect along the ray, and Eq. 6·18*b* is to be replaced by the integral

$$N = \frac{\beta}{\lambda\rho_s}\int_0^L (\rho - \rho_1)\,ds \tag{6·18c}$$

taken along the light path. The integral here is written only for the test section, but any effects of density difference at other points, for example, outside the test section, will also be integrated and will contribute to an error that may be lumped under the term "noise." Thus it is desirable that the two rays should travel through media as nearly identical as possible, except for the portion through the test section. In many installations, the reference ray 1, is conducted through a compensating chamber, in which the density may be adjusted to compensate for spurious differences in path length. An even simpler method is to pass the reference ray through the wind tunnel, at a section where the density is known. The density changes indicated by the fringe pattern are then relative to this reference density. In addition, spurious flow variations are automatically compensated. In particular, both beams pass through the side wall boundary layers,† which are otherwise rather troublesome to compensate.

6·15 Mach-Zehnder Interferometer

For wind tunnel work the Mach-Zehnder interferometer (Fig. 6·13) is now employed almost universally.‡ It has the advantage that the arrange-

†If the reference beam is not too far from the exploring beam, the change of boundary layer between the two sections is negligible, or easily estimated.

‡Its application to wind tunnel work was developed by Ladenburg and co-workers and by Zobel.

ment of the two arms is quite flexible and that it may be focused on any section. The principal components are the four optical plates, set up parallel to each other on the corners of a rectangle. Two of these, P_1 and P_2, which are fully reflecting, are fixed in position, once initial adjustments are complete. The other two, P_3 and P_4, which are half-silvered "splitter plates," may be rotated independently about horizontal or vertical axes. Monochromatic light from the effective source a is converted into a parallel

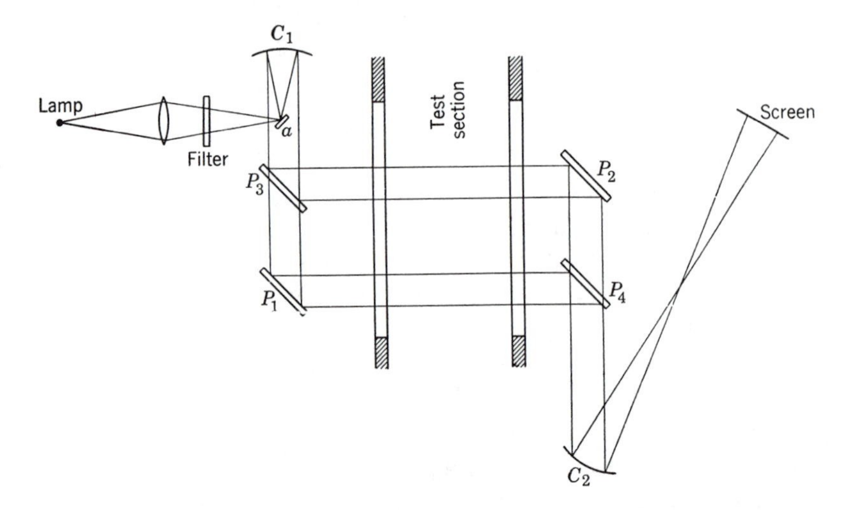

FIG. 6·13 Mach-Zehnder interferometer. a = plane mirror; C_1 = paraboloidal mirror; P_3, P_4 = half-silvered "splitter" plates; P_1, P_2 = plane mirrors; C_2 = spherical mirror.

beam by the concave mirror C_1. At plate P_3, this beam is split into two beams by partial reflection and transmission. These beams are then *coherent*. The reflected beam passes through the measuring part of the test section, while the transmitted beam, after reflection from P_1, passes through the reference part of the test section. They are both gathered by concave mirror C_2 and focused on the screen, where the fringes are formed. (A knife edge may be placed at the focal point of mirror C_2, if it is desired to also use the interferometer optical system for schlieren work.)

Since the rectangle on which the plates are set, as well as the parallel alignment of the plates themselves, must be accurate to a fraction of a wavelength, there is some delicacy in making the initial adjustments. Details of technique may be found in the literature.†

A point that warrants some discussion here is the problem of focusing both the test section and the fringes on the screen. A simplified diagram of the interferometer is given in Fig. 6·14. For clarity only two (coherent)

† Reference E·2; also Ashkenas and Bryson, *J. Aeronaut. Sci.*, *18* (1951), p. 82.

rays, r_1 and r_2, are shown instead of the full beams. If the plates are adjusted exactly parallel to each other, the two rays overlap again when they leave plate P_4. In this case both of them appear to have come from the same virtual source, S'. Now if the splitter plates P_3 and P_4 are rotated, the rays appear to come from *different* virtual sources. The method is illustrated in Fig. 6·14*b*.

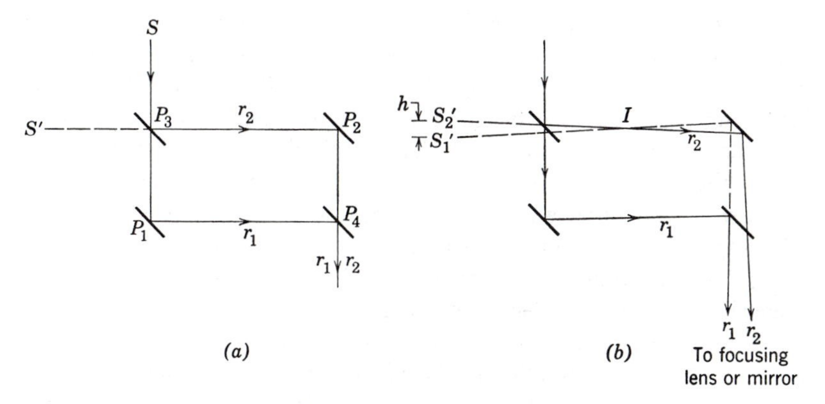

FIG. 6·14 Formation of two coherent virtual sources by rotation of plates P_1 and P_2.

A rotation of plate P_3 about an axis normal to the plane of the rays does not affect the transmitted ray r_1, but makes r_2 appear to come from S'_2. On the other hand, a rotation of plate P_4 does not affect r_2 but inclines r_1 in such a way that its virtual source is S'_1. The two rays no longer overlap when they leave P_4, but it is necessary only that they coincide at the screen, where the fringes are to be formed. In other words, they must be brought to focus there. On the other hand, the optical arrangement must be such that the test section is also focused on the screen. To accomplish both, the two rays r_1 and r_2 must appear to come from a single point I in the test section; that is, they must appear to *cross* there. Such an intersection may be arranged by giving the splitter plates suitable rotations; in fact, any arbitrary intersection angle, with corresponding spacing, h, between the two virtual sources, may be obtained.

According to interference theory, the fringe spacing at the screen will be given by

$$b = \text{const.}\,\frac{\lambda}{h} \tag{6·19}$$

where the constant depends on the details of the optical system; its order of magnitude is the path length from virtual source to screen.

6·16 Interferometer Techniques

Rotation of plates P_3 and P_4 about vertical axes (Fig. 6·14) produces vertical fringes, whereas rotation about horizontal axes produces horizontal ones. The latter case is illustrated in Fig. 6·15*a*, which shows horizontal fringes in the undisturbed flow ahead of a wedge. These were taken with monochromatic light, obtained by using a filter. In the disturbed flow, the fringes are displaced; to trace them through the shock wave, a *white* light picture, with no filter, is taken. The central fringe is then very prominent and may easily be traced through the shock. With this as reference, the displacements of the other fringes, relative to those ahead of the shock, are easily determined, and the entire density distribution is obtained. This extra white light picture is not always needed—a reference density may be obtained in some other way, for example, from a pressure measurement.

By superimposing the pictures taken with and without flow, the pattern in Fig. 6·15*b* is obtained. The faint lines that appear are lines of constant fringe shift and thus are density contours.

Density contours may also be obtained by starting with an "infinite fringe" in the no-flow condition. This is obtained by rotating the plates in such a way as to increase the fringe spacing (Eq. 6·19) until a single fringe fills the field. With flow on, the fringes that appear are density contours (Fig. 6·15*c*). The accuracy with this method is not usually as good as with counting of fringe shifts.

Figure 6·15*d* shows a white light interferogram of a shock wave propagating through CO_2 in a shock tube. The shock motion is from left to right; the density change is indicated by the fringe displacement, which can be easily traced by following the central fringe. The relatively long distance to reach the final equilibrium density is due to the long relaxation time for the vibrational mode in CO_2 (cf. Article 14·12).

An important consideration in the design of an interferometer is the problem of *mounting*, to isolate vibration. In the classical method, a massive supporting structure rests on a solid foundation. This is not usually practical in a wind tunnel installation. An entirely different arrangement,† using a light, damped structure on a very elastic, low-frequency mount gives quite satisfactory vibration isolation.

The sensitivity of the method may be defined as the number of fringe shifts per unit change of density (or any related flow parameter). From Eq. 6·18*b*, this is simply

$$s = \frac{N}{\Delta\rho} = \frac{\beta L}{\lambda \rho_s} \tag{6·20}$$

†Ashkenas and Bryson, *loc. cit.*, p. 166.

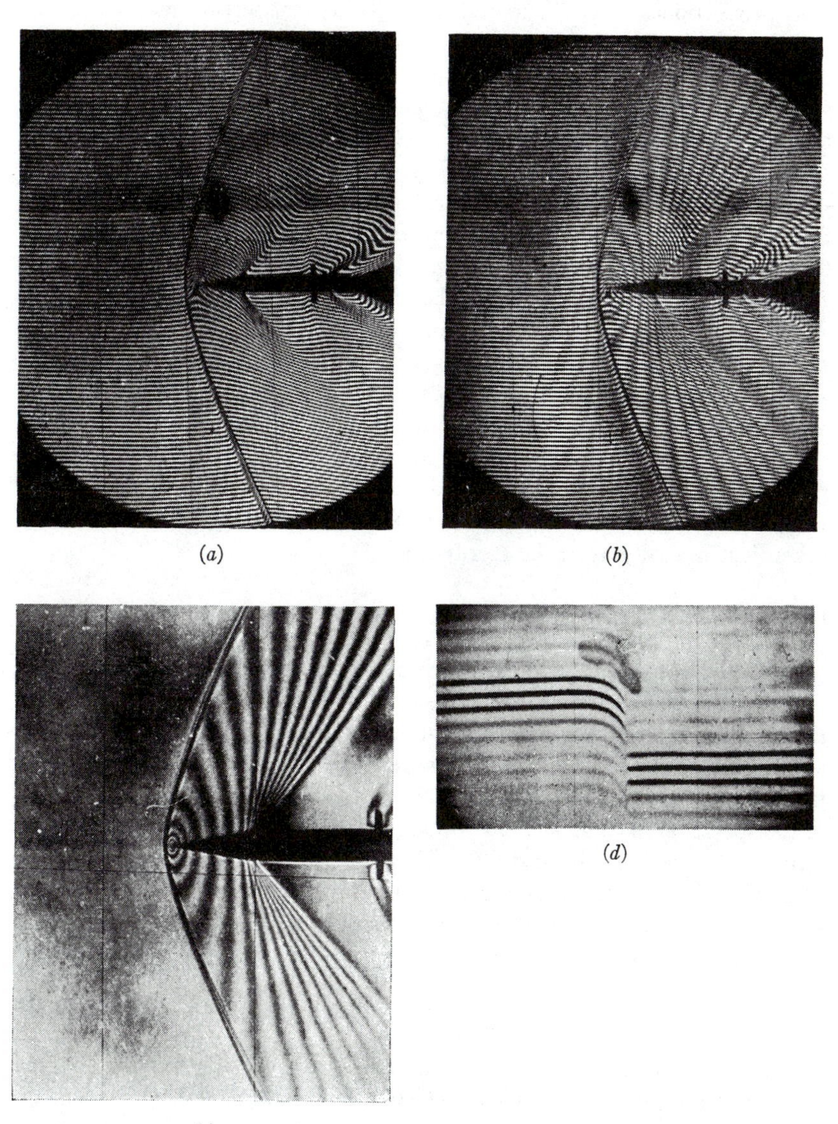

FIG. 6·15 Interference pictures. (*a*) Finite fringe interferogram. (*b*) Composite interferogram. (*c*) Infinite fringe interferogram. (*d*) White light interferogram. (*a*), (*b*), (*c*) from H. I. Ashkenas and A. E. Bryson, *J. Aeronaut. Sci.*, *18* (1951), p. 32; (*d*) from W. C. Griffith and Walker Bleakney. *Am. J. Phys.*, *22* (1954), p. 597.

It is proportional to the fluid refractivity and test section width, and inversely proportional to the wavelength and standard density of the fluid. As a typical example, for flow through a wind tunnel with $L = 4$ in. and with atmospheric stagnation conditions, the maximum sensitivity, based on Mach number, is about $N/\Delta M = 20$; that is, one fringe shift corresponds to $\Delta M = 0.05$. The sensitivity decreases directly with density.

In evaluating the *accuracy* of the method the following factors should be considered. (1) The accuracy of fringe shift determination is about 0.1 fringe. (2) Inhomogeneity in the flow and "noise" inside and outside the test section appear as errors in the final, *integrated* result (cf. Eq. 6·18*c*). A large portion of this may be eliminated by passing the reference ray through the test section. (3) In regions of large density gradient, such as boundary layers, *refraction* introduces a position error, since the path of a ray through the test section is not the same as in the no-flow condition.

It is possible to obtain accuracies comparable to those from surface pressure measurements. In *nonstationary* cases, the interferometric technique has an invaluable advantage, for a picture of the whole flow field can be taken just as easily as in the steady case.

6·17 X-Ray Absorption and Other Methods

In the operating ranges of most wind tunnels, down to about 0.1 atmosphere at the test section, the schlieren and shadow techniques for observation and the interferometer for measurement are at present the most useful methods. At lower densities, however, the resolutions of the optical methods become seriously low. Techniques using radiation in other ranges of the spectrum may then be advantageous. So far, however, none of them have become well established.

In the X-ray method, soft X-ray radiation is passed through the test section and measured by a Geiger counter or ionization chamber. The intensity of the emergent radiation follows the absorption law

$$I = I_0 e^{-\mu L} \qquad (6·21)$$

where the absorption coefficient μ depends, amongst other things, on the density, wavelength, etc. Thus I/I_0 furnishes a measure of the density. At low densities the time of measurement at the receiver is from several seconds to a few minutes. An advantage of X-rays is that refraction and diffraction errors are negligible, owing to the small wavelengths.

Other proposed techniques depend on the selective absorption at particular wavelengths. For instance, oxygen has a strong absorption band near $\lambda = 1450$ A. In a tunnel using water vapor as the working fluid, infrared absorption techniques might be employed, since water vapor has strong absorption bands in the infrared. Absorption techniques have been successfully applied to the study of chemical reactions in shock tubes.†

The *emission* of light which may be obtained in electrically excited gases at *low densities* has been used in *visualization* techniques for very low densities. The high temperatures obtainable in a shock tube have been measured by spectroscopic study of the radiation.‡

6·18 Direct Measurement of Skin Friction

A quantity of primary importance in aerodynamic problems is the force due to skin friction. In engineering applications, its *overall* value is usually estimated from measurements of the gross drag force. This involves an estimate of the pressure drag, which must be subtracted, and which itself may not be easily determinable.

Of more interest to the researcher and to the theoretical aerodynamicist is the *local* friction force, or shearing stress τ_w. At low speeds, this is ordinarily determined from a measurement of the velocity profile near the surface and an application of the Newtonian friction law,

$$\tau_w = \left(\mu \frac{du}{dy}\right)_w \tag{6·22}$$

where μ_w and $(du/dy)_w$ are the coefficient of viscosity and the velocity gradient at the wall, respectively. The measurement of the velocity profile at high speeds becomes very difficult, particularly in turbulent boundary layers, in which the measurements must be made very close to the surface. But it is just for turbulent boundary layers that measurements are most needed, since theory is not available.

An instrument that avoids these difficulties by measuring the local friction force *directly* is the skin friction meter developed by Dhawan§ (Fig. 6·16). The principle is very simple and had previously been applied on a much grosser scale. Its adaptation to a precision instrument for the small scales encountered in high-speed wind tunnels is dependent on the availability of high-precision machining methods and on modern techniques for measuring or detecting small displacement.

A small segment of the surface, separated from the remainder by very small gaps, is allowed to deflect under the action of the skin friction. Its movement, resisted by the flexure links on which it is mounted, is transmitted to the armature of a differential transformer. The movement is calibrated against standard weights, using a simple pulley arrangement.

†D. Britton, N. Davidson, G. Schott, *Faraday Soc. Discussion, 17* (1954), p. 58.

‡H. E. Petschek, P. H. Rose, H. S. Glick, A. Kane, A. Kantrowitz, *J. App. Phys., 26* (1955), p. 83.

§Satish Dhawan, "Direct Measurements of Skin Friction", *N.A.C.A. Rep.* 1121 (1953).

In the Dhawan instrument, the size of the "floating" element was only 0.08 in. × 0.8 in., the gaps separating it from the main surface being about 0.005 in.

There are various adaptations and modifications of this basic instrument. Measurements of turbulent skin friction at high speeds, obtained with two different versions of the instrument, by Coles and Korkegi, are shown in Fig. 13·10.

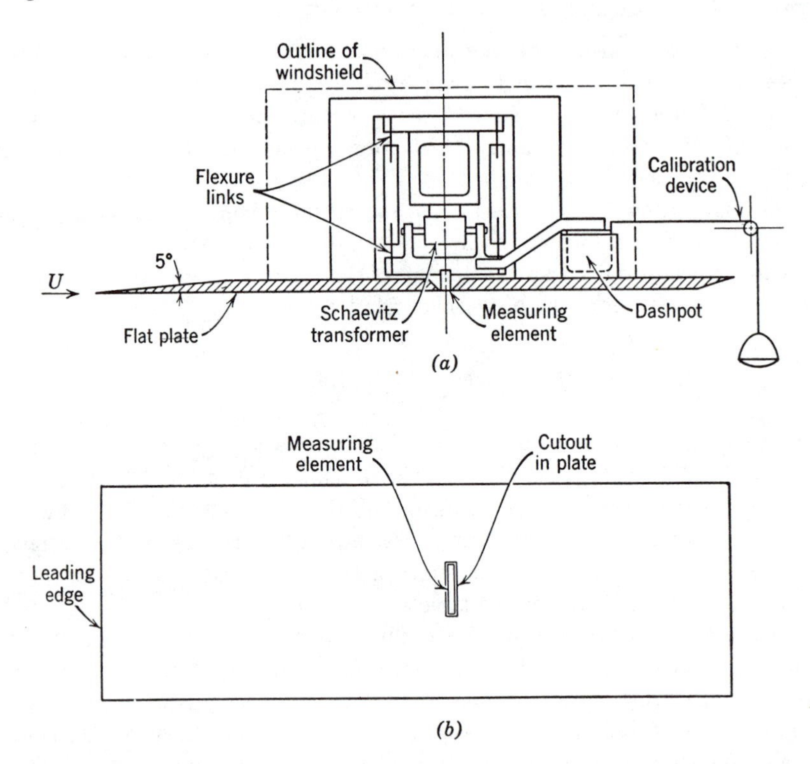

FIG. 6·16 Skin friction meter. (*a*) Side view of plate, balance, and calibration device. (*b*) Plan view of aerodynamic surface, showing measuring element in cutout. [From R. J. Hakkinen, "Measurements of Turbulent Skin Friction on a Flat Plate at Supersonic Speeds," *N.A.C.A.T.N.* 3486 (1955).]

6·19 Hot-Wire Probe

The cooling of a heated, fine wire in an airstream depends mainly on conduction to the fluid passing over it, and thus on the mass flow, ρu. This effect is the basis of the hot-wire probe, sketched in Fig. 6·17. The heating of the wire and the determination of its temperature (to which the resistance is related) are accomplished electrically.

In *incompressible flow*, the relation between heat transfer Q from the wire, the wire temperature T_w, and the flow speed U fits a formula known as King's equation

$$Q = (T_w - T)(a + b\sqrt{U}) \tag{6·23}$$

where a and b are calibration constants and T is the flow temperature.

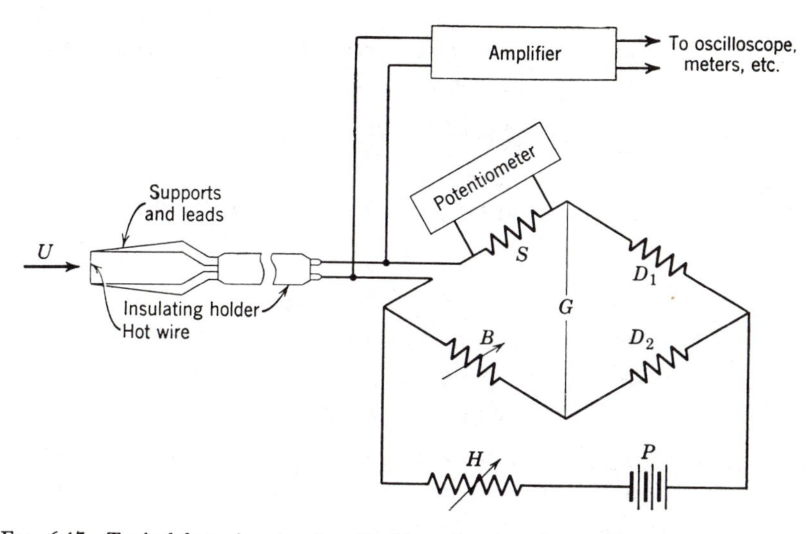

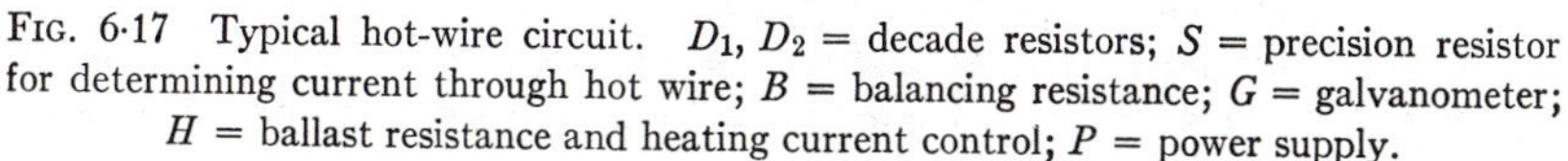
FIG. 6·17 Typical hot-wire circuit. D_1, D_2 = decade resistors; S = precision resistor for determining current through hot wire; B = balancing resistance; G = galvanometer; H = ballast resistance and heating current control; P = power supply.

The rate of heat transfer Q out of the wire is just equal, in *steady* conditions, to the heat supplied by electrical dissipation,

$$Q = i^2 R_w$$

The current i is measured by an ammeter (usually a potentiometer across a standard resistor in series with the wire), and the wire resistance R_w is measured by a Wheatstone bridge, as shown schematically in Fig. 6·17. The measurement of R_w also determines the wire temperature, from the relation

$$R_w = R'[1 + \alpha(T_w - T')] \tag{6·24}$$

where α is the coefficient of resistance and T' is the reference temperature.

The constants a, b of Eq. 6·23 are determined by calibration in a known flow. For measuring mean speed, it is usually convenient to operate at *constant temperature* by adjusting the current at each velocity just enough to keep R_w constant.

In Eq. 6·23, a and b are different for every wire; a more general form of the equation is obtained by using the dimensionless variables

$$Nu = \frac{Q}{\pi k l (T_w - T)} \quad \text{Nusselt number}$$

$$Re = \frac{\rho U d}{\mu} \quad \text{Reynolds number}$$

where k and μ are the conductivity and viscosity of the fluid, and l and d are the length and diameter of the wire. Then King's law becomes

$$Nu = A + B\sqrt{Re}$$

The universal "constants" A and B depend on parameters such as the ratio l/d, which determines "end losses," and the *temperature loading*†

$$\tau = \frac{T_w - T}{T_e}$$

For *compressible flow*, still other parameters such as the Mach number, Prandtl number, and γ must be considered. In general, one has

$$Nu = Nu(Re, M, Pr, \gamma, \tau, l/d)$$

Furthermore, in defining the Nusselt number and temperature loading for compressible flow it is necessary to use for T the recovery temperature T_e of the *unheated* wire; that is, $T_w = T_e$ when $Q = 0$.

Figure 6·18a gives an experimental determination of Nusselt number for wires at $l/d = 500$ in supersonic flow. The Nusselt number Nu_2 and Reynolds number Re_2 have been computed by evaluating the fluid properties at the conditions (2) behind the detached shock wave; that is,

$$Nu_2 = \frac{Q}{\pi k_2 l (T_w - T_e)}$$

$$Re_2 = \frac{\rho_2 u_2 d}{\mu_2} = \frac{\rho_1 U d}{\mu_2}$$

(The last equality follows from the fact that $\rho_2 u_2 = \rho_1 U$ across a normal shock.) With this choice of parameters, the calibration is independent of M_1 for values greater than 1.3, as shown in Figs. 6·18a and 6·18b. The latter gives the variation of equilibrium temperature ratio, T_e/T_0. An interesting fact is that for low Reynolds number the recovery temperature T_e of the unheated wire is higher than the total temperature T_0.

A Mach number effect appears at transonic and subsonic Mach numbers. The data for $M = 0$ and $\tau = 0$ is shown by a broken line in Fig. 6·18a, for

†The symbol τ is also used for the shear stress (Eq. 6·22).

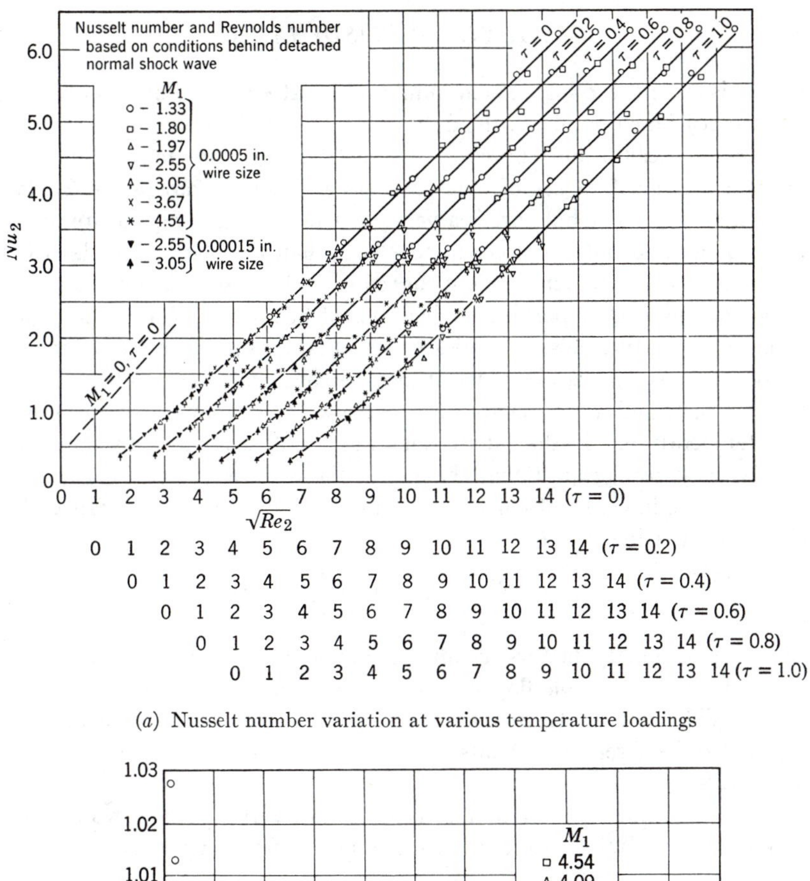

(*a*) Nusselt number variation at various temperature loadings

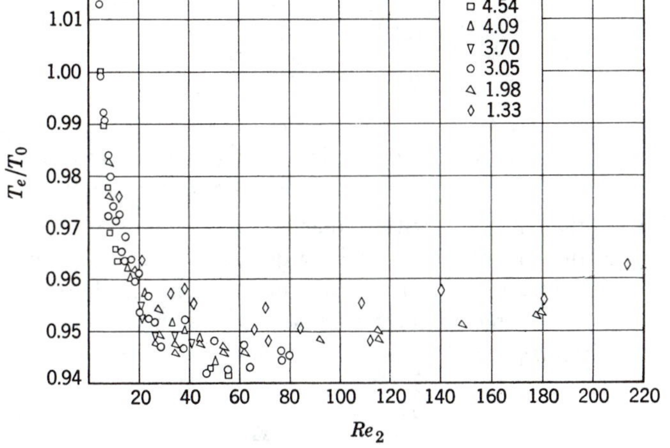

(*b*) Variation of equilibrium temperature

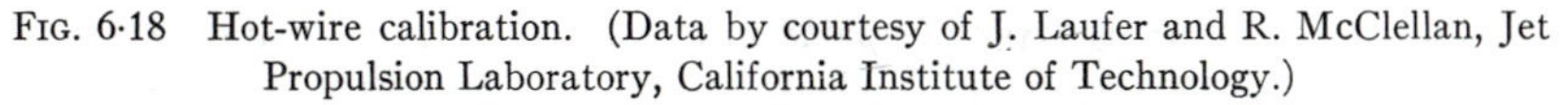
FIG. 6·18 Hot-wire calibration. (Data by courtesy of J. Laufer and R. McClellan, Jet Propulsion Laboratory, California Institute of Technology.)

comparison with the high Mach number data at $\tau = 0$. For intermediate (transonic) Mach numbers, the data falls between these two lines.

With the calibration curves of Fig. 6·18, a hot wire may be used to measure the local values of mass flow ρu and total temperature T_0 (see Exercise 6·11). The hot wire is a very convenient total temperature probe; it is smaller and much simpler to construct than the instrument described in Article 6·7. However, it is not rugged enough for some applications.

One of the most important applications of hot wires is for the measurement of fluctuating flow quantities.† Its suitability for such measurements is due to the small sizes in which it may be constructed (diameters of 10^{-5} to 10^{-3} in.), and which enable it to follow rapid changes of mass flow and temperature. If the fluctuations are small, compared to the mean values, they give proportional changes in the wire resistance (or current). For instance, in constant-current operation, the fluctuating resistance, R'_w gives a fluctuating voltage

$$e' = iR'_w = c_1 \frac{u'}{U} + c_1 \frac{\rho'}{\rho} + c_2 \frac{T_0'}{T_0}$$

The proportionality factors, c_1 and c_2, are obtained from the calibration data. In incompressible flow only the velocity fluctuation occurs,‡ but in compressible flow all three may be present. Resolution of the hot-wire signal into the three components requires readings at three different operating conditions.

Even these fine wires cannot *precisely* follow very fast changes and the proportionality factors decrease with fluctuation frequency n as

$$\frac{c(\omega)}{c(0)} = \frac{1}{\sqrt{1 + M^2\omega^2}}$$

where $\omega = 2\pi n$, and M is the *time constant* of the wire. It is possible to take this attenuation into account in making the measurements, but a more convenient arrangement is to use a *compensating amplifier*, with an amplification that increases with frequency by the factor $\sqrt{1 + M^2\omega^2}$.

The lag of the wire is almost entirely due to the time for thermal readjustment, the time for flow readjustment being much shorter (of order $d/10U$). Thus the time constant may be estimated by comparing the heat capacity of the wire with the rate of heat flow,

$$M \sim \frac{\rho_w(\pi l\, d^2)c\, \Delta T}{\pi k l Nu\, \Delta T}$$

†L. S. G. Kovasznay, *J. Aeronaut. Sci.*, *17* (1950), p. 565 and *20* (1953), p. 657. Also *NACA Rep.* 1209.

‡With the single straight wire illustrated in Fig. 6·17 only the fluctuation u', in the stream direction, can be measured. v'- and w'-probes may be constructed by using two wires set at an angle to the flow and to each other.

ρ_w and c are the density and specific heat of the wire, and ΔT is a typical temperature change. The exact theory gives an additional factor, R_w/R_e, and thus

$$M = \frac{R_w}{R_e} \frac{\rho_w c\, d^2}{kNu}$$

For platinum and tungsten, the usual hot-wire materials, operating in air, M is of the order of 1 millisecond for a 0.1-mil (0.0001 in.) wire.

6·20 Shock Tube Instrumentation

Shock tubes and, to a lesser extent, intermittent wind tunnels require special instrumentation since the flow durations are very short. In a shock tube they are of the order of milliseconds.

Instruments used for measuring pressure and temperature must have response times considerably shorter than the flow times. Small, pressure-sensitive crystals and temperature-sensitive resistance wires or films are typical elements for pressure and temperature gages. The outputs are obtained as voltage signals which may be observed on an oscilloscope, and photographically recorded. Suitable triggering devices and time delays must be provided, to synchronize the oscilloscope sweep with the portion of flow to be observed.

Wave speed is measured by timing the arrival of the wave at two trigger points along the tube.

Densities are measured by interferometry, using a triggered light source.† Similarly, schlieren and shadow techniques may be employed for flow visualization.

A useful instrument for flow visualization is the *drum camera*.‡ Using a slit window along the axis of the tube, the flow is photographed on a film which moves *perpendicular* to the slit while the flow moves *along* the slit. The film is on a rotating drum whose axis is parallel to the slit. The resulting picture is an x-t diagram of the flow (Article 3·2), x being along the slit and t in the direction of motion of the film.

†W. C. Griffith and Walker Bleakney, *Amer. J. Phys.*, *22* (1954), p. 567.

‡I. I. Glass and G. N. Patterson, *J. Aeronaut. Sci.*, *22* (1955), p. 73.

CHAPTER 7

The Equations of Frictionless Flow

7·1 Introduction

The one-dimensional theory of the earlier chapters took us a remarkably long way in the study of gasdynamics. It was even possible to extend it, in the supersonic case, to two-dimensional flows. For subsonic and transonic flow, however, and for three-dimensional flow in general, it is necessary to obtain the general equations of motion. This will be done in the present chapter.

The flow will again be considered nonviscous, that is, frictionless and nonconducting. It may be expected then that the general features will be similar to those already described by the one-dimensional results. The more general equations, however, will make possible the discussion of a much larger variety of configurations in the following chapters.

They will also bring out some features that are essentially two- or three-dimensional in nature. Thus there will be introduced the concept of vorticity, which does not exist in the one-dimensional case.

For convenience in writing the general three-dimensional equations of motion, we shall adopt the so-called *Cartesian tensor notation*, which is outlined in the following article. This is little more than another vector notation, having the advantage, however, of being applicable not only to vector quantities but also to quantities requiring more than three components for their specification, for example, stress.

7·2 Notation

The general three-dimensional equations will be derived for a Cartesian coordinate system, in which a position is given by the coordinates (x_1, x_2, x_3). In vector notation this is denoted by $\mathbf{x}$, but in the Cartesian tensor notation,† which we shall adopt, it is denoted by x_i, where the index i may be 1, 2, or 3. Similarly the velocity vector, $\mathbf{u} = (u_1, u_2, u_3)$ may be represented by u_j where j may be 1, 2, or 3. In general, different indices are used for different vectors, unless it is intended explicitly to indicate equality of indices. Thus a_i and b_j are two arbitrary vectors, whereas a_i and b_i are related by the equation

$$a_i = cb_i \tag{7·1}$$

†H. Jeffreys, *Cartesian Tensors*, Cambridge, 1952.

where c is a constant (scalar). That is,

$$a_1 = cb_1, \quad a_2 = cb_2, \quad a_3 = cb_3$$

which shows that the two vectors differ only in magnitude, not in direction.

The inner product (or so-called scalar product) of two vectors,

$$\mathbf{a} \cdot \mathbf{b} = a_1 b_1 + a_2 b_2 + a_3 b_3$$

may be written in index notation as

$$\mathbf{a} \cdot \mathbf{b} = \sum_{i=1}^{3} a_i b_i$$

Since inner products occur quite frequently, and result in summations like that in Eq. 7·2, it is convenient to adopt the convention that, whenever an index occurs twice in any term, a summation over that index is indicated, without the summation symbol having to be written. Thus

$$a_i b_i = a_1 b_1 + a_2 b_2 + a_3 b_3 \tag{7·2}$$

Such repeated indices, which indicate a summation, are often called *dummy* indices.

So far, the Cartesian tensor notation offers no apparent advantages over the vector notation. Its real advantage will appear later, when it will be necessary to treat quantities that cannot be represented conveniently in the vector notation, but for which the Cartesian tensor notation is easily generalized. Such a quantity is the *stress*, which has nine components (τ_{11}, $\tau_{12}, \tau_{13}, \ldots, \tau_{33}$), and may be represented by τ_{ik}. The nine components are obtained by taking all the possible combinations of i and k, each of which may have the values 1, 2, 3.

A quantity like τ_{ik} is called a tensor of second order;† a vector b_i is a tensor of first order; and a scalar c is simply a tensor of zero order. We shall have occasion, in Chapter 13, to use tensors like d_{ijkl} of fourth order.

Differentiation leads to a tensor of order one higher. For instance, the gradient of a scalar quantity

$$\nabla p = \frac{\partial p}{\partial x_i} \tag{7·3}$$

is a vector. Similarly the gradient of a vector

$$\nabla \mathbf{u} = \frac{\partial u_i}{\partial x_j} \tag{7·4}$$

is a tensor of second order.

†A tensor is more than a matrix of numbers. It has physical meaning; for instance, a second order tensor operating on a vector transforms it into another vector, $\tau_{ik} n_k = f_i$. A tensor also has certain invariants independent of the coordinate system.

On the other hand, the divergence of a vector is a scalar,

$$\nabla \cdot \mathbf{u} = \frac{\partial u_i}{\partial x_i} = \left(\frac{\partial u_1}{\partial x_1} + \frac{\partial u_2}{\partial x_2} + \frac{\partial u_3}{\partial x_3}\right) \tag{7·5}$$

The utility of the summation convention is evident here.†

A useful "operator" is the *Kronecker delta*, δ_{ij}, which is defined by

$$\begin{aligned} \delta_{ij} &= 1 \quad \text{if } i = j \\ \delta_{ij} &= 0 \quad \text{if } i \neq j \end{aligned} \tag{7·6}$$

For instance,

$$\delta_{ij} b_j = b_i$$

In developing the equations of motion, we shall need the *theorem of Gauss*, which in vector notation is written

$$\int_A \mathbf{b} \cdot \mathbf{n} \, dA = \int_V \nabla \cdot \mathbf{b} \, dV \tag{7·7}$$

This states that, for any volume V in a vector field $\mathbf{b}$, the normal component $\mathbf{b} \cdot \mathbf{n}$, integrated over the enclosing surface A, is equal to the divergence $\nabla \cdot \mathbf{b}$ integrated over the volume. In index notation, this is

$$\int_A b_j n_j \, dA = \int_V \frac{\partial b_k}{\partial x_k} \, dV \tag{7 7a}$$

Gauss's theorem may be generalized to any tensor field. Thus

$$\int_A g n_j \, dA = \int_V \frac{\partial g}{\partial x_j} \, dV \tag{7·7b}$$

where g is any tensor, including scalars and vectors.

7·3 The Equation of Continuity

The general three-dimensional equation of continuity may be derived by the method that was used for one-dimensional flow (Article 2·2), that is, by observing the *flux of matter through a fixed "control surface"* (Fig. 7·1). The surface may be of arbitrary shape and area A, but must be *closed*; it encloses a volume V of the space through which the fluid flows. Part of the surface may consist of physical boundaries, such as walls.

An element of surface area is denoted by $\mathbf{n} \, dA$, where $\mathbf{n}$ is a unit vector, normal to the surface, positive when directed outward. The component of velocity which carries matter through the surface is $\mathbf{u} \cdot \mathbf{n}$, so that the rate of flow of mass through the surface element is $\rho \mathbf{u} \cdot \mathbf{n} \, dA$ or $\rho u_j n_j \, dA$. When

†Since the repeated index is a "dummy," the choice of letter is optional; of course it should not duplicate an index used in some other sense in the same equation.

the contributions to the flux are added over the whole surface, the net flux *out* of the fixed volume is obtained,

$$\int_A \rho u_j n_j \, dA$$

The rate at which the enclosed mass increases is simply

$$\frac{\partial}{\partial t} \int_V \rho \, dV = \int_V \frac{\partial \rho}{\partial t} \, dV$$

the differentiation and integration being interchangeable since a fixed volume is considered. By the law of conservation of mass, this rate of increase must

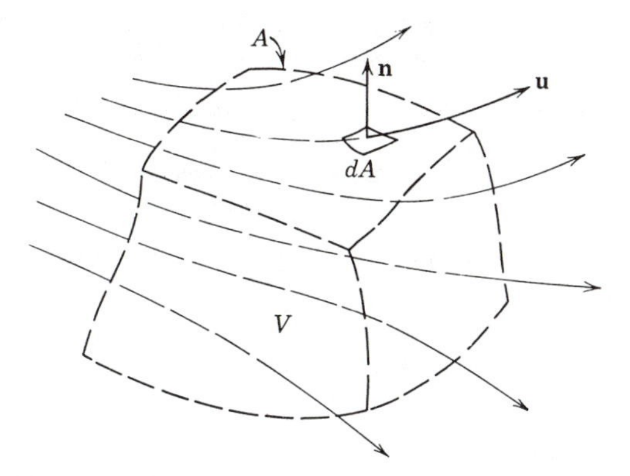

FIG. 7·1 Fluid flow through a "control surface." Surface area = A; volume = V.

be due entirely to the *inflow* of mass through the surface (assuming that there are no internal sources), that is,

$$\int_V \frac{\partial \rho}{\partial t} \, dV = -\int_A \rho u_j n_j \, dA \qquad \blacktriangleright(7{\cdot}8)$$

This is one form of the law of conservation of mass.

In one-dimensional flow (Article 2·2), the relation becomes

$$\int_V \frac{\partial \rho}{\partial t} \, dV = -(\rho_2 u_2 A_2 - \rho_1 u_1 A_1)$$

The volume term on the left side of Eq. 7·8, called the *nonstationary* term, is due to the fact that the density in V is changing, if the flow is nonstationary. The surface term on the right-hand side, called the *convective* term, expresses the fact that the flow carries mass into and out of V.

The fluid which flows into and out of V *transports* not only mass, but also various characteristics associated with the fluid, such as momentum, energy, entropy, etc. If such a characteristic be denoted by g, *per unit volume*, the rate of change of the amount of g enclosed in V will always be expressed in terms of a nonstationary volume term

$$\int_V \frac{\partial g}{\partial t}\, dV \tag{7·9a}$$

and a convective surface term

$$\int_A g u_j n_j \, dA \tag{7·9b}$$

The characteristic g may be any (transportable) scalar, vector, or general tensor quantity.

Returning now to the continuity equation, we may use Gauss's theorem (Eq. 7·7*a*) to rewrite the surface term as a volume integral. The vector b_j in Gauss's theorem is replaced here by ρu_j, giving the result

$$\int_A \rho u_j n_j \, dA = \int_V \frac{\partial}{\partial x_j} (\rho u_j)\, dV$$

Bringing it to the left-hand side of Eq. 7·8, we obtain the continuity equation in the form

$$\int_V \left[\frac{\partial \rho}{\partial t} + \frac{\partial}{\partial x_j} (\rho u_j)\right] dV = 0$$

Since this equation must be true for any volume V, the integral can vanish only if the quantity in brackets vanishes everywhere.† Thus we find the differential form of the continuity equation

$$\frac{\partial \rho}{\partial t} + \frac{\partial}{\partial x_j} (\rho u_j) = 0 \tag{7·10}$$

For one-dimensional flow, it reduces to Eq. 3·7.

The surface integral in the general expression (Eq. 7·9*b*) may similarly be transformed into a volume integral by Gauss's theorem (7·7*b*), as we shall see in later applications.

7·4 The Momentum Equation

We now apply the law of momentum to the fluid flowing through the volume V (Fig. 7·1). If F_i is the net force acting on the fluid in V, then

$$F_i = \text{rate of change of momentum of the fluid in } V$$

†For example, an integral $\int_a^b F(x)\,dx$ might be zero for special values of a, b, i.e., whenever "negative areas" cancel positive ones, but it can be zero for all *arbitrary* values of a, b only if $F(x) \equiv 0$ identically.

But what if the flow is *steady*, that is, conditions at each point of the field do not change in time? How can we compute the rate of change of momentum of the fluid in V, that is, of the fluid passing through V? Clearly it must be just equal to the net rate of outflow, since with steady conditions momentum cannot accumulate in V. Thus, for steady flow,

$$F_i = \int_A (\rho u_i) u_j n_j \, dA$$

We have computed the outflow here by replacing g in expression 7·9*b* by ρu_i, the momentum per unit volume.

If the flow is *nonstationary*, we must add the nonstationary term, thus obtaining the complete momentum equation

$$F_i = \int_V \frac{\partial}{\partial t} (\rho u_i) \, dV + \int_A \rho u_i u_j n_j \, dA \tag{7·11}$$

Next, we compute the force F_i in terms of more specific flow parameters. The forces are of two kinds, surface forces and volume forces. The surface forces are due to whatever medium is adjacent to the surface A, for example, a solid wall or simply the adjacent fluid. In the *nonviscous* flow which we are considering, they can only be due to *normal pressure*, on the surface; there are no tangential friction forces. Thus the normal force per unit area is p. It is directed *inward*, so that, if n_i is the *outward* unit normal, the force acting on element area dA is equal to $(-p n_i \, dA)$. The total pressure force acting on the fluid inside A is

$$\int_A (-p n_i) \, dA$$

Examples of the volume or "body" force are inertial forces, gravity forces and electromagnetic forces. Such a force is proportional to the mass, and may be represented by the vector f_i, per unit mass, or by ρf_i per unit volume. Summed over the volume, it contributes the force

$$\int_V \rho f_i \, dV$$

Replacing F_i in Eq. 7·11 by the sum of the surface and body forces, we have the integral form of the momentum equation,

$$\int_V \frac{\partial}{\partial t} (\rho u_i) \, dV + \int_A \rho u_i u_j n_j \, dA = -\int_A p n_i \, dA + \int_V \rho f_i \, dV \quad ▶(7·12)$$

Again the surface integrals may be transformed to volume integrals by Gauss's theorem. The same argument as before then gives the differential form of the momentum equation,

$$\frac{\partial}{\partial t}(\rho u_i) + \frac{\partial}{\partial x_j}(\rho u_i u_j) = -\frac{\partial p}{\partial x_i} + \rho f_i \tag{7·13}$$

In this equation the four terms are, respectively, the nonstationary and the convective rates of change of momentum, per unit volume, the net pressure force acting on the surface of unit volume, and the body force per unit volume.

Both the integral and differential forms of the momentum equation are very important. Most of the problems treated in this book start from the

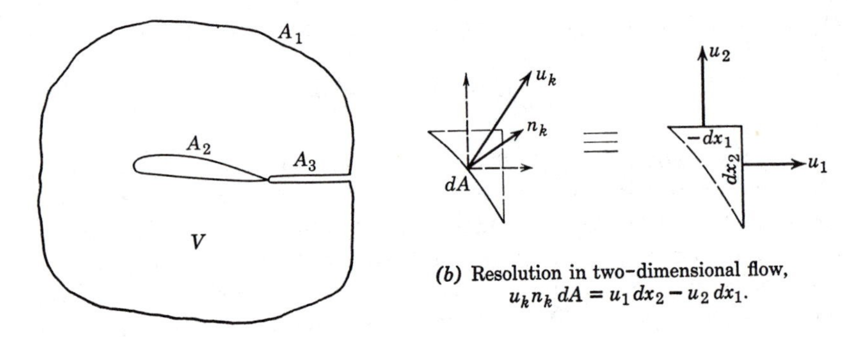

(a) Control surface, $A \equiv A_1 + A_2 + A_3$.

(b) Resolution in two-dimensional flow, $u_k n_k\, dA = u_1\, dx_2 - u_2\, dx_1$.

FIG. 7·2 Application of momentum theorem.

differential equations, but there are many cases in which application of the integral form may be more direct. We have had one example in flow through a shock.

In applying the integral form, it is important to remember that the enclosing "control surface" A must be *closed*. As an example, we shall work out the expressions for lift and drag on a body in terms of the momentum flux through an enclosing control surface. We shall assume steady flow with no body forces, in which case the momentum equation becomes

$$-\int_A p n_i\, dA = \int_A (\rho u_i) u_k n_k\, dA\dagger \tag{7·14}$$

The area A, which is made up of three parts,

$$A = A_1 + A_2 + A_3$$

is shown in Fig. 7·2*a*. A_1 is an arbitrary control surface, A_2 is the area of

†Here we adopt k for the dummy index. Cf. footnote on p. 180.

the body, and A_3 is the area of the surface of a slit, or cut, which makes A a simple closed surface.

The net contribution from A_3 to the integrals is zero, since both pressures and flows cancel each other on the two sides of the cut. Also, A_2 gives no contribution to the flux integral, since there the velocity component normal to A_2 is zero. Thus Eq. 7·14 becomes

$$\int_{A_2} (-pn_i)\, dA = \int_{A_1} (\rho u_i) u_k n_k \, dA + \int_{A_1} (pn_i)\, dA$$

The left-hand integral represents the force exerted on the fluid surface A_2 by the body. Hence, by changing the sign, we have the force of the *fluid on the body*, that is,

$$F_i = -\int_{A_1} (\rho u_i) u_k n_k \, dA - \int_{A_1} (pn_i)\, dA \tag{7·15}$$

If the drag D and lift L are defined as the components of F_i in the x_1- and x_2-directions, respectively, then

$$\begin{aligned} D &= -\int_{A_1} (\rho u_1) u_k n_k \, dA - \int_{A_1} pn_1 \, dA \\ L &= -\int_{A_1} (\rho u_2) u_k n_k \, dA - \int_{A_1} pn_2 \, dA \end{aligned} \tag{7·16}$$

For example, in two-dimensional flow, Fig. 7·26*b*, integrating counterclockwise,

$$\begin{aligned} D &= -\int_{A_1} \rho u_1 (u_1 \, dx_2 - u_2 \, dx_1) - \int_{A_1} p \, dx_2 \\ L &= -\int_{A_1} \rho u_2 (u_1 \, dx_2 - u_2 \, dx_1) - \int_{A_1} p \, dx_1 \end{aligned} \tag{7·16a}$$

The form of the control surface A_1 is arbitrary; it may be chosen to be a rectangular "box," or circle, or whatever is most convenient.

7·5 The Energy Equation

It may be expected that the energy law, a scalar equation, will be essentially the same as that already obtained in Chapter 2. Here, we shall obtain a little more generality by including the nonstationary terms. The derivation will also be a useful reference for Chapter 13, where friction and heat transfer terms will be included.

The energy law applied to the fluid in V (Fig. 7·1) states that

Heat added + work done on the fluid = increase of energy

For a flowing fluid it is convenient to consider *rate* of change of energy. Thus

Rate of heat addition + rate of work on the fluid
= rate of increase of energy of the fluid

The rate of heat addition, per unit mass, will be denoted by q. This includes only the heat that is added *externally* and is not already latent in the fluid. Thus, heat released by a transformation of the fluid is not included in q, but heat absorbed from external radiation is included. q is a *volume* term, and does not include heat that is transferred by conduction from one part of the fluid to another. The latter is a *surface* term; it will be included in the discussion of viscous fluids in Chapter 13.

The *rate of work* on the fluid is due to the volume forces and the pressure p. It was shown in the last article that these contribute the forces $\rho f_i\,dV$, on the volume dV, and $-pn_i\,dA$, on the surface area dA. The corresponding rates of work are obtained by forming a scalar product with u_i, giving $\rho f_i u_i\,dV$ and $-pn_i u_i\,dA$, respectively. In a viscous fluid there will be additional surface terms due to friction.

The *energy of the fluid*, consisting of internal energy and kinetic energy, is $\rho e + \frac{1}{2}\rho u^2$ per unit volume, where $u^2 \equiv u_1{}^2 + u_2{}^2 + u_3{}^2 \equiv u_i u_i$. The rate at which the energy is changing inside V has a nonstationary and a convective part. It may be written by replacing g in Eqs. 7·9 by $\rho e + \frac{1}{2}\rho u^2$

If we collect the various terms, the energy equation in integral form is

$$\int_V \rho q\,dV + \int_V \rho f_i u_i\,dV - \int_A p n_i u_i\,dA = \int_V \frac{\partial}{\partial t}\left(\rho e + \frac{1}{2}\rho u^2\right) dV + \int_A \left(\rho e + \frac{1}{2}\rho u^2\right) u_j n_j\,dA \qquad \blacktriangleright(7\cdot17)$$

with a corresponding differential expression

$$\rho q + \rho f_i u_i - \frac{\partial}{\partial x_i}(p u_i) = \frac{\partial}{\partial t}\left(\rho e + \frac{1}{2}\rho u^2\right) + \frac{\partial}{\partial x_j}\left[\left(\rho e + \frac{1}{2}\rho u^2\right) u_j\right] \qquad \blacktriangleright(7\cdot17a)$$

7·6 The Eulerian Derivative

The equations of motion are often derived from a somewhat different, but equivalent, point of view, which we shall now compare briefly with the integral method of the preceding articles.

Any characteristic or property associated with the fluid may be expressed as a *field*. For example, there is a temperature field, $T(x_1, x_2, x_3)$, which might be specified by temperature contours, as in Fig. 7·3. If the field is *nonstationary*,

$$T = T(x_1, x_2, x_3, t) = T(x_i, t)$$

The expression means simply that a fluid particle that is at position x_i at time t will have the temperature T. Similarly, we can specify fields for density, pressure, velocity, etc.

The *rate of change* of any of these characteristics for a particle of fluid is due to two effects, a *convective* effect and a *nonstationary* effect. The con-

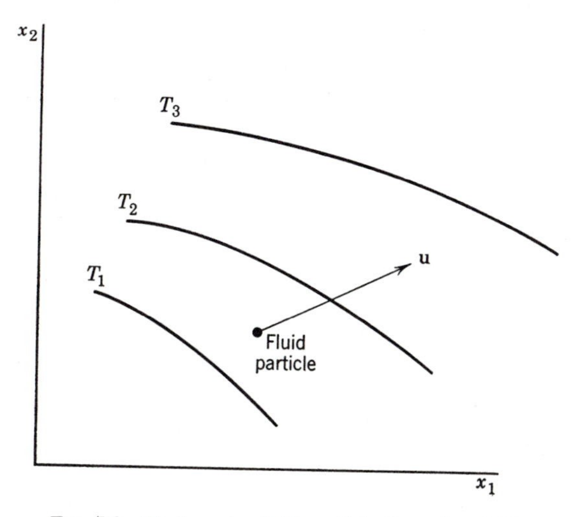

FIG. 7·3 Motion of a fluid particle through a field.

vective rate of change of temperature is equal to the *gradient* of T in the field times the *velocity* with which the particle moves through it,

$$u_k \frac{\partial T}{\partial x_k} = u_1 \frac{\partial T}{\partial x_1} + u_2 \frac{\partial T}{\partial x_2} + u_3 \frac{\partial T}{\partial x_3}$$

The nonstationary effect depends on the rate at which T changes *locally*, i.e., $\partial T/\partial t$. Thus the net rate of change of temperature of a flowing particle is

$$\frac{\partial T}{\partial t} + u_k \frac{\partial T}{\partial x_k} \equiv \frac{DT}{Dt} \tag{7·18}$$

The special notation, D/Dt, for this rate of change is called the *Eulerian derivative.*

The Eulerian derivative may be used to compute the rate of change of any fluid characteristic which is expressible as a field, whether it be a scalar, vector, or more general tensor quantity. Furthermore, the characteristic may be an "intensive" one, such as pressure or velocity, or an "extensive" one, such as energy. In the latter case, the field should express the amount of characteristic *per unit mass*, since in following a particle we follow a definite mass.

To illustrate the field method, we shall write Newton's equation for a flowing particle of fluid. The acceleration, or rate of change of velocity, is

$$\frac{Du_i}{Dt} \equiv \frac{\partial u_i}{\partial t} + u_k \frac{\partial u_i}{\partial x_k}$$

The last term here sometimes seems perplexing, because the velocity appears in two roles, as the convecting agent and as the field characteristic in question!

Newton's law states that mass times acceleration equals the net force. If we consider a unit volume, the mass is ρ, and the force per unit volume is given by the expression already obtained in Article 7·4, namely, $\left(-\frac{\partial p}{\partial x_i} + \rho f_i\right)$. Application of Newton's law then gives us

$$\rho \frac{Du_i}{Dt} \equiv \rho \frac{\partial u_i}{\partial t} + \rho u_k \frac{\partial u_i}{\partial x_k} = -\frac{\partial p}{\partial x_i} + \rho f_i \qquad \blacktriangleright(7\cdot 19)$$

This is often called Euler's equation. It is equivalent to the differential momentum equation, (7·13),

$$\frac{\partial}{\partial t}(\rho u_i) + \frac{\partial}{\partial x_k}(\rho u_i u_k) = -\frac{\partial p}{\partial x_i} + \rho f_i$$

as may be seen by writing out the left-hand side of the latter, that is,

$$\begin{aligned} &\rho \frac{\partial u_i}{\partial t} + u_i \frac{\partial \rho}{\partial t} + \rho u_k \frac{\partial u_i}{\partial x_k} + u_i \frac{\partial}{\partial x_k}(\rho u_k) \\ &\equiv \rho \frac{\partial u_i}{\partial t} + \rho u_k \frac{\partial u_i}{\partial x_k} + u_i \left[\frac{\partial \rho}{\partial t} + \frac{\partial}{\partial x_k}(\rho u_k)\right] \end{aligned} \qquad (7\cdot 20)$$

The expression in brackets is zero, by the continuity equation.

This example, in fact, exhibits the equivalence between the transport method, of the preceding articles, and the field method, of this article. Whenever an extensive property, such as the momentum ρu_i per unit volume, is associated with an intensive property, such as the velocity u_i, the continuity equation may be "subtracted out" in the same way as in Eq. 7·20. In the next article we shall apply it to the energy equation.

7·7 Splitting the Energy Equation

If the continuity equation is multiplied by $(e + \frac{1}{2}u^2)$ and then subtracted from the energy equation (7·17*a*), the latter is put into the form

$$\rho \frac{De}{Dt} + \rho \frac{D}{Dt}\left(\frac{1}{2}u^2\right) = \rho q + \rho f_i u_i - \frac{\partial}{\partial x_i}(p u_i) \qquad (7\cdot 21)$$

This can be further reduced by introducing Euler's equation (7·19), first multiplying the latter by u_i, to transform it into a scalar equation,

$$\rho u_i \frac{Du_i}{Dt} = \rho \frac{D}{Dt}\left(\frac{1}{2} u^2\right) = -u_i \frac{\partial p}{\partial x_i} + \rho f_i u_i \tag{7·22}$$

Subtracting this from Eq. 7·21 leaves

$$\rho \frac{De}{Dt} = \rho q - p \frac{\partial u_i}{\partial x_i}$$

which may be put into more familiar form by noting that, from the continuity equation,

$$\frac{\partial u_i}{\partial x_i} = \rho \frac{D}{Dt}\left(\frac{1}{\rho}\right)$$

so that

$$\frac{De}{Dt} + p \frac{D}{Dt}\left(\frac{1}{\rho}\right) = q \tag{7·23}$$

Thus we have effectively split the energy equation (7·21) into two parts, Eqs. 7·22 and 7·23. Equation 7·22 shows that the kinetic energy is interchangeable with the work due to pressure and body forces, as is usual in "conservative" systems, whereas Eq. 7·23 is simply the first law of thermodynamics for a system in equilibrium. The fact that the latter is written as a *rate* equation does not contradict the idea of equilibrium. It simply states that the system (which is the fluid particle) passes through equilibrium states only, as may be expected when dissipation is absent.

We can also use Eq. 7·23 to compute the rate of change of entropy† of a fluid particle; referring to Eq. 1·44*a*, we have

$$\frac{DS}{Dt} = \frac{1}{T}\left[\frac{De}{Dt} + p \frac{D}{Dt}\left(\frac{1}{\rho}\right)\right] = \frac{q}{T} \tag{7·24}$$

The entropy may increase or decrease, depending on whether external heat is added or removed. If there is no heat addition, that is, $q = 0$, the changes of state of the fluid particle are *isentropic*

$$\frac{DS}{Dt} = 0 \tag*{▶(7·24a)}$$

In Chapter 13, we shall see that the effect of friction and conduction is to change the equilibrium thermal equation (7·23) to a nonequilibrium equation which has *dissipation terms* on the right-hand side; these are always *positive*, and appear as entropy production terms on the right-hand side of Eq. 7·24.

†Cf. footnote on p. 192.

It should be noted that the equations are written for a fluid particle. For instance, the isentropic result of Eq. 7·24*a* applies only to the particle; it does not imply that the entropy is the same in every part of the flow. We may extend the result to *steady* flow by noting that in this case the particle paths coincide with streamlines, and thus the entropy along streamlines must be constant. It may be different on different streamlines, which is a point we shall investigate in Article 7·9.

7·8 The Total Enthalpy

In Chapter 2, it was shown that for steady, adiabatic flow the quantity $h_0 = h + \frac{1}{2}u^2$, the total enthalpy, is the same at all equilibrium sections of a streamline. To show this quantity explicitly in the present derivation we go back to the energy equation (7·21) and rewrite the pressure term in the form

$$-\frac{\partial}{\partial x_i}(pu_i) = \frac{\partial p}{\partial t} - \rho\frac{D}{Dt}\left(\frac{p}{\rho}\right)$$

which may be checked with the help of the continuity equation. Putting this into Eq. 7·21 and dividing through by ρ gives

$$q + f_i u_i + \frac{1}{\rho}\frac{\partial p}{\partial t} = \frac{D}{Dt}\left(\frac{p}{\rho}\right) + \frac{De}{Dt} + \frac{D}{Dt}\left(\frac{1}{2}u^2\right)$$

Introducing the enthalpy, $h = e + p/\rho$, the result is

$$\frac{D}{Dt}\left(h + \frac{1}{2}u^2\right) = q + f_i u_i + \frac{1}{\rho}\frac{\partial p}{\partial t} \qquad ▶(7·25)$$

This equation shows that changes in the quantity $h + \frac{1}{2}u^2$ (which is $c_pT + \frac{1}{2}u^2$ for a perfect gas) may be due to heat addition, the work of volume forces, or a nonstationary effect associated with the pressure. In the absence of external heat addition (see Article 7·5), the first term is zero. Also, in most aerodynamic problems the volume forces are zero. The last term, however, is important in many *nonstationary* problems. For instance, the large temperature differences observed in the Hilsch tube, and in vortex formation behind bluff bodies in compressible flow, are attributed to this nonstationary term.†

For steady, adiabatic flow, without volume forces, all three terms on the right-hand side of Eq. 7·25 are zero, giving

$$h + \tfrac{1}{2}u^2 = \text{const.} = h_0 \qquad (7·26)$$

for a fluid particle. In steady flow, particle paths coincide with streamlines, and then Eq. 7·26 is also valid for each streamline. This result was obtained

†L. F. Ryan, *Mitteilungen aus dem Inst. für Aerodynamik*, Nr. 18, Zürich, 1954.

in Chapter 2. The constant may be different on different streamlines; for example, if the streamlines have originated in different reservoirs.

7·9 Natural Coordinates. Crocco's Theorem

We saw in the two preceding articles that both entropy and total enthalpy are conserved along streamlines if the flow is *steady*, provided that it is also frictionless, nonconducting, and adiabatic. We now have to investigate the

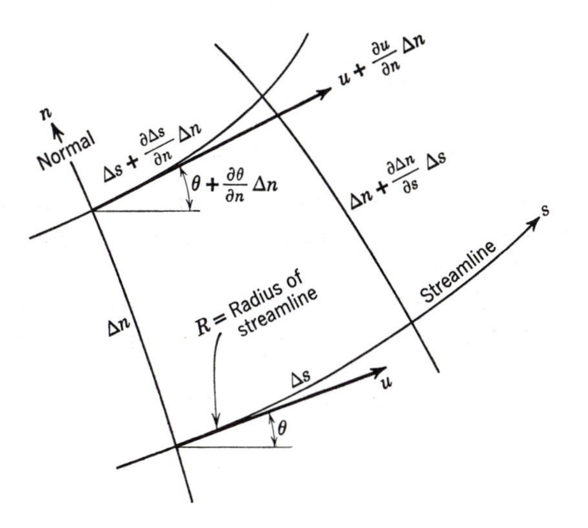

FIG. 7·4 Natural coordinates.

variation of the entropy and the total heat from one streamline to another. This also provides a good opportunity to introduce the so-called *natural coordinate system*, in which one coordinate is along the streamline and the other two are normal to it. As shown in Fig. 7·4, the coordinates are simply the mesh formed by the streamlines and the normals to them. It will be sufficient to consider only *two-dimensional* flow, as shown; in the general case, there would be an additional network normal to this one.

In this coordinate system, s is measured along the streamline in the direction of the velocity vector, whereas n is measured normal to it. The velocity is denoted by its magnitude u and direction θ. Thus,

$$(u, \theta) \text{ is a function of } (s, n)$$

Since the coordinates are curvilinear, the equations derived for a Cartesian system are not directly applicable. However, it is easy to write down the equations of motion, since most of the results of the one-dimensional channel flow of Chapter 2 may be applied to the streamtube which is formed by the two streamlines at n and $n + \Delta n$. The difference from the one-dimensional case is that now we want to consider the momentum balance normal to the flow direction as well as in the direction of flow. The equations of motion, for *steady flow*, are as follows.

$$\text{Continuity} \qquad \rho u \, \Delta n = \text{const.} \tag{7·27}$$

$$s\text{-momentum} \qquad \rho u \frac{\partial u}{\partial s} = -\frac{\partial p}{\partial s} \tag{7·28}$$

$$n\text{-momentum} \qquad \rho \frac{u^2}{R} = -\frac{\partial p}{\partial n} = \rho u^2 \frac{\partial \theta}{\partial s} \tag{7·29}$$

$$\text{Energy} \qquad h + \tfrac{1}{2}u^2 = h_0 \tag{7·30}$$

In Eq. 7·27, the area of the stream tube, for unit width, is Δn (cf. Eq. 2·2). Equation 7·28 gives the balance between acceleration and pressure gradient along the streamline, whereas Eq. 7·29 gives it normal to the streamline. In the latter case, the acceleration depends on the streamline curvature, $1/R$ or $\partial\theta/\partial s$. In the energy equation, h_0 is constant on each streamline but may be different on different ones.

The entropy† is related to the other thermodynamic variables by

$$T\,dS = dh - \frac{1}{\rho}\,dp$$

and thus may be related to the velocity by using the energy equation to replace dh, which is equal to $dh_0 - u\,du$. Thus

$$T\,dS = -\left(u\,du + \frac{1}{\rho}\,dp\right) + dh_0 \tag{7·31}$$

From this differential relation between the variables, the variations along and normal to the streamlines are

$$T\frac{\partial S}{\partial s} = -\left(u\frac{\partial u}{\partial s} + \frac{1}{\rho}\frac{\partial p}{\partial s}\right)$$

$$T\frac{\partial S}{\partial n} = -\left(u\frac{\partial u}{\partial n} + \frac{1}{\rho}\frac{\partial p}{\partial n}\right) + \frac{dh_0}{dn}$$

($\partial h_0/\partial s$ is zero because h_0 is constant along streamlines.) The pressure gradient terms are now replaced by using the momentum equations, Eqs.

†In this chapter we shall use capital S for specific entropy (cf. Article 1·3) to distinguish it from the streamline coordinate s.

7·28 and 7·29, which gives

$$T\frac{\partial S}{\partial s} = 0$$

$$T\frac{\partial S}{\partial n} = -u\left(\frac{\partial u}{\partial n} - \frac{u}{R}\right) + \frac{dh_0}{dn} \tag{7·32}$$

The first equation shows that entropy is constant *along* streamlines, the result previously obtained in Article 7·7. The second equation shows how entropy varies *normal* to the streamlines. It is known as *Crocco's theorem*, and is usually written in the form

$$T\frac{dS}{dn} = \frac{dh_0}{dn} + u\zeta \tag{7·33}$$

The quantity

$$\zeta \equiv \frac{u}{R} - \frac{\partial u}{\partial n} \equiv u\frac{\partial \theta}{\partial s} - \frac{\partial u}{\partial n} \tag{7·34}$$

is the *vorticity* of the flow. Its kinematic meaning will be discussed in the next article.

Equation 7·33 shows what is necessary for a frictionless flow to have the same entropy on different streamlines, that is, to be *isentropic throughout* (called "homentropic" in the British literature). The conditions are:

(1) $h_0 = \text{const.}$ throughout

(2) $\zeta = 0$ throughout

The first condition is satisfied in most aerodynamic problems. Then the vorticity is related directly to the entropy gradient across streamlines. It may be recalled (Article 2·4) that, for a perfect gas, $h_0 = \text{const.}$ implies $T_0 = \text{const.}$, in which case the entropy is related directly to the stagnation pressure (Eq. 2·15*a*). Under these conditions, the vorticity is a measure of the variation of p_0 across streamlines,

$$\zeta = \frac{T}{u}\frac{dS}{dn} = -\frac{RT_0}{up_0}\frac{dp_0}{dn} \tag{7·35}$$

In summary, *zero vorticity implies uniform entropy, provided h_0 is uniform.* In general form, Crocco's theorem is

$$T \operatorname{grad} S + \mathbf{u} \times \operatorname{curl} \mathbf{u} = \operatorname{grad} h_0 + \frac{\partial \mathbf{u}}{\partial t} \tag{7·36}$$

which holds in nonstationary flow.

7·10 Relation of Vorticity to Circulation and Rotation

We may associate with any *velocity field* a function Γ which is defined by

$$\Gamma \equiv \oint_C \mathbf{u} \cdot d\mathbf{l} \equiv \oint_C u_i \, dx_i \tag{7·37}$$

The method of forming the integral is illustrated in Fig. 7·5*a*. An element of length, $d\mathbf{l}$, along an arbitrary curve C, is multiplied by the component of velocity along the curve. The sum of these infinitesimal scalar products

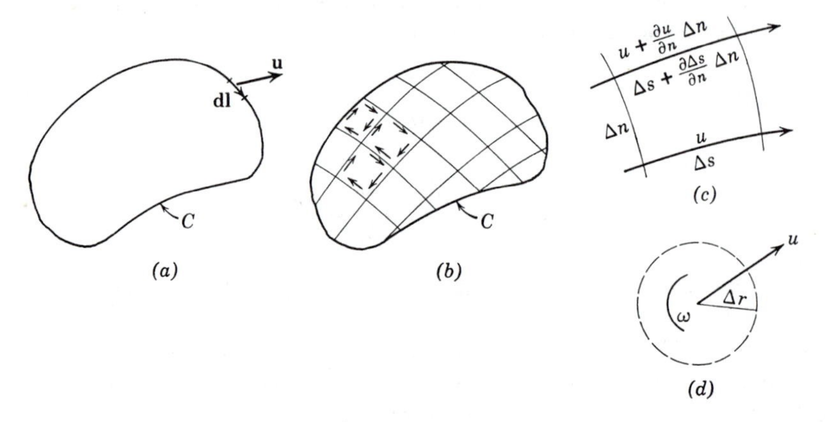

FIG. 7·5 Calculation of circulation. (*a*) Typical length element and velocity on the contour C; (*b*) subdivision of C into smaller meshes; (*c*) elements for calculating circulation around a mesh in natural coordinates; (*d*) a circular mesh element in a flow with rotation.

$\mathbf{u} \cdot d\mathbf{l}$, over a *closed curve* C, defines Γ, which is called the *circulation* around C. The circulation is assigned a definite direction, say, positive when clockwise.

In the same way, we may calculate the circulation around any one of the meshes shown in Fig. 7·5*b*. Let us call a typical one $\Delta\Gamma_i$. The sum of these over all the meshes enclosed by C is equal to the circulation around C,

$$\Gamma = \sum_i \Delta\Gamma_i \tag{7·38}$$

since the contributions from adjacent meshes cancel on their common boundaries, leaving only the contributions from the enclosing curve.

If the sides of the meshes are chosen to lie along streamlines and normals, as shown in Fig. 7·5*c*, the value of ΔΓ is

$$\Delta\Gamma = -(u \, \Delta s) + \left(u + \frac{\partial u}{\partial n} \Delta n\right)\left(\Delta s + \frac{\partial \, \Delta s}{\partial n} \Delta n\right)$$

there being no contribution along the normals. From Fig. 7·4, it may be seen that $\partial\, \Delta s/\partial n = -\Delta s/R$, and thus the above expression may be written

$$\Delta\Gamma = \left(\frac{\partial u}{\partial n} - \frac{u}{R}\right) \Delta s\, \Delta n - \left(\frac{\partial u}{\partial n}\frac{1}{R}\right) \Delta s\, (\Delta n)^2$$

Dividing by the element area, $\Delta A = \Delta s\, \Delta n$, we have

$$\frac{d\Gamma}{dA} = \lim_{\Delta A \to 0} \frac{\Delta\Gamma}{\Delta A} = \frac{\partial u}{\partial n} - \frac{u}{R} = -\zeta \tag{7·39}$$

We may now rewrite Eq. 7·38, neglecting the higher-order term in Δn,

$$\Gamma = \sum \Delta\Gamma = -\sum \zeta\, \Delta A$$

that is,

$$\Gamma \equiv \oint_C \mathbf{u} \cdot d\mathbf{l} = -\iint_A \zeta\, dA \tag{7·40}$$

Equation 7·40 relates the circulation around each mesh element to the vorticity of the fluid within the element.

Finally, the vorticity is related to the *angular velocity*, or "spin," of the fluid. To show this, we may choose a circular mesh element (Fig. 7·5*d*) and calculate the circulation around it. If the fluid is rotating with angular velocity ω, this being the local value in the area covered by the small circle, the circulation is

$$\Delta\Gamma = (-\omega\, \Delta r)(2\pi\, \Delta r) = -2\omega\, \Delta A$$

(Any translational velocity of the fluid gives no net contribution to the line integral.) Thus, the vorticity of the fluid is related to the angular velocity by

$$\zeta = \lim_{\Delta A \to 0} \frac{\Delta\Gamma}{\Delta A} = 2\omega \tag{7·41}$$

Flow with vorticity is called *rotational*; flow with zero vorticity throughout is called *irrotational*.

We have been considering so far only two-dimensional flow, for which the angular velocity (and the vorticity) has only one component, normal to the two-dimensional plane. In the general case there are three components of vorticity, related to the components of angular velocity by

$$\zeta_i = 2\omega_i$$

For this case, *Stokes' theorem* furnishes the result

$$\oint_C \mathbf{u} \cdot d\mathbf{l} = \iint_A \operatorname{curl} \mathbf{u}\, dA$$

where A is any surface bounded by C and curl $\mathbf{u}$ has the components

$$\zeta_1 = \frac{\partial u_3}{\partial x_2} - \frac{\partial u_2}{\partial x_3}$$

$$\zeta_2 = \frac{\partial u_1}{\partial x_3} - \frac{\partial u_3}{\partial x_1} \tag{7·42}$$

$$\zeta_3 = \frac{\partial u_2}{\partial x_1} - \frac{\partial u_1}{\partial x_2}$$

A flow is irrotational only if all three components of vorticity are zero, i.e., curl $\mathbf{u} = 0$.

Sometimes it is convenient to represent the vorticity by the tensor notation

$$\zeta_{ij} = \frac{\partial u_j}{\partial x_i} - \frac{\partial u_i}{\partial x_j} = 2\omega_{ij}$$

In fact, vorticity (and rotation) is a second-order tensor, rather than a true vector. However, it is of a special kind, having only three independent components, as may be seen from the above relation. The three components in which both indices are the same are evidently zero, whereas of the remaining six components only three are independent because $\zeta_{12} = -\zeta_{21}$, etc. Such a tensor is called a pseudo-vector.

A final point of interest is the *tensor gradient*, $\partial u_i/\partial x_j$, which contains all the first-order derivatives of the velocity components. It may be formally split into two parts,

$$\frac{\partial u_i}{\partial x_j} = \frac{1}{2}\left(\frac{\partial u_i}{\partial x_j} + \frac{\partial u_j}{\partial x_i}\right) + \frac{1}{2}\left(\frac{\partial u_i}{\partial x_j} - \frac{\partial u_j}{\partial x_i}\right) = \epsilon_{ij} - \omega_{ij} \tag{7·43}$$

The "symmetrical" part of this gradient, ϵ_{ij}, which is called the *rate of strain tensor*, measures the rate of strain of the fluid. It is related to the viscous stresses, and will be discussed in Chapter 13. The antisymmetrical part ω_{ij} is the *rotation.*

7·11 The Velocity Potential

If the vorticity is zero,

$$-\zeta_{ij} = \frac{\partial u_i}{\partial x_j} - \frac{\partial u_j}{\partial x_i} = 0$$

the velocity is related to a certain function Φ by the relation

$$u_i = \frac{\partial \Phi}{\partial x_i} = \text{grad } \Phi \tag{▶7·44}$$

as may be checked by noting that

$$\frac{\partial u_i}{\partial x_j} - \frac{\partial u_j}{\partial x_i} = \frac{\partial^2 \Phi}{\partial x_j \partial x_i} - \frac{\partial^2 \Phi}{\partial x_i \partial x_j} \equiv 0$$

Φ is called the *velocity potential* and, consequently, *irrotational flow is called potential flow.* We shall see in the following article how to put Φ to good use in the equations of motion.

A simple example of a velocity potential is the one for uniform flow in the x_1-direction,

$$\Phi = Ux_1$$

The velocities are given by

$$u_1 = \frac{\partial \Phi}{\partial x_1} = U$$

$$u_2 = \frac{\partial \Phi}{\partial x_2} = 0$$

$$u_3 = \frac{\partial \Phi}{\partial x_3} = 0$$

The flow may be nonstationary $U = U(t)$, and so, in general, $\Phi = \Phi(x_i, t)$.

7·12 Irrotational Flow

Most of the remaining chapters will deal with adiabatic, irrotational flows. According to Crocco's theorem, such flows are isentropic. It will be useful to summarize the equations that describe them.

For *Cartesian coordinates,* they are as follows.

Continuity $$\frac{\partial \rho}{\partial t} + \frac{\partial}{\partial x_j}(\rho u_j) = 0 \tag{7·45}$$

Momentum $$\rho \frac{\partial u_i}{\partial t} + \rho u_j \frac{\partial u_i}{\partial x_j} = -\frac{\partial p}{\partial x_i} \tag{7·46}$$

Isentropic relation $$\frac{p}{p_0} = \left(\frac{\rho}{\rho_0}\right)^{\gamma} \tag{7·47}$$

Instead of the energy equation, we have the simpler, isentropic relation (Eq. 7·47). It has been written for a perfect gas; the more general form, if needed, is simply $S =$ const.

The above system of five equations is the basic system for the five unknowns, p, ρ, u_i $(i = 1, 2, 3)$. The equations may be recombined in various ways, it being often convenient to introduce the auxiliary equations that are implied by the above conditions. Some of these are:

$$\text{Irrotationality} \quad \frac{\partial u_i}{\partial x_j} - \frac{\partial u_j}{\partial x_i} = 0 \tag{7·48}$$

or its equivalent,

$$\text{Velocity potential} \quad u_i = \frac{\partial \Phi}{\partial x_i} \tag{7·49}$$

Often it is convenient to use the energy equation,

$$\frac{u_1{}^2 + u_2{}^2 + u_3{}^2}{2} + h = h_0 \tag{7·50}$$

or, for a perfect gas,

$$\frac{u_1{}^2 + u_2{}^2 + u_3{}^2}{2} + \frac{a^2}{\gamma - 1} = \frac{a_0{}^2}{\gamma - 1} = \frac{1}{2}\frac{\gamma + 1}{\gamma - 1}a^{*2} \tag{7·50a}$$

If any of these auxiliary equations are introduced, they simply replace one or more of the basic equations above; they are not independent.

As an example of how the equations may be rearranged, consider *steady flow*. In the momentum equation, the pressure term may be rewritten in the form†

$$\frac{\partial p}{\partial x_i} = \left(\frac{\partial p}{\partial \rho}\right)_S \frac{\partial \rho}{\partial x_i} = a^2 \frac{\partial \rho}{\partial x_i}$$

and the momentum equation may be transformed into a scalar equation by multiplying it by u_i. Thus

$$u_i u_j \frac{\partial u_i}{\partial x_j} = -\frac{u_i}{\rho}\frac{\partial p}{\partial x_i} = -\frac{a^2}{\rho} u_i \frac{\partial \rho}{\partial x_i}$$

Combined with the steady continuity equation, this gives

$$u_i u_j \frac{\partial u_i}{\partial x_j} = a^2 \frac{\partial u_k}{\partial x_k} \tag{▶7·51}$$

Written out, the equation is

$$(u_1{}^2 - a^2)\frac{\partial u_1}{\partial x_1} + (u_2{}^2 - a^2)\frac{\partial u_2}{\partial x_2} + (u_3{}^2 - a^2)\frac{\partial u_3}{\partial x_3} + u_1 u_2\left(\frac{\partial u_1}{\partial x_2} + \frac{\partial u_2}{\partial x_1}\right)$$
$$+ u_2 u_3\left(\frac{\partial u_2}{\partial x_3} + \frac{\partial u_3}{\partial x_2}\right) + u_3 u_1\left(\frac{\partial u_3}{\partial x_1} + \frac{\partial u_1}{\partial x_3}\right) = 0 \tag{7·51a}$$

a^2 is related to the velocity components by the energy equation (7·50).

†Cf. Articles 2·9 and 3·3.

If the velocity components in Eq. 7·51 are replaced by the appropriate derivatives of the potential, the equation becomes

$$\frac{1}{a^2}\frac{\partial\Phi}{\partial x_i}\frac{\partial\Phi}{\partial x_j}\frac{\partial^2\Phi}{\partial x_i\,\partial x_j} = \frac{\partial^2\Phi}{\partial x_j\,\partial x_j} \tag{7·52}$$

Thus the equations may be reduced to a single differential equation for the scalar function Φ. Once the solution is found, the velocities are calculated from Eq. 7·49.

For incompressible flow, $a^2 \to \infty$, Eq. 7·52 reduces to the familiar Laplace equation,

$$\frac{\partial^2\Phi}{\partial x_j\,\partial x_j} \equiv \nabla^2\Phi = 0$$

The above equations are for Cartesian coordinates. For other coordinate systems, the equations of motion may be worked out by applying the conservation laws directly to the system in question, or they may be obtained from the Cartesian equations by formal transformation of coordinates.

For the *natural coordinate system*, the equations may be summarized from Article 7·9.

$$\text{Continuity} \quad \rho u\,\Delta n = \text{const.}$$

$$\text{Momentum} \quad \rho u\frac{\partial u}{\partial s} = -\frac{\partial p}{\partial s} \tag{7·53}$$

$$\text{Irrotationality} \quad \frac{\partial u}{\partial n} - u\frac{\partial\theta}{\partial s} = 0 \tag{7·54}$$

Instead of the momentum equation for the n-component, we have written the irrotationality condition. These three equations, together with the isentropic relation, determine the variables p, ρ, u, θ. The continuity equation here may be written in the more convenient form

$$\frac{1}{\rho}\frac{\partial\rho}{\partial s} + \frac{1}{u}\frac{\partial u}{\partial s} + \frac{1}{\Delta n}\frac{\partial\Delta n}{\partial s} = 0$$

or

$$\frac{1}{\rho}\frac{\partial\rho}{\partial s} + \frac{1}{u}\frac{\partial u}{\partial s} + \frac{\partial\theta}{\partial n} = 0 \tag{7·55}$$

The last term may be obtained from the geometry in Fig. 7·4.

Some further rearrangement is possible. By eliminating the pressure as before, the momentum equation becomes

$$u\frac{\partial u}{\partial s} = -\frac{a^2}{\rho}\frac{\partial\rho}{\partial s}$$

and, combined with the continuity equation (7·55), gives

$$\left(\frac{u^2}{a^2} - 1\right)\frac{1}{u}\frac{\partial u}{\partial s} - \frac{\partial \theta}{\partial n} = 0 \tag{7·56}$$

This corresponds to Eq. 7·51 in the Cartesian coordinates.

7·13 Remarks on the Equations of Motion

To solve the equations of motion, one has to have the *boundary conditions,* which define a particular problem. They will be discussed in the following chapters. Here, we need only mention the general condition, that at a solid boundary the velocity must be tangential to the boundary. Since the flow is frictionless, the velocity there is not required to be zero.

The difficulty in solving the equations arises from the fact that they are nonlinear, as may be illustrated by the following simple examples,

$$x\frac{\partial f}{\partial x} + \frac{\partial f}{\partial y} = 0 \tag{7·57}$$

$$f\frac{\partial f}{\partial x} + \frac{\partial f}{\partial y} = 0 \tag{7·58}$$

Equation 7·57 is *linear,* because the dependent variable f occurs only linearly; Eq. 7·58 is nonlinear because f multiplies a derivative of f. The linear equation is simpler to treat for the following reason: if we find two solutions, f_1 and f_2, which satisfy Eq. 7·57, a third one can be constructed by simple superposition, $f_3 = f_1 + f_2$, or $f_3 = af_1 + bf_2$, since f_3 *also satisfies the differential equation.* This method of superposition does not work for the nonlinear equation; f_3 does not satisfy the equation, because it has "cross-terms," like $f_1\,(\partial f_2/\partial x)$ and $f_2\,(\partial f_1/\partial x)$.

The method of building up solutions by *superposition* is the very basis of the general theory of linear differential equations, for example, the methods of Fourier series, Fourier integral, Laplace transform, etc. As yet, there is no general theory for nonlinear partial differential equations. Unfortunately, the equations of fluid motion belong to the latter class, since they contain the typical nonlinear terms $u_k(\partial u_i/\partial x_k)$.

Lacking a general method, there remain the following alternative possibilities.

(1) Obtain solutions numerically.

(2) Find *transformations* of the variables which will make the equations linear, but still exact. (The hodograph method mentioned in Article 4·20 is an example.)

(3) Find *linear* equations which are approximations to the exact nonlinear ones.

A large part of aerodynamic theory is based on method (3), because, once the equations are linearized, a wide variety of configurations may be treated, so long as they fulfill the assumptions of the linearization. The approximations made in the linearization may be quite consistent with the accuracy which is needed in experimental or engineering practice, or which may be imposed by other limitations. But one must be aware of the approximations and their range of validity.

Method (1), using numerical solutions, does not usually give results in a form from which general trends and rules may be formulated, but it is sometimes necessary if one must obtain greater accuracy than is obtainable from the linearized (or higher-order) solutions. Furthermore it provides standards against which the approximate ones may be compared. Examples of numerical methods based on the theory of characteristics are given in Chapter 12.

Some problems, such as transonic flow, are inherently nonlinear, and to linearize them is to destroy their sense entirely. For these problems it is very difficult to obtain general results and rules. Some of the available methods are discussed in Chapter 11 in connection with transonic flow. A very useful approach to nonlinear problems is through the method of *similarity*, discussed in Chapter 10.

CHAPTER 8

Small-Perturbation Theory

8·1 Introduction

In a great number of aerodynamic problems one is interested in the perturbation of a known fluid motion. The most common and obvious case is that of a uniform, steady flow (Fig. 8·1). Let U denote the uniform velocity, and choose a coordinate system in which U is parallel to the

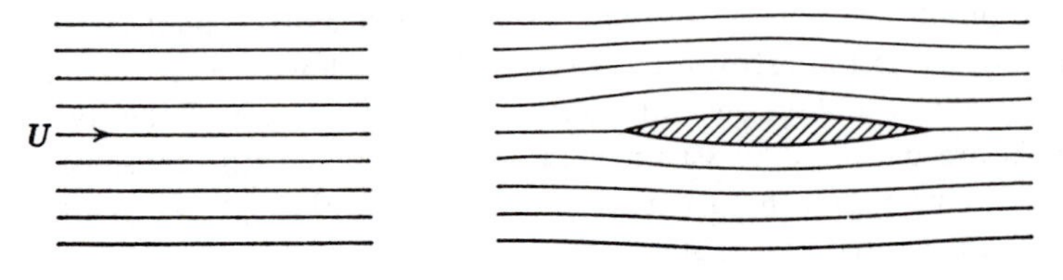

(a) Uniform flow. (b) Perturbed flow.

FIG. 8·1 Perturbation of a uniform flow by a thin body.

x_1-axis. Density, pressure, and temperature are also uniform in this basic motion and will be denoted by ρ_∞, p_∞, T_∞, respectively. The corresponding velocity of sound is a_∞, and the Mach number is $U/a_\infty = M_\infty$. The velocity field of this basic flow is given by

$$u_1 = U$$

$$u_2 = 0$$

$$u_3 = 0$$

Assume now that a solid body, for example, an airfoil, is placed in this uniform stream. The body disturbs the basic motion, and changes its velocity field, which, in the presence of the body, may be written,

$$\begin{aligned} u_1 &= U + u \\ u_2 &= v \\ u_3 &= w \end{aligned} \tag{8·1}$$

u, v, w are called "induced" or "perturbation" velocity components.

The object of this chapter is to study the case for which these perturbation velocities are small compared with the mean velocity U. We shall assume that

$$\frac{u}{U}, \frac{v}{U}, \frac{w}{U} \ll 1 \tag{8·2}$$

and shall simplify the equations of motion by neglecting small terms in the perturbation velocities. In this way we shall be able to arrive at equations which, though not always linear, are still much simpler than the full equations, and which form the basis for by far the largest part of airfoil theory, wing theory, flow past slender bodies, wind tunnel interference problems, transonic flow, etc.

According to convenience, we shall use the notation x_1, x_2, x_3 or x, y, z for the coordinate system. x_1 or x will usually be in the direction of the undisturbed flow; in two-dimensional problems, it is customary to use y as the normal coordinate, whereas, in problems involving wing-like (planar) bodies, it is conventional to let z be the coordinate normal to the wing plane, and y the spanwise coordinate.

8·2 Derivation of the Perturbation Equations

The equations of motion for steady frictionless flow were obtained in Article 7·12, in the form

$$u_i u_j \frac{\partial u_i}{\partial x_j} = a^2 \frac{\partial u_k}{\partial x_k} \tag{8·3}$$

Writing this out in full, and substituting the velocity field defined in Eq. 8·1, we obtain the equation in terms of perturbation velocities

$$\begin{aligned} a^2\left(\frac{\partial u}{\partial x_1} + \frac{\partial v}{\partial x_2} + \frac{\partial w}{\partial x_3}\right) &= (U+u)^2 \frac{\partial u}{\partial x_1} + v^2 \frac{\partial v}{\partial x_2} + w^2 \frac{\partial w}{\partial x_3} \\ &+ (U+u)v\left(\frac{\partial u}{\partial x_2} + \frac{\partial v}{\partial x_1}\right) + vw\left(\frac{\partial v}{\partial x_3} + \frac{\partial w}{\partial x_2}\right) + w(U+u)\left(\frac{\partial w}{\partial x_1} + \frac{\partial u}{\partial x_3}\right) \end{aligned} \tag{8·3a}$$

a^2 may be obtained in terms of the perturbation velocities, from the energy equation (7·50a) for a perfect gas,

$$\frac{(U+u)^2 + v^2 + w^2}{2} + \frac{a^2}{\gamma - 1} = \frac{U^2}{2} + \frac{a_\infty{}^2}{\gamma - 1}$$

or

$$a^2 = a_\infty{}^2 - \frac{\gamma - 1}{2}(2uU + u^2 + v^2 + w^2) \tag{8·4}$$

Substituting this in Eq. 8·3a, dividing by a_∞^2, and rearranging the terms, we obtain

$$
\begin{aligned}
(1 - M_\infty^2)\frac{\partial u}{\partial x_1} &+ \frac{\partial v}{\partial x_2} + \frac{\partial w}{\partial x_3} \\
&= M_\infty^2\left[(\gamma+1)\frac{u}{U} + \frac{\gamma+1}{2}\frac{u^2}{U^2} + \frac{\gamma-1}{2}\frac{v^2+w^2}{U^2}\right]\frac{\partial u}{\partial x_1} \\
&+ M_\infty^2\left[(\gamma-1)\frac{u}{U} + \frac{\gamma+1}{2}\frac{v^2}{U^2} + \frac{\gamma-1}{2}\frac{w^2+u^2}{U^2}\right]\frac{\partial v}{\partial x_2} \\
&+ M_\infty^2\left[(\gamma-1)\frac{u}{U} + \frac{\gamma+1}{2}\frac{w^2}{U^2} + \frac{\gamma-1}{2}\frac{u^2+v^2}{U^2}\right]\frac{\partial w}{\partial x_3} \\
&+ M_\infty^2\left[\frac{v}{U}\left(1+\frac{u}{U}\right)\left(\frac{\partial u}{\partial x_2} + \frac{\partial v}{\partial x_1}\right) + \frac{w}{U}\left(1+\frac{u}{U}\right)\left(\frac{\partial u}{\partial x_3} + \frac{\partial w}{\partial x_1}\right)\right. \\
&\qquad \left. + \frac{vw}{U^2}\left(\frac{\partial w}{\partial x_2} + \frac{\partial v}{\partial x_3}\right)\right]
\end{aligned}
\tag{8·5}
$$

This is the full equation in terms of the perturbation velocities. It is still exact. On the left-hand side, the terms are linear, but on the right they are nonlinear, because they involve products of the perturbation velocities with their derivatives.

If the perturbation velocities are small, as expressed in Eq. 8·2, it becomes possible to neglect many of the terms, and thus simplify the equation. For instance, the coefficients of the derivatives on the right-hand side all contain the perturbation velocities. As a first step, we may neglect the terms containing squares of the perturbation velocities, in comparison to those containing first powers. Thus we obtain the simpler equation

$$
\begin{aligned}
(1 - M_\infty^2)\frac{\partial u}{\partial x_1} &+ \frac{\partial v}{\partial x_2} + \frac{\partial w}{\partial x_3} \\
&= M_\infty^2(\gamma+1)\frac{u}{U}\frac{\partial u}{\partial x_1} + M_\infty^2(\gamma-1)\frac{u}{U}\left(\frac{\partial v}{\partial x_2} + \frac{\partial w}{\partial x_3}\right) \\
&+ M_\infty^2\frac{v}{U}\left(\frac{\partial u}{\partial x_2} + \frac{\partial v}{\partial x_1}\right) + M_\infty^2\frac{w}{U}\left(\frac{\partial u}{\partial x_3} + \frac{\partial w}{\partial x_1}\right)
\end{aligned}
\tag{8·6}
$$

Even further simplification is possible, for we may neglect all the terms on the right-hand side, in comparison to those on the left, which contain no perturbation velocities. This gives the *linear* equation

$$(1 - M_\infty^2)\frac{\partial u}{\partial x_1} + \frac{\partial v}{\partial x_2} + \frac{\partial w}{\partial x_3} = 0 \tag{8·7}$$

One must, however, compare the terms a little more critically. For instance, in transonic flow, where $M_\infty \to 1$, the coefficient of $\partial u/\partial x_1$, on the left-hand side, becomes very small, and it is then not possible to neglect the first term on the right-hand side of Eq. 8·6. However, the condition $M_\infty \to 1$ does not affect the terms $\partial v/\partial x_2$ and $\partial w/\partial x_3$, and so the other terms on the right-hand side may still be neglected in comparison. Thus we arrive at the small-perturbation equation which is *valid for subsonic, transonic, and supersonic flow,*

$$(1 - M_\infty^2)\frac{\partial u}{\partial x_1} + \frac{\partial v}{\partial x_2} + \frac{\partial w}{\partial x_3} = M_\infty^2(\gamma + 1)\frac{u}{U}\frac{\partial u}{\partial x_1} \qquad \blacktriangleright(8{\cdot}8)$$

Another case in which certain terms on the right-hand side must be retained arises if M_∞ is very large, for, even though the perturbation velocities may be small, their product with M_∞^2 may not be negligible. This is the case of *hypersonic flow,* to which we shall return in Article 10·7.

Equation 8·3, from which the others have been derived, is for frictionless flow. We shall assume, in addition, that it is irrotational, so that a perturbation velocity potential ϕ exists, giving

$$u = \frac{\partial \phi}{\partial x_1}, \quad v = \frac{\partial \phi}{\partial x_2}, \quad w = \frac{\partial \phi}{\partial x_3}$$

(To the order of accuracy of Eqs. 8·7 and 8·8, this is not really an additional assumption, for variations of stagnation pressure from one streamline to another may be neglected with no greater error than in the other approximations. For instance, to the order of these equations, entropy changes across shock waves are negligible.)

In terms of the velocity potential, Eqs. 8·8 and 8·7 become

$$(1 - M_\infty^2)\frac{\partial^2 \phi}{\partial x_1^2} + \frac{\partial^2 \phi}{\partial x_2^2} + \frac{\partial^2 \phi}{\partial x_3^2} = \frac{M_\infty^2(\gamma + 1)}{U}\frac{\partial \phi}{\partial x_1}\frac{\partial^2 \phi}{\partial x_1^2} \qquad \blacktriangleright(8{\cdot}9a)$$

and

$$(1 - M_\infty^2)\frac{\partial^2 \phi}{\partial x_1^2} + \frac{\partial^2 \phi}{\partial x_2^2} + \frac{\partial^2 \phi}{\partial x_3^2} = 0 \qquad \blacktriangleright(8{\cdot}9b)$$

For transonic flow, Eq. 8·9*a* must be used; outside the transonic range, in subsonic or supersonic flow, Eq. 8·9*b* is applicable. Since the latter is linear, it is much simpler to treat, and may be solved in great generality. In principle, any problem can be solved, for example, a wing of any shape. In the transonic case, on the other hand, there are only a few special solutions, which cannot be generalized because the equation is nonlinear.

In this and the following chapter, we shall be mainly concerned with the linear equation (8·9*b*). We shall return to Eq. 8·9*a* in Chapter 10, where we shall see that it forms the basis of the *similarity rules* of subsonic, transonic, and supersonic motion.

8·3 Pressure Coefficient

The pressure coefficient is defined by

$$C_p = \frac{p - p_\infty}{\frac{1}{2}\rho_\infty U^2} = \frac{2}{\gamma M_\infty^2}\frac{p - p_\infty}{p_\infty}$$

Using the form given in Eq. 2·40*b*, it may be expressed in terms of the perturbation velocities,

$$C_p = \frac{2}{\gamma M_\infty^2}\left\{\left[1 + \frac{\gamma - 1}{2} M_\infty^2\left(1 - \frac{(U+u)^2 + v^2 + w^2}{U^2}\right)\right]^{\gamma/(\gamma-1)} - 1\right\}$$

$$= \frac{2}{\gamma M_\infty^2}\left\{\left[1 - \frac{\gamma - 1}{2} M_\infty^2\left(\frac{2u}{U} + \frac{u^2 + v^2 + w^2}{U^2}\right)\right]^{\gamma/(\gamma-1)} - 1\right\} \tag{8·10}$$

Using the binomial expansion on the expression inside the square brackets, we obtain the expression

$$C_p = -\left[\frac{2u}{U} + (1 - M_\infty^2)\frac{u^2}{U^2} + \frac{v^2 + w^2}{U^2}\right] \tag{8·11}$$

in which cubes and higher powers of the perturbation velocities are neglected. For two-dimensional and planar flows, it is consistent with the first-order perturbation equations of the previous article to retain only the first term in Eq. 8·11, and use the expression

$$C_p = -\frac{2u}{U} \tag*{▶(8·12)}$$

For flow over axially symmetric or elongated bodies, it is necessary to retain the third term, the consistent approximation then being

$$C_p = -\frac{2u}{U} - \frac{v^2 + w^2}{U^2} \tag*{▶(8·13)}$$

For large Mach numbers, in the hypersonic range, a suitable approximate expression is given in Eq. 10·40.

8·4 Boundary Conditions

At the surface of a body the direction of the flow must be tangential to the solid surface. Expressed differently, the velocity vector has to be at right angles to the normal on the solid surface. Let the surface of the body be described by an equation of the form

$$f(x_1, x_2, x_3) = 0 \tag{8·14}$$

and write the velocity vector $\mathbf{u}$. The boundary condition can then be written

$$\mathbf{u} \cdot \operatorname{grad} f = 0 \tag{8·15}$$

since grad f is always normal to $f = 0$ and the vanishing of the scalar product expresses the condition $\mathbf{u} \perp \operatorname{grad} f$. In the Cartesian tensor notation, Eq. 8·15 is written

$$u_i \frac{\partial f}{\partial x_i} = 0$$

or, introducing the perturbation velocities,

$$(U + u) \frac{\partial f}{\partial x_1} + v \frac{\partial f}{\partial x_2} + w \frac{\partial f}{\partial x_3} = 0 \tag{8·16}$$

In the first term, u may be neglected compared to U and we have

$$U \frac{\partial f}{\partial x_1} + v \frac{\partial f}{\partial x_2} + w \frac{\partial f}{\partial x_3} = 0 \tag{8·17}$$

Equation 8·17 has to be satisfied on the surface of the body. To make the next step clearer, consider the two-dimensional case, for which $w = 0$ and $\partial f/\partial x_3 = 0$. Then Eq. 8·17 becomes

$$\frac{v}{U} = -\frac{\partial f/\partial x_1}{\partial f/\partial x_2} = \frac{dx_2}{dx_1} \tag{8·18}$$

where dx_2/dx_1 is the slope of the body and v/U is approximately the slope of the streamline. Hence, Eq. 8·18 tells us that we have to find a velocity field such that, if we insert into $v(x_1, x_2)$ the coordinates of a point on the body, v should be equal to U times the slope of the body at this point.

Now the body has to be thin in order to satisfy our assumptions of small perturbations. Therefore the ordinate x_2 on the surface of the body differs little from zero (this statement will be made more precise later), and we may try to simplify Eq. 8·18 further by developing $v(x_1, x_2)$ in powers of x_2. That is, we put

$$v(x_1, x_2) = v(x_1, 0) + \left(\frac{\partial v}{\partial x_2}\right)_{x_2=0} x_2 + \cdots \tag{8·19}$$

Within the frame of small-perturbation theory it is possible to neglect all terms after the first in Eq. 8·19 and apply the boundary condition in the form

$$v(x_1, 0) = U \left(\frac{dx_2}{dx_1}\right)_{\text{body}} \tag{8·20}$$

The step from Eq. 8·18 to Eq. 8·20 is always possible for two-dimensional flow. The corresponding step in three dimensions is possible for so-called "planar" systems, that is, configurations that are essentially flat, like three-dimensional wings, etc. For the *planar* case, $\partial f/\partial x_3 \doteq 0$, and the boundary condition becomes

$$v(x_1, 0, x_3) = U\left(\frac{\partial x_2}{\partial x_1}\right)_{\text{body}} \tag{8·20a}$$

One says: "The boundary condition is applied in the plane of the wing." On the other hand, for axially symmetric slender bodies the step leading from Eq. 8·18 to Eq. 8·20*a* breaks down, because, in the neighborhood of the axis, the perturbation velocity normal to the mean flow cannot be developed in a power series. This case will be considered in Chapter 9.†

Finally, it is necessary to specify a boundary condition at infinity, for instance, that the perturbation velocities be zero, or at least finite. The condition depends on the nature of the particular problem.

8·5 Two-Dimensional Flow Past a Wave-Shaped Wall

The following simple example, due to Ackeret, serves to apply and clarify the concepts of the previous paragraphs. In addition, the example is of considerable value for general considerations later.

We have flow past a boundary of sinusoidal shape, the so-called "wavy wall" (Fig. 8·2). Equation 8·14, specifying the boundary, becomes

$$x_2 - \epsilon \sin \alpha x_1 = 0 \tag{8·21}$$

where ϵ denotes the "amplitude" of the waves of the wall and $l = 2\pi/\alpha$, the wavelength.

For subsonic or supersonic flow, we may use the linear equation (8·9*b*). For plane flow, this is

$$(1 - M_\infty^2)\frac{\partial^2\phi}{\partial x_1^2} + \frac{\partial^2\phi}{\partial x_2^2} = 0 \tag{8·22}$$

subject to the boundary conditions

$$\frac{\partial\phi}{\partial x_1},\ \frac{\partial\phi}{\partial x_2}\quad \text{finite at infinity}$$

and

$$v(x_1, 0) \equiv \left(\frac{\partial\phi}{\partial x_2}\right)_{x_2=0} = U\left(\frac{dx_2}{dx_1}\right)_{\text{wall}} = U\epsilon\alpha\cos\alpha x_1 \tag{8·23}$$

†Experience has shown that the consideration of boundary conditions and the significance of Eqs. 8·20 and 8·20*a* is difficult for the beginner to grasp. If difficulties are encountered here, it is suggested that the reader first study the examples in the following articles and then return to this article.

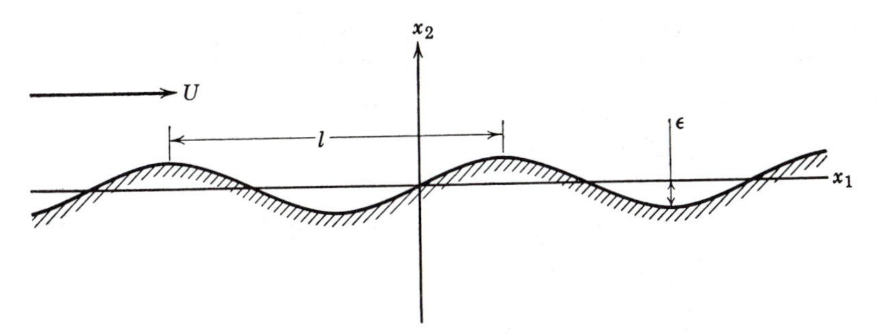

FIG. 8·2 Wave-shaped wall.

We first work out the solution for *subsonic motion*, for which $1 - M_\infty^2 \equiv m^2 > 0$. Equation 8·22 is then of the elliptic type,

$$\frac{\partial^2\phi}{\partial x_1^2} + \frac{1}{m^2}\frac{\partial^2\phi}{\partial x_2^2} = 0 \tag{8·22a}$$

We try separation of variables by putting

$$\phi(x_1, x_2) = F(x_1)G(x_2)$$

Equation 8·22*a* becomes

$$m^2F''G + FG'' = 0$$

or

$$\frac{1}{F}F'' + \frac{1}{m^2G}G'' = 0$$

Here the first term is a function of x_1 only, the second of x_2 only. Since the equation must hold for arbitrary x_1 and x_2, the sum can be zero only if

$$\frac{1}{F}F'' = \text{const.} = -k^2 \tag{8·24a}$$

$$\frac{1}{m^2G}G'' = +k^2 \tag{8·24b}$$

The sign of k^2 is chosen for convenience in later steps, and the *square* of a real constant k is used to make the sign definite. This choice stems from the consideration that on the boundary we want a *harmonic* variation in x_1; hence the sign is chosen so that Eq. 8·24*a* has harmonic solutions, that is,

$$F = A_1 \sin kx_1 + A_2 \cos kx_1 \tag{8·25a}$$

Then the solution of Eq. 8·24*b* is

$$G = B_1e^{-mkx_2} + B_2e^{mkx_2} \tag{8·25b}$$

The boundary condition for $x_2 \to \infty$ requires that $B_2 = 0$ to assure that the velocity components remain finite. The boundary condition on the wall (Eq. 8·23) is

$$\left(\frac{\partial \phi}{\partial x_2}\right)_{x_2=0} = F(x_1)\left(\frac{dG}{dx_2}\right)_{x_2=0} = U\epsilon\alpha \cos \alpha x_1 \dagger$$

The condition is satisfied if $A_1 = 0$, $k = \alpha$, and $-A_2B_1mk = U\epsilon\alpha$. Hence the solution of the problem is

$$\phi(x_1, x_2) = -\frac{U\epsilon}{\sqrt{1 - M_\infty^2}} e^{-x_2\alpha\sqrt{1-M_\infty^2}} \cos \alpha x_1 \tag{8·26a}$$

$$u = \frac{U\epsilon\alpha}{\sqrt{1 - M_\infty^2}} e^{-x_2\alpha\sqrt{1-M_\infty^2}} \sin \alpha x_1 \tag{8·26b}$$

$$v = U\epsilon\alpha e^{-x_2\alpha\sqrt{1-M_\infty^2}} \cos \alpha x_1 \tag{8·26c}$$

$$C_p = -2\frac{u}{U} = -\frac{2\epsilon\alpha}{\sqrt{1 - M_\infty^2}} e^{-x_2\alpha\sqrt{1-M_\infty^2}} \sin \alpha x_1 \tag{8·26d}$$

Discussion of the Subsonic Solution. From the exponential term it is evident that the largest perturbation occurs at the boundary, as should be expected. The pressure coefficient on the boundary is

$$(C_p)_{\text{boundary}} = (C_p)_{x_2=0} = -\frac{2\epsilon\alpha}{\sqrt{1 - M_\infty^2}} \sin \alpha x_1 \tag{8·27}$$

From Eq. 8·27 it follows immediately that: (1) there is no drag force, since the pressure is "in phase" with the wall and hence symmetrical about the crest of the waves; (2) the pressure coefficient increases with Mach number proportionally to $1/\sqrt{1 - M_\infty^2}$. We encounter here for the first time the well-known Prandtl-Glauert factor. Furthermore, it is evident from Eq. 8·26 that the attenuation of the perturbation away from the wall becomes weaker as the Mach number is increased, because of the factor $\sqrt{1 - M_\infty^2}$ in the exponent.

The two latter statements can be combined into a single *similarity law*, the nature of which will be discussed in full generality in Chapter 10. It is evidently possible to write Eq. 8·26*d* in the form

$$\frac{C_p\sqrt{1 - M_\infty^2}}{\epsilon\alpha} = fn(x_1\alpha; x_2\alpha\sqrt{1 - M_\infty^2}) \tag{8·28}$$

In this form the relation is between *three* variables instead of *six*; these are a

†The usefulness of the simplified boundary condition appears at this step.

modified pressure coefficient and a modified coordinate system. The factors that make this reduction possible incorporate the effects of Mach number, wave amplitude, and wavelength in such a way that the one relation is valid for all combinations of these three latter variables.

We may now return to the important question of the applicability of the approximations.

(1) It was assumed to begin with that $\frac{u}{U}, \frac{v}{U} \ll 1$. It is evident from the solution that this means

$$\frac{\epsilon\alpha}{\sqrt{1 - M_\infty^2}} \ll 1 \tag{8·29}$$

(2) In using the linearized equation (8·7) instead of Eq. (8·8), it was assumed that

$$(1 - M_\infty^2) \gg M_\infty^2(\gamma + 1)\frac{u}{U}$$

This is equivalent to

$$(1 - M_\infty^2) \gg \frac{M_\infty^2(\gamma + 1)\epsilon\alpha}{\sqrt{1 - M_\infty^2}}$$

or

$$\frac{M_\infty^2(\gamma + 1)\epsilon\alpha}{(1 - M_\infty^2)^{3/2}} \ll 1 \tag{8·30}$$

It is left to the reader to show that

$$\frac{M_\infty^2(\gamma + 1)\epsilon\alpha}{(1 - M_\infty^2)^{3/2}} = 1$$

is the condition for the occurrence of local sonic velocity. Equation 8·30 states that the linearized equation (8·9) is adequate as long as M_∞ is low enough so that the local velocity of sound is not reached anywhere in the field. It should be noted that this transonic parameter contains the gas constant, γ, whereas it does not appear explicitly in the linearized range. Another characteristic appearing in the transonic range is the occurrence of the 3/2 exponent on the parameter $(1 - M_\infty^2)$.

(3) Finally, we are now able to show explicitly the meaning of the simplified boundary condition (Eq. 8·20). From Eq. 8·26*c* we have

$$\frac{v}{U} = \epsilon\alpha e^{-\alpha x_2\sqrt{1-M_\infty^2}} \cos \alpha x_1$$

If we introduce for x_2 the coordinates of the boundary, we have

$$\left(\frac{v}{U}\right)_{\text{boundary}} = \epsilon\alpha e^{-\epsilon\alpha\sqrt{1-M_\infty^2}\sin \alpha x_1} \cos \alpha x_1$$

Developing the exponential, we have

$$\left(\frac{v}{U}\right)_{\text{boundary}} = \epsilon\alpha \cos \alpha x_1[1 - \epsilon\alpha\sqrt{1 - M_\infty^2} \sin \alpha x_1 + \cdots]$$

The second term is always smaller than or at most equal to $\epsilon\alpha\sqrt{1 - M_\infty^2}$; it should be negligible, compared to the first, for the small-perturbation theory to be valid. Note that *this* approximation becomes even better as the Mach number is increased. The amplitude of the permissible perturbation, for linearized theory to be accurate, is given by

$$\epsilon\alpha\sqrt{1 - M_\infty^2} \ll 1$$

or, in terms of the maximum inclination, θ, of the wall,

$$\theta\sqrt{1 - M_\infty^2} \ll 1$$

This, together with Eq. 8·29, determines the permissible limits on $\epsilon\alpha$ (or θ) and $\sqrt{1 - M_\infty^2}$.

8·6 Wavy Wall in Supersonic Flow

For supersonic flow, $M_\infty^2 - 1 \equiv \lambda^2 > 0$, Eq. 8·22 becomes of hyperbolic type,

$$\frac{\partial^2\phi}{\partial x_1^2} - \frac{1}{\lambda^2}\frac{\partial^2\phi}{\partial x_2^2} = 0 \tag{8·22b}$$

In principle the equation could be solved by the same method of separating variables that was used in the subsonic case (Eq. 8·22*a*). However, Eq. 8·22*b* allows a much simpler approach. It is the simple wave equation discussed in Article 3·4, where it was verified that the general solution is the sum of two arbitrary functions, $f(x_1 - \lambda x_2)$ and $g(x_1 + \lambda x_2)$,

$$\phi(x_1, x_2) = f(x_1 - \lambda x_2) + g(x_1 + \lambda x_2) \tag{8·31}$$

The boundary conditions are the same as in the subsonic case.

For reasons that will become clear presently, only the function f is needed, and g is set equal to zero. (This choice has to do with the direction of flow, or the distinction between upstream and downstream regions of flow.) The boundary conditions on the wall are then written

$$\left(\frac{\partial\phi}{\partial x_2}\right)_{x_2=0} = -\lambda[f'(x_1 - \lambda x_2)]_{x_2=0} = -\lambda f'(x_1) = U\epsilon\alpha \cos \alpha x_1$$

where f' denotes the derivative of f with respect to its argument. Consequently

$$f(x_1) = -\frac{U\epsilon}{\lambda} \sin \alpha x_1$$

and hence

$$\phi(x_1, x_2) = f(x_1 - \lambda x_2) = -\frac{U\epsilon}{\lambda}\sin\alpha(x_1 - \lambda x_2)$$

that is,

$$\phi(x_1, x_2) = -\frac{U\epsilon}{\sqrt{M_\infty^2 - 1}}\sin\alpha[x_1 - x_2\sqrt{M_\infty^2 - 1}] \tag{8·32a}$$

and

$$u = -\frac{U\epsilon\alpha}{\sqrt{M_\infty^2 - 1}}\cos\alpha[x_1 - x_2\sqrt{M_\infty^2 - 1}] \tag{8·32b}$$

$$v = U\epsilon\alpha\cos\alpha[x_1 - x_2\sqrt{M_\infty^2 - 1}] \tag{8·32c}$$

$$C_p = \frac{2\epsilon\alpha}{\sqrt{M_\infty^2 - 1}}\cos\alpha[x_1 - x_2\sqrt{M_\infty^2 - 1}] \tag{8·32d}$$

Discussion of the Supersonic Solution. (1) The solution does not involve an exponential attenuation factor like the subsonic solution, and hence the perturbation, for example, in pressure, does not decrease with x_2. Instead, the same value of the perturbation exists all along the straight lines

$$x_1 - x_2\sqrt{M_\infty^2 - 1} = \text{const.} \tag{8·33}$$

These lines are inclined at the *Mach angle* with respect to the undisturbed flow. They are the Mach lines, or *characteristics*. The existence of these characteristics is independent of the specific boundary conditions, being already contained in the form of the solution (Eq. 8·31). It is clear from Eq. 8·31 that f = const. along lines $x_1 - \lambda x_2$ = const. and g = const. along lines $x_1 + \lambda x_2$ = const. The former characteristics are inclined downstream; they "originate" at the wall. The latter characteristics are inclined upstream; they "originate" at infinity, and so carry no perturbation. It is for this reason that $g = 0$ when the fluid above the wall is unlimited.

(2) C_p on the wall has the value

$$C_p = \frac{2\epsilon\alpha}{\sqrt{M_\infty^2 - 1}}\cos\alpha x_1$$

Comparing with Eq. 8·27, it may be seen that the maxima and minima of the pressure are now shifted by the phase $\pi/2$ from the maxima and minima of the wall coordinate x_2. Hence the pressure distribution is now antisymmetrical around the crests and troughs of the wall, and a drag force exists (Fig. 8·3). The magnitude of the drag coefficient per wavelength is

$$C_D = \int_0^l \frac{C_p(dx_2/dx_1)}{l}\,dx_1 \tag{8·34}$$

obtained by replacing the sine of the slope of the wall by the tangent, dx_2/dx_1, valid within the small-perturbation approximation. But C_p can be written

$$C_p = \frac{2}{\sqrt{M_\infty^2 - 1}} \frac{dx_2}{dx_1} \tag{8·35}$$

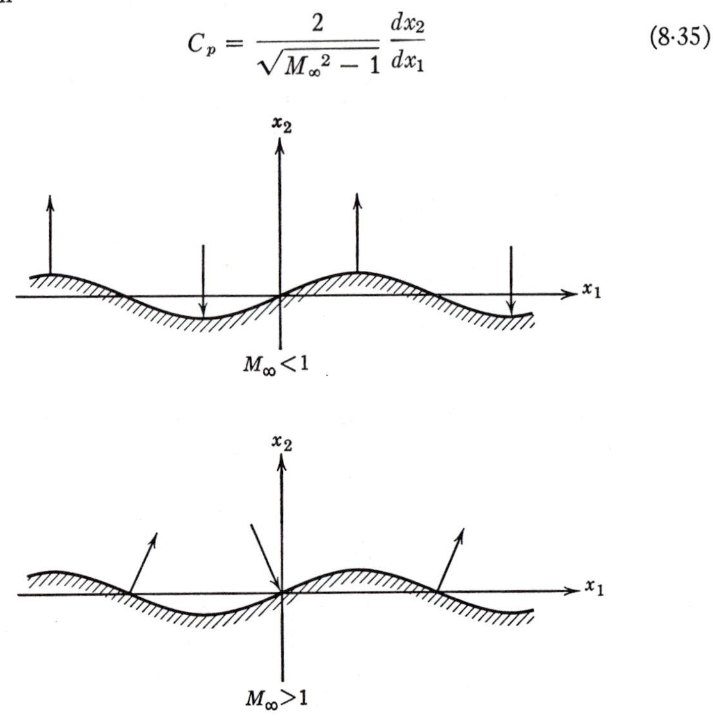

FIG. 8·3 Forces on the wave-shaped wall in subsonic and supersonic flow.

and hence

$$C_D = \frac{2}{\sqrt{M_\infty^2 - 1}} \overline{\left(\frac{dx_2}{dx_1}\right)^2} \tag{8·36}$$

where the bar over $\overline{(dx_2/dx_1)^2}$ denotes the mean value defined by

$$\overline{\left(\frac{dx_2}{dx_1}\right)^2} = \frac{1}{l} \int_0^l \left(\frac{dx_2}{dx_1}\right)^2 dx_1$$

The expressions for C_p and C_D written in the form Eqs. 8·35 and 8·36 are independent of the special boundary conditions, and apply in general to all steady, two-dimensional, supersonic flow patterns within the frame of small-perturbation theory.

(3) The discussion of the range of validity of the approximations follows exactly the same line as in the subsonic case.

The wave-shaped wall examples of this and the preceding article bring out some of the main features of linearized subsonic and supersonic flow, and the differences between them. These features are typical for any boundary shape, since other solutions can be built up from these by superposition, for instance, by the technique of Fourier analysis. Usually, however, one attacks specific problems more directly, as in the example in the next article.

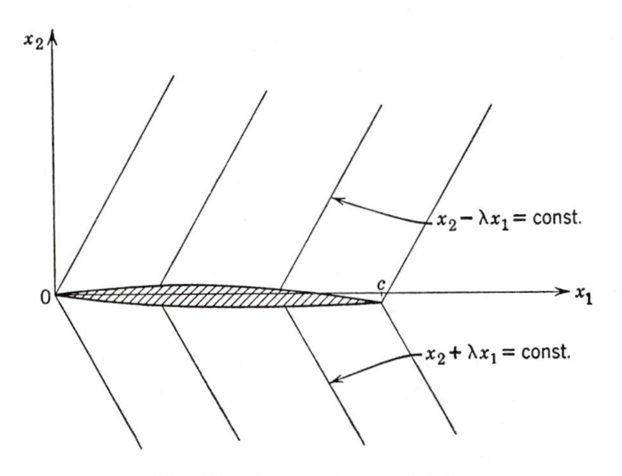

FIG. 8·4 Supersonic thin airfoil.

8·7 Supersonic Thin Airfoil Theory

The general solution of the wave equation

$$\phi(x_1, x_2) = f(x_1 - \lambda x_2) + g(x_1 + \lambda x_2)$$

may also be applied to the problem of a two-dimensional supersonic airfoil, illustrated in Fig. 8·4. Since disturbances are propagated only along downstream-running Mach lines, we need only the function f for the upper surface and g for the lower surface. Thus

$$\phi(x_1, x_2) = f(x_1 - \lambda x_2) \qquad x_2 > 0$$

$$\phi(x_1, x_2) = g(x_1 + \lambda x_2) \qquad x_2 < 0$$

On the upper surface, the boundary condition is

$$U\left(\frac{dx_2}{dx_1}\right)_U = v_U(x_1, 0) = \left(\frac{\partial\phi}{\partial x_2}\right)_{x_2=0+} = -\lambda f'(x_1)$$

and thus

$$f'(x_1) = -\frac{U}{\lambda}\left(\frac{dx_2}{dx_1}\right)_U$$

Similarly,

$$g'(x_1) = \frac{U}{\lambda}\left(\frac{dx_2}{dx_1}\right)_L$$

The pressure coefficient on the surface is then obtained from

$$C_p = -2\left(\frac{u}{U}\right)_{\text{body}} = -\frac{2}{U}\left(\frac{\partial\phi}{\partial x_1}\right)_{x_2=0} = \begin{cases} -\dfrac{2}{U}f'(x_1) & \text{Upper surface} \\[2ex] -\dfrac{2}{U}g'(x_1) & \text{Lower surface} \end{cases}$$

Thus

$$C_{pU} = \frac{2}{\sqrt{M_\infty^2 - 1}}\left(\frac{dx_2}{dx_1}\right)_U$$
$$C_{pL} = \frac{2}{\sqrt{M_\infty^2 - 1}}\left(-\frac{dx_2}{dx_1}\right)_L \tag{8·37}$$

This is the same result obtained in Article 4·17, where the pressure on thin airfoils was obtained by an approximation to the shock-expansion method. In fact, for purposes of calculating the velocity and pressure perturbations on the surface, the linearized theory is equivalent to the "weak wave" approximations of Chapter 4.

The remaining details for calculating lift and drag of thin airfoils are the same as those already given in Article 4·17, and again in the preceding article.

It should be noted that in these examples, as well as the supersonic wavy wall, the pressure coefficient on the surface is related to the local surface slope $\theta(x)$ by the relation

$$C_p = \frac{2\theta}{\sqrt{|M_\infty^2 - 1|}} \tag{8·38}$$

This is the basic relation for linearized, two-dimensional, supersonic flow. For subsonic flow the constant is not 2, but depends on the particular geometry, and on the relation of the point in question to other parts of the boundary.

8·8 Planar Flows

Three-dimensional bodies that are wing-like are called planar systems. The condition is that one dimension (in the x_3-direction, normal to the "plane of the wing") should be small compared to the other dimensions (x_2, spanwise, and x_1, streamwise), that all surface slopes be small relative to the free-stream direction, and that the system be "near" the plane $x_3 = 0$.

For a planar system, the three-dimensional equation (8·7 or 8·8) is required, but the boundary conditions and pressure coefficient are similar to those for two-dimensional flow. Thus, if $x_3 = f(x_1, x_2)$ defines the body surface, then, for supersonic flow,

$$(M_\infty^2 - 1)\frac{\partial^2\phi}{\partial x_1^2} - \frac{\partial^2\phi}{\partial x_2^2} - \frac{\partial^2\phi}{\partial x_3^2} = 0$$

$$\left(\frac{\partial\phi}{\partial x_3}\right)_{x_3=0} = U\frac{\partial f(x_1, x_2)}{\partial x_1} \tag{8·39}$$

$$C_p = -\frac{2}{U}\left(\frac{\partial\phi}{\partial x_1}\right)$$

These are the basic equations of supersonic wing theory for steady flow. A more general formulation of the problem is outlined in Article 9·19.

A systematic treatment of the theory and applications would be too extensive to be included in this book. A very brief discussion of one example has been given in Article 4·18.

CHAPTER 9

Bodies of Revolution Slender Body Theory

9·1 Introduction

The class of small-perturbation flows is divided into the following types of problem (Fig. 9·1).

(*a*) *Two-Dimensional.* This flow, in which conditions are identical in sections parallel to the x-z plane, was studied in the last chapter. In

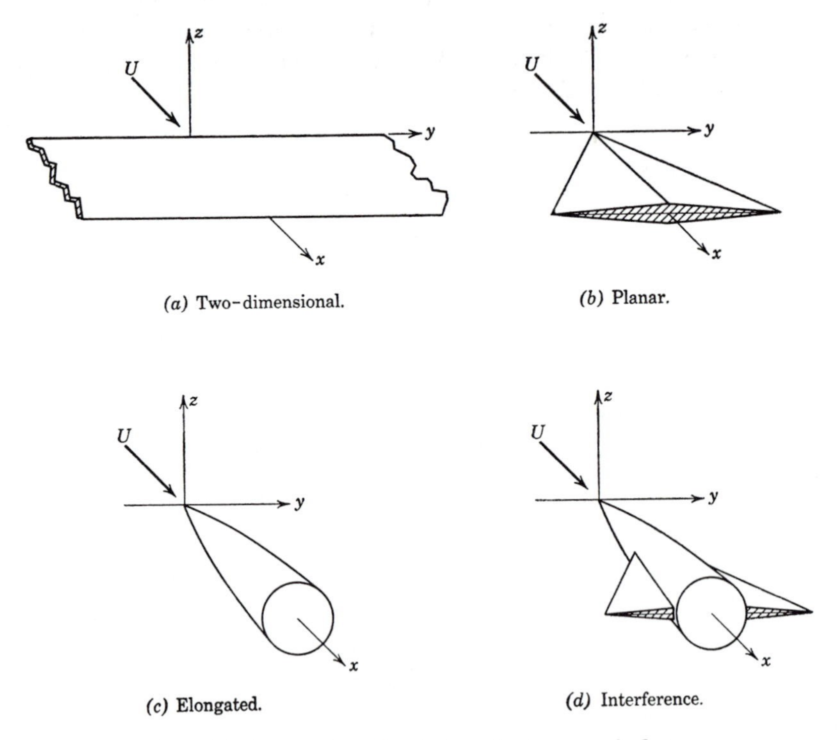

FIG. 9·1 Types of problem in linearized supersonic flow.

many supersonic problems these results are directly applicable to certain *portions* of wings, where two-dimensional conditions exist, as shown in Article 4·18.

(*b*) *Planar System.* The surface may have any plan form, but its slope (measured relative to the x-y plane) is everywhere small enough to ensure that the velocity perturbation is small. Furthermore, all parts of the surface are near enough to the "plane of the wing" ($z = 0$), so that the boundary conditions may be applied on this plane instead of on the actual surface. Two-dimensional flow is a special case of planar flow.

(*c*) *Elongated Body.* The body is elongated in the x-direction. The surface slope (relative to the free stream) is everywhere small. A special case is the *slender* body, which is elongated so much that (a special form of) the boundary conditions may be applied *on the axis.*

(*d*) *Interference Problem.* Combinations of simple systems (planar or elongated or both) form more complex ones. It is not possible simply to superimpose the results of the separate systems, for the interference of their flow fields requires that adjustments be made to satisfy the boundary conditions.

In this chapter, we shall discuss the problem of elongated bodies, mainly bodies of revolution. Although the governing equation is linear, as for two-dimensional flow, the peculiarities of the geometry here introduce some new problems and lead to results that could not be anticipated from the two-dimensional study.

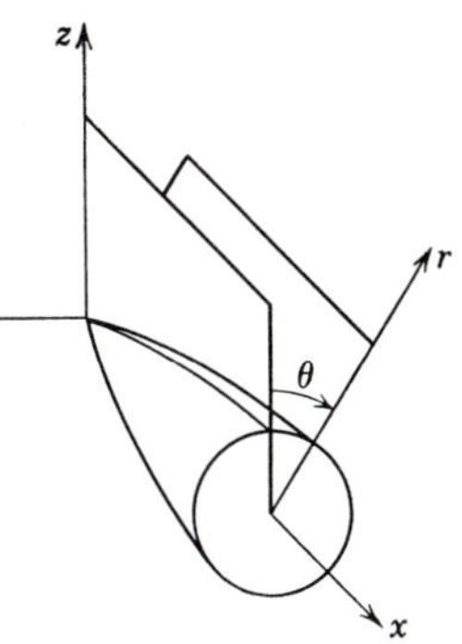

Fig. 9·2 Cylindrical coordinates.

9·2 Cylindrical Coordinates

For bodies of revolution, and elongated bodies in general, it is convenient to adopt cylindrical coordinates (x, r, θ) aligned with the body axis (Fig. 9·2). The angular coordinate θ locates the meridian plane, x-r, relative to some convenient reference, such as the x-z plane. The velocity components corresponding to x, r, θ are u_1, v, and w, respectively. They may be derived from the velocity potential function,

$$u_1 = U + u = \frac{\partial \Phi}{\partial x}, \quad v = \frac{\partial \Phi}{\partial r}, \quad w = \frac{1}{r}\frac{\partial \Phi}{\partial \theta}$$

The potential equation in cylindrical coordinates could be obtained from the one in Cartesian coordinates (Eq. 8·9a) by straightforward transformation of coordinates. However, it will be instructive to derive it more primitively, by starting with the continuity equation for the cylindrical coordinates.

The elementary volume elements in Cartesian and cylindrical coordinates are shown in Fig. 9·3. The continuity equation for *steady flow* states that the excess of mass outflow over inflow through the volume is zero.

On Fig. 9·3, the mass flow through only one face of each element has been marked. Similar expressions for the other faces, added up with due regard to sign, give, for the Cartesian element,

$$\frac{\partial}{\partial x_1}(\rho u_1\,\Delta x_2\,\Delta x_3)\Delta x_1 + \frac{\partial}{\partial x_2}(\rho u_2\,\Delta x_3\,\Delta x_1)\Delta x_2 + \frac{\partial}{\partial x_3}(\rho u_3\,\Delta x_1\,\Delta x_2)\Delta x_3 = 0$$

and, for the cylindrical element,

$$\frac{\partial}{\partial x}(\rho u_1 \cdot \Delta r \cdot r\,\Delta\theta)\Delta x + \frac{\partial}{\partial r}(\rho v \cdot r\,\Delta\theta \cdot \Delta x)\Delta r + \frac{\partial}{\partial \theta}(\rho w \cdot \Delta x\,\Delta r)\Delta\theta = 0$$

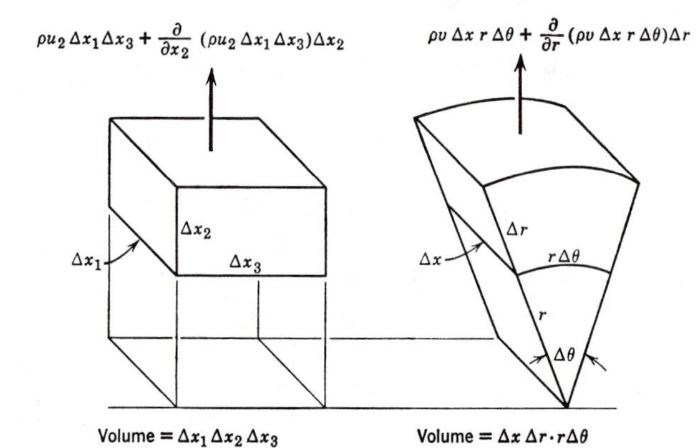

FIG. 9·3 Volume elements in Cartesian and cylindrical coordinates.

When written for unit volume, these give the continuity equation in the forms

$$\frac{\partial}{\partial x_1}(\rho u_1) + \frac{\partial}{\partial x_2}(\rho u_2) + \frac{\partial}{\partial x_3}(\rho u_3) = 0 \tag{9·1a}$$

$$\frac{\partial}{\partial x}(\rho u_1) + \frac{1}{r}\frac{\partial}{\partial r}(\rho v r) + \frac{1}{r}\frac{\partial}{\partial \theta}(\rho w) = 0 \tag{9·1b}$$

These are to be combined with the Bernoulli equation, which is similar in the two coordinate systems:

$$\rho(u_1\,du_1 + u_2\,du_2 + u_3\,du_3) = -dp = -a^2\,d\rho \tag{9·2a}$$

and

$$\rho(u_1\,du_1 + v\,dv + w\,dw) = -dp = -a^2\,d\rho \tag{9·2b}$$

The basic step is to use the Bernoulli equation to rewrite the derivatives of ρ which appear in each of the terms of the continuity equation.

A typical term is

$$\frac{\partial}{\partial x_1}(\rho u_1) = \rho \frac{\partial u_1}{\partial x_1} + u_1 \frac{\partial \rho}{\partial x_1}$$

$$= \rho \frac{\partial u_1}{\partial x_1} - u_1 \frac{\rho}{a^2}\left(u_1 \frac{\partial u_1}{\partial x_1} + u_2 \frac{\partial u_2}{\partial x_1} + u_3 \frac{\partial u_3}{\partial x_1}\right)$$

Summation with the other two terms, application of the small-perturbation procedure of Chapter 8, and neglect of all but first-order terms gives the Cartesian potential equation

$$(1 - M_\infty{}^2)\frac{\partial^2 \phi}{\partial x_1{}^2} + \frac{\partial^2 \phi}{\partial x_2{}^2} + \frac{\partial^2 \phi}{\partial x_3{}^2} = 0$$

In the cylindrical system, the only difference is due to the second term of the continuity equation,

$$\frac{1}{r}\frac{\partial}{\partial r}(\rho v r) = \frac{\partial}{\partial r}(\rho v) + \frac{\rho v}{r}$$

Compared with the Cartesian system, there is the extra term $(\rho v)/r$. Now, even though v is a perturbation velocity, this term cannot be neglected in the first-order equation, since r may also be very small. It contributes the extra term $\frac{v}{r} = \frac{1}{r}\frac{\partial \phi}{\partial r}$ to the linearized perturbation potential equation,

$$(1 - M_\infty{}^2)\frac{\partial^2 \phi}{\partial x^2} + \frac{\partial^2 \phi}{\partial r^2} + \frac{1}{r}\frac{\partial \phi}{\partial r} + \frac{1}{r^2}\frac{\partial^2 \phi}{\partial \theta^2} = 0 \qquad \blacktriangleright(9\cdot 3)$$

9·3 Boundary Conditions

In two-dimensional flow, the boundary condition, that the slope of the surface be equal to the slope of the streamline, is

$$\left(\frac{dx_2}{dx_1}\right)_{\text{body}} = \left(\frac{u_2}{U + u}\right)_{\text{body}} \doteq \frac{u_2(0)}{U + u(0)} \doteq \frac{u_2(0)}{U}$$

The second term is the exact condition, whereas the third term is the approximate condition, applied on the axis. The last term is a further simplification, whose accuracy is usually consistent with the previous step.

For a body of revolution, aligned in cylindrical coordinates, the component of velocity w is automatically tangent to the surface. It is necessary only to consider the boundary condition in the meridian plane, in which the body contour is given by $r = R(x)$. The exact boundary condition is

$$\frac{dR}{dx} = \left(\frac{v}{U + u}\right)_R \qquad (9\cdot 4)$$

However, it is not possible to use the same approximation as in the two-dimensional case. The reason can easily be seen from the following example.

Figure 9·4*a* shows a profile of the longitudinal section of either a two-dimensional body or a body of revolution. Figure 9·4*b* shows radial velocities near the surface, in the cross-sections of the two respective cases.

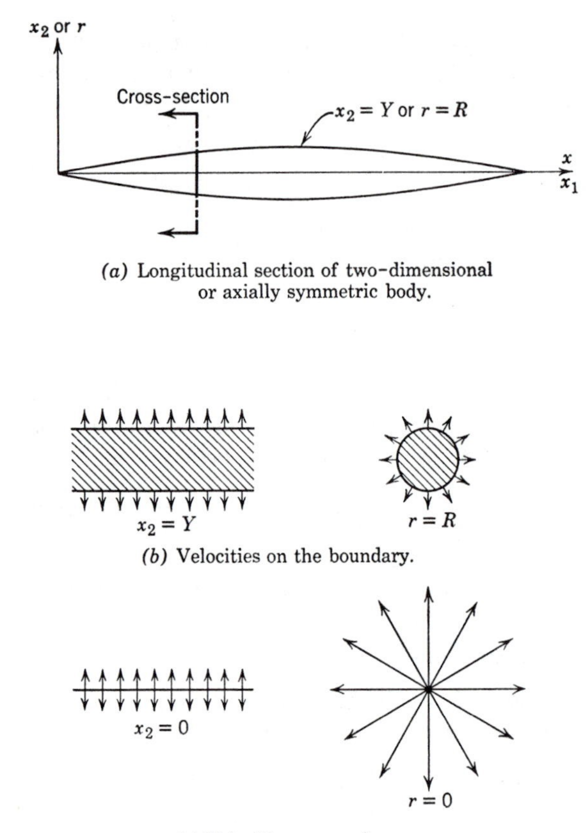

(*a*) Longitudinal section of two-dimensional or axially symmetric body.

(*b*) Velocities on the boundary.

(*c*) Velocities near axis

FIG. 9·4 Velocity fields near the axes of two-dimensional and axially symmetric bodies.

Figure 9·4*c* shows the velocity near $x_2 = 0$ or $r = 0$ in a flow field which is intended to approximate to the exact flow field of (*b*). (It may be obtained by suitable distributions of sources on $x_2 = 0$ or $r = 0$.) In the two-dimensional flow the required velocity near the axis is nearly the same as at the position of the boundary, whereas in the axially symmetric flow the radial velocity at the axis must be infinite if it is to be finite at the position of the boundary.

In the two-dimensional case, the velocities near the axis may be written as power series expansions,

$$\begin{aligned} u_1(x_1, x_2) &= U + u(x_1, 0) + a_1 x_2 + a_2 x_2^2 + \cdots \\ u_2(x_1, x_2) &= u_2(x_1, 0) + b_1 x_2 + b_2 x_2^2 + \cdots \end{aligned} \tag{9·5}$$

where $a_1, a_2 \cdots, b_1, b_2 \cdots$ are functions of x_1. For the approximate boundary condition, only the first term of each series is used.

In the axially symmetric case, on the other hand, velocity gradients near the axis are very large, and a power series expansion is not possible. This is a consequence of the radial term, $\frac{1}{r}\frac{\partial}{\partial r}(vr)$, in the continuity equation (9·1*b*). To obtain an estimate of the velocity near the axis, we note that this term must be of the same order as the other terms, for example,

$$\frac{1}{r}\frac{\partial}{\partial r}(vr) \sim \frac{\partial u}{\partial x}$$

or

$$\frac{\partial}{\partial r}(vr) \sim r\frac{\partial u}{\partial x}$$

In general, the velocity gradient $\partial u/\partial x$ is not infinite, so that, for $r \rightarrow 0$,

$$\frac{\partial}{\partial r}(vr) \sim 0$$

$$vr = a_0(x)$$

That is, near the axis, v is of the order $1/r$.† Instead of a power series like Eq. 9·5, the correct expansion is

$$vr = a_0 + a_1 r + a_2 r^2 + \cdots$$

or

$$v = a_0/r + a_1 + a_2 r + \cdots \tag{9·6}$$

Thus the correct form for an approximate boundary condition, *on the axis*, in the case of an elongated body, is

$$R\frac{dR}{dx} = R\left(\frac{v}{U+u}\right)_R \doteq \frac{(vr)_0}{U} \tag{9·7}$$

We may also obtain an estimate of the behavior of the other component, u near the axis, by using the irrotationality relation

$$\frac{\partial u}{\partial r} = \frac{\partial v}{\partial x}$$

†The result is still the same if $r(\partial u/\partial x)$ remains finite for $r \rightarrow 0$.

This gives

$$\frac{\partial u}{\partial r} = \frac{a'_0}{r} + a'_1 + \cdots$$

where $a'_0 = \partial a_0/\partial x$, etc. Thus

$$u = a'_0 \log r + a'_1 r + \cdots \tag{9·8}$$

In summary, to "apply boundary conditions at the axis" means to specify a velocity at $x_2 = 0$ or $r = 0$ which will give a velocity field having the (nearly) correct velocity at the position of the actual boundary, $x_2 = Y$ or $r = R$.

9·4 Pressure Coefficient

To evaluate the pressures, we have the exact pressure coefficient, written in terms of perturbation velocities (Eq. 2·40*b*),

$$C_p = \frac{2}{\gamma M_\infty^2}\left\{\left[1 - \frac{\gamma - 1}{2} M_\infty^2 \left(\frac{2u}{U} + \frac{u^2}{U^2} + \frac{v^2}{U^2} + \frac{w^2}{U^2}\right)\right]^{\gamma/(\gamma-1)} - 1\right\} \tag{9·9}$$

For small perturbations, this may be expanded in series, as in Article 8·3. To second order,

$$C_p = -\left(2\frac{u}{U} + (1 - M_\infty^2)\frac{u^2}{U^2} + \frac{v^2}{U^2} + \frac{w^2}{U^2}\right) \tag{9·10}$$

In the two-dimensional case, it was possible to neglect all but the first term, in the first-order theory. Now, however, the radial component v is of different order from u, near the axis, as shown by Eqs. 9·6 and 9·8. To evaluate the pressure coefficient near the axis (that is, on the surface of a slender body) the correct approximation for the *first-order* pressure coefficient is†

$$C_p = -\frac{2u}{U} - \left(\frac{v}{U}\right)^2 \tag{9·10a}$$

9·5 Axially Symmetric Flow

The flow over a body of revolution aligned parallel to the free stream (Fig. 9·2) is axially symmetric; that is, conditions are the same in every meridian plane. There is no variation with θ, and so the potential equation (9·3) may be written

$$\frac{\partial^2\phi}{\partial r^2} + \frac{1}{r}\frac{\partial\phi}{\partial r} + (1 - M_\infty^2)\frac{\partial^2\phi}{\partial x^2} = 0 \tag{9·11}$$

†M. J. Lighthill, "Supersonic Flow Past Bodies of Revolution," *Aero. Research Council (Britain), Rept. and Memo.* No. 2003 (1945).

We recall that ϕ is the *perturbation potential* for the perturbation velocities, u, v, w. (Here $w = 0$.)

For incompressible flow, $M_\infty = 0$, the equation becomes Laplace's equation,

$$\frac{\partial^2\phi}{\partial r^2} + \frac{1}{r}\frac{\partial\phi}{\partial r} + \frac{\partial^2\phi}{\partial x^2} = 0 \tag{9·12}$$

It has the basic solution

$$\phi = \frac{\text{const.}}{\sqrt{x^2 + r^2}}$$

as may be checked by substitution in Eq. 9·12. Furthermore, it becomes vanishingly small at large values of r, i.e., at infinity, as the perturbation potential should. This solution represents a source or sink, depending on the sign of the constant, situated at the origin of coordinates. For a source, the sign is negative, and we may represent a source of arbitrary strength by the relation

$$\phi = \frac{-A}{\sqrt{x^2 + r^2}}$$

If the source is at the position $x = \xi$ on the x-axis, the potential is

$$\phi(x, r) = \frac{-A}{\sqrt{(x - \xi)^2 + r^2}} \tag{9·13}$$

Owing to the linearity of the differential equation, it is possible to superimpose solutions. Thus

$$\phi(x, r) = -\frac{A_0}{\sqrt{x^2 + r^2}} - \frac{A_1}{\sqrt{(x - \xi_1)^2 + r^2}} - \frac{A_2}{\sqrt{(x - \xi_2)^2 + r^2}} - \cdots \tag{9·14}$$

represents the flow due to a series of sources placed along the x-axis.

Finally, instead of sources of finite strength, we may introduce source *distributions*. If $f(\xi)$ is the source strength *per unit length*, then $f(\xi)\,d\xi$ represents the (infinitesimal) source strength at $x = \xi$. The effect of such sources distributed along the x-axis may be added up as before, but instead of the sum (Eq. 9·14) we now have the integral

$$\phi(x, r) = -\int_0^l \frac{f(\xi)\,d\xi}{\sqrt{(x - \xi)^2 + r^2}} \tag{9·15}$$

which gives the potential at x, r due to a source distribution on the x-axis, from $x = 0$ to $x = l$. Differentiation of ϕ gives corresponding integral expressions for the velocity components.

Equation 9·15 is an *integral equation* for the "airship problem" of incompressible flow theory. In a given problem, $f(\xi)$ is determined by satisfying the boundary conditions. The solution is usually a numerical one, and often amounts to nothing more than approximating the integral by a sum like Eq. 9·14, with a finite number of sources and sinks. Application of the boundary conditions, at the same number of boundary points, leads to a set of simultaneous equations for the source strengths.

9·6 Subsonic Flow

For compressible, subsonic flow,

$$m^2 \equiv 1 - M^2 > 0$$

The potential equation may be written

$$\frac{1}{m^2}\frac{\partial^2\phi}{\partial r^2} + \frac{1}{m^2 r}\frac{\partial\phi}{\partial r} + \frac{\partial^2\phi}{\partial x^2} = 0$$

If we introduce a new coordinate system by the relation

$$r' = mr \tag{9·16}$$

the equation becomes

$$\frac{\partial^2\phi}{\partial r'^2} + \frac{1}{r'}\frac{\partial\phi}{\partial r'} + \frac{\partial^2\phi}{\partial x^2} = 0$$

This is identical with the incompressible flow equation, whose solution is given in Eq. 9·15. Transforming back to the original coordinates, we thus obtain the basic solution

$$\phi(x, r) = \frac{-A}{\sqrt{(x - \xi)^2 + m^2 r^2}} \tag{9·17}$$

and the general solution

$$\phi(x, r) = -\int_0^l \frac{f(\xi)\, d\xi}{\sqrt{(x - \xi)^2 + m^2 r^2}} \tag{▶9·17a}$$

The method of solving this integral equation for a given body shape is the same as that outlined above.

Another method is first to compute the *incompressible* flow past a similar body, which is more slender, in the ratio given in Eq. 9·16. The pressures on the original body are then computed by a rule that will be given in Chapter 10. It is called the Göthert similarity rule.

9·7 Supersonic Flow

For supersonic flow, we shall use the notation

$$\lambda^2 \equiv M^2 - 1 > 0$$

Equation 9·11 then becomes the *wave equation,*

$$\frac{\partial^2\phi}{\partial r^2} + \frac{1}{r}\frac{\partial\phi}{\partial r} - \lambda^2\frac{\partial^2\phi}{\partial x^2} = 0 \tag{9·18}$$

By formal analogy with the subsonic solution, Eq. 9·17, we try the basic solution

$$\phi(x, r) = \frac{-A}{\sqrt{(x-\xi)^2 - \lambda^2 r^2}} \tag{9·19}$$

This satisfies the wave equation, but to use it for representing flows is now more problematic. In some parts of the field, the denominator in Eq. 9·19 is zero or negative, giving values of ϕ there that are infinite or imaginary. Indeed, the method of solving the equation has to be completely re-examined, for the solutions to the supersonic equation are fundamentally different from those of the subsonic case, a fact already demonstrated in Chapter 8.

The mathematical theory of the wave equation is well developed, since it describes a large number of physical phenomena that involve wave propagation, for example, acoustical and electromagnetic phenomena. The theory was first adapted to the present problem by von Kármán and Moore, who showed that the potential may again be represented as an integral over a distribution of "sources,"

$$\phi(x, r) = -\int_0^{x-\lambda r} \frac{f(\xi)\,d\xi}{\sqrt{(x-\xi)^2 - \lambda^2 r^2}} \tag{9·20}$$

It will be noted that an imaginary value for the denominator does not appear, since the integral is evaluated only for values of ξ given by

$$\xi \leq x - \lambda r$$

For suitable source distributions, this integral gives solutions that have no singularities off the axis.

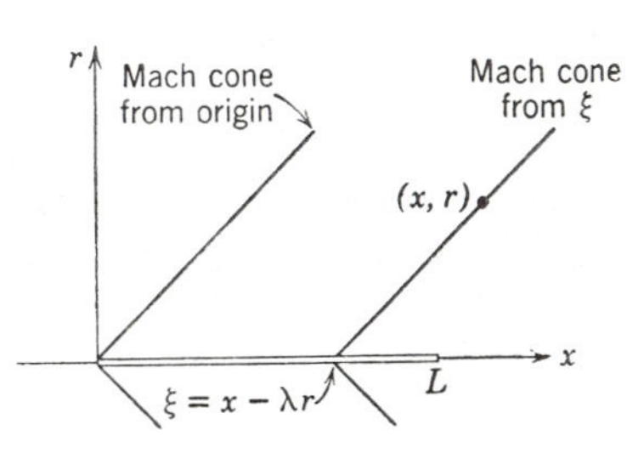

FIG. 9·5 Region of dependence of a point (x, r).

The meaning of the upper limit of integration is illustrated in Fig. 9·5. The sources are distributed along the x-axis, from 0 to L. However, to obtain the value of ϕ at (x, r) it is necessary to include only the sources up to $\xi = x - \lambda r$, the sources downstream of this point having no effect on the conditions at (x, r).

The physical meaning of this limit may be appreciated by noting that the angle

$$\tan^{-1}\frac{1}{\lambda} = \tan^{-1}\frac{1}{\sqrt{M_\infty{}^2 - 1}} = \mu \tag{9·21}$$

is the Mach angle. Thus, the upper limit of integration means that a source has no influence *ahead* of its Mach cone.

We learned earlier that in linearized, two-dimensional supersonic flow a disturbance has no effect either upstream or downstream of its Mach lines, the effect being confined entirely to the Mach lines. In the axially

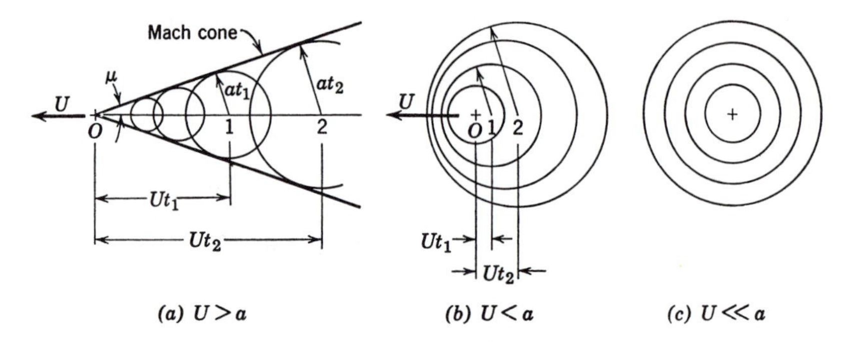

FIG. 9·6 Wave propagation from a moving source.

symmetric case, however, there is an effect over the whole region downstream of the Mach cone. Thus (x, r) is affected by all the disturbances ahead of $\xi = x - \lambda r$, and hence the *integral* expression for ϕ.

In an often-quoted example to illustrate the limited upstream influence, the sources are considered to be moving into undisturbed fluid which is at rest. This corresponds to the superposition of a uniform velocity, $-U$, on the flow problem considered above. The disturbance due to each source element spreads outward, from the source as center, with the acoustic speed a. Figure 9·6a shows such a source element at the position O and moving to the left with velocity U. At an earlier time t_1 it was at the position 1. The disturbance created at that position has now spread outward into the spherical region centered at 1 and having radius at_1. On the other hand, the source itself has moved beyond this region. The envelope of all the spherical disturbance regions due to the previous history of the source forms an envelope which is the Mach cone with vertex at the source, and which has the semi-vertex angle

$$\mu = \sin^{-1}\frac{at}{Ut} = \sin^{-1}\frac{1}{M} = \tan^{-1}\frac{1}{\sqrt{M^2 - 1}}$$

The effect of the source is not felt ahead of this Mach cone.

By contrast, if $U < a$ (Fig. 9·6*b*), no such envelope is formed, since the spreading disturbance fronts do not overtake one another. If the motion is considered to have been established a long time, so that a steady state has been reached, the disturbance extends to upstream infinity, as well as downstream.

If $U/a \rightarrow 0$ the flow is incompressible, and the basic source pattern appears as in Fig. 9·6*c*.

It is clear that the effects of the source are more "concentrated" in the supersonic case.

9·8 Velocities in the Supersonic Field

To find the velocities it is necessary to obtain the derivatives of ϕ in Eq. 9·20. Since the variable of differentiation, x or r, appears not only in the integrand but also in the upper limit, the differentiation must follow the Leibnitz rule.† However, there are some further problems, since the integrand in Eq. 9·20 becomes infinite at the upper limit. The rules for treating such integrals were given by Hadamard.‡ Sometimes the need for an explicit application of Hadamard's method may be avoided by introducing a suitable change of the variable of integration. This is possible in the present case, by putting

$$\xi = x - \lambda r \cosh \sigma \tag{9·22}$$

Then

$$d\xi = -\lambda r \sinh \sigma \, d\sigma = -\sqrt{(x-\xi)^2 - \lambda^2 r^2}\, d\sigma$$

The corresponding changes in the limits of integration are

$$\xi = 0 \rightarrow \sigma = \cosh^{-1} \frac{x}{\lambda r}$$

$$\xi = x - \lambda r \rightarrow \sigma = \cosh^{-1}(1) = 0$$

In the new variable, then, the potential may be written

$$\phi(x, r) = -\int_0^{\cosh^{-1}(x/\lambda r)} f(x - \lambda r \cosh \sigma)\, d\sigma \tag{9·23}$$

Application of the Leibnitz rule now gives the velocity components

$$u = \frac{\partial \phi}{\partial x} = -\int_0^{\cosh^{-1}(x/\lambda r)} f'(x - \lambda r \cosh \sigma)\, d\sigma - f(0)\left(\frac{1}{\sqrt{x^2 - \lambda^2 r^2}}\right) \tag{9·24a}$$

† $\dfrac{d}{ds}\displaystyle\int_0^{g(s)} F(\xi; s)\, d\xi = \int_0^{g(s)} \frac{\partial F}{\partial s}\, d\xi + F\{g(s); s\}\frac{dg}{ds}$.

‡See discussion by M. A. Heaslet and H. Lomax in *General Theory of High Speed Aerodynamics,* W. R. Sears (Ed.), Princeton, 1954.

$$v = \frac{\partial \phi}{\partial r} =$$

$$-\int_0^{\cosh^{-1}(x/\lambda r)} f'(x-\lambda r \cosh \sigma)(-\lambda \cosh \sigma)\, d\sigma + f(0)\left(\frac{x}{r\sqrt{x^2-\lambda^2 r^2}}\right) \tag{9·24b}$$

where f' denotes differentiation of f with respect to its argument.

We shall consider only bodies for which $f(0) = 0$ (for example, pointed bodies as will be shown later). In the original variable of integration, Eqs. 9·24 become

$$u = -\int_0^{x-\lambda r} \frac{f'(\xi)}{\sqrt{(x-\xi)^2 - \lambda^2 r^2}}\, d\xi \tag{9·25a}$$

$$v = \frac{1}{r}\int_0^{x-\lambda r} \frac{f'(\xi)(x-\xi)}{\sqrt{(x-\xi)^2 - \lambda^2 r^2}}\, d\xi \tag{9·25b}$$

9·9 Solution for a Cone

The direct problem of finding $f(\xi)$ for given boundary conditions is, in general, quite difficult, involving the solution of an integral equation. An alternate procedure is to assume a definite function for $f(\xi)$ and then find the corresponding flow. A particularly simple and useful one is obtained from

$$f(\xi) = a\xi \tag{9·26}$$

which gives

$$f(0) = 0 \quad \text{and} \quad f'(\xi) = a$$

The velocity potential is then easily found from Eq. 9·23, using Eq. 9·22 for ξ. The result is

$$\phi(x, r) = -\int_0^{\cosh^{-1}(x/\lambda r)} (ax - a\lambda r \cosh \sigma)\, d\sigma$$

$$= -ax\left[\cosh^{-1}\frac{x}{\lambda r} - \sqrt{1 - \left(\frac{\lambda r}{x}\right)^2}\right] \tag{9·27}$$

The perturbation velocity components may be found by direct differentiation here, or from the integral expressions (9·24). They are

$$u = -a \cosh^{-1}\frac{x}{\lambda r} \tag{9·27a}$$

$$v = a\lambda\sqrt{\left(\frac{x}{\lambda r}\right)^2 - 1} \tag{9·27b}$$

It will be observed that both u and v are functions of the parameter $x/\lambda r$. Thus u and v are invariant along lines for which x/r is constant, that is, along radial lines, or *rays*, from the origin. The flow field is said to be *conical*. This is illustrated in Fig. 9·7, where typical velocities are indicated by arrows on several rays. On ray a, which is in the Mach cone

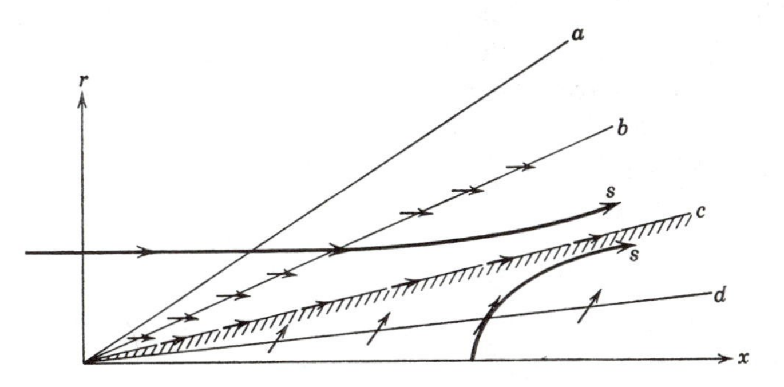

FIG. 9·7 Conical velocity field. Velocities constant on each ray. a = ray in Mach cone; b, c, d = typical rays in velocity field; c = ray along which velocity vector is parallel to ray; s = typical streamlines.

from the vertex, the perturbation velocities are zero, and the flow is still at the free-stream condition. For rays nearer the axis, the magnitude of the perturbations increases and the flow direction changes as shown. Two typical streamlines, marked s, are also shown, one originating in the free stream and one in a source on the axis. Any streamline may be chosen to represent a solid surface.

There is one streamline, c, that lies along a ray. The cone that contains this ray may therefore represent a solid cone in supersonic flow. Its flow field is given by the above solution. The flow field "inside" the cone is of no interest here.

The vertex angle of the cone depends on the constant a in Eq. 9·27. Conversely, to find the solution for a cone of *given* semi-vertex angle, δ, it is necessary to determine a by satisfying the boundary condition

$$\left(\frac{v}{U + u}\right)_{\text{cone}} = \tan \delta \tag{9·28}$$

From Eq. 9·27 the velocity components on the cone surface are

$$u = -a \cosh^{-1} \frac{\cot \delta}{\lambda} \tag{9·29a}$$

$$v = a\sqrt{\cot^2 \delta - \lambda^2} \tag{9·29b}$$

giving, with the boundary condition, the result

$$a = \frac{U \tan \delta}{\sqrt{\cot^2 \delta - \lambda^2} + \tan \delta \cosh^{-1}\left(\frac{\cot \delta}{\lambda}\right)} \tag{9·29c}$$

To find the pressure on the cone (it is uniform over the surface) the results of Eqs. 9·29 are substituted into Eq. 9·9 for the pressure coefficient. There is no need to write it out here.

An *exact* solution, such as this one, of the linear differential equation has been called a first-order solution by Van Dyke.†

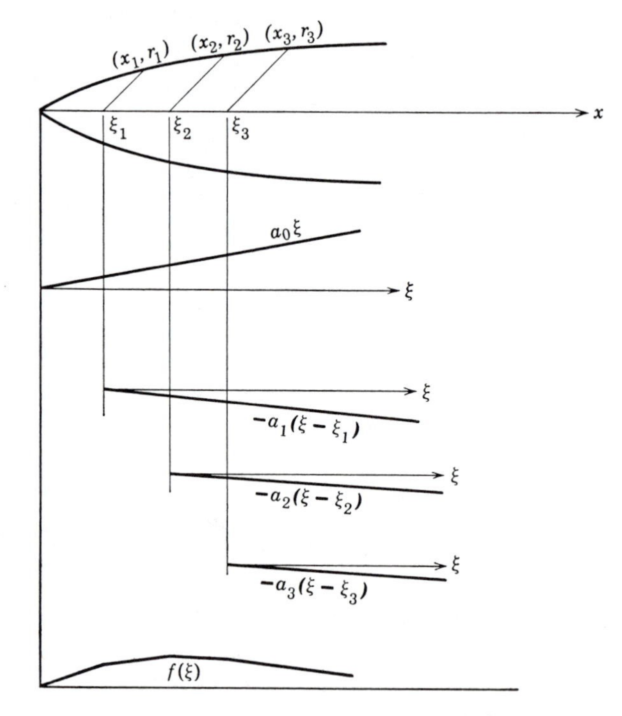

FIG. 9·8 Superposition of conical solutions.

9·10 Other Meridian Shapes

In a numerical method given by Kármán and Moore, the flow around an arbitrary (pointed) body of revolution can be obtained by superposition of the conical solutions obtained in the preceding article.

†M. Van Dyke, "Supersonic Flow Past Bodies of Revolution," *J. Aeronaut. Sci.*, *18* (1951), p. 161.

For the ogive shape shown in Fig. 9·8, the solution is obtained by using the distribution

$$f(\xi) = a_0\xi - a_1(\xi - \xi_1) - a_2(\xi - \xi_2) + \cdots$$

in which $a_0, a_1, a_2, \cdots$ are positive and $\xi_1 = x_1 - \lambda r_1$, $\xi_2 = x_2 - \lambda r_2$, etc.

The constants are evaluated as follows: a_0 corresponds to the solution for a cone having the same vertex angle as the ogive, and gives the flow near the nose. At (x_1, r_1) this gives a velocity that does not satisfy the boundary condition there, and so it is corrected by superimposing another conical flow, $-a_1(\xi - \xi_1)$, with a_1 appropriately chosen. Having its origin at $\xi_1 = x_1 - \lambda r_1$, it does not affect the nose flow, ahead of (x_1, r_1).

In this way, all the constants are evaluated, step by step. Successive steps do not affect those portions of the flow, upstream, which have already been constructed. This convenience is a feature of solutions to the wave equation, in contrast to the relaxation type of numerical solutions needed for problems governed by Laplace's equation. It will appear again in the numerical method of characteristics (Chapter 12).

It may be observed that for pointed bodies the source distribution will always be cone-like near the nose, that is,

$$f'(0) = \text{const.}, \quad f(0) = 0$$

This method is applicable only if the meridional section is smooth. For bodies with shoulders one must add to the solution certain terms which give the proper "jump" at each shoulder.†

9·11 Solution for Slender Cone

If the nose angle, 2δ, of the cone described in Article 9·9 is very small, the solution for a, in Eq. 9·29*c*, may be simplified. For, if M is not too large, then $\cot\delta \gg \lambda$, and $\cosh^{-1}(\cot\delta/\lambda) \doteq \log(2/\delta\lambda)$. Furthermore, $\tan\delta \log(2/\delta\lambda) \to 0$, and thus

$$a \doteq U\delta/\cot\delta \doteq U\delta^2 \tag{9·30}$$

To discuss conditions near the surface of such a slender cone, it is also possible to simplify the expression for ϕ, given in Eq. 9·27, since r/x will then be small. The result is

$$\phi = -U\delta^2 x\left(\log\frac{2x}{\lambda r} - 1\right) \tag{9·31}$$

The velocities near the cone surface, which may be obtained by differentiat-

†Cf. Van Dyke, *loc. cit.*, where additional refinements are given.

ing this (or by making the approximations in Eq. 9·27), are

$$\frac{u}{U} = -\delta^2 \log \frac{2x}{\lambda r} \tag{9·31a}$$

$$\frac{v}{U} = \delta^2 \frac{x}{r} \tag{9·31b}$$

On the cone surface, where $r/x \doteq \delta$, these are

$$\left(\frac{u}{U}\right)_{\text{cone}} = -\delta^2 \log \frac{2}{\lambda\delta} \tag{9·32a}$$

$$\left(\frac{v}{U}\right)_{\text{cone}} = \delta \tag{9·32b}$$

The corresponding pressure coefficient is

$$C_p = -\frac{2u}{U} - \left(\frac{v}{U}\right)^2 = 2\delta^2\left(\log\frac{2}{\lambda\delta} - \frac{1}{2}\right) \tag{9·32c}$$

Values from this approximate formula may be compared with the results of exact theory, which are given in Fig. 4·27.

The result in Eq. 9·32*c* may also be compared to the pressure coefficient for a thin wedge of nose angle 2δ, which is

$$C_p = \frac{2\delta}{\sqrt{M^2 - 1}} = \frac{2\delta}{\lambda}$$

The pressure rise on the cone is much less than on the wedge, being of the order

$$\delta^2 \log (1/\delta) \quad \text{compared to} \quad \delta$$

The lower pressure rise is due to the three-dimensional effect, which gives "more room for adjustment" of the flow over the cone.

There is less dependence on Mach number in the case of the cone, the very slender cones being practically independent.

Yet another difference may be observed by calculating the pressure coefficient along a line parallel to the axis, $r =$ const. The distribution of pressure between the Mach cone and cone surface is shown in Fig. 9·9. In the wedge flow, all the pressure change occurs at the nose wave, whereas in the cone flow, the pressure rises continuously downstream of the Mach cone. This typical difference between the two cases was mentioned in Article 9·7. In the slender body theory there is no pressure jump at the nose wave (cf. Fig. 4·26).

The relation for the pressure coefficient on a slender cone (Eq. 9·32*c*) may be put in the functional form

$$\frac{C_p}{\delta^2} = f_n(\delta\sqrt{M_\infty^2 - 1}) \tag{9·32d}$$

This is a general result which may be obtained by the method of similarity, discussed in Chapter 10.

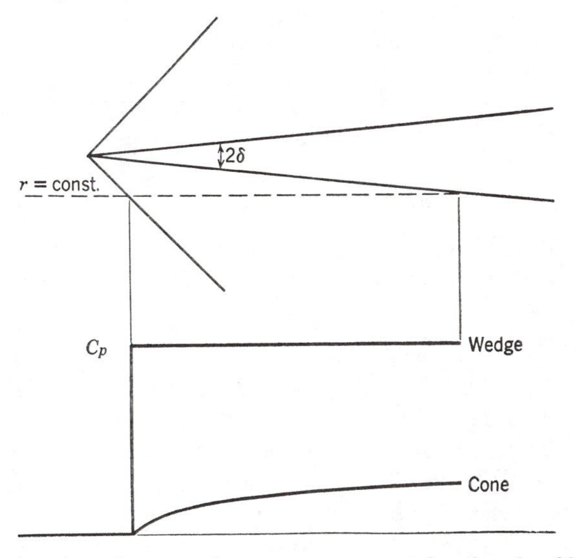

FIG. 9·9 Distributions of pressure between nose wave and surface for thin wedge and slender cone.

9·12 Slender Body Drag

The slender cone results of the preceding article were obtained by first obtaining an exact solution and then finding approximations for it. It is possible to reach the same results by making the approximations directly in the general, integral solution (Eq. 9·20),

$$\phi(x, r) = -\int_0^{x-\lambda r} \frac{f(\xi)\,d\xi}{\sqrt{(x-\xi)^2 - \lambda^2 r^2}}$$

The approximation will then apply not only to the cone but to bodies of arbitrary profile.

The slender body approximation is made by evaluating the integral for small values of r or, more precisely, for small values of $\lambda r/x$. Owing to the singular nature of the integrand near the upper limit, some care is required in the evaluation. The integral may be broken into two parts,

$$\phi = -I_1 - I_2 = -\int_0^{x-\lambda r-\epsilon} \frac{f(\xi)\,d\xi}{\sqrt{(x-\xi)^2-\lambda^2 r^2}} - \int_{x-\lambda r-\epsilon}^{x-\lambda r} \frac{f(\xi)\,d\xi}{\sqrt{(x-\xi)^2-\lambda^2 r^2}}$$

In the first integral, the integrand may be expanded in a power series of $\lambda^2 r^2$,

$$\frac{f(\xi)}{\sqrt{(x-\xi)^2-\lambda^2 r^2}} = \frac{f(\xi)}{x-\xi} + \frac{1}{2}\lambda^2 r^2 \frac{f(\xi)}{(x-\xi)^3} + \cdots$$

and integrated term by term; then, for $\lambda r \to 0$,

$$I_1 = f(0)\log x - f(x)\log\epsilon + \int_0^{x-\epsilon} f'(\xi)\log(x-\xi)\,d\xi + \epsilon f'(x)\log\epsilon$$

The second integral may be rewritten using the substitution in Eq. 9·22 and expanded in a power series of λr,

$$I_2 = \int_0^{\cosh^{-1}[(\lambda r+\epsilon)/\lambda r]} f(x-\lambda r\cosh\sigma)\,d\sigma$$

$$= f(x)\int_0^{\cosh^{-1}[(\lambda r+\epsilon)/\lambda r]} d\sigma - \lambda r\int_0^{\cosh^{-1}[(\lambda r+\epsilon)/\lambda r]} f'(x)\cosh\sigma\,d\sigma + \cdots$$

Then, for $\lambda r \to 0$,

$$I_2 = f(x)\log\frac{2}{\lambda r} + f(x)\log\epsilon - \epsilon f'(x)$$

For pointed bodies, the first term in I_1 is zero, since $f(0) = 0$ (cf. Article 9·10). The second term cancels the second one in I_2. Finally, ϵ may be made arbitrarily small, giving, for $r \to 0$, the result

$$\phi = -f(x)\log\frac{2}{\lambda r} - \int_0^x f'(\xi)\log(x-\xi)\,d\xi \qquad \blacktriangleright(9\cdot33)$$

This result not only applies to the cone but also is valid for all axially symmetric slender bodies of arbitrary profile.

To find $f(x)$ for a given body, we have to apply the boundary condition (Eq. 9·7). The radial component of velocity is

$$v = \frac{\partial\phi}{\partial r} = \frac{f(x)}{r} \quad \text{or} \quad vr = f(x) \qquad (9\cdot33a)$$

On the body surface, $r = R$, and so

$$f(x) = (v)_{\text{body}}R$$

On the other hand, the condition of tangency is

$$\left(\frac{v}{U}\right)_{\text{body}} = \frac{dR}{dx}$$

Thus the solution for the source strength is

$$f(x) = UR\frac{dR}{dx} = \frac{U}{2\pi}\frac{dS}{dx} \qquad \blacktriangleright(9\cdot33b)$$

where $S(x) = \pi R^2$ is the cross-sectional area of the body at x. We see that $f(0) = 0$ if $R = 0$, that is, the nose of the body is closed. The condition $f(0) = 0$ is not restricted to pointed bodies. However, there is a certain restriction on the allowable bluntness. (See Exercise 9·1.)

Equation 9·33*b* gives the interesting result that the source strength is proportional only to the *local* rate of change of area of the body. For a very slender body, those portions of the body that are "far away" do not influence the local conditions. The rate at which the fluid is "pushed outward," locally, depends entirely on the local rate of area change.

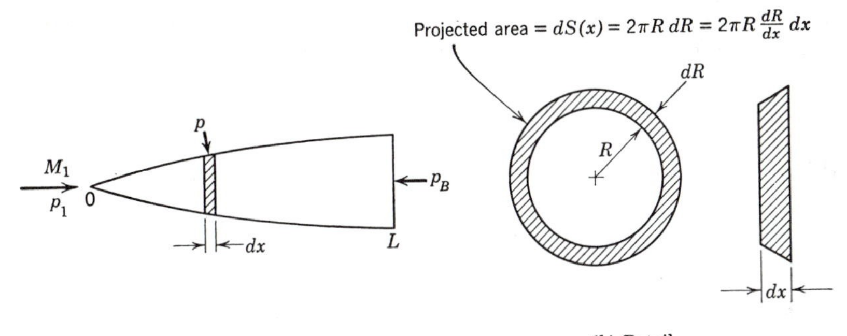

(*a*) Element of surface area. (*b*) Detail.

FIG. 9·10 Pressure on axially symmetric body.

Thus the solution for axial flow over a slender body of revolution with closed nose and arbitrary (smooth) meridional section is

$$\phi(x, r) = -\frac{U}{2\pi} S'(x) \log \frac{2}{\lambda r} - \frac{U}{2\pi} \int_0^x S''(\xi) \log (x - \xi)\, d\xi \qquad \blacktriangleright(9\cdot34)$$

$$\frac{u}{U} = -\frac{S''(x)}{2\pi} \log \frac{2}{\lambda r} - \frac{1}{2\pi} \frac{d}{dx} \int_0^x S''(\xi) \log (x - \xi)\, d\xi \qquad \blacktriangleright(9\cdot34a)$$

$$\frac{v}{U} = \frac{S'(x)}{2\pi r} = \frac{R}{r} \frac{dR}{dx} \qquad \blacktriangleright(9\cdot34b)$$

On the body, $r = R$, the pressure coefficient is

$$C_p = \frac{S''(x)}{\pi} \log \frac{2}{\lambda R} + \frac{1}{\pi} \frac{d}{dx} \int_0^x S''(\xi) \log (x - \xi)\, d\xi - \left(\frac{dR}{dx}\right)^2 \qquad (9\cdot34c)$$

This may now be used to compute the drag, as follows.

In Fig. 9·10*a*, the pressure p, at an arbitrary section x, is uniform over the circumference of the section. It contributes to the drag because the surface is inclined, so that the pressure acts on the projected area $dS = 2\pi R\, dR$, as shown in Fig. 9·10*b*. Thus the drag is

$$D = \int_0^L p\,dS - p_B S(L) = \int_0^L (p - p_1)\,dS + (p_1 - p_B)S(L)$$

or, in dimensionless form,

$$C_D = \frac{D}{q_\infty S(L)} = \frac{1}{S(L)} \int_0^L C_p \frac{dS}{dx}\,dx + C_{pB} = C_{D1} + C_{pB}$$

The last term is the contribution to the drag of the base pressure p_B. This is determined by the mechanics of the *wake*, for which there is as yet no complete theory. Values of the base pressure coefficient C_{pB} must be obtained experimentally. The integral may be evaluated by inserting for C_p the expression given in Eq. 9·34*c*. Thus

$$\int_0^L C_p S'(x)\,dx = \frac{1}{\pi} \int_0^L S'(x)S''(x) \log \frac{2}{\lambda R(x)}\,dx - \int_0^L \left(\frac{dR}{dx}\right)^2 S'(x)\,dx$$
$$+ \frac{1}{\pi} \int_0^L S'(x) \frac{d}{dx} \int_0^x S''(\xi) \log (x - \xi)\,d\xi\,dx$$

The first and last integrals on the right-hand side are, respectively,

$$I_1 = -\frac{1}{2\pi} \int_0^L \log \frac{\lambda R}{2}\,d[S'(x)]^2$$
$$= -\frac{1}{2\pi} \left[(S')^2 \log \frac{\lambda R}{2} \right]_0^L + \int_0^L S' \left(\frac{dR}{dx}\right)^2 dx$$

$$I_3 = \frac{1}{\pi} \left[S'(x) \int_0^x S''(\xi) \log (x - \xi)\,d\xi \right]_{x=0}^L$$
$$- \frac{1}{\pi} \int_0^L S''(x) \int_0^x S''(\xi) \log (x - \xi)\,d\xi\,dx$$

$$= \frac{1}{\pi} S'(L) \int_0^L S''(\xi) \log (L - \xi)\,d\xi$$
$$- \frac{1}{\pi} \int_0^L \int_0^x S''(x)S''(\xi) \log (x - \xi)\,d\xi\,dx$$

Combining these with the second integral, we have for the forebody drag coefficient

$$C_{D1} = \frac{[S'(L)]^2}{2\pi} \log \frac{2}{\lambda R(L)} + \frac{S'(L)}{\pi} \int_0^L S''(\xi) \log (L - \xi)\,d\xi$$
$$- \frac{1}{\pi} \int_0^L \int_0^x S''(x)S''(\xi) \log (x - \xi)\,d\xi\,dx \quad (9{\cdot}35)$$

If $S'(L) = 0$, the first two terms in this expression are zero. Since $S' = 2\pi RR'$, this special case occurs for

(1) $R(L) = 0$ the body is closed at the base

or

(2) $R'(L) = 0$ the slope of the body at the base is zero

The drag coefficient is then given by the expression

$$S(L)C_{D1} = -\frac{1}{\pi}\int_0^L\int_0^x S''(\xi)S''(x)\log(x-\xi)\,d\xi\,dx \tag{9·35a}$$

This integral is taken over the triangular area bounded by $\xi = 0$, $x = \xi$, and $x = L$. The integrand may be made symmetrical about the line $x = \xi$ by replacing $\log(x-\xi)$ by $\log|x-\xi|$, and it may then be written

$$S(L)C_{D1} = -\frac{1}{2\pi}\int_0^L\int_0^L S''(\xi)S''(x)\log|x-\xi|\,d\xi\,dx \tag{9·35b}$$

It was first given by von Kármán.

9·13★ Yawed Body of Revolution in Supersonic Flow

To study the flow over a body of revolution with angle of attack, or yaw, it is convenient to align the body along the x-axis (Fig. 9·11), and allow

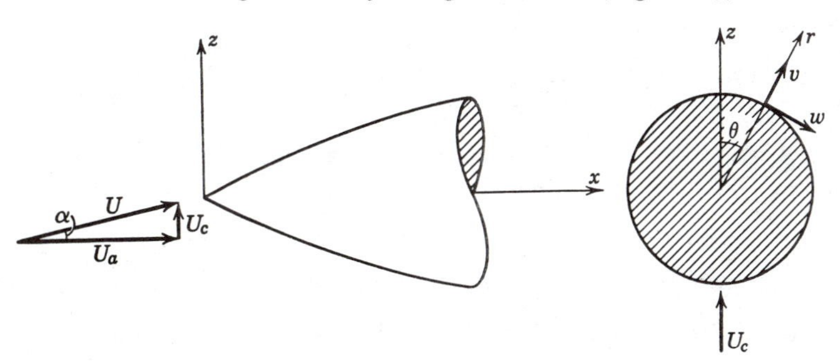

FIG. 9·11 Alignment of axes for axial flow and cross flow. U_c is the free stream cross-flow velocity. $v = \partial\phi/\partial r$ and $w = \dfrac{1}{r}\dfrac{\partial\phi}{\partial\theta}$ are radial and tangential perturbation velocities.

the flow to be inclined. Evidently the flow is not axially symmetric, and so the full equation (9·3) for the potential ϕ must be used. However, owing to the *linearity* of the equation it is possible to write the solution as the sum of two potentials,

$$\phi(x, r, \theta) = \phi_a(x, r) + \phi_c(x, r, \theta) \tag{9·36}$$

where ϕ_a is an *axial* flow, not dependent on θ, and ϕ_c is a *cross* flow. This is the superposition of two flows, in one of which the velocity far from the body is parallel to the axis

$$U_a = U \cos \alpha$$

whereas in the other it is normal to the axis,

$$U_c = U \sin \alpha$$

The axial problem is the same as the one that has been worked out in the preceding articles, only with the modification that the reference velocity is U_a, provided that the boundary conditions can also be "split" appropriately.

The cross-flow problem, on the other hand, must satisfy Eq. 9·3, that is,

$$\frac{\partial^2 \phi_c}{\partial r^2} + \frac{1}{r}\frac{\partial \phi_c}{\partial r} + \frac{1}{r^2}\frac{\partial^2 \phi_c}{\partial \theta^2} - \lambda^2 \frac{\partial^2 \phi_c}{\partial x^2} = 0 \qquad (9\cdot37)$$

in which $\lambda^2 = M^2 - 1 > 0$ for supersonic flow. The Mach number M refers to the full velocity U. The fact that the component U_c may be "subsonic" is irrelevant, the "cross flow" here being simply one component of the full solution, Eq. 9·36. Similarly, the fact that the velocity perturbations in the "cross flow" are large does not invalidate the result, so long as the perturbations in the combined solution are small.

A solution of Eq. 9·37 may be obtained by comparing with the equation for axial flow,

$$\frac{\partial^2 \phi_a}{\partial r^2} + \frac{1}{r}\frac{\partial \phi_a}{\partial r} - \lambda^2 \frac{\partial^2 \phi_a}{\partial x^2} = 0$$

If ϕ_a is a solution of this equation, it is also a solution of the equation that is obtained by differentiating with respect to r,

$$\frac{\partial^2}{\partial r^2}\left(\frac{\partial \phi_a}{\partial r}\right) + \frac{1}{r}\frac{\partial}{\partial r}\left(\frac{\partial \phi_a}{\partial r}\right) - \frac{1}{r^2}\frac{\partial \phi_a}{\partial r} - \lambda^2 \frac{\partial^2}{\partial x^2}\left(\frac{\partial \phi_a}{\partial r}\right) = 0 \qquad (9\cdot37a)$$

This is still true, if $(\partial\phi_a/\partial r)$ is replaced by $\cos\theta\, \partial\phi_a/\partial r$, but then the third term may be rewritten

$$-\frac{\cos\theta}{r^2}\frac{\partial \phi_a}{\partial r} = \frac{1}{r^2}\frac{\partial^2}{\partial \theta^2}\left(\cos\theta \frac{\partial \phi_a}{\partial r}\right)$$

With this substitution, Eq. 9·37a is the same as Eq. 9·37. Thus the basic cross-flow solution may be obtained from the basic axially symmetric solution by putting

$$\phi_c(x, r, \theta) = \cos\theta \frac{\partial \phi_a}{\partial r} \equiv \frac{\partial \phi_a}{\partial z}$$

This is, in fact, the rule for deriving a doublet from a source, the doublet axis being in the z-direction.

9·14★ Cross-Flow Boundary Conditions

The superposition of axial flow and cross flow satisfies the differential equation and the boundary condition far from the body. It is still necessary to satisfy the boundary condition at the body. This may also be "split" into two parts, corresponding to axial and cross flow, as follows.

The axial and cross-flow velocity perturbations are obtained from the respective potentials by differentiation. The radial velocity, in any cross-section (Fig. 9·11), is

$$U_c \cos\theta + \frac{\partial\phi}{\partial r}$$

and the axial velocity is

$$U_a + \frac{\partial\phi}{\partial x}$$

When these are placed in the exact, full boundary condition (Eq. 9·4), there is obtained

$$\left[\frac{\partial\phi}{\partial r} + U_c \cos\theta\right]_{\text{body}} = \frac{dR}{dx}\left[U_a + \frac{\partial\phi}{\partial x}\right]_{\text{body}}$$

that is,

$$\left[\frac{\partial\phi_a}{\partial r} + \frac{\partial\phi_c}{\partial r} + U_c \cos\theta\right]_{\text{body}} = \frac{dR}{dx}\left[U_a + \frac{\partial\phi_a}{\partial x} + \frac{\partial\phi_c}{\partial x}\right]_{\text{body}} \tag{9·38}$$

This may be split in two parts,

$$\left[\frac{\partial\phi_a}{\partial r}\right]_{\text{body}} = \frac{dR}{dx}\left[U_a + \frac{\partial\phi_a}{\partial x}\right]_{\text{body}} \tag{9·38a}$$

$$\left[\frac{\partial\phi_c}{\partial r}\right]_{\text{body}} + U_c \cos\theta = \frac{dR}{dx}\left[\frac{\partial\phi_c}{\partial x}\right]_{\text{body}} \tag{9·38b}$$

It will be observed that Eq. 9·38*a* is the exact boundary condition for axial flow with velocity U_a (see Eq. 9·4). Similarly, Eq. 9·38*b* involves only the cross-flow velocities, and is the exact boundary condition for that problem. In certain cases an approximate boundary condition may be written by noting that the product on the right-hand side of Eq. 9·38*b* is of lower order than the terms on the left, and may be neglected, giving

$$\left[\frac{\partial\phi_c}{\partial r}\right]_{\text{body}} + U_c \cos\theta \doteq 0 \tag{9·39}$$

9·15★ Cross-Flow Solutions

It was shown in Article 9·13 that cross-flow solutions may be obtained from axial flow solutions by the relation

$$\phi_c(x, r, \theta) = \cos\theta \frac{\partial \phi_a}{\partial r} \tag{9·40}$$

The derivative here has already been evaluated (see Eq. 9·25*b*). It is

$$\frac{\partial \phi_a}{\partial r} = \frac{1}{r}\int_0^{x-\lambda r} \frac{f'(\xi)(x-\xi)}{\sqrt{(x-\xi)^2 - \lambda^2 r^2}}\, d\xi \tag{9·41}$$

This may be put into another form by integrating by parts,

$$\frac{\partial \phi_a}{\partial r} = -\lambda^2 r\int_0^{x-\lambda r} \frac{f(\xi)\, d\xi}{[(x-\xi)^2 - \lambda^2 r^2]^{3/2}} + \frac{1}{r}\left[\frac{f(\xi)(x-\xi)}{\sqrt{(x-\xi)^2 - \lambda^2 r^2}}\right]_0^{x-\lambda r}$$

In the last bracket, the lower limit gives a term containing $f(0)$, which will again be zero for pointed bodies, but the upper limit gives a term that is infinite. The rules for handling this singularity are given in Hadamard's method. The above result may be written formally

$$\frac{\partial \phi_a}{\partial r} = -\lambda^2 r \left|\overline{\int_0^{x-\lambda r} \frac{f(\xi)\, d\xi}{[(x-\xi)^2 - \lambda^2 r^2]^{3/2}}}\right. \tag{9·42}$$

where the bar around the integral denotes the *finite part* of the integral, and prescribes Hadamard's method for operations on the integral.†

Both forms, (Eq. 9·41) and (Eq. 9·42), appear in the literature. The undetermined functions $f'(\xi)$ and $f(\xi)$ which appear in these integrals are so far arbitrary, to be determined by the cross-flow boundary condition. To avoid confusion with the axial flow solutions, it is convenient to use a different notation. Thus the two forms for the solution may be written

$$\phi_c(x, r, \theta) = \frac{\cos\theta}{r}\int_0^{x-\lambda r} \frac{m(\xi)(x-\xi)}{\sqrt{(x-\xi)^2 - \lambda^2 r^2}}\, d\xi \tag{9·41a}$$

$$\phi_c(x, r, \theta) = \lambda^2 r \cos\theta \left|\overline{\int^{x-\lambda r} \frac{\sigma(\xi)\, d\xi}{[(x-\xi)^2 - \lambda^2 r^2]^{3/2}}}\right. \tag{9·42a}$$

These results and their application were first given by Tsien and Ferrari.

9·16 Cross Flow for Slender Bodies of Revolution

For a very slender body it is possible to simplify the integral solutions, by evaluating them for small values of $\lambda r/x$, as in Article 9·12. A more

†Cf. M. A. Heaslet and H. Lomax, *loc. cit.*, p. 229.

direct procedure is to use the axially symmetric slender body solution obtained in Eq. 9·33 and apply the rule given in Eq. 9·40 for deriving a cross flow. Thus,

$$\phi_c = \cos\theta \frac{\partial \phi_a}{\partial r} = \frac{\sigma(x)}{r} \cos\theta \qquad \blacktriangleright(9{\cdot}43)$$

where $\sigma(x)$ is used instead of $f(x)$, to distinguish the two cases. This will be recognized as the perturbation potential for plane flow normal to an infinitely long cylinder, well known from incompressible fluid theory. The result means that the conditions at any section are the same as if the section were part of an infinite cylinder, normal to a uniform flow. It expresses the fact that for a very slender body the conditions at each section are not influenced by portions of the body that are "far away."

The strength of the *doublet* $\sigma(x)$ is related to the section radius. The relation is determined by the boundary conditions, which, for slender bodies, may be used in the approximate form given in Eq. 9·39. The result is

$$\sigma(x) = U_c R^2(x) = \frac{U_c}{\pi} S(x) \qquad \blacktriangleright(9{\cdot}44)$$

The doublet strength is proportional to the local section area. Thus the solution for the cross flow may be written

$$\phi_c = U_c \frac{R^2(x)}{r} \cos\theta = U \sin\alpha \frac{R^2(x)}{r} \cos\theta \qquad (9{\cdot}43a)$$

9·17 Lift of Slender Bodies of Revolution

To find the pressures on the body it is necessary, of course, to obtain the velocities from the full solution, $\phi = \phi_a + \phi_c$. Since these appear as squares in the exact pressure coefficient, it is not possible, in general, to split C_p into axial and cross-flow components. However, in the slender body approximation it may be done as follows.

The exact square of the velocity (cf. Fig. 9·11) is

$$(\text{vel})^2 = \left(U_a + \frac{\partial \phi}{\partial x}\right)^2 + \left(U_c \cos\theta + \frac{\partial \phi}{\partial r}\right)^2 + \left(U_c \sin\theta - \frac{1}{r}\frac{\partial \phi}{\partial \theta}\right)^2$$

It will be remembered that $U_a = U\cos\alpha$ and $U_c = U\sin\alpha$, and that α is small, as are the perturbation velocities compared to U. Retaining only terms up to second order, the expression in the first parenthesis may be approximated as follows, for points on the body surface:

$$\left(U_a + \frac{\partial \phi}{\partial x}\right)^2_{\text{body}} = U_a^2 + 2U_a \left(\frac{\partial \phi}{\partial x}\right)_{\text{body}} + \left(\frac{\partial \phi}{\partial x}\right)^2_{\text{body}}$$

$$\doteq U^2(1 - \alpha^2) + 2U \left(\frac{\partial \phi}{\partial x}\right)_{\text{body}} + \left(\frac{\partial \phi}{\partial x}\right)^2_{\text{body}}$$

The second parenthesis may be rewritten with the help of the boundary condition (Eq. 9·38),

$$\left(U_c \cos\theta + \frac{\partial\phi}{\partial r}\right)^2_{\text{body}} \doteq \left(U\frac{dR}{dx}\right)^2$$

Finally, in the last parenthesis, $\partial\phi/\partial\theta$ may be obtained from Eq. 9·43*a*, giving

$$\left(U_c \sin\theta - \frac{1}{R}\frac{\partial\phi}{\partial\theta}\right)^2_{\text{body}} = (2U_c \sin\theta)^2 \doteq (2U\sin\theta)^2\alpha^2$$

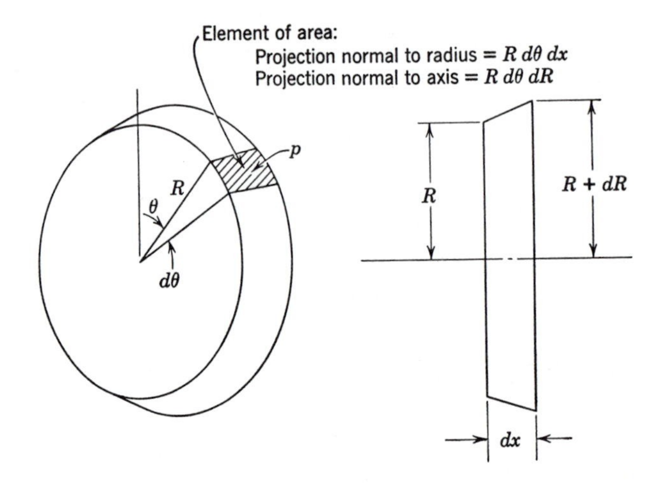

Fig. 9·12 Pressure on element of surface area.

The pressure coefficient to second order (Eq. 9·10) is

$$C_p \doteq 1 - \frac{\text{vel}^2}{U^2} + \frac{M_\infty{}^2}{U^2}\left(\frac{\partial\phi}{\partial x}\right)^2$$

Thus,

$$(C_p)_{\text{body}} = -\frac{2}{U}\left(\frac{\partial\phi}{\partial x}\right)_{\text{body}} + \alpha^2 - \left(\frac{dR}{dx}\right)^2 - 4\alpha^2\sin^2\theta + \lambda^2\left(\frac{\partial\phi}{\partial x}\right)^2_{\text{body}}$$

The last term may be neglected, as indicated in the next article. The expression for $(C_p)_{\text{body}}$ may be separated into two parts, one of which has axial symmetry,

$$C_{pa} = -\frac{2}{U}\left(\frac{\partial\phi_a}{\partial x}\right)_{\text{body}} - \left(\frac{dR}{dx}\right)^2$$

and one which depends on α and θ,

$$C_{pc} = -\frac{2}{U}\left(\frac{\partial \phi_c}{\partial x}\right)_{\text{body}} + (1 - 4\sin^2\theta)\alpha^2$$

The expression for C_{pa} is the one given in Eq. 9·34*c*. It evidently contributes the same axial force as before, and gives no contribution to the cross force. The expression for C_{pc} may be evaluated from the cross-flow potential (Eq. 9·43*a*), which gives

$$C_{pc} = -4\alpha \frac{dR}{dx} \cos\theta + (1 - 4\sin^2\theta)\alpha^2$$

The *cross-force* is determined entirely by C_{pc}, and may be computed as follows. The radial component of the force acting on an element of surface area is $C_{pc} q R\, d\theta\, dx$ as shown in Fig. 9·12. (q is the dynamic pressure.) To find the component of cross force in the direction of U_c, this must be multiplied by $(-\cos\theta)$ and integrated over the body surface:

$$N = q_\infty \int_0^L \int_0^{2\pi} \cos\theta(-C_{pc}) R\, d\vartheta\, dx$$

$$= 4\alpha q_\infty \int_0^L \int_0^{2\pi} \cos^2\theta \frac{R\, dR}{dx}\, d\theta\, dx - \alpha^2 q_\infty \int_0^L \int_0^{2\pi} \cos\theta (1 - 4\sin^2\theta) R\, d\theta\, dx$$

$$= 2\alpha q_\infty S(L)$$

where $S(L) = \pi R^2(L)$ is the area of the base. Referred to this base area, the normal force coefficient is thus

$$C_N = 2\alpha$$

If the base area is zero, that is, the body is closed, the *cross force is zero*, in the slender body approximation.

The cross flow contributes to the *axial* force (cf. Fig. 9·12) the term

$$A_2 = q_\infty \int_0^{R(L)} \int_0^{2\pi} C_{pc} R\, d\theta\, dR = -q_\infty \pi [R(L)]^2 \alpha^2 = -q_\infty S(L) \alpha^2$$

and so the corresponding contribution to the axial force coefficient is

$$C_2 = -\alpha^2$$

The net axial force is $A = A_1 + A_2$, where $A_1 = qS(L)C_{D1}$ may be computed for a specific body from the axially symmetric relation given in Eq. 9·35. We are omitting the base pressure drag (Article 9·12).

Finally, from the cross force N and the axial force A, we may compute the lift and drag,

$$L = N \cos\alpha - A \sin\alpha$$

$$D = N \sin\alpha + A \cos\alpha$$

or

$$C_L \doteq C_N\left(1 - \frac{\alpha^2}{2}\right) - (C_{D1} + C_2)\alpha$$

$$C_D \doteq C_N\alpha + (C_{D1} + C_2)\left(1 - \frac{\alpha^2}{2}\right)$$

Inserting the expressions for C_N and C_2, and retaining only the dominant terms, assuming $C_{D1} \ll 2$, we obtain

$$C_L = 2\alpha$$

$$C_D = C_{D1} + \alpha^2 \tag{9.45}$$

This is the slender body result for the lift and drag coefficients, referred to the *base area*.

The drag increment due to angle of attack is called the *induced drag;* the induced drag coefficient is

$$C_{Di} = \alpha^2$$

A point of interest is the value of the ratio

$$\frac{C_{Di}}{C_L} = \frac{D_i}{L} = \frac{\alpha}{2}$$

This shows that the force vector due to angle of attack is midway between the normal to the body axis and the normal to the flight path. This result, which is the same for all slender bodies, may be compared to the well-known result for incompressible flow over an elliptic wing of high aspect ratio, for which $C_{Di}/C_L = C_L/\pi Æ$.

The induced drag of the slender body, like that of the high aspect ratio wing, is related to the appearance of trailing vortices.

9·18 Slender Body Theory

In the preceding articles some general results were obtained for bodies of circular cross-section and arbitrary shape in the meridian plane, by assuming them to be so slender that certain approximations, valid near the axis, can be used. A further consequence is that the cross flow in each cross-section is independent of the flow at other sections (Eq. 9·43), and is governed by the two-dimensional equation

$$\frac{\partial^2 \phi_c}{\partial y^2} + \frac{\partial^2 \phi_c}{\partial z^2} = 0 \tag{9.46}$$

instead of the full equation

$$(1 - M_\infty)^2 \frac{\partial^2 \phi}{\partial x^2} + \frac{\partial^2 \phi}{\partial y^2} + \frac{\partial^2 \phi}{\partial z^2} = 0$$

This simplification was first introduced by Munk, in connection with airship theory, and later was extended by R. T. Jones to slender bodies of arbitrary cross-section at arbitrary Mach number. The underlying physical idea is that, for a very elongated body, variations in the x-direction are much smaller than those in the other directions, at least near the body. More precisely, the term $(1 - M_\infty^2)\partial^2\phi/\partial x^2$ is assumed to be negligible compared to the others.

Equation 9·46 is the same equation that governs plane incompressible flow, and for which the general methods of conformal mapping are available. The connection between this two-dimensional flow and the *actual* three-dimensional problem is through the boundary condition. Thus solutions of Eq. 9·46 have the form

$$\phi_c = \phi_c(y, z; x) \tag{9·47}$$

x appears as a parameter, corresponding to the fact that at each section the boundary condition is a function of x, since it depends on the geometry of the body.

According to slender body theory, the lift coefficient for a given body depends only on the cross flow and hence is independent of Mach number. The accuracy of the theory for practical application is rather low. Its usefulness lies in its generality, and for complicated configurations it may provide the only means for obtaining an estimate of the lift. An example of its application to a slender airplane is shown in Fig. 9·13.

The simple cross-flow theory does not give the correct thickness drag, since this has terms that depend on interaction between different sections of the body (cf. Eq. 9·35); a general slender body theory for drag is not as simple as for lift. A generalization of the results of Article 9·12 to bodies of noncircular cross-section has been given by Ward.† Cole‡ has given a systematic expansion procedure in terms of suitable slenderness parameters; this gives a systematic ordering of successive terms of the expansion and an estimate of the accuracy, and provides for a consistent extension to higher-order (nonlinear) approximations to the exact equations of motion.

9·19★ Rayleigh's Formula

In closing this chapter we shall briefly discuss the motion of a body in fluid at rest. In Chapters 3 and 4, some remarks were made to the effect that each part of the body acts like a piston on the fluid which it encounters, setting up a wave motion and establishing a corresponding "piston pressure." The local "piston velocity" or particle velocity which the body

†G. N. Ward, *Linearized Theory of Steady High-Speed Flow*, Cambridge (1955).

‡J. D. Cole and A. F. Messiter, "Expansion Procedures and Similarity Laws for Transonic Flow," Ninth International Congress Applied Mechanics, Brussels (1956).

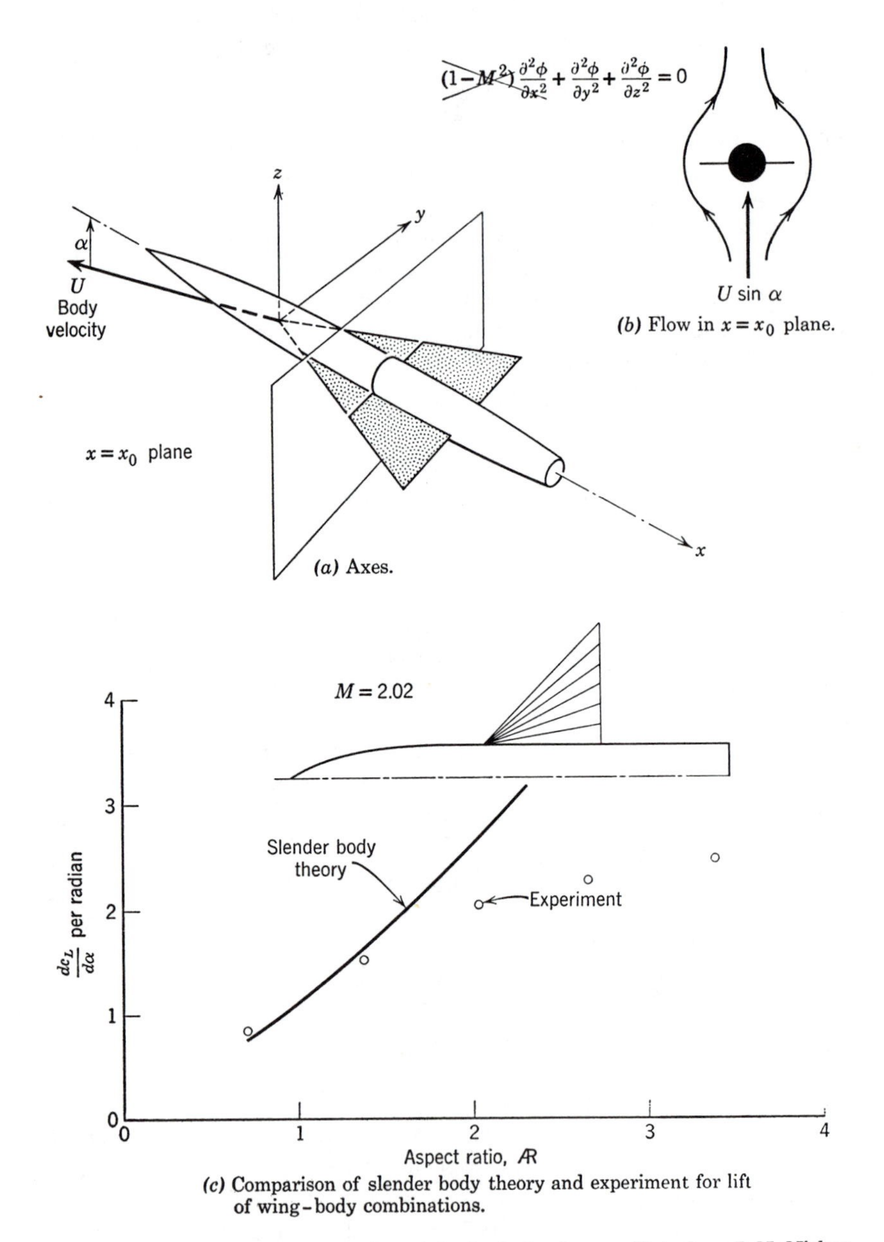

(b) Flow in $x = x_0$ plane.

(a) Axes.

(c) Comparison of slender body theory and experiment for lift of wing-body combinations.

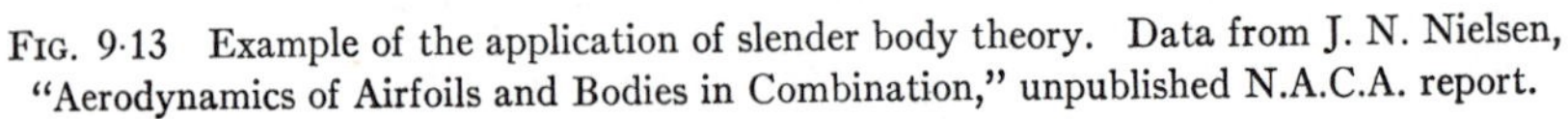

FIG. 9·13 Example of the application of slender body theory. Data from J. N. Nielsen, "Aerodynamics of Airfoils and Bodies in Combination," unpublished N.A.C.A. report.

imparts to the fluid is $U \tan \theta \doteq U\theta$, where θ is the body surface slope along the direction of motion. Within linearized theory, the piston velocity is normal to the direction of motion. This also gives some insight into the method of representing a body by distributions of sources; the sources also act on the fluid like pistons, establishing a velocity normal to the direction of motion. This "piston velocity" is usually called the *upwash.*

The piston pressure corresponding to the piston velocity cannot be obtained from the simple one-dimensional formula of Chapter 3, because in three-dimensional space there is a spreading of the effect from one part of the body to another. In linearized theory, the disturbances propagate with the constant, acoustic speed a_∞, and the motion of the fluid is described by the *acoustic* equation†

$$\nabla^2\phi - \frac{1}{a_\infty^{\,2}}\frac{\partial^2\phi}{\partial t^2} = 0 \tag{9·48}$$

The fluid velocity is derived from the potential,

$$u_i = \frac{\partial\phi}{\partial x_i} \equiv \nabla\phi$$

and the perturbation pressure is obtained from

$$p' = p - p_\infty = -\rho_\infty \frac{\partial\phi}{\partial t}$$

The acoustic equation describes the fluid motion for any arbitrary motion of the body, so long as the conditions of linearization are satisfied; that is, so long as the disturbances or "perturbations" are small. The speed of the body may be changing, $U = U(t)$, and the body may be oscillating or vibrating. The motion may be subsonic or supersonic, and even the transonic range may be treated in cases of accelerated motion, provided the body accelerates through transonic speed fast enough so that large perturbations do not build up.‡

A *planar body*, such as a thin wing, is represented by a surface distribution of sources $f(x, y, t)$ moving in the plane $z = 0$. The solution of the acoustic equation is given by Rayleigh's formula§

$$\phi(x, y, z, t) = -\iint_{\substack{\text{Plane}\\ z=0}} \frac{f(\xi, \eta, t - h/a_\infty)}{h}\, d\xi\, d\eta \tag{▶9·49}$$

†H. Lamb, *Hydrodynamics*, 6th Ed., Dover, N.Y., 1945, p. 493.

‡J. D. Cole, "Acceleration of Slender Bodies of Revolution through Sonic Velocity," *J. Appl. Phys., 26* (1955), p. 322.

§Rayleigh, *Theory of Sound (II)*, Dover, N.Y. (1945), p. 107.

where

$$h = \sqrt{(x - \xi)^2 + (y - \eta)^2 + z^2}$$

This formula states that the potential at a point $P(x, y, z)$ and time t is obtained by integrating the contributions from the sources. A typical source, of strength $f\, d\xi\, d\eta$, at the point $Q(\xi, \eta)$ is at a distance h from P. The formula shows that for the source strength at Q we must use the value at an earlier time $t - h/a_\infty$, since the effect takes a finite time, h/a_∞, to reach P. Furthermore, the effect is attenuated with distance, as $1/h$.

An *axially symmetric* body may similarly be represented by a line distribution of sources $f(x, t)$ along the x-axis. Rayleigh's formula for the axially symmetric acoustic field is

$$\phi(x, r, t) = -\int_{\substack{\text{Line} \\ r=0}} \frac{f(\xi, t - h/a_\infty)}{h}\, d\xi \qquad \blacktriangleright(9\cdot50)$$

where

$$h = \sqrt{(x - \xi)^2 + r^2}$$

The appropriate source distribution for a given body is found by computing the upwash, $(\partial\phi/\partial z)_{z=0}$ or $(\partial\phi/\partial r)_{r=0}$, respectively, from Eqs. 9·49 or 9·50, and setting this equal to $U\theta = U\partial Z/\partial X$ or $U\partial R/\partial X$, respectively. $Z(X, Y)$ and $R(X)$ are the respective equations for the body surface, the coordinates X, Y, which are fixed in the body, being used to distinguish from the space coordinates x, y. For a planar body, moving in the negative x-direction with speed $U(t)$, the resulting relation between source strength and upwash is

$$f(x, y, t) = \frac{1}{2\pi} U(t) \frac{\partial Z}{\partial X} \qquad \blacktriangleright(9\cdot51)$$

Thus the source strength for a planar system depends only on the *local* upwash, and is proportional to it.

In the axially symmetric case the relation is not so simple; an integral equation is obtained. It may be shown, however, that for a very *slender body* the source strength again depends only on the local body geometry, according to the simple relation†

$$f(x, t) = \frac{1}{4\pi} \frac{\partial S(x, t)}{\partial t}$$

Thus, the source strength is equal to the local rate of increase of cross-

†Due to F. I. Frankl. See J. D. Cole, "Note on Non-Stationary Slender Body Theory," *J. Aeronaut. Sci.*, *20* (1953), p. 798.

sectional area, that is, to the outward "pushing" of the body as it moves through the fluid. For a rigid body, $S = S(X)$, the relation may be written

$$f(x, t) = \frac{1}{4\pi} U(t) \frac{dS}{dX} \tag{9·52}$$

As noted above, these results are very general. They apply also to the special case $U =$ const. It may be shown that in this case the transformation $x' = x + Ut$ reduces the results to those of steady-flow conditions in the conventional coordinate system (now labeled x', y, z), in which the body is stationary and the flow is in the positive x-direction. For instance, the acoustic equation (9·48) becomes

$$\frac{\partial^2\phi}{\partial x'^2} + \frac{\partial^2\phi}{\partial y^2} + \frac{\partial^2\phi}{\partial z^2} - \frac{U^2}{a_\infty^2}\frac{\partial^2\phi}{\partial x'^2} = 0$$

(cf. Eq. 8·9*b*). The integral solution (9·49) reduces to the form

$$\phi(x', y, z) = -\frac{1}{2\pi}\int_{-\infty}^{\infty}\int_{-\infty}^{\infty}\frac{w(\xi, \eta, 0)\, d\xi\, d\eta}{\sqrt{(x' - \xi)^2 + (1 - M_\infty^2)[(y - \eta)^2 + z^2]}} \qquad M_\infty < 1 \tag{9·53a}$$

$$\phi(x', y, z) = -\frac{1}{\pi}\iint_{\text{hyp}}\frac{w(\xi, \eta, 0)\, d\xi\, d\eta}{\sqrt{(x' - \xi)^2 - (M_\infty^2 - 1)[(y - \eta)^2 + z^2]}} \qquad M_\infty > 1 \tag{9·53b}$$

where $w = U(dz/dx)_{\text{body}}$ is the upwash normal to the plane of the wing. "hyp" denotes the region of integration for the supersonic case. It includes only those sources which lie ahead of the forward running Mach lines from the point x', y, z (cf. Article 9·7). These form a Mach cone whose intersection with the x', y plane defines a hyperbolic region for the integration.

Equations 9·53 are the basic equations for the source method of solution, for steady flow past planar bodies, such as wings.

Application of the same procedure to Eq. 9·50 gives the corresponding steady-state equations, 9·17*a* and 9·20, for axially symmetric flow.

CHAPTER 10

The Similarity Rules of High-Speed Flow

10·1 Introduction

In the special examples of linearized two-dimensional and axially symmetric flows we noticed that the parameters could be arranged into certain functional groups, so that a *single curve* represents the solution for a whole family of shapes and range of Mach numbers. For example, for flow past wave-shaped walls (Article 8·5) the relation involving the pressure coefficient is of the form

$$\frac{C_p\sqrt{|1 - M_\infty^2|}}{\epsilon\alpha} = fn(\alpha x_1, \alpha x_2\sqrt{|1 - M_\infty^2|}) \qquad (10\cdot1a)$$

whereas for supersonic flow over slender cones (Article 9·11) it is of the form

$$\frac{C_p}{\delta^2} = fn(\delta\sqrt{M_\infty^2 - 1}) \qquad (10\cdot1b)$$

These are examples of *similarity relations*. In these special cases, the explicit form of the functions is known (Eqs. 8·26*d* and 9·32*c*), since the solution of the equations of motion can be found. It is in cases where solutions are not readily obtainable, for example, the nonlinear case of transonic flow, that a similarity analysis is especially useful. But even where explicit solutions are available, as in linearized flow, the rearrangement into similarity groups provides valuable insight into the solution and is a useful concept in engineering applications.

In this chapter we shall discuss the idea of similarity in the sense of the above examples, aiming mainly for an extension to transonic flow. We shall be particularly interested in the pressure coefficient, C_p, on the surface of the body, and in the corresponding forces.

The steady, two-dimensional or axially symmetric flow of a perfect gas over a body of chord (or length) c and maximum thickness t is characterized by the following parameters,

$$C_p = C_p(x_1/c, M_\infty, \gamma, t/c) \qquad (10\cdot2)$$

This is simply the result of dimensional analysis. The problem now is to find the functional form which will represent C_p in such a way that the five dimensionless variables in Eq. 10·2 are grouped into a *smaller* number of "similarity parameters." It will be found possible to reduce the number to three. The pressure coefficient parameter at a given station, x_1/c, can then be represented by a *single* curve for all Mach numbers and gases and a whole family of shapes.

The difference between *dimensional analysis* and *similarity considerations* is the following. Dimensional analysis simply lists for us the dimensionless parameters that are involved, whereas similarity analysis goes much further, showing how to group these dimensionless quantities in such a way as to reduce the number of independent variables. For dimensional analysis, we need only a certain amount of "bookkeeping," that is, we have to know, or guess, the variables involved in a problem. For similarity analysis, we need to know more, for example, the differential equations and boundary conditions, or possibly some integral relations. Sometimes similarity rules come from a set of experiments.

In our particular application, we know the differential equations and boundary conditions, and hence we shall derive the similarity rules from these. It should be emphasized that for the *linearized* equations it is not essential to make a separate study of similarity. We could abstract the rules from specific examples, like the wave-shaped wall, since the general solutions can be built up from the special ones by superposition. The main importance of the similarity rules is in their extension to transonic and hypersonic flow, where explicit solutions are rare and superposition is not possible.

10·2 Two-Dimensional Linearized Flow. Prandtl-Glauert and Göthert Rules

The linearized equations for the perturbation potential $\phi(x, y)$ in plane, steady flow may be written

$$\frac{\partial^2 \phi}{\partial x^2} + \frac{1}{1 - M_1^2} \frac{\partial^2 \phi}{\partial y^2} = 0 \tag{10·3}$$

The subscript 1, used on the free stream Mach number M_1, is to distinguish it from a second flow at Mach number M_2, with which it will be presently compared.

The shape of the boundary may be given in the form

$$y = t_1 f\left(\frac{x}{c}\right) = \tau_1 c f\left(\frac{x}{c}\right) \tag{10·4}$$

c is the chord, and τ_1 is the thickness ratio t_1/c, of whatever body shape is being considered. c and t_1 are the characteristic dimensions in the x and y directions, respectively. Equation 10·4 is in a form suitable for a similarity argument. More explicitly

$$\frac{y}{c} = \tau_1 f\left(\frac{x}{c}\right) \tag{10·4a}$$

The boundary condition which ϕ must satisfy is

$$\left(\frac{\partial\phi}{\partial y}\right)_{y=0} = U_1\left(\frac{dy}{dx}\right)_{\text{body}} = U_1\tau_1 f'\left(\frac{x}{c}\right) \tag{10·5}$$

where U_1 is the free-stream velocity.

The pressure coefficient, C_{p1}, on the boundary is given by

$$C_{p1} = -\frac{2}{U_1}\left(\frac{\partial\phi}{\partial x}\right)_{y=0} \tag{10·6}$$

Now consider the potential function $\Phi(\xi, \eta)$ of a second flow in the coordinates (ξ, η), which we assume is related to ϕ by the relation

$$\phi(x, y) = A\frac{U_1}{U_2}\Phi(\xi, \eta) = A\frac{U_1}{U_2}\Phi\left(x, \sqrt{\frac{1 - M_1^2}{1 - M_2^2}}\,y\right) \tag{10·7}$$

where A is a constant to be determined later. This implies that the correspondence between the two coordinate systems is

$$\xi = x; \quad \eta = \sqrt{\frac{1 - M_1^2}{1 - M_2^2}}\,y$$

Introducing Eq. 10·7 into the differential equation (10·3), we find that Φ satisfies the equation

$$\frac{\partial^2\Phi}{\partial\xi^2} + \frac{1}{1 - M_2^2}\frac{\partial^2\Phi}{\partial\eta^2} = 0 \tag{10·8}$$

Hence, if ϕ is a solution corresponding to a Mach number M_1, then Φ is a solution corresponding to M_2. The boundary condition (10·5) yields

$$\left(\frac{\partial\phi}{\partial y}\right)_{y=0} = A\frac{U_1}{U_2}\sqrt{\frac{1 - M_1^2}{1 - M_2^2}}\left(\frac{\partial\Phi}{\partial\eta}\right)_{\eta=0} = U_1\tau_1 f'\left(\frac{x}{c}\right) \tag{10·9}$$

The only variable in Eq. 10·9 is $x/c(=\xi/c)$; hence Eq. 10·9 will also satisfy the boundary condition

$$\left(\frac{\partial\Phi}{\partial\eta}\right)_{\eta=0} = U_2\tau_2 f'\left(\frac{\xi}{c}\right)$$

provided that f' is the same function in both cases, and that τ_2 is related to τ_1 by

$$A\sqrt{\frac{1-M_1^2}{1-M_2^2}}\,\tau_2 = \tau_1 \tag{10·10}$$

The fact that the function f must be the same in both flows simply means that we can compare only bodies of the same "family," whose family shape is determined by the shape function f.

The pressure coefficient C_{p1} may also be rewritten,

$$C_{p1} = -\frac{2}{U_1}\left(\frac{\partial\phi}{\partial x}\right)_{y=0} = -\frac{2}{U_2}A\left(\frac{\partial\Phi}{\partial\xi}\right)_{\eta=0}$$

But the pressure coefficient in the second flow must be

$$C_{p2} = -\frac{2}{U_2}\left(\frac{\partial\Phi}{\partial\xi}\right)_{\eta=0}$$

Thus the relation between the pressure coefficients in the two flows is

$$C_{p1} = AC_{p2} \tag{10·11}$$

Equations 10·10 and 10·11 express the following similarity rules: Two members of a family of shapes characterized by the thickness ratios τ_1 and τ_2 have pressure distributions given by coefficients C_{p1} and C_{p2}. If the Mach numbers of the flows are M_1 and M_2, respectively, then $C_{p1} = AC_{p2}$, provided that

$$\tau_1 = A\sqrt{\frac{1-M_1^2}{1-M_2^2}}\,\tau_2$$

This somewhat clumsy statement may be written in the general form

$$\frac{C_p}{A} = fn\left(\frac{\tau}{A\sqrt{1-M_\infty^2}}\right) \tag*{▶(10·12)†}$$

So far the factor A remains arbitrary. This is due to the fact that the linear Eq. 10·3 is homogeneous in ϕ, making it possible to multiply by any constant factor, without changing the equation. In the next article it will be seen that in the transonic equation the nonlinearity of one of the terms precludes an arbitrary choice of A. In fact, the condition imposed on A there will determine the transonic similarity rules.

Equation 10·12 includes both the Prandtl-Glauert rule and the Göthert rule, which may be obtained as follows.

(1) Choosing $A = 1$ yields $C_p = fn\left(\dfrac{\tau}{\sqrt{1-M_\infty^2}}\right)$ (10·13*a*)

†It may be helpful to the reader to first put $A = A_1/A_2$ in Eqs. 10·10 and 10·11.

(2) Choosing $A = \dfrac{1}{\sqrt{1 - M_\infty^2}}$ yields $C_p = \dfrac{1}{\sqrt{1 - M_\infty^2}} fn(\tau)$ (10·13b)

(3) Choosing $A = \tau$ yields $C_p = \tau fn(\sqrt{1 - M_\infty^2})$ (10·13c)

(4) Choosing $A = \dfrac{1}{1 - M_\infty^2}$ yields $C_p = \dfrac{1}{1 - M_\infty^2} fn(\tau\sqrt{1 - M_\infty^2})$ (10·13d)

Methods (1), (2), and (3) are the standard forms of the Prandtl-Glauert rule. Method (1) states that C_p remains invariant with M if the thickness ratio τ is reduced as M is increased, by an amount which will keep $\tau/\sqrt{1 - M_\infty^2} =$ const. Method (2) states that for a given member of the family of shapes, C_p increases with M_∞ as $(1 - M_\infty^2)^{-1/2}$, and method (3) states that C_p is proportional to τ for a fixed value of M_∞.

Method (4) gives the rule due to Göthert, which applies to the axially symmetrical as well as the two-dimensional case (as will be shown in Article 10·4). It states that C_p increases with Mach number as $(1 - M_\infty^2)^{-1}$ if the thickness ratio increases as $(1 - M_\infty^2)^{-1/2}$.

Applications of these rules will be given in Article 10·6.

In Eqs. 10·12 and 10·13 the rules are written for linearized *subsonic flow*. Evidently, Eq. 10·12 could have been written with $\sqrt{M_\infty^2 - 1}$, since all the preceding equations could have been written with $\sqrt{(M_1^2 - 1)/(M_2^2 - 1)}$ instead of $\sqrt{(1 - M_1^2)/(1 - M_2^2)}$. Hence, to extend the rules to supersonic flow one has only to replace $\sqrt{1 - M_\infty^2}$ everywhere by $\sqrt{|1 - M_\infty^2|}$ or else rewrite the rules in forms where the square roots do not appear. For example, one can rewrite Eq. 10·12 in the form

$$\frac{C_p}{A} = fn\left(\frac{\tau^2}{A^2(1 - M_\infty^2)}\right)$$

which is valid for subsonic and supersonic motion.

Figure 10·1 shows an application of the Prandtl-Glauert rule to the pressure distribution on an airfoil. Values of C_p at $M_\infty = 0.60$, 0.70 and 0.80 are calculated from the experimental data at $M_\infty = 0.40$. For example $(C_p)_{0\cdot 80} = (C_p)_{0\cdot 40}\sqrt{[1 - (0.40)^2]/[1 - (0.80)^2]} = 1.38(C_p)_{0\cdot 40}$. The agreement with experiment becomes less satisfactory as the transonic range is approached. For $M_\infty = 0.80$, the values of C_p have exceeded the *critical* value, which corresponds to sonic conditions, and there is a supersonic region between 10 and 30 percent chord.

10·3 Two-Dimensional Transonic Flow. von Kármán's Rules

As the free-stream Mach number approaches unity, the linearized equa-

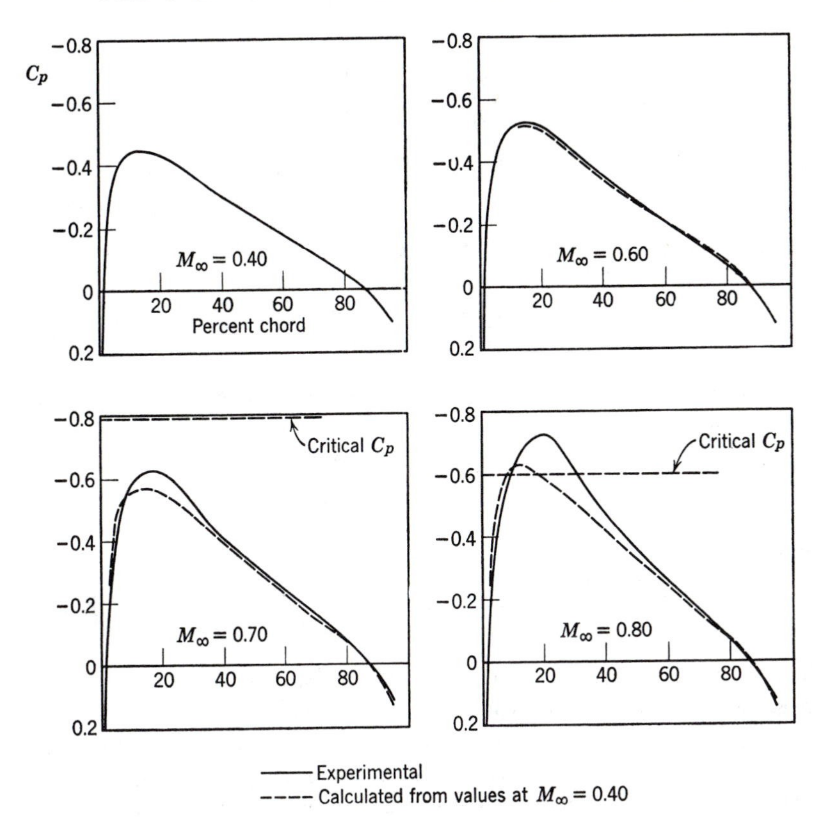

FIG. 10·1 Comparison of Prandtl-Glauert similarity rule with experiment. (Experimental data for NACA 0012 airfoil, taken from *NACA Tech. Note* 2174 by J. L. Amick.)

tion becomes inaccurate and has to be replaced by (cf. Eq. 8·9*a*)

$$\frac{\partial^2\phi}{\partial x^2} + \frac{1}{1 - M_1{}^2}\frac{\partial^2\phi}{\partial y^2} = \frac{(\gamma_1 + 1)M_1{}^2}{1 - M_1{}^2}\frac{1}{U_1}\frac{\partial\phi}{\partial x}\frac{\partial^2\phi}{\partial x^2} \tag{10·14}$$

As before, the equation is written for a flow with free-stream Mach number M_1 and velocity U_1. If we again introduce Φ by the relation 10·7 we find that $\Phi(\xi, \eta)$ satisfies the equation

$$\frac{\partial^2\Phi}{\partial \xi^2} + \frac{1}{1 - M_2{}^2}\frac{\partial^2\Phi}{\partial \eta^2} = \frac{(\gamma_1 + 1)M_1{}^2}{1 - M_1{}^2}\frac{A}{U_2}\frac{\partial\Phi}{\partial \xi}\frac{\partial^2\Phi}{\partial \xi^2} \tag{10·15}$$

We notice now that, if Φ is to satisfy the same equation as ϕ, that is, the transonic equation 10·14, then A must be chosen to give the correct coefficient in the right-hand term of Eq. 10·15. The required condition is

$$\frac{(\gamma_1 + 1)M_1{}^2}{1 - M_1{}^2}A = \frac{(\gamma_2 + 1)M_2{}^2}{1 - M_2{}^2}$$

Hence

$$A = \frac{\gamma_2 + 1}{\gamma_1 + 1} \frac{M_2^2}{M_1^2} \frac{1 - M_1^2}{1 - M_2^2} \tag{10·16}$$

For transonic flow, Eq. 10·16 supplements Eqs. 10·10 and 10·11. Thus, if we write the expression for C_p in the form of Eq. 10·12, then A has to be replaced by $\frac{1 - M_\infty^2}{(\gamma + 1)M_\infty^2}$, and instead of Eq. 10·12 we now have

$$\frac{C_p(\gamma + 1)M_\infty^2}{1 - M_\infty^2} = fn\left(\frac{\tau(\gamma + 1)M_\infty^2}{(1 - M_\infty^2)^{3/2}}\right) \tag{10·17}$$

This relation may be rewritten in various ways. For instance, if both sides are multiplied by

$$\chi = \frac{1 - M_\infty^2}{[\tau(\gamma + 1)M_\infty^2]^{2/3}}$$

the relation becomes

$$\frac{C_p[(\gamma + 1)M_\infty^2]^{1/3}}{\tau^{2/3}} = fn\left(\frac{1 - M_\infty^2}{[\tau(\gamma + 1)M_\infty^2]^{2/3}}\right) = fn(\chi) \tag{10·18}$$

This is von Kármán's transonic similarity rule, in a form due to Spreiter; it is slightly different from von Kármán's original one.

Equation 10·18 is valid from subsonic through sonic to supersonic velocities. It includes the Prandtl-Glauert and Göthert rules as special cases, valid in the regions where the linearized equations hold.

The fact that the constant A could not be chosen arbitrarily now appears in the circumstance that it is impossible to compare the same body at different Mach numbers, or different bodies at a given Mach number, as, for example, in the Prandtl-Glauert rules of Eqs. 10·13*b* and *c*. One can only compare bodies of *different* thickness ratios, τ_1 and τ_2, at *different* Mach numbers, M_1 and M_2, which must be chosen so that $\chi_1 = \chi_2$, that is,

$$\frac{1 - M_1^2}{[\tau_1(\gamma_1 + 1)M_1^2]^{2/3}} = \frac{1 - M_2^2}{[\tau_2(\gamma_2 + 1)M_2^2]^{2/3}} \tag{10·19}$$

10·4 Linearized Axially Symmetric Flow

The linearized equation for axially symmetric flow, $\phi(x, r)$, with free-stream velocity U_1 and Mach number M_1, is

$$\frac{\partial^2\phi}{\partial x^2} + \frac{1}{1 - M_1^2}\left(\frac{\partial^2\phi}{\partial r^2} + \frac{1}{r}\frac{\partial\phi}{\partial r}\right) = 0 \tag{10·20}$$

Again, as in Article 10·2, $\phi(x, r)$ may be related to a second flow $\Phi(\xi, R)$ by the transformation

$$\phi(x, r) = A\,\frac{U_1}{U_2}\,\Phi\left(x, r\sqrt{\frac{1 - M_1^2}{1 - M_2^2}}\right) \tag{10·21}$$

With this choice, Φ satisfies the same equation as ϕ, namely, Eq. 10·20, for a Mach number M_2. The procedure so far is identical with the one followed in two-dimensional flow. But there is a difference in the boundary condition. If $r = \tau_1 c f(x/c)$ is the shape of the axially symmetric body, then the exact boundary condition is

$$\left(\frac{\partial \phi}{\partial r}\right)_{\text{body}} = U_1 \tau_1 f'\left(\frac{x}{c}\right)$$

In two-dimensional flow, the boundary condition could be approximately satisfied by applying it at $r = 0$ but this is not possible in axially symmetric flow, as was shown in Article 9·3. Hence it is necessary to use the exact boundary condition

$$\left(\frac{\partial \phi}{\partial r}\right)_{r=\tau_1 c f(x/c)} = U_1 \tau_1 f'\left(\frac{x}{c}\right)$$

The left-hand side may be written in terms of Φ, from Eq. 10·21,

$$\left(\frac{\partial \phi}{\partial r}\right)_{r=\tau_1 c f(x/c)} = A\,\frac{U_1}{U_2}\sqrt{\frac{1 - M_1^2}{1 - M_2^2}}\left(\frac{\partial \Phi}{\partial R}\right)_{R=\sqrt{\frac{1-M_1^2}{1-M_2^2}}\,\tau_1 c f(x/c)} \tag{10·22}$$

On the other hand, the boundary condition which must be satisfied by Φ is

$$\left(\frac{\partial \Phi}{\partial R}\right)_{R=\tau_2 c F(\xi/c)} = U_2 \tau_2 c F'\left(\frac{\xi}{c}\right) \tag{10·23}$$

In order to compare Eqs. 10·22 and 10·23, it is required that $f(x/c) = F(\xi/c)$, which is the same condition as before, and in addition that

$$\tau_1\sqrt{\frac{1 - M_1^2}{1 - M_2^2}} = \tau_2$$

Putting these conditions back into Eq. 10·22 gives the equation

$$\tau_1 f'\left(\frac{x}{c}\right) = A\sqrt{\frac{1 - M_1^2}{1 - M_2^2}}\,\tau_1\sqrt{\frac{1 - M_1^2}{1 - M_2^2}}\,f'\left(\frac{x}{c}\right)$$

Thus A is determined by the relation

$$A = \frac{1 - M_2^2}{1 - M_1^2} \tag{10·24}$$

Next we must work out the relation between the pressure coefficients.

Again, because of the singular conditions near the axis, it is not accurate to use the linearized pressure coefficient (Eq. 10·6). The consistent pressure coefficient for axially symmetric flow (Article 9·4) is

$$C_{p1} = -\frac{2}{U_1}\left(\frac{\partial \phi}{\partial x}\right)_{r=0} - \frac{1}{U_1^2}\left(\frac{\partial \phi}{\partial r}\right)_{r=0}^2$$

Using Eq. 10·21, we may write this in terms of Φ,

$$C_{p1} = -\frac{2}{U_2} A \left(\frac{\partial \Phi}{\partial \xi}\right)_{R=0} - \frac{A^2}{U_2^2}\frac{1 - M_1^2}{1 - M_2^2}\left(\frac{\partial \Phi}{\partial R}\right)_{R=0}^2$$

Using Eq. 10·24, we may factor out the constant A, leaving

$$C_{p1} = AC_{p2}$$

as in the two-dimensional case. Thus, the similarity law may again be expressed in the form

$$\frac{C_p}{A} = fn\left(\frac{\tau}{A\sqrt{1 - M_\infty^2}}\right) \tag{10·25}$$

But now it is not possible to choose A arbitrarily since it must satisfy Eq. 10·24. That is, it is required that $A = (1 - M_\infty^2)^{-1}$. This gives Göthert's rule,

$$C_p(1 - M_\infty^2) = fn(\tau\sqrt{1 - M_\infty^2}) \tag{10·26}$$

It is clear from the derivation, and from Article 10·2, that Eq. 10·26 may also be used for two-dimensional flow.

Dividing both sides by $\tau^2(1 - M_\infty^2)$ gives the alternate form

$$\frac{C_p}{\tau^2} = fn(\tau\sqrt{1 - M_\infty^2}) \tag{10·27}$$

Since it has been necessary to use the free parameter A to adjust the "non-linear" boundary condition, it would not be possible to use it for the similarity condition in the *transonic* equation. There is no corresponding similarity relation for axially symmetric transonic flow.†

10·5 Planar Flow

The same method that was used for two-dimensional flow can be extended to three-dimensional planar flow.

†A transonic similarity rule for slender bodies can be obtained if one subtracts out a source term. Such a similarity rule has been obtained by Oswatitsch and Berndt and by J. D. Cole. E.g., for a cone of nose angle 2θ it is

$$\frac{C_p}{\theta^2} + 4 \log \theta = fn\left(\frac{|1 - M_\infty^2|}{\theta^2}\right)$$

The differential equation and the boundary conditions for this case are

$$\frac{\partial^2 \phi}{\partial x} + \frac{1}{1 - M_1^2}\left(\frac{\partial^2 \phi}{\partial y^2} + \frac{\partial^2 \phi}{\partial z^2}\right) = 0 \tag{10·28}$$

$$\left(\frac{\partial \phi}{\partial z}\right)_{z=0} = U_1 \tau_1 c \frac{\partial f}{\partial x} \tag{10·29}$$

where f is the shape function of the planar boundary†

$$z = \tau_1 c f\left(\frac{x}{c}, \frac{y}{b}\right)$$

A planar shape is one whose dimension in the y-direction, as well as in the x-direction, is large compared to the thickness. This characteristic dimension, b, is called the span, since planar shapes are essentially wing shapes.

In analogy with the two-dimensional treatment, a new flow $\Phi(\xi, \eta, \zeta)$ is introduced by the transformation

$$\phi(x, y, z) = A \frac{U_1}{U_2} \Phi\left(x, y\sqrt{\frac{1 - M_1^2}{1 - M_2^2}}, z\sqrt{\frac{1 - M_1^2}{1 - M_2^2}}\right) \tag{10·30}$$

With this choice, Φ satisfies the differential equation (10·28), for M_2. The boundary conditions are then related by

$$\left(\frac{\partial \phi}{\partial z}\right)_{z=0} = A \frac{U_1}{U_2}\sqrt{\frac{1 - M_1^2}{1 - M_2^2}}\left(\frac{\partial \Phi}{\partial \zeta}\right)_{\zeta=0}$$

The left-hand side may be replaced by Eq. 10·29, and there is an analogous expression for $(\partial \phi/\partial \zeta)_{\zeta=0}$. Thus

$$\tau_1 \frac{\partial}{\partial x} f\left(\frac{x}{c}, \frac{y}{b_1}\right) = A\sqrt{\frac{1 - M_1^2}{1 - M_2^2}}\, \tau_2 \frac{\partial}{\partial \xi} f\left(\frac{\xi}{c}, \frac{\eta}{b_2}\right)$$

Just as before, it is necessary that the shape function f be the same in both cases, giving the equation

$$\frac{\tau_1}{\tau_2} = A\sqrt{\frac{1 - M_1^2}{1 - M_2^2}}$$

In addition, it is necessary that

$$\frac{y}{b_1} = \frac{\eta}{b_2} = \frac{y}{b_2}\sqrt{\frac{1 - M_1^2}{1 - M_2^2}}$$

†In the conventional notation for planar systems, x is streamwise, y is spanwise, and z is normal to the plane of the wing.

or

$$\frac{b_2}{b_1} = \sqrt{\frac{1 - M_1^2}{1 - M_2^2}} \tag{10·31}$$

The relations for C_p are obtained just as in the two-dimensional case, and the pressure coefficient on any point of the planar surface satisfies a similarity law of the form

$$\frac{C_p}{A} = fn\left(\frac{\tau}{A\sqrt{1 - M_\infty^2}};\quad b\sqrt{1 - M_\infty^2}\right) \tag{10·32}$$

Since the chord c was kept unaltered, the wing area S is proportional to b for given plan form; hence the aspect ratio $AR \equiv b^2/S$ is proportional to b, and hence Eq. 10·32 may be written

$$\frac{C_p}{A} = f\left(\frac{\tau}{A\sqrt{1 - M_\infty^2}};\quad AR\sqrt{1 - M_\infty^2}\right) \tag{10·32a}$$

For transonic flow, a similar procedure gives the relation

$$\frac{C_p[(\gamma + 1)M_\infty^2]^{1/3}}{\tau^{2/3}} = \mathcal{P}\left(\frac{1 - M_\infty^2}{\tau^{2/3}[(\gamma + 1)M_\infty^2]^{2/3}};\quad AR\sqrt{1 - M_\infty^2}\right) \tag{10·33}$$

10·6 Summary and Application of the Similarity Laws

The similarity laws for planar flow are contained in Eq. 10·33 where $\mathcal{P}$ is a function that depends on the family shape. The two-dimensional case, as well as the linearized forms of the similarity laws, are special cases of this one. The pressure coefficient, C_p, given by this relation is for a specific position on the body surface. For example, one might be interested in the value of C_p on a wing at a specific value of x/c and y/b.

Now the local lift coefficient, C_l, is also equal to C_p. This may be seen from the fact that $C_l = C_p \cos\theta \doteq C_p$, where θ is the local inclination of the surface, relative to free stream. Within the frame of small-perturbation theory, the body thickness and angle of attack are small enough so that θ is always small and allows the above approximation. Since the overall body lift coefficient C_L is obtained by simply taking the mean value of C_l, it will follow the same type of similarity law,

$$\frac{C_L[(\gamma + 1)M_\infty^2]^{1/3}}{\tau^{2/3}} = \mathcal{L}\left(\frac{1 - M_\infty^2}{\tau^{2/3}[(\gamma + 1)M_\infty^2]^{2/3}};\quad AR\sqrt{|M_\infty^2 - 1|}\right) \tag{10·34}$$

Similarly, the overall drag coefficient C_D is the mean value of the local drag coefficient C_d. The latter is related to C_p by $C_d = C_p \sin\theta \doteq C_p\theta$. For a given family of shapes, $\theta \sim \tau$, and so $C_d \sim C_p\tau$. Hence, the similarity law for the drag coefficient (local or overall) has the form

$$\frac{C_D[(\gamma + 1)M_\infty^2]^{1/3}}{\tau^{5/3}} = \mathfrak{D}\left(\frac{1 - M_\infty^2}{\tau^{2/3}[(\gamma + 1)M_\infty^2]^{2/3}};\quad Æ\sqrt{|M_\infty^2 - 1|}\right) \tag{10·35}$$

An application of the similarity rule to experimental measurements is given in Fig. 10·2*a*, where the drag coefficients of three wedges of different thicknesses are given. They are plotted in terms of the ordinary parameters $C_D = C_D(M_\infty)$, and appear as three curves. When replotted, in Fig. 10·2*b*, in the parameters of the similarity rule (10·35), they follow a single curve. Outside the transonic range, the curve approaches the linearized theory.

The similarity rules discussed in this chapter are valid through the whole range of Mach numbers from subsonic through supersonic. In the linearized subsonic and supersonic ranges, they may be used in the simpler, Prandtl-Glauert and Göthert forms, which apply when

$$\chi \equiv \frac{|1 - M_\infty^2|}{\tau^{2/3}[(\gamma + 1)M_\infty^2]^{2/3}} > 1 \tag{10·36}$$

In all cases, the laws derived here apply only to *small perturbations*, that is, to thin sections and slender bodies, at small angles of attack. Any improvement of the similarity rules must remove this restriction. An extension to thicker sections has been accomplished in the von Kármán-Tsien rule and in the Ringleb rule,† but with some limitations. It is difficult to estimate the range of applicability, and it is not possible to extend them to three-dimensional cases, since they are derived from hodograph considerations. Modern aerodynamics of necessity is concerned mainly with *thin* wings, and thus the similarity rules given here actually cover a wide range of practical interest.

At very high values of M_∞, these rules are no longer valid, and have to be replaced by the *hypersonic* similarity rules, which are outlined in the following article.

10·7 High Mach Numbers. Hypersonic Similarity

The subsonic, transonic, and supersonic similarity rules of the preceding paragraphs were obtained from the small-perturbation equation (8·9*a*)

$$(1 - M_\infty^2)\frac{\partial^2\phi}{\partial x_1^2} + \frac{\partial^2\phi}{\partial x_2^2} + \frac{\partial^2\phi}{\partial x_3^2} = \frac{M_\infty^2(\gamma + 1)}{U}\frac{\partial\phi}{\partial x_1}\frac{\partial^2\phi}{\partial x_1^2}$$

and the approximate expressions for the pressure coefficient,

$$C_p = -\frac{2}{U}\frac{\partial\phi}{\partial x_1}$$

†For example, see Oswatitsch, *Gasdynamik*, Springer, Vienna, 1952, pp. 254 ff.

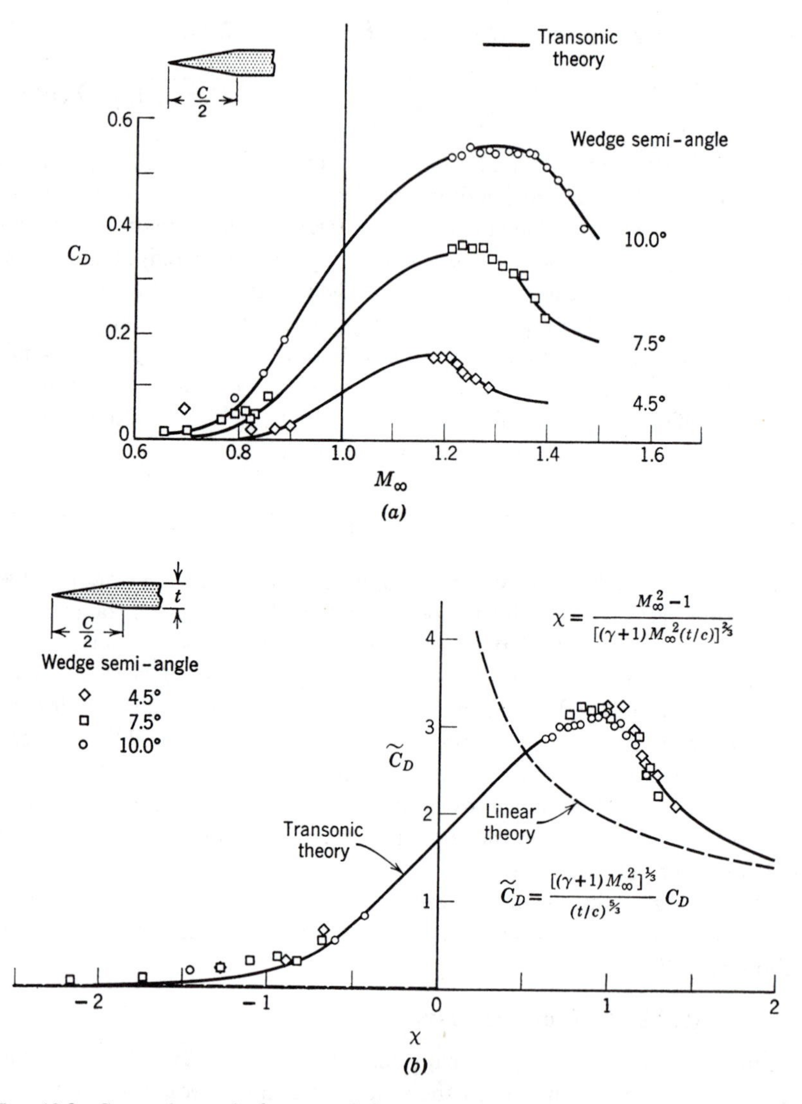

FIG. 10·2 Comparison of the extended transonic similarity law with experiment. (*a*) Plotted in conventional coordinates. (*b*) Plotted in transonic similarity coordinates. (After J. R. Spreiter. NACA.)

for planar flow and

$$C_p = -\frac{2}{U}\frac{\partial \phi}{\partial x_1} - \frac{1}{U^2}\left(\frac{\partial \phi}{\partial r}\right)^2$$

for axially symmetric flow. As pointed out in Articles 8·2 and 8·3, the validity of these equations must fail at sufficiently high Mach numbers, because some of the neglected terms become important. The approximations used in deriving Eq. 8·9*a* imply that the shock waves on the body are weak. In supersonic flow, these shock waves lie close to a characteristic (that is, a Mach line), whereas in transonic flow they are nearly normal. At very high Mach numbers, the Mach angle μ may be of the same order or less than the maximum deflection angle, θ, of the body. Since $\mu \doteq \sin \mu = 1/M_\infty$, this case occurs when

$$\frac{1}{M_\infty} \leq \theta \quad \text{that is,} \quad M_\infty \theta \geq 1$$

This new regime is thus characterized by the parameter $M_\infty \theta$, or rather $M_\infty \tau$, where τ is the body thickness ratio. It is called the hypersonic similarity parameter

$$K = M_\infty \tau \tag{10·37}$$

For large values of the hypersonic parameter, the difference between β, the shock angle, and μ, the Mach angle, is important. This is true also in transonic cases. In both cases, the equations are nonlinear, but the nonlinearity comes from different terms, since the nature of the flow fields is radically different. In the transonic case, the *lateral* extent of the field is large and the changes *along* the direction of flow are of main importance. In the hypersonic regime, the flow field is narrow, since the shock waves lie close to the body, and the changes *normal* to the direction of flow are the most important ones. Typical flow fields from subsonic through hypersonic flow are illustrated in Fig. 10·3.

Furthermore, in hypersonic flow it is not possible, in general, to assume irrotational flow, with a corresponding velocity potential. The shock waves are strong even for thin bodies, and the entropy gradients in the wakes of shocks are not always negligible.

The derivation of the small-perturbation equation for hypersonic flow was first given by Tsien.† Here we shall use another approach, formulating the similarity rules from some simple relations based on the behavior of shock waves and Prandtl-Meyer expansions.

The relation between Mach number M, wave angle β, and deflection angle θ for an oblique shock wave, obtained in Chapter 4, may be written in the form

$$M^2 \sin^2 \beta - 1 = \frac{\gamma + 1}{2} M^2 \frac{\sin \beta \sin \theta}{\cos (\beta - \theta)}$$

For small values of θ, and large values of M, such that $M\theta > 1$, β must also

†H. S. Tsien, "Similarity Laws of Hypersonic Flows," *J. Math. Phys.*, *25* (1946), p. 247.

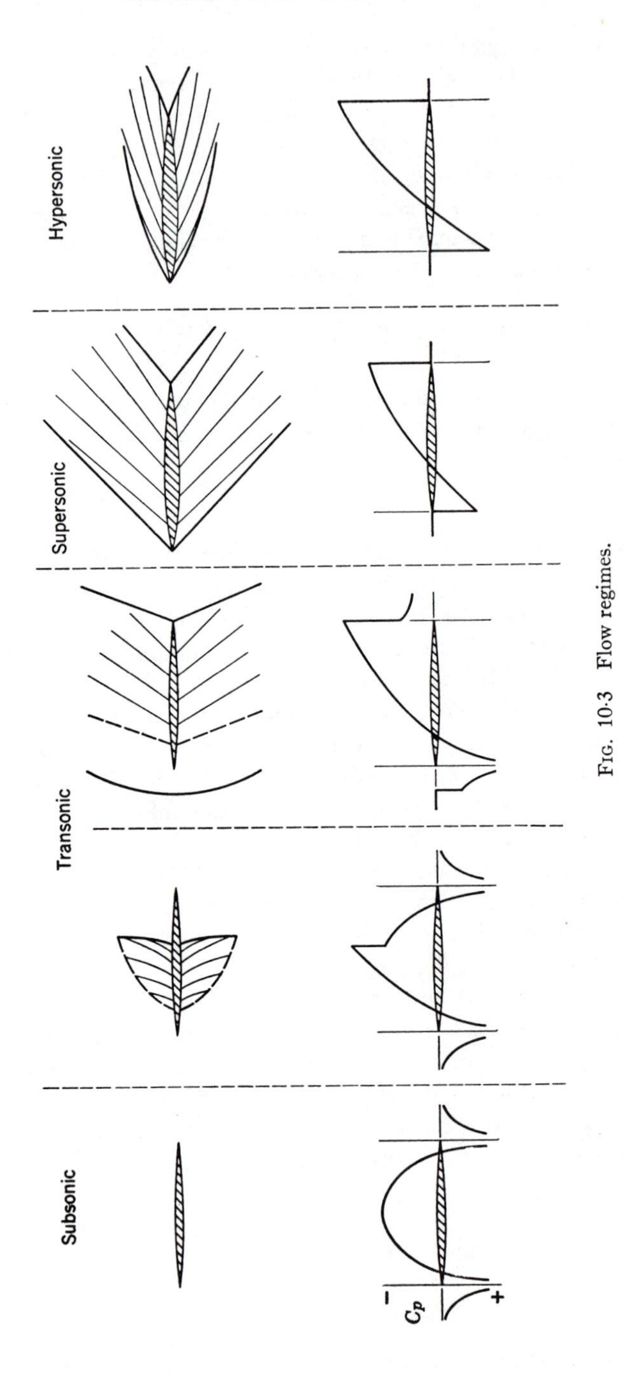

FIG. 10·3 Flow regimes.

be small, and hence we may use the approximations

$$\sin \beta \doteq \beta; \quad \sin \theta \doteq \theta; \quad \cos (\beta - \theta) \doteq 1$$

Thus, up to terms of order θ^2 or β^2, we have

$$M^2\beta^2 - 1 = \frac{\gamma + 1}{2} M^2\beta\theta$$

This may be solved to give

$$\frac{\beta}{\theta} = \frac{\gamma + 1}{4} + \sqrt{\left(\frac{\gamma + 1}{4}\right)^2 + \frac{1}{(M\theta)^2}}$$

The pressure ratio across the oblique shock wave is

$$\frac{p_2 - p_1}{p_1} = \frac{2\gamma}{\gamma + 1} (M^2 \sin^2 \beta - 1) \doteq \frac{2\gamma}{\gamma + 1} (M^2\beta^2 - 1)$$

and hence the pressure coefficient is

$$C_p = \frac{2}{\gamma M^2}\left(\frac{p_2 - p_1}{p_1}\right) = 2\beta\theta = 2\theta^2\left[\frac{\gamma + 1}{4} + \sqrt{\left(\frac{\gamma + 1}{4}\right)^2 + \frac{1}{(M\theta)^2}}\right] \tag{10·38}$$

Thus the pressure coefficient is of the form

$$C_p = \theta^2 f(M\theta)$$

and satisfies the similarity law

$$\frac{C_p}{\theta^2} = f(M\theta) \tag{10·38a}$$

Next, we show that the pressure coefficient in a Prandtl-Meyer expansion (cf. Article 4·10) has the same form. Let M denote the local Mach number, M_∞ the free-stream Mach number, and θ the expansion angle from the free-stream direction. The Prandtl-Meyer relation may then be written

$$\theta = \sqrt{\frac{\gamma + 1}{\gamma - 1}}\left[\tan^{-1}\sqrt{\frac{\gamma - 1}{\gamma + 1}}\sqrt{M_\infty^2 - 1} - \tan^{-1}\sqrt{\frac{\gamma - 1}{\gamma + 1}}\sqrt{M^2 - 1}\right]$$
$$- [\tan^{-1}\sqrt{M_\infty^2 - 1} - \tan^{-1}\sqrt{M^2 - 1}]$$

Now if M_∞ is large, then so is M, and thus $\sqrt{M^2 - 1} \doteq M$. Then, using the series expansion,

$$\tan^{-1} x = \frac{\pi}{2} - \frac{1}{x} + \cdots$$

for large x, the above expression reduces to

$$\theta = \frac{2}{\gamma - 1}\left(\frac{1}{M} - \frac{1}{M_\infty}\right)$$

or

$$\frac{M_\infty}{M} = 1 + \frac{\gamma - 1}{2} M_\infty\theta \tag{10·39}$$

The pressure coefficient, as given in Eq. 2·40*a*, is

$$C_p = \frac{2}{\gamma M_\infty^2}\left\{\left[\frac{2 + (\gamma - 1)M_\infty^2}{2 + (\gamma - 1)M^2}\right]^{\gamma/(\gamma-1)} - 1\right\}$$

For large values of M and M_∞ it reduces to

$$C_p = \frac{2}{\gamma M_\infty^2}\left[\left(\frac{M_\infty}{M}\right)^{2\gamma/(\gamma-1)} - 1\right] \tag{10·40}$$

Combining this with Eq. 10·39, we have

$$C_p M_\infty^2 = \frac{2}{\gamma}\left[\left(1 + \frac{\gamma - 1}{2} M_\infty\theta\right)^{2\gamma/(\gamma-1)} - 1\right] \tag{10·40a}$$

Dividing both sides by $M_\infty^2\theta^2$, we have for Prandtl-Meyer flow at high Mach number the similarity law

$$\frac{C_p}{\theta^2} = f(M_\infty\theta) \tag{10·40b}$$

Thus both the shock wave and the isentropic expansion wave have the same form of similarity law for C_p. Consequently, we may expect this law to apply also to all flows that can be constructed from shock waves and expansion fans. Actually, it has been shown that the validity is quite general,† and that the general form of the hypersonic similarity law, in terms of a typical thickness parameter τ, is

$$\frac{C_p}{\tau^2} = \mathcal{P}(M_\infty\tau) \tag{10·41a}$$

Hence we also have

$$\frac{C_D}{\tau^3} = \mathcal{D}(M_\infty\tau) \tag{10·42a}$$

where the drag coefficient C_D is based on the chord.

In Fig. 10·4, the hypersonic similarity law has been used to correlate the pressures on two inclined ogives of thickness ratio 1/3 and 1/5. They are compared at values of $M_\infty = 3$ and 5, respectively, in order that $M_\infty\tau$ be the same. It is necessary, for similarity, that the parameter $M_\infty\alpha$ also

†W. D. Hayes, "On Hypersonic Similitude," *Quart. Appl. Math.*, *5* (1947), p. 105.
M. D. Van Dyke, "Hypersonic Small Disturbance Theory." *J. Aeronaut. Sci.*, *21* (1954), p. 179.

be kept constant by adjusting α, for reasons similar to those outlined in Article 10·6 concerning C_L.

We may now compare the form of the hypersonic similarity law for C_p with the one obtained for linearized supersonic flow,

$$\frac{C_p}{A} = fn\left(\frac{\tau}{A\sqrt{M_\infty^2 - 1}}\right)$$

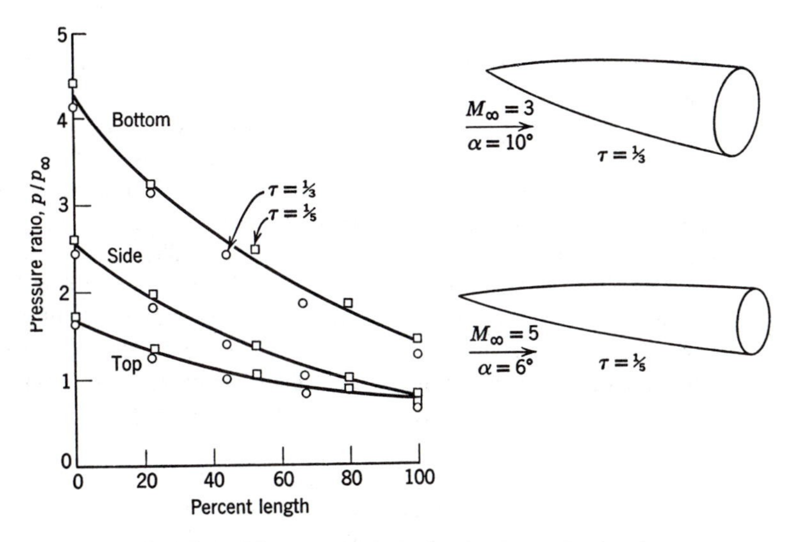

FIG. 10·4 Applicability of hypersonic similarity law to inclined ogives. $M_\infty \tau = 1.0$. $M_\infty \alpha = 30°$. (After A. J. Eggers and C. A. Syvertson, NACA.)

Choosing $A = \tau^2$ yields

$$\frac{C_p}{\tau^2} = fn\left(\frac{1}{\tau\sqrt{M_\infty^2 - 1}}\right) = fn(\tau\sqrt{M_\infty^2 - 1}) \qquad (10·43)$$

In hypersonic flow, the difference between M_∞^2 and $M_\infty^2 - 1$ is negligible, and consequently Eq. 10·43 is a combined rule which is valid over the whole supersonic-hypersonic range. This formulation is due to Van Dyke.

By specializing Eq. 10·32*a*, we may also obtain an extended, supersonic-hypersonic law for planar systems,

$$\frac{C_p}{\tau^2} = \mathcal{P}(\tau\sqrt{M_\infty^2 - 1};\quad ɶ\sqrt{M_\infty^2 - 1}) \qquad ▶(10·44)$$

CHAPTER 11

Transonic Flow

11·1 Introduction

The efforts of aeronautical research were focused on the transonic range by the practical problem of passing through the speed of sound, and of flight in the transonic range. The flow pattern and thus the forces exerted on a moving body in this range differ considerably from those in the low-speed range. Shock waves are an essential feature of transonic flow, as we know today, and their appearance on a moving body leads to a rapid increase in drag coefficient with increasing Mach number. The new phenomena encountered in this range were, in fact, so baffling to many aerodynamicists used to low-speed, incompressible flow, that the myth of the "sonic barrier" arose. This in spite of the fact that there had already been almost a century of experience with artillery shells, which reach supersonic muzzle speeds and have to decelerate through the speed of sound during their flight.

The *qualitative* features of transonic flow are now well understood, and the experimental methods of investigating transonic flow problems are continuously being improved. The difficulties that remain are questions of detail design and of *quantitative* theoretical prediction. For the former, a sufficient reservoir of test data is needed; the latter is due to the *inherent nonlinearity of the equation governing transonic flow.*

Furthermore, in transonic flow, and also in hypersonic flow, it is often impossible to separate clearly the nonviscous flow from the effects of viscosity. One encounters here what is known as "shock-wave boundary-layer interaction"; that is, an interplay between the strength and position of shock waves on a body and of boundary layer character, separation, etc.

The present chapter is intended to describe and discuss the essential physical features of transonic flow for simple boundary conditions, without getting involved in much computation, and to build up a feeling for the problems encountered in the transonic range.

11·2 Definition of the Transonic Range

For the present we shall assume the fluid to be frictionless; the flow may then slip past solid boundaries. The flow is called transonic if *both* subsonic and supersonic regions are present in the field. If the free-stream Mach number is increased continuously from zero, the transonic range begins when the *highest* local Mach number reaches unity, and ends when the *lowest* local Mach number reaches unity.

This condition can be made more specific if we restrict ourselves to thin or slender bodies in the sense of small-perturbation theory. If the characteristic body thickness ratio is called τ, the flow past such a body is transonic if the transonic parameter (Article 10·3) falls within the range

$$-1 \leq \frac{M_\infty^2 - 1}{[(\gamma + 1)\tau M_\infty^2]^{2/3}} \leq 1 \tag{11·1}$$

11·3 Transonic Flow Past Wedge Sections

As a first and very instructive example, consider flow past a simple wedge of opening angle 2θ and length c followed by a straight section. Figure 11·1 shows the essentials of the flow pattern at various free-stream Mach numbers. As M_∞ increases from subsonic values, a shock wave system appears near the shoulder. The main (nearly normal) shock grows and moves downstream as M_∞ increases toward unity. At $M_\infty = 1$ it has moved to "downstream infinity" and a second shock wave has appeared at "minus infinity." With further increase of M_∞ this bow wave approaches the wedge vertex, finally attaches, and becomes straight. The supersonic range has now been reached. Thus for $M_\infty < 1$ there is a local supersonic region ahead of the main shoulder shock, while for $M_\infty > 1$ there is a local subsonic region behind the detached bow wave.

The measured local Mach number distributions for such a wedge are shown in Fig. 11·2. The change-over from a typical incompressible distribution to the characteristic constant Mach number at supersonic speeds is evident.

It may be seen that no discontinuous changes occur in the whole range. The change-over from subsonic to supersonic flow is smooth and continuous. The behavior of the drag coefficient C_D, which was shown in Fig. 10·2, is, of course, also smooth. C_D varies "rapidly" only if plotted as a function of M_∞, which is indeed an improper choice of variable for this range. The proper independent variable, as we have seen in Chapter 10, is

$$\chi = \frac{1 - M_\infty^2}{[(\gamma + 1)\tau M_\infty^2]^{2/3}}$$

the dependent one

$$\tilde{C}_D = \frac{C_D[(\gamma + 1)M_\infty^2]^{1/3}}{\tau^{5/3}}$$

One can give some intuitive arguments concerning the general features of the flow, as follows.†

(*a*) *Sonic velocity, that is* $M = 1$, *occurs at the shoulder throughout the*

†H. W. Liepmann and A. E. Bryson, *J. Aeronaut. Sci.*, *17* (1950), p. 745.

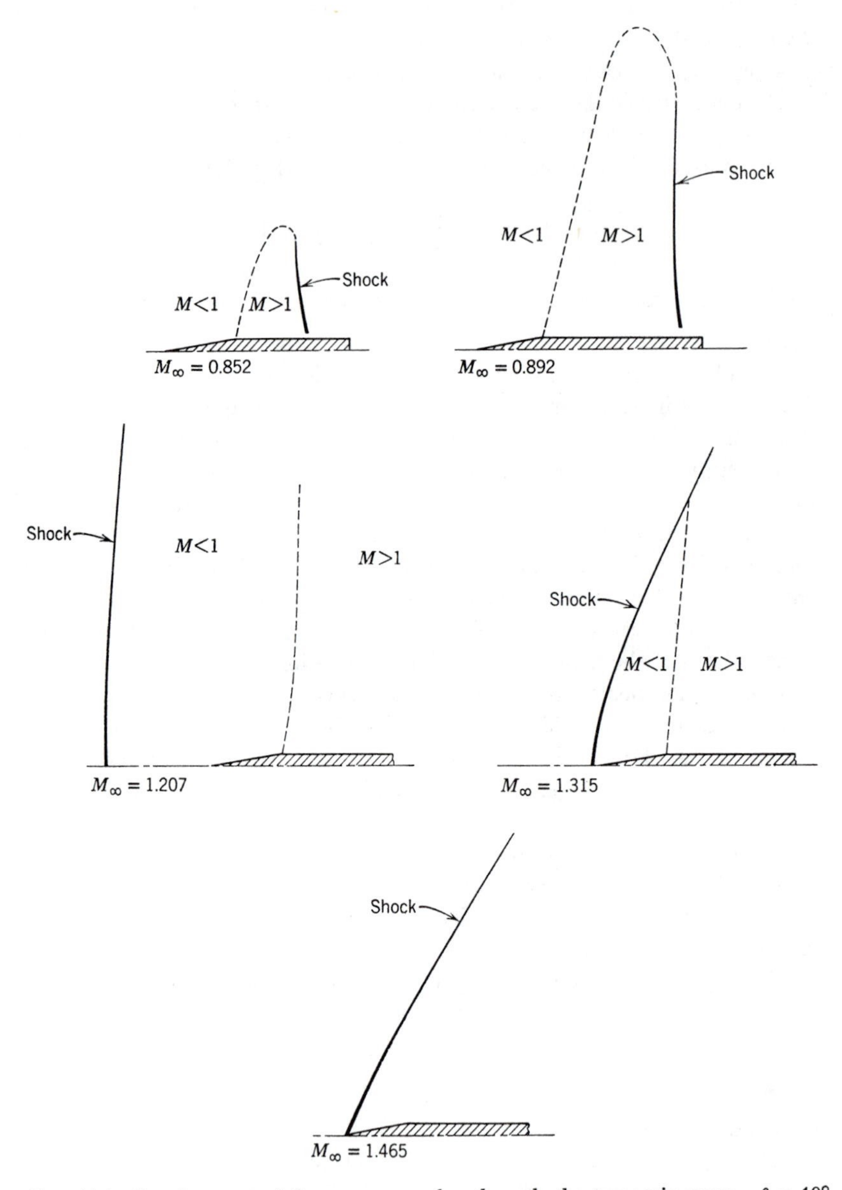

Fig. 11·1 Development of flow over a wedge through the transonic range. $\theta = 10°$. [Drawings were traced from interferograms by A. E. Bryson, "An Experimental Investigation of Transonic Flow Past Two-Dimensional Wedge and Circular-Arc Sections Using a Mach-Zehnder Interferometer," *NACA Tech. Note* 2560 (1951).]

whole transonic range. This fact seems intuitively plausible since incompressible flow past a sharp corner will have infinite velocity at the corner. Hence compressible flow will reach at least sonic velocity there. Hence

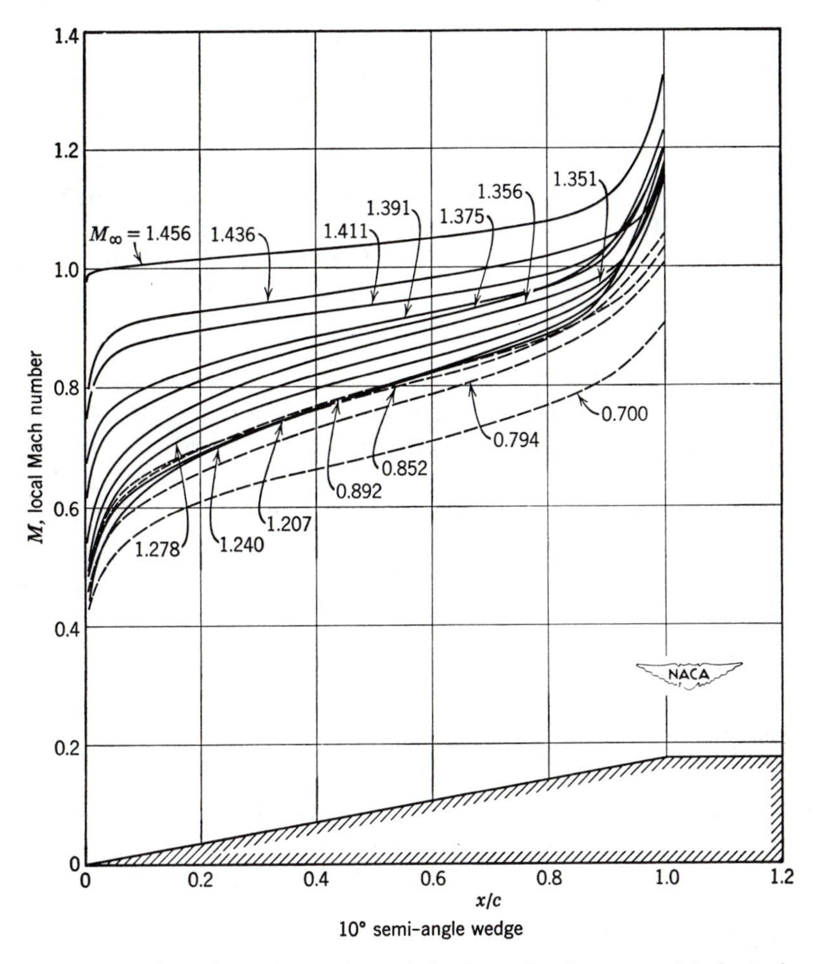

FIG. 11·2 Local Mach number against x/c for increasing free-stream Mach number. [From A. E. Bryson, *NACA Tech. Note* 2560 (1951).]

in the range where $M_\infty < 1$ we expect sonic velocity to exist along a line starting at the corner and terminating in a shock wave downstream of the shoulder. On the other hand, for $M_\infty > 1$ the sonic line must end at the point on the *bow shock wave* for which the Mach number downstream of the shock is just unity. For this case again we can argue that the sonic line must pass through the shoulder corner. If it should terminate *downstream* of the shoulder, the flow over the wedge would remain subsonic and the former argument would apply. If, on the other hand, the sonic line should terminate *upstream* of the corner, the shoulder would have no influence upstream, the wedge would act like one with an infinite chord, and no finite detachment distance for the shock wave could exist.

Actually, with a little more care these plausible arguments can be made rigorous. It is also possible to demonstrate analytically that sonic velocity cannot be reached on any flat part of the body.

(*b*) *The local Mach number M becomes stationary as M_∞ passes through unity. That is,* $(dM/dM_\infty)_{M_\infty=1} = 0$. For this consider M_∞ a little larger than unity. The detached bow wave will stand far ahead of the body, and nearly normal to the flow. The Mach number M_2 downstream of a weak normal shock is related to the upstream Mach number M_1 by

$$M_1 - 1 = 1 - M_2 \tag{11·2}$$

In our case $M_1 \equiv M_\infty$. But as far as the wedge is concerned the upstream Mach number is M_2. As $M_\infty \to 1$ from supersonic values, $M_2 \to 1$ from subsonic values. Consequently, as far as the local Mach number on the wedge is concerned, the conditions just below and just above $M_\infty = 1$ are identical. Hence, if M denotes the local Mach number at a point on the wedge, then

$$\left(\frac{dM}{dM_\infty}\right)_{M_\infty=1} = 0 \tag{11·3}$$

The pressure coefficient C_p, expressed in terms of M and M_∞, is given by Eq. 2·40*a*,

$$C_p = \frac{2}{\gamma M_\infty^2}\left[\left(\frac{2 + (\gamma - 1)M_\infty^2}{2 + (\gamma - 1)M^2}\right)^{\gamma/(\gamma-1)} - 1\right] \tag{11·4}$$

If we now form dC_p/dM_∞ and make use of Eq. 11·3 we can obtain an expression for the slope of pressure coefficient versus Mach number curve at sonic velocity. To do this, it is convenient to rewrite Eq. 11·4 slightly:

$$\log\left(\frac{\gamma}{2} M_\infty^2 C_p + 1\right)$$
$$= \frac{\gamma}{\gamma - 1}\{\log[2 + (\gamma - 1)M_\infty^2] - \log[2 + (\gamma - 1)M^2]\} \tag{11·5}$$

Differentiating with respect to M_∞ (remembering Eq. 11·3) and letting $M_\infty \to 1$ yields

$$\frac{\gamma C^*_p + \dfrac{\gamma}{2}\left(\dfrac{dC_p}{dM_\infty}\right)_{M_\infty=1}}{1 + \dfrac{\gamma}{2}C^*_p} = \frac{2\gamma}{\gamma + 1}$$

where C^*_p is written for C_p at $M_\infty = 1$. Thus we have

$$\left(\frac{dC_p}{dM_\infty}\right)_{M_\infty=1} = \frac{4}{\gamma + 1}\left(1 - \frac{1}{2}C^*_p\right) \qquad \blacktriangleright(11\cdot6)$$

A point on the wedge where the pressure coefficient is C_p contributes $C_p\theta$ to the drag coefficient of the wedge. Hence from Eq. 11·6 it follows that the slope of the C_D vs. M_∞ curve at sonic velocity is

$$\left(\frac{dC_D}{dM_\infty}\right)_{M_\infty=1} = \frac{4\theta}{\gamma + 1} - \frac{2C^*_D}{\gamma + 1} \qquad \blacktriangleright(11\cdot7)$$

If the wedge angle is small, in the sense of small-perturbation theory, then, in Eq. 11·6, it is permissible to neglect $C^*_p/2$ compared to 1 (that is, $u/U \ll 1$). Similarly, the second term in Eq. 11·7 is also negligible. Within small-perturbation theory, then,

$$\left(\frac{dC_p}{dM_\infty}\right)_{M_\infty=1} = \frac{4}{\gamma + 1} \qquad (11\cdot8)$$

$$\left(\frac{dC_D}{dM_\infty}\right)_{M_\infty=1} = \frac{4\theta}{\gamma + 1} \qquad (11\cdot9)$$

These may now be rewritten in terms of the transonic similarity parameters,

$$\tilde{C}_p = C_p \frac{[(\gamma + 1)M_\infty^2]^{1/3}}{\theta^{2/3}}$$

$$\tilde{C}_D = C_D \frac{[(\gamma + 1)M_\infty^2]^{1/3}}{\theta^{5/3}}$$

$$\chi = \frac{M_\infty^2 - 1}{[(\gamma + 1)M_\infty^2\theta]^{2/3}}$$

In these variables, Eqs. 11·8 and 11·9 become simply

$$\left(\frac{d\tilde{C}_p}{d\chi}\right)_{\chi=0} = \left(\frac{d\tilde{C}_D}{d\chi}\right)_{\chi=0} = 2 \qquad \blacktriangleright(11\cdot10)$$

The agreement of experiments (Fig. 10·2) with Eqs. 11·8, 11·9, or 11·10 is very good.

The flow past wedges in the transonic range has been worked out theoretically by Cole,† Guderley and Yoshihara,‡ and Vincenti and Wagoner,§ using the transonic equation. (The last two papers apply also to double-wedge sections.) The wedge flow thus represents the best-known case of

†J. D. Cole, *J. Math. and Phys.*, *30* (1951), p. 79.

‡G. Guderley and H. Yoshihara, *J. Aeronaut. Sci.*, *17* (1950), p. 723.

§W. Vincenti and C. Wagoner, *NACA Tech. Notes* 2339 and 2588 (1951).

transonic flow. The excellent agreement between theory and experiment demonstrates the validity of the approximations made in the transonic small-perturbation theory and in the derivation of the transonic similarity laws.

The lift of a double-wedge† section in the transonic range has also been investigated theoretically and experimentally but not nearly as completely as the nonlifting section. Figure 11·3 shows some representative data. The "transonic small disturbance theory" plotted in this figure is shock expan-

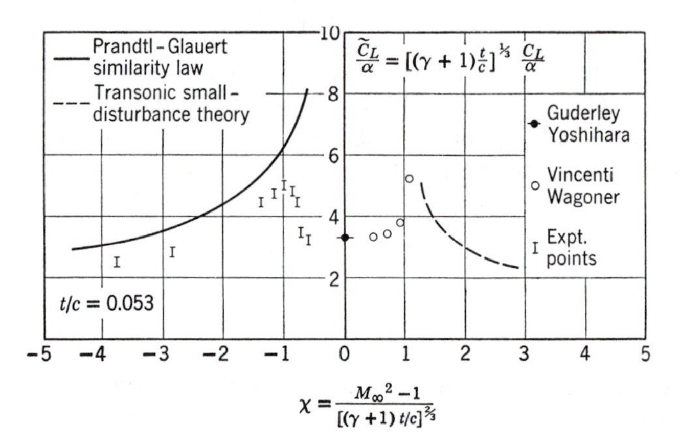

FIG. 11·3 Lift curve slope of a double wedge at small angles of attack. [From J. D. Cole, G. E. Solomon, W. W. Willmarth, "Transonic Flow Past Simple Bodies," *J. Aeronaut. Sci.*, *20* (1953), p. 627.]

sion theory in which the shock wave relations and Prandtl-Meyer function have been simplified for small disturbances near sonic conditions.

More important for theoretical application are wings of finite span, and wing-body combinations. A rigorous theory of transonic flow past such configurations is very difficult and usually requires considerable numerical work. Fortunately, good use can be made of the similarity rules and of slender body theory. The similarity rules are used to correlate experimental data with each other and with the few theoretically known results. Slender body theory can be used to obtain the *lift*, and the *drag due to lift*, of low aspect ratio configurations, right through the transonic regime.

11·4 Transonic Flow Past a Cone

The second simple example of transonic flow to be discussed here is flow past a conical tip. Some features of this problem are similar to the transonic wedge flow of the preceding paragraph. For example, the sonic line passes

†I.e., diamond airfoil.

through the shoulder point, and the principle of stationarity at $M_\infty = 1$, that is, $(dM/dM_\infty)_{M_\infty=1} = 0$, applies as well for a conical tip.

However, the axially symmetrical cone flow differs in an interesting way from the two-dimensional wedge flow. Consider first the supersonic case. Both the wedge and the cone have the property that flow conditions along a ray through the apex are constant (cf. Article 4·21). This means, for

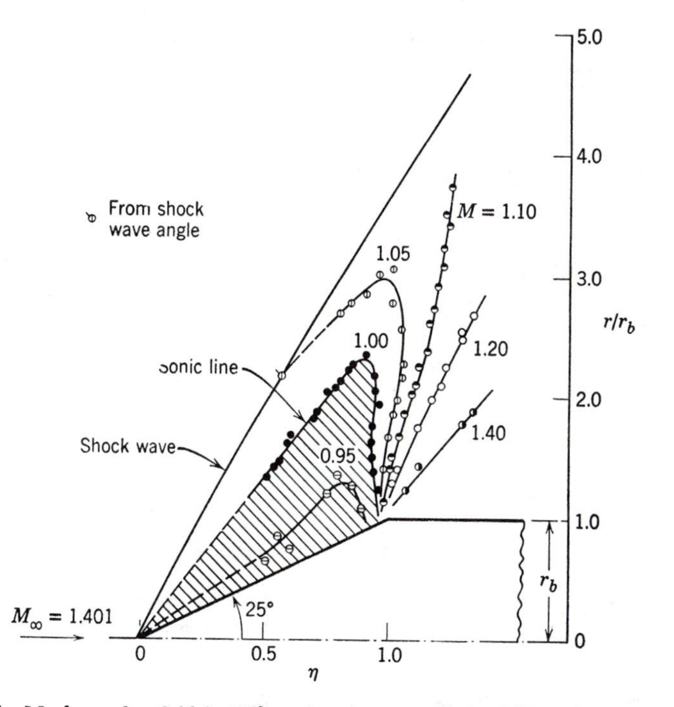

Fig. 11·4 Mach number field for 25° semi-angle cone-cylinder. [From J. D. Cole, G. E. Solomon, and W. W. Willmarth, *J. Aeronaut. Sci.*, *20* (1953), p. 627.]

example, that the pressure along the surface of a wedge or cone is constant. In the case of flow past a wedge, the flow properties are not only the same on one ray but on all rays between the shock wave and the surface. For the cone this is not so. For example, the pressure on a ray just downstream of the shock wave is *lower* than the pressure on the cone surface. The flow downstream of the shock wave undergoes an isentropic compression. This is simply due to the geometry of the flow field and is formally expressed in the extra v/r term in the continuity equation.

Consequently there should exist a low enough supersonic Mach number of the free stream such that the flow downstream of the shock wave is still *supersonic* while the Mach number on the surface of the cone is already *subsonic*. That is, for cone flow a smooth, isentropic compression from supersonic to subsonic flow is to be expected, as was pointed out already by Taylor and Maccoll.

It is not obvious that the flow field in the neighborhood of a conical tip of *finite* length will exhibit the same transonic features as an infinite cone. Experiments, however, have shown that a *subsonic* zone embedded in a *supersonic region* does indeed exist even for conical tips. Figure 11·4 shows these experimental results. The experimental point marked on the shock wave is the point at which the shock wave inclination β corresponds to $M_2 = 1$ (cf. Chart 1 at the back of the book).

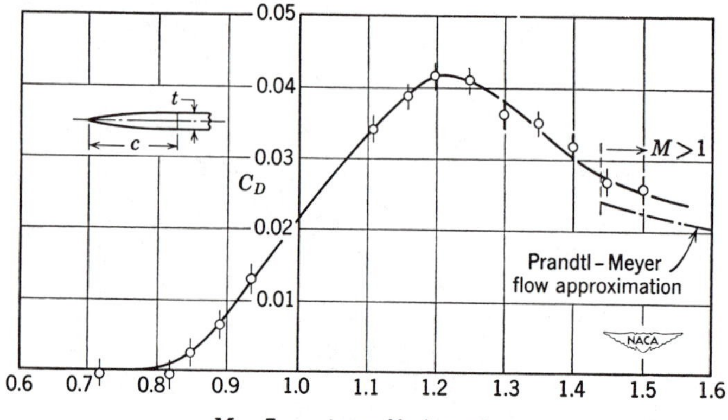

FIG. 11·5 Drag coefficient against Mach number for 8.8 percent circular arc section. (From A. E. Bryson, *NACA Tech. Note* 2560.)

11·5 Transonic Flow Past Smooth Two-Dimensional Shapes. The Question of Shock-Free Flow

Experience has shown that the flow past wedge sections (or double-wedge sections) is typical for transonic two-dimensional flow. The existence of the corner at the shoulder point makes the analysis simpler—because the sonic point can be located immediately—but does not alter the essential features of the flow in cases where there is no shoulder. Results of experiments with circular arc sections, shown in Fig. 11·5, are typical.

Thus for transonic flow past a two-dimensional, semi-infinite body we expect the C_D versus M_∞ curve to be qualitatively similar to the wedge sections, and for a full body to be qualitatively similar to the double-wedge profile. These remarks refer to flow with zero viscosity. A semi-infinite body actually demonstrates transonic flow effects, for nonviscous flow, better than a finite chord body, because of the differences in boundary-layer separation and interaction problems. On the semi-infinite body, for example a wedge with afterbody, these viscous problems are negligible since the pressure is continuously falling in the direction of flow. The boundary-layer shock-wave interaction problem will be briefly discussed in Chapter 13.

Here a few remarks will be added concerning the possibility of shock-free transonic flow. We have seen that the drag in the transonic and supersonic zone is due to shock waves (in the absence of skin friction). We know that shock waves usually appear if the free-stream Mach number anywhere is larger than unity. However, we also know that it is possible in purely supersonic flow to cancel waves by interference. The Busemann biplane (Article 4·19) is the classical example. In the transonic regime with $M_\infty < 1$, a supersonic zone forms in the vicinity of the profile. Shock-free transonic flow would exist on a profile shape for which this supersonic zone, within the subsonic flow field, would remain smooth, that is, all compression waves would be canceled by expansion waves. The correspondence with the Busemann biplane or shock-free nozzles and diffusers is clear.

A rigorous proof of the impossibility of shock-free supersonic zones on airfoils has not been given. In fact, there do exist a few exact solutions of the potential equation—for odd shapes, to be sure—that yield smooth transonic flow. However, smooth transonic flow is almost certainly *singular* in the sense that, if it exists for a specific shape at a specific free-stream Mach number, the flow will not be smooth if the profile or the free-stream Mach number is altered. A detailed discussion of this point would lead far beyond the scope of this book, and the reader is referred to the literature, especially to a survey paper by Guderley.†

However, it should be emphasized that the problem is not the question of whether a smooth deceleration through sonic flow is possible. The problem is rather whether it is possible to *have a smooth supersonic zone enclosed within a subsonic field.* This is probably impossible because a supersonic, that is, hyperbolic, type of field is in general overconstrained if boundary conditions are given at every point of a closed boundary. The opposite case, namely, the *existence of a subsonic zone within a supersonic field,* is quite different. For an elliptic zone, boundary conditions along the complete, closed boundary are quite natural. Hence the existence of locally subsonic flow, as in the case of transonic cone flow, is not too surprising.

For application to airfoil sections, this problem of smooth transonic flow is of little interest today. The appearance of shock waves and of shock drag has to be accepted. The sections are designed to delay and reduce the drag rise by reducing the thickness or by the use of sweepback. Other

†G. Guderley, *Advances of Applied Mechanics* III, Academic Press, New York, 1953, p. 145.

important problems are posed by secondary shock-wave effects, such as transonic buffeting, control effectiveness, etc. These problems are strongly influenced by boundary-layer effects.

11·6★ The Hodograph Transformation of the Equations

The difficulty in dealing theoretically with transonic flow is evidently due to the essential nonlinearity of the equations of motion. However, for two-dimensional flow it is possible to transform the equations into linear ones. This is accomplished by the *hodograph transformation* in which the dependent and independent variables are interchanged. By far the larger part of the theoretical results of transonic flow has been obtained by the use of hodograph methods.

The transonic equations can be written in the form of Eq. 8·8:

$$(1 - M_\infty^2)\frac{\partial u}{\partial x} + \frac{\partial v}{\partial y} = \frac{(\gamma + 1)M_\infty^2}{U} u \frac{\partial u}{\partial x} \tag{11·11}$$

The x_1 and x_2 of Eq. 8·8 have been replaced by x and y, for convenience.

In transonic flow within the small-perturbation theory we deal with irrotational flow and hence we have a second equation, the condition of irrotationality

$$\frac{\partial u}{\partial y} - \frac{\partial v}{\partial x} = 0 \tag{11·12}$$

In Chapters 8 and 9 we usually satisfied the second equation identically by introducing the velocity potential ϕ. Here it is more convenient to keep the two first-order equations.

Equations 11·11 and 11·12 are differential equations for the dependent variables u and v as functions of the independent variables x, y. One says that Eqs. 11·11 and 11·12 are differential equations in the "physical plane," that is, in x and y (Fig. 11·6). One can transform Eqs. 11·11 and 11·12 into equations in which the variables x, y become dependent, that is, functions of the *independent variables* u, v. The resulting equations are in the "hodograph plane" (Article 4·20), that is, in a plane where u and v are the coordinates. We write

$$u = u(x, y)$$

$$v = v(x, y)$$

and hence†

$$du = u_x\, dx + u_y\, dy$$

$$dv = v_x\, dx + v_y\, dy$$

†Here we use the compact notation $u_x = \partial u/\partial x$, etc.

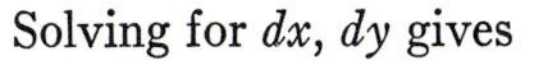

Solving for dx, dy gives

$$dx = \frac{1}{\Delta}[v_y\,du - u_y\,dv] \tag{11·13}$$

$$dy = \frac{1}{\Delta}[-v_x\,du + u_x\,dv]$$

with

$$\Delta = \begin{vmatrix} u_x & u_y \\ v_x & v_y \end{vmatrix}$$

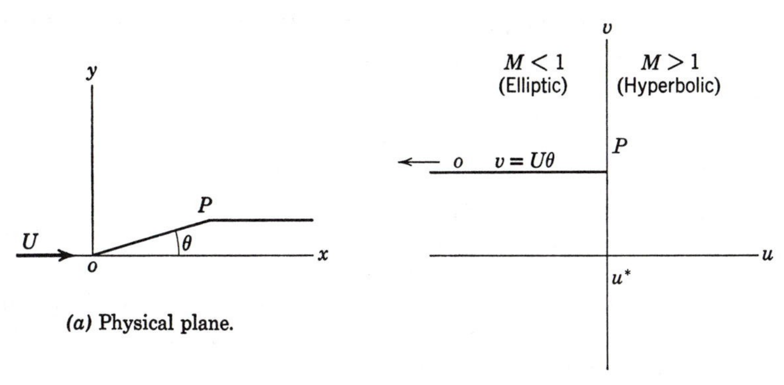

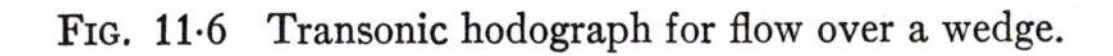

FIG. 11·6 Transonic hodograph for flow over a wedge.

We now consider x and y to be functions of u and v, that is,

$$x = x(u, v)$$

$$y = y(u, v)$$

and thus

$$dx = x_u\,du + x_v\,dv$$

$$dy = y_u\,du + y_v\,dv \tag{11·14}$$

Comparing Eq. 11·14 with Eq. 11·13, we find the transformation scheme

$$x_u = \frac{1}{\Delta}v_y \qquad x_v = -\frac{1}{\Delta}u_y$$

$$y_u = -\frac{1}{\Delta}v_x \qquad y_v = \frac{1}{\Delta}u_x \tag{11·15}$$

With Eq. 11·15 the differential equations (11·11 and 11·12) can be transformed into the hodograph plane by simply inserting for u_x, v_y, etc., the corresponding derivatives of x and y from Eq. 11·15. The functional determinant Δ evidently occurs in every term and hence cancels out (provided it is not zero). Thus we obtain

$$(1 - M_\infty^2)\frac{\partial y}{\partial v} + \frac{\partial x}{\partial u} = \frac{(\gamma + 1)M_\infty^2}{U} u \frac{\partial y}{\partial v} \qquad \blacktriangleright(11\cdot16a)$$

$$\frac{\partial x}{\partial v} - \frac{\partial y}{\partial u} = 0 \qquad \blacktriangleright(11\cdot16b)$$

Equations 11·16*a* and 11·16*b* are the *transonic hodograph equations.* These equations are *linear* because the *dependent variables* x and y occur linearly. Equation 11·16*a* changes type, from elliptic to hyperbolic, for a specific value of the perturbation velocity u, that is, for

$$(1 - M_\infty^2) - \frac{\gamma + 1}{U} M_\infty^2 u = 0$$

or

$$\frac{u}{U} = \frac{1 - M_\infty^2}{(\gamma + 1)M_\infty^2} = \frac{u^*}{U} \quad \text{(say)} \qquad (11\cdot17)$$

If u/U is smaller than the value given by Eq. 11·17, y_v in Eq. 11·16*a* has the *same sign* as x_u and the equation is "elliptic." If u/U is larger than u^*/U, y_v has the *opposite* sign from x_u and the equation is "hyperbolic." But since u and v are now the *independent* variables, Eq. 11·17 is a relation that corresponds to a given straight line in the hodograph plane (Fig. 11·6*b*). To the left of $u = u^*$, the equation is elliptic; to the right, it is hyperbolic.

Equations 11·16 are of a type of differential equation studied by Tricomi and thus often called the "Tricomi equation." To solve a transonic problem in the hodograph plane, one thus has to solve the Tricomi equation. One difficulty in the application always arises from the boundary conditions, which inherently are given in the physical plane, that is, for flow past a given shape in x and y. The reader can convince himself easily of this point by sketching the streamlines of flow past an arbitrary airfoil in the hodograph plane, and by referring to the examples in Article 4·20.

The straight wedge is one case for which the boundary condition can be formulated in the hodograph plane. The physical and the hodograph planes for flow with $M_\infty < 1$ are compared in Fig. 11·6. Along *o-P* the flow has to be tangential to the wedge, and hence, within small-perturbation theory,

$$\frac{v}{U} = \theta$$

At P the flow becomes sonic. Consequently, oP transforms into the horizontal line $v = U\theta$ in the hodograph plane, with P lying on the intersection of $v = U\theta$ and $u = u^*$, where the equation changes type because the flow passes locally through sonic velocity. The point o, a stagnation point, maps into $u = -\infty$, in the small-perturbation theory. All streamlines are thus confined to the strip between $v = 0$ and $v = U\theta$ in the hodograph plane. Some of them, next to the corner P, will extend to the right of $u = u^*$; that is, the flow will be locally supersonic.

The hodograph transformation becomes singular if $\Delta = 0$. It can be shown that this is possible only in the supersonic range. The locus of all points for which $\Delta = 0$ is called the "limiting line." For a time a great deal of work was devoted to its study, in the belief that it held the clue to the appearance of shocks in transonic flow. However, it now appears that the limiting line does not have the physical significance that was anticipated.

A detailed discussion of these problems would lead far beyond the scope of this book. The preceding sketch may serve to indicate the general purpose of the hodograph transformation. For further study, the reader is referred to the literature.

CHAPTER 12

The Method of Characteristics

12·1 Introduction

In preceding chapters, it has been shown that general solutions for frictionless flow over slender bodies may be obtained by using the approximate, linearized equations. If the accuracy of these is not sufficient, it is necessary to work out improved solutions, by including higher-order terms in the approximate equations, or by applying the exact equations. In the latter case, however, it is rarely possible to obtain solutions in analytical form, because *the equations are nonlinear.* One must then resort to numerical methods.

The full, nonlinear equations of motion for two-dimensional, nonviscous, irrotational flow are

$$(u_1^2 - a^2)\frac{\partial u_1}{\partial x_1} + u_1 u_2\left(\frac{\partial u_1}{\partial x_2} + \frac{\partial u_2}{\partial x_1}\right) + (u_2^2 - a^2)\frac{\partial u_2}{\partial x_2} = 0$$

$$\frac{\partial u_2}{\partial x_1} - \frac{\partial u_1}{\partial x_2} = 0 \qquad (12{\cdot}1)$$

They are easily extended to flow with vorticity by including the appropriate terms in the right-hand side of the second equation (see Article 7·9).

The forms of the numerical solutions for these two cases are fundamentally different for subsonic and supersonic flow. If $(u_1^2 + u_2^2)/a^2 < 1$, the equations are of a type that is called *elliptic,* for which the *relaxation method of* solution is appropriate, whereas, if $(u_1^2 + u_2^2)/a^2 > 1$, the equations are of *hyperbolic type* and the numerical solution is obtained by the *method of characteristics.* The latter type will be discussed in this chapter. Transonic flow is a mixed case, in which both subsonic and supersonic regions occur. Here even numerical methods present problems, because the boundary between the two regions is not known a priori.

It may be noted at the outset that numerical methods are very much slower and more tedious than the linearized methods, or even than higher-order solutions. They should be used only where great accuracy is needed and is consistent with the limitations imposed by other idealizations. However, apart from providing an exact method of solution, a study of characteristics is important for further insight into the structure of supersonic flow.

12·2 Hyperbolic Equations

It is not possible here to go into the mathematical theory of hyperbolic equations, which may, however, be found in several excellent works.† We shall simply borrow the main results needed for this chapter. They are as follows:

(1) An equation is "hyperbolic" if a certain relation is satisfied by the coefficients of its highest-order derivatives. For Eqs. 12·1 this leads to the condition $(u_1^2 + u_2^2)/a^2 > 1$.

(2) The distinguishing property of hyperbolic equations is the existence of certain characteristic directions or lines in the x_1-x_2 plane, usually simply called *characteristics*. On a characteristic the normal derivative of the dependent variables (u_1, u_2) may be *discontinuous*. For Eqs. 12·1 *the characteristics are the Mach lines*. It must be noted that the Mach-line network is not known a priori, a result of the nonlinearity of the equations.

Because the normal velocity *derivatives* on Mach lines may be discontinuous, it is possible to "patch" different flows together at these lines. The only restriction is that the velocity itself must be continuous.‡ This is unlike the elliptic or subsonic case, for which all derivatives are continuous, so that a change in any part of the field affects every other part.

(3) On the characteristics, or Mach lines, the dependent variables satisfy a certain relation known as the *compatibility relation*. It provides the key to the method of computation.

12·3 The Compatibility Relation

The characteristics method may be applied directly to Eqs. 12·1, which are for Cartesian coordinates. However, it is more convenient, for both derivation and application, to use the *natural coordinate system*, in which the velocity is expressed in terms of its magnitude and direction (w, θ) and the independent variables are the streamline coordinates (s, n). The equations (Article 7·12) are

$$\frac{\cot^2 \mu}{w} \frac{\partial w}{\partial s} - \frac{\partial \theta}{\partial n} = 0 \tag{12·2a}$$

$$\frac{1}{w} \frac{\partial w}{\partial n} - \frac{\partial \theta}{\partial s} = 0 \tag{12·2b}$$

The characteristic, or Mach, direction is explicitly introduced here by the expression

$$\cot^2 \mu = M^2 - 1 \tag{12·3}$$

†For example, Courant and Hilbert, *Methods of Mathematical Physics*, Vol. II, Interscience, New York, 1956; Courant and Friedrichs, *Supersonic Flow and Shock Waves*, Interscience, New York, 1948.

‡Characteristics, or Mach lines, must not be confused with finite *waves*, across which the velocity may also be discontinuous.

The form of these equations lends itself naturally to introduction of the Prandtl-Meyer function, ν, which is a dimensionless measure of the speed defined by the relation (Article 4·10)

$$\nu = \int \frac{\cot \mu}{w}\, dw$$

or (12·4)

$$d\nu = \cot \mu \frac{dw}{w}$$

It will be seen presently that, for the characteristics method, the function ν is the most "natural" one of the many functions that are related to w (or to the Mach number M). With this substitution, Eqs. 12·2 become

$$\frac{\partial \nu}{\partial s} - \tan \mu \frac{\partial \theta}{\partial n} = 0$$

$$\tan \mu \frac{\partial \nu}{\partial n} - \frac{\partial \theta}{\partial s} = 0 \qquad (12 \cdot 5)$$

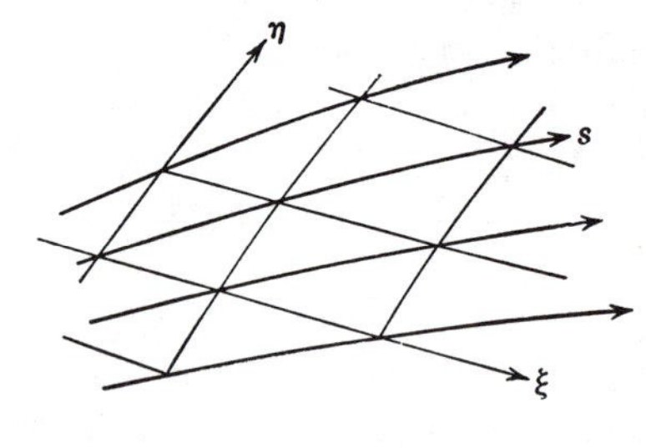

(a) Characteristic network.

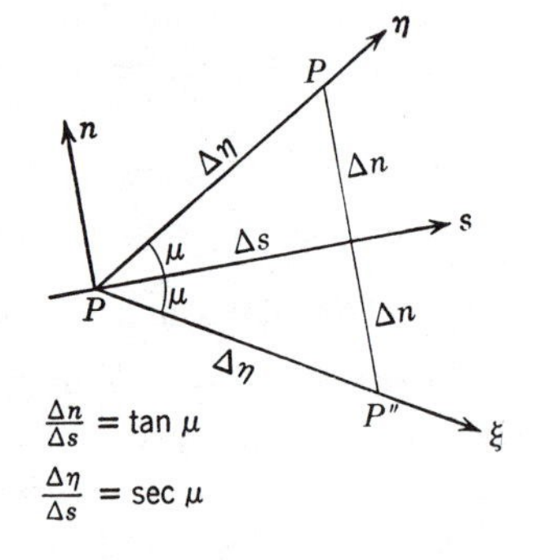

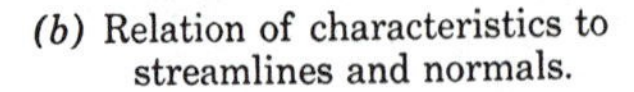
(b) Relation of characteristics to streamlines and normals.

Fig. 12·1. Natural and characteristic coordinate systems.

We must now find the *compatibility relation* between ν and θ, which according to the theory of hyperbolic equations must exist on the characteristics, or Mach lines. The theory gives rules for finding this relation, but here we shall obtain it simply by "inspection." We may expect this to be facilitated by rewriting the above equations in a coordinate system (ξ, η) which consists of the network of Mach lines shown in Fig. 12·1*a*. The relation between the two sets of coordinates is established by the fact that the Mach lines are inclined at the angles $\pm\mu$ to the streamlines.

To write derivatives in the new coordinate system, note that the change in any function f, in going from P to P' (Fig. 12·1b), may be written

$$\Delta f = \frac{\partial f}{\partial \eta} \Delta \eta$$

On the other hand, Δf may also be calculated by going along the streamline coordinate system,

$$\Delta f = \frac{\partial f}{\partial s} \Delta s + \frac{\partial f}{\partial n} \Delta n = \left(\frac{\partial f}{\partial s} + \frac{\partial f}{\partial n} \frac{\Delta n}{\Delta s} \right) \Delta s$$

Comparing the two equations,

$$\frac{\partial f}{\partial \eta} \frac{\Delta \eta}{\Delta s} = \frac{\partial f}{\partial s} + \frac{\partial f}{\partial n} \frac{\Delta n}{\Delta s}$$

From the geometry of Fig. 12·1b, this may be written in the form

$$\sec \mu \frac{\partial f}{\partial \eta} = \frac{\partial f}{\partial s} + \tan \mu \frac{\partial f}{\partial n} \tag{12·6}$$

The only difference in the ξ-derivative, that is, from P to P'', is that $\Delta n/\Delta s = -\tan \mu$, and thus

$$\sec \mu \frac{\partial f}{\partial \xi} = \frac{\partial f}{\partial s} - \tan \mu \frac{\partial f}{\partial n} \tag{12·7}$$

Equations 12·6 and 12·7 give the rules that relate the derivatives of any function f, in the two coordinate systems. We could solve them for $\partial f/\partial s$ and $\partial f/\partial n$ explicitly and then apply them directly to Eqs. 12·5. However, a shorter procedure is to first add and subtract Eqs. 12·5, which gives respectively,

$$\frac{\partial}{\partial s} (\nu - \theta) + \tan \mu \frac{\partial}{\partial n} (\nu - \theta) = 0$$

$$\frac{\partial}{\partial s} (\nu + \theta) - \tan \mu \frac{\partial}{\partial n} (\nu + \theta) = 0$$

Comparing with our rule for derivatives, we see that these may be written

$$\frac{\partial}{\partial \eta} (\nu - \theta) = 0$$

$$\frac{\partial}{\partial \xi} (\nu + \theta) = 0$$

or

$$\left.\begin{aligned} \nu - \theta &= R \quad \text{const. along an } \eta\text{-characteristic} \\ \nu + \theta &= Q \quad \text{const. along a } \xi\text{-characteristic} \end{aligned}\right\} \tag{12·8}$$

These are the *compatibility relations* between ν and θ. They give the simple result that the functions $Q = \nu + \theta$ and $R = \nu - \theta$ are invariant on the ξ- and η-characteristics, respectively. Q and R are called the *Riemann invariants.*

It should be noted that the compatibility relations are not always obtained in such convenient form. In general, they are obtained in differential form, and cannot always be integrated in this way, independently of the specific flow field to be solved. We shall have an example of this in Article 12·6.

12·4 The Computation Method

The preceding article shows that the characteristics problem is "naturally" formulated in terms of the Prandtl-Meyer function ν. Thus, in computing solutions by this method it is convenient to work with ν instead of w. Once

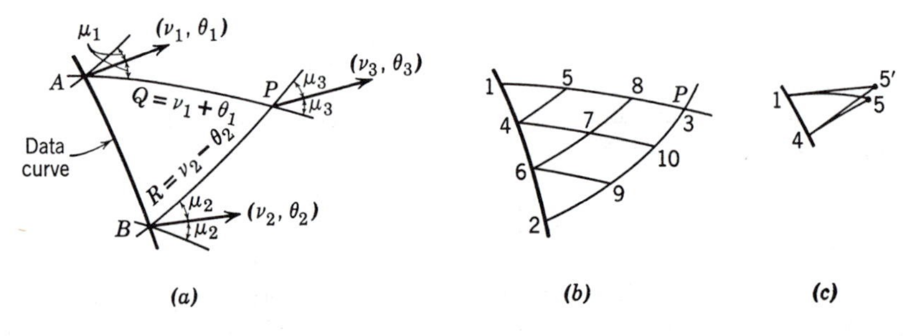

FIG. 12·2 Characteristic networks for computation.

the problem is solved, the values of ν may be converted into M, μ, w/a^*, p/p_0, or any other related supersonic variable. Some of these are tabulated in Table V at the back of the book.

The method of computation is illustrated in Fig. 12·2. The data, or boundary conditions, are given on the data curve. It is required to find the conditions at an arbitrary point P. It will be noted that the velocity vector is specified by ν and θ, rather than u and θ or M and θ.

Through the point P there are two characteristics, one of ξ-type and one of η-type, which intersect the data curve at the points A and B, respectively. A and B lie on the data curve, and so the values of the functions ν and θ (and thus Q and R) are known there. The values of Q and R at the point P are then easily written down by choosing the appropriate one for each of the two characteristics. Thus, from Eqs. 12·8,

$$Q_3 = Q_1$$

$$R_3 = R_2$$

or

$$\nu_3 + \theta_3 = \nu_1 + \theta_1$$
$$\nu_3 - \theta_3 = \nu_2 - \theta_2$$

From these two simple algebraic equations the solution is easily found to be

$$\nu_3 = \tfrac{1}{2}(\nu_1 + \nu_2) + \tfrac{1}{2}(\theta_1 - \theta_2)$$
$$\theta_3 = \tfrac{1}{2}(\nu_1 - \nu_2) + \tfrac{1}{2}(\theta_1 + \theta_2)$$

It may also be written directly in terms of the invariants,

$$\nu = \tfrac{1}{2}(Q + R)$$
$$\theta = \tfrac{1}{2}(Q - R) \qquad \blacktriangleright(12{\cdot}9)$$

Thus the solution for the conditions at an arbitrary point P is very simple and elegant. However the solution is not yet complete, *for the location of the characteristics is not known a priori.*

... this point the solution becomes a numerical one: to locate the characteristics a step-by-step procedure must be used. By subdividing the region into a characteristics network, as shown in Fig. 12·2*b*, the "mesh" may be made small enough so that the mesh sides may be approximated by straight-line segments. For instance, point 5 is located by using the known Mach angles and flow directions at 1 and 4 to draw the characteristic segments. Conditions at 5 are determined from the data at 1 and 4. Similarly point 7 is found, and then 8 is found from 7 and 5. Thus the computation proceeds *outward from the data curve.*

The accuracy of the construction depends on the mesh size. In Fig. 12·2*c* a typical, constructed mesh element is compared, with exaggeration, to the true mesh which is defined by the true, curved characteristics. The calculated conditions, ν_5 and θ_5, strictly apply to point 5 on the true characteristics but the construction gives them for 5′, and the solution is in error to this extent.

Various methods for improving the construction will occur to the reader. One of these is an iteration procedure. However, usually the best method is simply to use smaller mesh size, for it is desirable to keep the number of operations in the computing sequence to a minimum.

The typical "working outward" from a data curve gives some indication of the nature of the boundary conditions which may be imposed, and of their limited regions of influence. This is in contrast to the Laplacian or elliptic type of field, in which the region of computation must be completely bounded and in which each point is influenced by all other points in the region. Unfortunately it is not possible here to go further into these interesting questions. In practice, the correct boundary conditions will usually be evident.

An example of a characteristics construction is given in Fig. 12·3. A

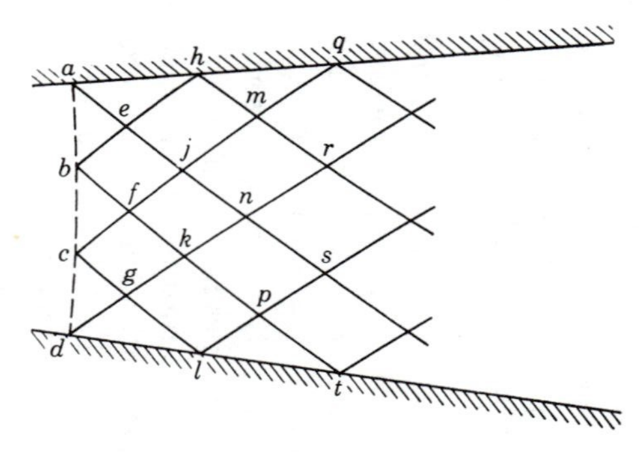

$$Q = \nu + \theta$$
$$R = \nu - \theta$$

$$\nu = \tfrac{1}{2}(Q + R)$$
$$\theta = \tfrac{1}{2}(Q - R)$$

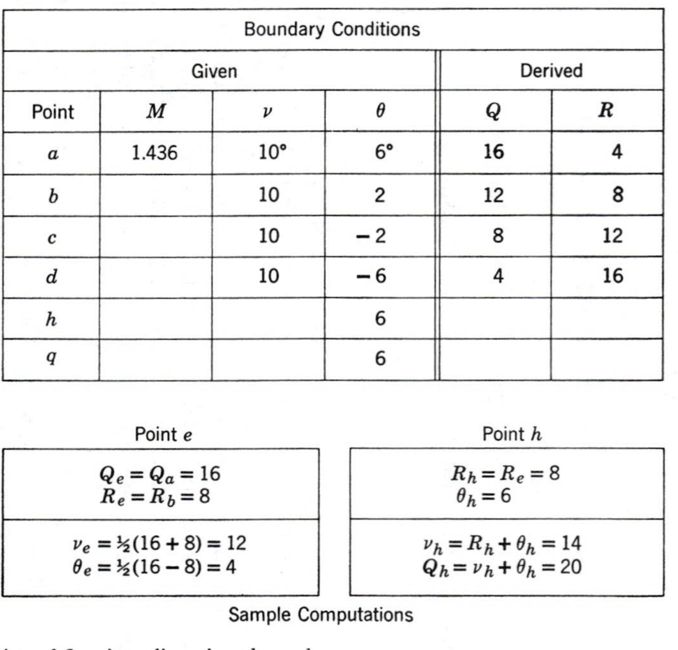

Boundary Conditions					
Given				Derived	
Point	M	ν	θ	Q	R
a	1.436	10°	6°	16	4
b		10	2	12	8
c		10	−2	8	12
d		10	−6	4	16
h			6		
q			6		

Point e

$$Q_e = Q_a = 16$$
$$R_e = R_b = 8$$

$$\nu_e = \tfrac{1}{2}(16 + 8) = 12$$
$$\theta_e = \tfrac{1}{2}(16 - 8) = 4$$

Point h

$$R_h = R_e = 8$$
$$\theta_h = 6$$

$$\nu_h = R_h + \theta_h = 14$$
$$Q_h = \nu_h + \theta_h = 20$$

Sample Computations

FIG. 12·3 Computation of flow in a diverging channel.

channel with straight walls diverging by 12° produces radial flow, with a Mach number of 1.436 on the arc *ad*. It is required to compute the flow downstream of this arc. Here it could be done very simply, of course, from area relations, since the flow is radial, but the example serves well to illustrate the characteristics method, and it furnishes a check on the accuracy.

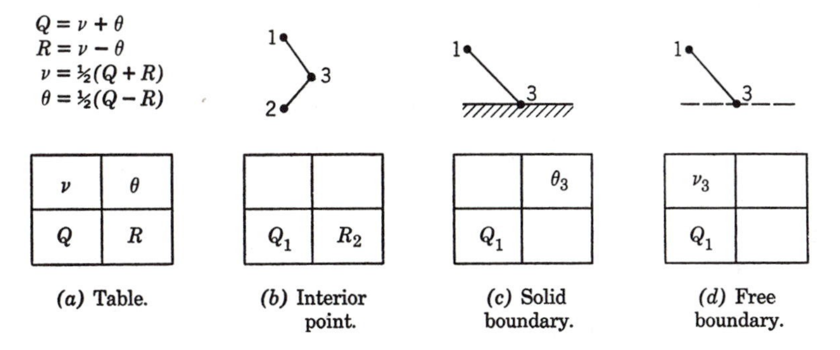

FIG. 12·4 Known data for computing toward a point 3.

The table lists the boundary conditions, which are given on the arc *ad* and on the channel walls. They are given in terms of ν and θ, from which the corresponding values of Q and R are easily calculated, where needed.

Typical computations are shown for a point *e* inside the channel and a point *h* on the wall.

12·5 Interior and Boundary Points

The above example shows that boundary conditions fit into the computation quite readily. At a boundary one of the two invariants is not available, but one of the other two variables is determined instead. Thus, at a solid boundary, θ is given, whereas, at a free boundary, such as the edge of a jet, the pressure ratio p/p_0, and thus ν, are given.

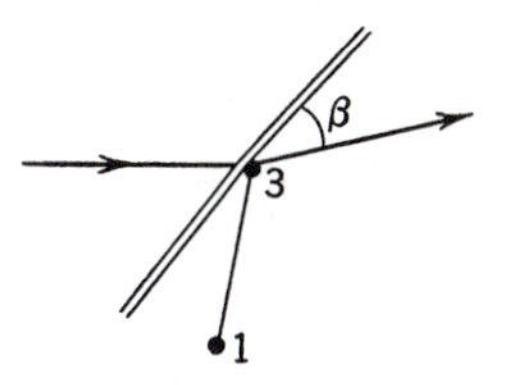

FIG. 12·5 Computation of a point 3 on the downstream side of a shock.

The possibilities may be classified into a tabular scheme as shown in Fig. 12·4. In (*a*) the scheme of tabulation is shown. In (*b*) is shown an "interior point 3, for which the values of Q and R are obtained from 1 and 2, respectively. In (*c*) the value of Q is obtained from 1 and the value of θ is given; in (*d*) the value of ν is given. In each case there are two quantities, so that the other two may be computed from the equations listed in (*a*).

It is sometimes necessary to compute flows in which shocks appear, for example, the airfoil of Fig. 4·17*b*. The method is illustrated in Fig. 12·5, where it is required to compute point 3 just behind the shock. One quantity,

R, is obtained from point 1. The other is determined by the shock equations. It is not given explicitly, but as a relation between ν_3 and θ_3. Thus the condition at point 3 may be solved. These then determine the shock wave angle β, which is used to draw the next shock segment. If the shock is strongly curved, the flow downstream of it has vorticity and the equations of Article 12·7 must be used.

12·6★ Axially Symmetric Flow

The treatment of plane flow in the preceding articles demonstrates the main features of the characteristics method. There is also a theory of characteristics for general three-dimensional flow, but the computations become

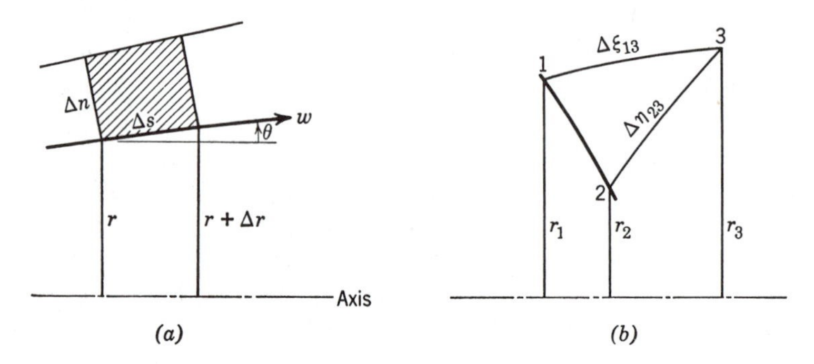

FIG. 12·6 Coordinates for axially symmetric flow. (*a*) Natural coordinates; (*b*) characteristic network.

quite complicated. However, for axially symmetric flow, which has only two independent and two dependent variables, the method is easily extended from the plane flow case.

The equations of motion (Exercise 7·1) are

$$\frac{\cot^2 \mu}{w} \frac{\partial w}{\partial s} - \frac{\partial \theta}{\partial n} = \frac{\sin \theta}{r}$$
$$\frac{1}{w} \frac{\partial w}{\partial n} - \frac{\partial \theta}{\partial s} = 0 \qquad (12\cdot10)$$

where (w, θ) defines the velocity in a meridian plane (Fig. 12·6*a*.) The first equation differs from the two-dimensional case only in the last term. At points whose distance, r, from the axis is large, this term is small, and the flow is practically two dimensional. The second equation, of irrotationality, is unchanged.

By multiplying through the first equation by $\tan \mu$ and through the second one by $\tan\mu \cot\mu$, the equations are obtained in the form

$$\frac{\partial\nu}{\partial s} - \tan\mu \frac{\partial\theta}{\partial n} = \tan\mu \frac{\sin\theta}{r}$$

$$\tan\mu \frac{\partial\nu}{\partial n} - \frac{\partial\theta}{\partial s} = 0$$

which correspond to Eqs. 12·5 for the two-dimensional case By adding and subtracting them as before, and applying the transformations (12·6 and 12·7) to characteristic coordinates, there is obtained,

$$\frac{\partial}{\partial\eta}(\nu - \theta) = \sin\mu \frac{\sin\theta}{r}$$

$$\frac{\partial}{\partial\xi}(\nu + \theta) = \sin\mu \frac{\sin\theta}{r} \qquad (12\cdot11)$$

It is not possible to integrate these, as before, because *the geometry of the flow field is now involved*, through the variable r. The integration must now be done numerically, step by step, simultaneously with the construction of the characteristics network.

Figure 12·6*b*, shows a typical mesh element, for which point 3 is to be solved from the known data at 2 and 1. From Eqs. 12·11 we may write, along the characteristic segments,

$$\int_2^3 d(\nu - \theta) = \int_2^3 \left(\sin\mu \frac{\sin\theta}{r}\right) d\eta$$

$$\int_1^3 d(\nu + \theta) = \int_1^3 \left(\sin\mu \frac{\sin\theta}{r}\right) d\xi$$

Since the mesh is small, the quantities in parentheses on the right-hand side may be assumed to be approximately constant, over the interval of integration, and to have the known values at 1 and 2, respectively. The result is

$$(\nu_3 - \theta_3) - (\nu_2 - \theta_2) = \sin\mu_2 \frac{\sin\theta_2}{r_2} \Delta\eta_{23}$$

$$(\nu_3 + \theta_3) - (\nu_1 + \theta_1) = \sin\mu_1 \frac{\sin\theta_1}{r_1} \Delta\xi_{13}$$

where $\Delta\eta_{23}$ and $\Delta\xi_{13}$ are the segment lengths along η- and ξ-characteristics. The solution of the two equations is

$$\nu_3 = \frac{1}{2}(\nu_1 + \nu_2) + \frac{1}{2}(\theta_1 - \theta_2) + \frac{1}{2}\left(\sin\mu_1 \frac{\sin\theta_1}{r_1} \Delta\xi_{13} + \sin\mu_2 \frac{\sin\theta_2}{r_2} \Delta\eta_{23}\right)$$

$$\theta_3 = \frac{1}{2}(\nu_1 - \nu_2) + \frac{1}{2}(\theta_1 + \theta_2) + \frac{1}{2}\left(\sin\mu_1 \frac{\sin\theta_1}{r_1}\Delta\xi_{13} - \sin\mu_2 \frac{\sin\theta_2}{r_2}\Delta\eta_{23}\right) \tag{12·12}$$

These differ from the two-dimensional equations (12·8) only in the additional terms which depend on the geometry of the particular problem. In these terms, the radial distances r_1 and r_2, of the points in question, and

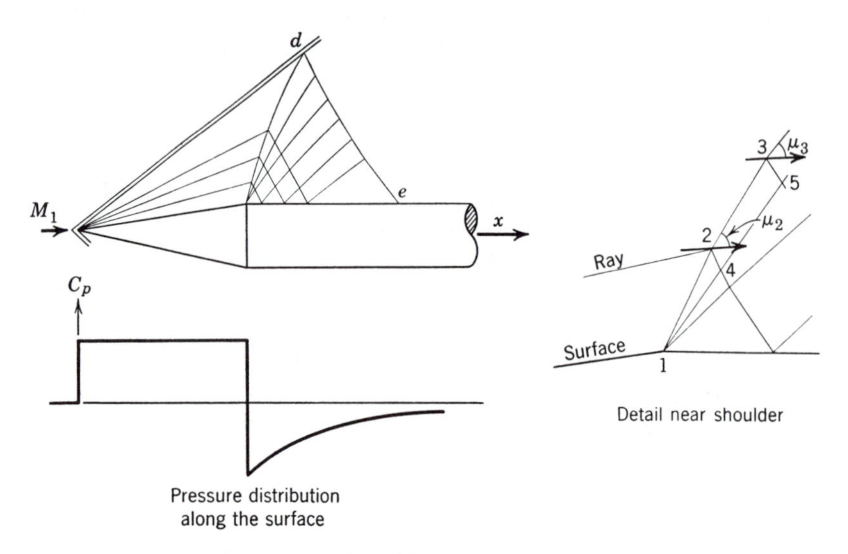

Fig. 12·7 Computation of flow over a cone with afterbody.

the lengths of the mesh sides, $\Delta\eta_{23}$ and $\Delta\xi_{13}$, must be obtained from the flow field by measurement on a drawing or by computation. These terms are large near the axis.

An example of an axially symmetric flow computation for a cone with afterbody is shown in Fig. 12·7. Some preliminary construction is needed before the actual characteristics solution is started. The flow up to the first Mach line from the shoulder is conical (Articles 4·21 and 9·9). This Mach line is located by simply drawing the segments at the local Mach angle relative to the local flow direction, starting the construction at the shoulder, as shown in the detail sketch. The Mach angle and flow direction at each point are obtained from the conical solution.

The other Mach line segments *at the shoulder* are also easily located because the flow is *locally* a two-dimensional Prandtl-Meyer expansion (Exercise 12·3). The characteristics construction and the computation for points 4, 5, etc., can now proceed.

From the pressure distribution shown in the figure, it will be seen that after expansion at the shoulder the pressure rises again toward the free-stream value.

The effect of the interaction with the shock is first felt downstream at the point *e*. For an exact computation it is necessary to include the effects of vorticity for the computations downstream of *de*, but usually the shock and expansion are not strong enough for the effect to be appreciable.

12·7★ Nonisentropic Flow

In the preceding example, the shock strength starts to decrease at *d*, where the shock begins to interact with the expansion. Downstream of the curved shock the entropy varies from one streamline to another, and the isentropic equations are not strictly valid in this region. They must be replaced by

$$\frac{\cot^2 \mu}{w}\frac{\partial w}{\partial s} - \frac{\partial \theta}{\partial n} = \frac{\sin \theta}{r}$$

$$\frac{1}{w}\frac{\partial w}{\partial n} - \frac{\partial \theta}{\partial s} = -\frac{T}{w^2}\frac{dS}{dn} + \frac{1}{w^2}\frac{dh_0}{dn}$$

The second equation was derived in Chapter 7 (see Eqs. 7·32 to 7·34).

The procedure for transforming these to characteristic coordinates is exactly the same as in the preceding article, and gives the result

$$\frac{\partial}{\partial \eta}(\nu - \theta) = \sin \mu \frac{\sin \theta}{r} - \frac{\cos \mu}{w^2}\left(T\frac{dS}{dn} - \frac{dh_0}{dn}\right)$$

$$\frac{\partial}{\partial \xi}(\nu + \theta) = \sin \mu \frac{\sin \theta}{r} + \frac{\cos \mu}{w^2}\left(T\frac{dS}{dn} - \frac{dh_0}{dn}\right)$$

The last terms in each equation may be written as derivatives along the characteristics by using the geometry of Fig. 12·1*b*,

$$\frac{\partial \eta}{\partial n} = \csc \mu, \quad \frac{\partial \xi}{\partial n} = -\csc \mu$$

Integration over a small-mesh element then gives the equations

$$\nu_3 - \theta_3 = \nu_2 - \theta_2 + \sin \mu_2 \frac{\sin \theta_2}{r_2}\Delta\eta_{23} - \frac{\cot \mu_2}{{w_2}^2}[T_2(S_3 - S_2) - (h_{03} - h_{02})]$$

$$\nu_3 + \theta_3 = \nu_1 + \theta_1 + \sin \mu_1 \frac{\sin \theta_1}{r_1}\Delta\xi_{13} - \frac{\cot \mu_1}{{w_1}^2}[T_1(S_3 - S_1) - (h_{03} - h_{01})] \tag{12·13}$$

They may be solved for ν_3 and θ_3.

It will be noted that the values of S_3 and h_{03} at point 3 are needed for the computation. These may be determined as follows (Fig. 12·8). Once 3 is located, the streamline through it may be approximately located by drawing a line with slope $\theta'_3 = \frac{1}{2}(\theta_1 + \theta_2)$, intersecting the data curve at 3′. Since S and h_0 are invariant along streamlines, their values at 3 are the same as at 3′ on the data curve, where they are known.

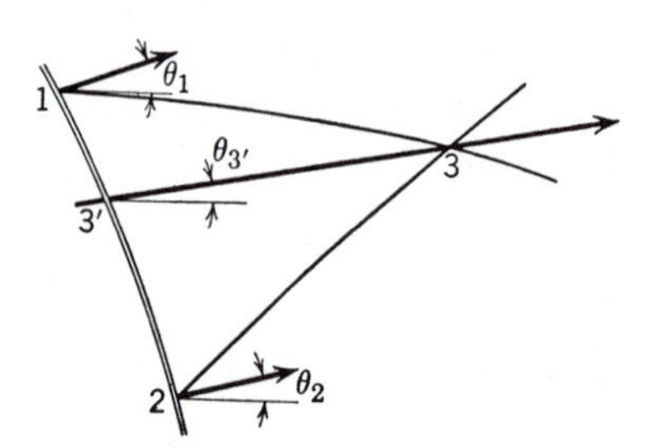

FIG. 12·8 Streamline through 3.

12·8 Theorems about Plane Flow

In the remainder of this chapter we shall consider some further aspects of plane, supersonic flow, for which we have the compatibility relations

$$\nu + \theta = Q \quad \text{on } \xi\text{-characteristics}$$

$$\nu - \theta = R \quad \text{on } \eta\text{-characteristics}$$

These relations, which are *independent of the specific flow geometry*, lead to the following useful theorems.

The flow may be classified into three kinds: (1) general or nonsimple region, (2) simple region, sometimes called simple wave, (3) uniform region, or constant state.

The *general region* is illustrated in Fig. 12·9. The characteristics are curved, and each one corresponds to one value of R or Q. The value of ν and θ at the intersection of any two characteristics is found from the solution of the above equations,

$$\nu = \tfrac{1}{2}(Q + R), \quad \theta = \tfrac{1}{2}(Q - R)$$

Now, in going *along an η-characteristic*, R is constant and so the changes in ν and θ depend only on the changes in Q, owing to the crossings of ξ-characteristics. Thus

$$\Delta\nu = \tfrac{1}{2}\,\Delta Q = \Delta\theta \tag{12·14a}$$

Similarly, in going *along a ξ-characteristic*,

$$\Delta\nu = \tfrac{1}{2}\,\Delta R = -\,\Delta\theta \tag{12·14b}$$

Thus, flow changes are easily determined once the values of R and Q on the characteristics are known.

Next we consider the *simple region* or simple wave, which is defined by the condition that one of the invariants, Q or R, *is constant throughout the region*. This is illustrated in Fig. 12·10, in which all the η-characteristics have the same value of R $(=R_0)$. Then, from Eq. 12·14*b*, ν and θ are *individually* constant along a ξ-characteristic, and it must therefore be *straight*. Thus, in a simple wave one set of characteristics consists of straight lines, with uniform conditions on each one. The flow changes encountered in crossing the straight characteristics are related by

$$\Delta\nu = \pm\,\Delta\theta \tag{12·15}$$

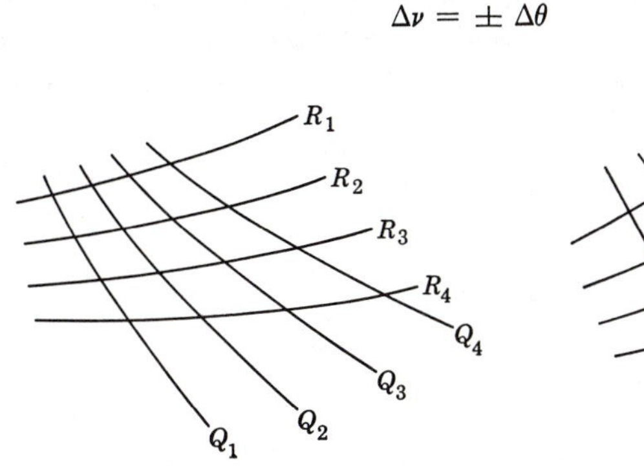

FIG. 12·9 Characteristics in a nonsimple region.

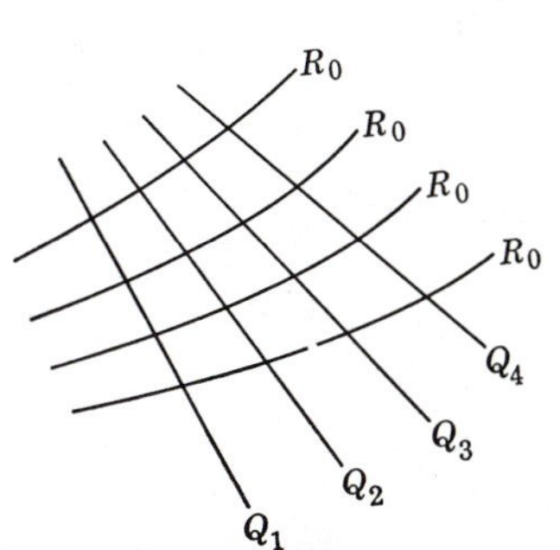

FIG. 12·10 Characteristics in a simple region.

the sign depending on whether they are of ξ- or η-type. The difference between the relation here and in Eqs. 12·14 is that now it holds on *any* line that crosses the straight characteristics, and in particular on a streamline.

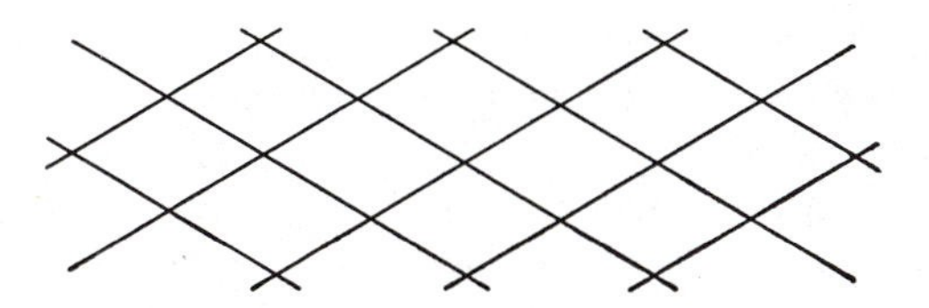

FIG. 12·11 Characteristics in a region of uniform flow.

Finally, a constant state or *uniform flow* is one in which $R = R_0$ and $Q = Q_0$ throughout. Then ν and θ are uniform, and both sets of characteristics consist of straight lines forming a parallel network, as shown in Fig. 12·11.

An example in which all three regions occur is given in Fig. 4·10. As shown there, the usual convention is to omit the Mach lines in the uniform region, to show only the straight lines in a simple wave, and both sets in the nonsimple region.

It will be noted that *the uniform region does not adjoin the nonsimple region* (except at one point). This is a general theorem, which may be easily proved by trying to construct a contrary case, if we remember the definitions given above. The theorem is useful in the construction of flow fields.

12·9 Computation with Weak, Finite Waves

The method of constructing two-dimensional, supersonic flows by using waves was outlined in Chapter 4. If the waves are weak, one can set up a computing procedure that is equivalent to the characteristics method. (See Article 12·12.) In fact, this is usually also called the characteristics method.

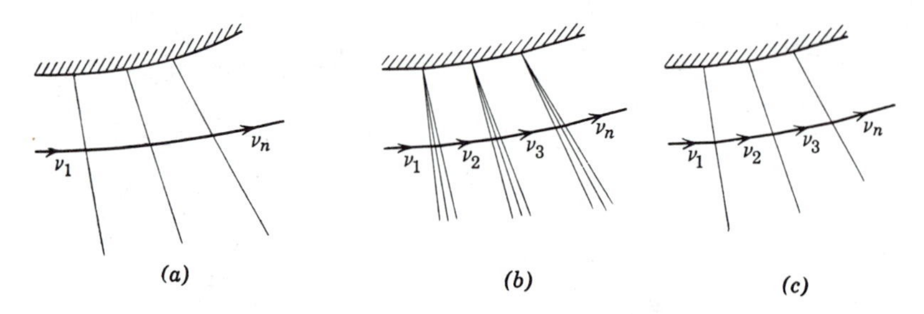

Fig. 12·12 Approximation of curved wall by straight segments. [In (c) the expansion wave from each corner is represented by a single line.]

However, the points of view are slightly different. The main point to remember is that *characteristics are not waves*, or vice versa.

The difference and similarity between them is illustrated by Fig. 12·12. In (a), the wall and flow are *continuous*. Several characteristics of the straight family are shown. We could draw as many of these as we like, from arbitrary points on the wall. The characteristics of the other family are not shown.

In (b), the wall is *approximated* by a series of straight segments, meeting at finite angles. The expansion fans, or *waves*, which originate at the corners, divide the field into *segments of uniform flow*. Since they are simple waves, the flow changes across them are related by Eq. 12·15, of the preceding article,

$$\Delta\nu = \pm\Delta\theta$$

The positive sign is for waves from the *upper* wall, as in Fig. 12·12, whereas the negative sign is for waves from the *lower* wall, $\Delta\theta$ being considered positive when counterclockwise.

In part (c) of Fig. 12·12, a further approximation is made by replacing each fan by a *single line*, using for this the central one. The changes at the "waves" are now discontinuous, but this approximation is no greater than in representing the wall by straight segments.

If the wall is *concave*, it gives a continuous *compression*, which may be approximated by a series of weak shocks as in Article 4·8. The above equations for the flow changes apply without modification. The position of each wave is midway between the characteristics of the uniform flow fields which it separates (Exercise 4·2). This is the same rule as for the expansion waves, above.

In summary, the method of weak waves is to replace a continuously varying flow by a series of uniform segments, with "steps" between them. These steps occur at the positions of the waves. So far we have considered only simple flows, in which all the waves are of one family. To extend them to nonsimple flows, we need the rules for intersection of waves of opposite family.

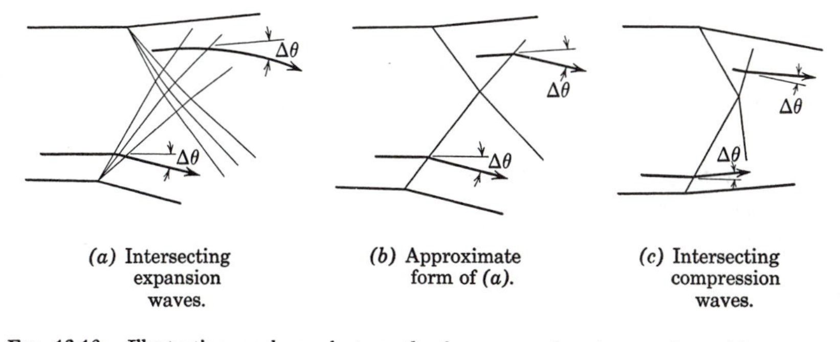

(*a*) Intersecting expansion waves.

(*b*) Approximate form of (*a*).

(*c*) Intersecting compression waves.

FIG. 12·13 Illustrating unchanged strength of a wave after intersection with another wave.

12·10 Interaction of Waves

The use of finite weak waves for the construction of plane flows depends on the following theorem:

> The strength of a weak wave is not affected by intersection with other waves.

Here the "strength" is defined as the *flow deflection* $\Delta\theta$ which the wave produces.

Figure 12·13 illustrates the theorem. (*a*) shows the intersection of two simple expansions, for which the theorem is easily proved by noting that the flow crosses the same characteristics on either side of the intersection. In (*b*) the expansion waves are represented by single lines. (*c*) illustrates the theorem for compression waves (Exercise 4·5).

It is now possible to set up a single, systematic computing method. We first consider an example in which the waves are *expansion* waves. They divide the flow into a number of *cells*, as shown in Fig. 12·14, in each of which the flow is uniform. It is required to calculate ν and θ in each cell.

It will be convenient to let δ_i = absolute value of flow deflection produced by a wave and to distinguish between waves from upper and lower walls by subscripts, $\delta_{\xi i}$ and $\delta_{\eta i}$. Expansion waves from the upper wall deflect the flow upward (positive), and increase ν, whereas those from the lower wall

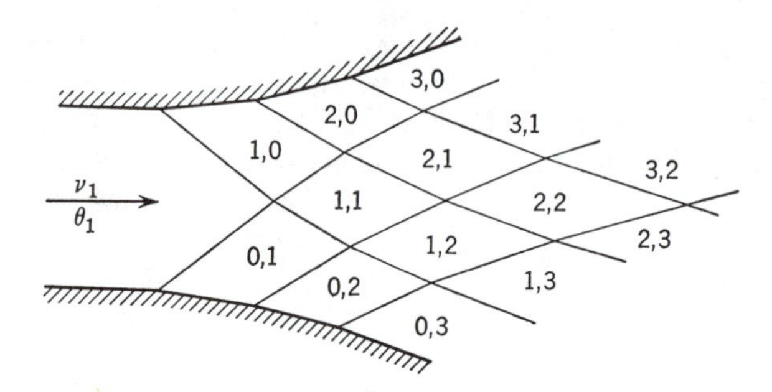

FIG. 12·14 Numbering of cells.

deflect the flow downward, and also increase ν. To reach a given cell the flow has crossed m waves of ξ-type and n waves of η-type. (This is true for any path from the initial point to the cell in question!) Thus the conditions in the (m, n) cell are given by

$$\theta = \theta_1 + \sum_{i=1}^{m} \delta_{\xi i} - \sum_{i=1}^{n} \delta_{\eta i}$$

$$\nu = \nu_1 + \sum_{i=1}^{m} \delta_{\xi i} + \sum_{i=1}^{n} \delta_{\eta i}$$

A particularly simple relation is obtained if all *the waves are of equal strength*, say $\delta_i = 1°$. Then

$$\begin{aligned} \theta - \theta_1 &= m - n \\ \nu - \nu_1 &= m + n \end{aligned} \qquad \blacktriangleright(12{\cdot}16)$$

that is, the conditions in a cell are obtained by simply adding up the number of waves of each kind that are crossed.

To include *compression* waves in the scheme it is necessary to introduce additional notation. If the flow crosses k compression waves from the upper wall and l compression waves from the lower wall, then the above result generalizes to

$$\begin{aligned} \theta - \theta_1 &= m - n - k + l \\ \nu - \nu_1 &= m + n - k - l \end{aligned} \qquad (12{\cdot}17)$$

One more result needed for flow construction concerns the *reflection* and *cancellation* of waves. Figure 12·15 shows that on reflection from a wall a wave of ξ-type is changed to one of η-type. The turning strength of the reflected wave is the same as that of the incident one, since the flow must return to the original direction, parallel to the wall.

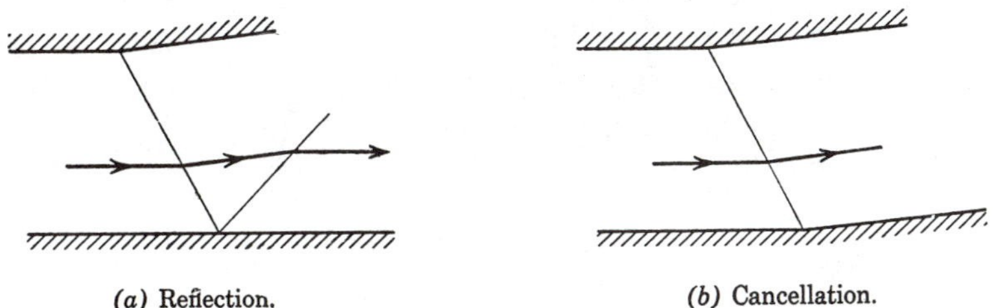

(a) Reflection. (b) Cancellation.

FIG. 12·15 Reflection and cancellation of a wave.

Figure 12·15b shows how the reflected wave may be "canceled" by accommodating the wall to the flow direction after the incident wave. The wall deflection is equal to the strength of the wave.

In designing supersonic nozzles by the method of waves, the idea of cancellation is applied in the following way. Since the uniform flow in the test

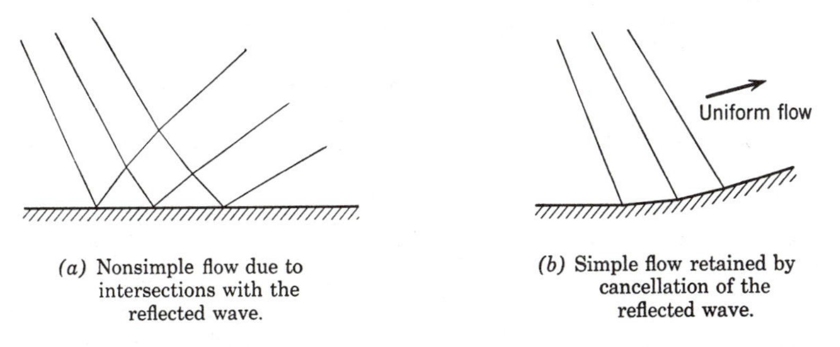

(a) Nonsimple flow due to intersections with the reflected wave.

(b) Simple flow retained by cancellation of the reflected wave.

FIG. 12·16 Simple and nonsimple regions.

section must be "wave-free," all the waves produced in the upstream part of the nozzle must be canceled before they reach the test section. The appearance of the wave field with and without cancellation is illustrated in Fig. 12·16. In the latter case, there is an obvious correspondence with the simple region of Article 12·8. We have here another example of the theorem that a uniform region must be adjacent to a simple wave.

12·11 Design of Supersonic Nozzles

In Fig. 12·17 the method of computation with weak waves is applied to the design of a supersonic nozzle. It serves to illustrate the method of application in general.

The problem here is to expand the flow from $M = 1$ at the throat to $M = M_T$ in the test section, where the flow is to be *uniform* and parallel to the direction at the throat. For this example, we have chosen $\nu_T = 16$, which corresponds to $M_T = 1.639$. The main steps, using values from Table V, are as follows.

(1) The *initial expansion*, whose length and shape are arbitrary, is divided into segments with equal deflection angles at the corners. Here we use 2° deflections, instead of the 1° suggested in the last article, so that

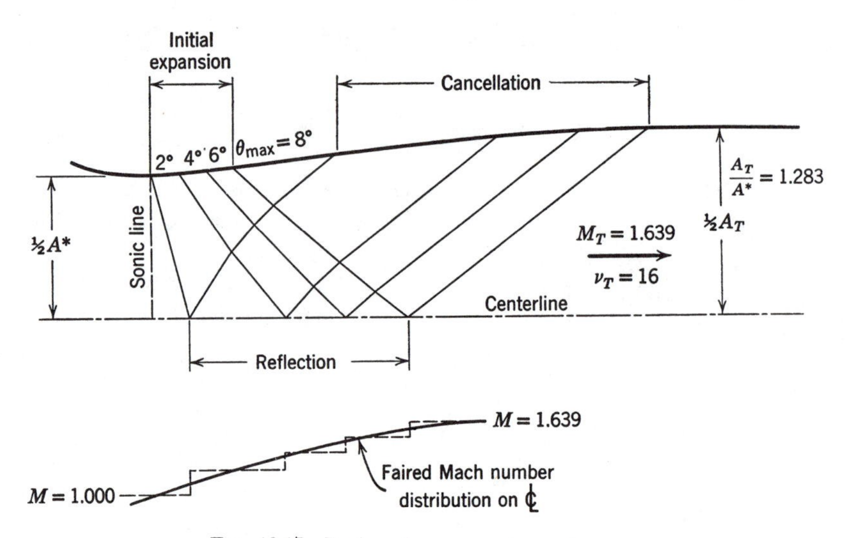

FIG. 12·17 Design of a supersonic nozzle contour.

all the waves are of strength 2°. There are four such waves, the corresponding maximum value of the wall deflection being $\theta_{max} = 8°$. The inclination of each wave is midway between the Mach lines on either side of it. For example, the inclination of the first wave, measured relative to the centerline, is $\frac{1}{2}[(90 - 0) + (62.07 - 2)] = 75.04$.

(2) The waves are *"reflected"* from the centerline of the symmetrical nozzle, of which only one half is shown. (Alternatively, the "centerline" might be the wall of an asymmetrical nozzle, complete as shown.)

(3) In the region of *cancellation* the reflected waves are canceled by deflecting the wall 2° at each wave intersection. After the last wave the wall is parallel to the centerline and the flow is uniform. It has the value $\nu_T = 16$, having crossed eight expansion waves of 2° each—four initial ones and four reflected ones. From this we get the rule that the value of the initial expansion (that is the maximum wall deflection) is given by

$$\theta_{max} = \tfrac{1}{2}\nu_T \tag{12·18}$$

(4) The final test section height must agree with the value A_T/A^* corresponding to $\nu_T = 16$. From Table V, this is $A_T/A^* = 1.283$.

The example of Fig. 12·17 contains the basic elements of a nozzle design, but many variations and modifications are possible. Here we can comment on only a few points, as follows.

(*a*) The length of the initial expansion contour is arbitrary. The *shortest nozzle* is obtained by having zero length for the initial expansion, that is, a Prandtl-Meyer expansion to $\theta_{\max}$. However, such rapid expansion usually gives troublesome effects in the side-wall boundary layers.

(*b*) In the example of Fig. 12·17, the construction has been started at the narrowest part of the throat, assuming the sonic line to be straight. Actually the sonic line is curved and meets the wall somewhat upstream of the narrowest section, the details depending on the throat shape. For great accuracy it may be necessary to use a transonic solution for the throat, in order to start the supersonic solution correctly.

(*c*) These throat problems are sometimes avoided by starting the construction well downstream in the initial expansion, where the flow may be assumed to be *radial*, as in Fig. 12·3. The assumption is satisfactory if the wall curvatures are not too great. The Mach number in the radial flow is obtained from the area relation A/A^* in Table III, with A being the area of the *curved* cross-section over which M is uniform.

(*d*) The rule given in Eq. 12·18 is valid only if all the waves are canceled after *one* reflection from the centerline. If, instead of being canceled, they are reflected back to the centerline and reflected there once more, then $\theta_{\max} = \frac{1}{4}\nu_T$, since the flow will then cross the initial expansion waves four times. *Partial cancellation* may also be used. The nozzles obtained are longer than the basic one.

(*e*) Figure 12·17 includes a sketch of the centerline distribution of Mach number, obtained from the construction. Another approach to nozzle design is to *specify the centerline distribution.* It should fair smoothly into M_T, as shown, and at $M = 1$ its slope should be consistent with the actual, transonic throat flow, which depends on throat shape. This approach is particularly suitable if the method of characteristics is used, instead of the method of waves. It is convenient in designing axially symmetric nozzles.

(*f*) The boundary layer on the nozzle and side walls has a displacing effect which reduces the effective height and width of the nozzle. Allowance for this is made by adding a correction for boundary-layer growth to the designed contour. The side walls should also diverge to allow for their boundary layers, but often this side-wall correction is taken up by additional correction of the nozzle contour.

12·12 Comparison of Characteristics and Waves

In closing this chapter we shall briefly compare the methods of characteristics and of waves.

(1) The method of characteristics deals with a *continuous* velocity field, the computations being made at the *lattice points* of a network of characteristics. The wave method deals with a patchwork field of *cells* of uniform flow, with discontinuities between them. The two cases are compared schematically in Fig. 12·18. Accuracies in the two methods are similar, being dependent in both cases on the fineness of the mesh.

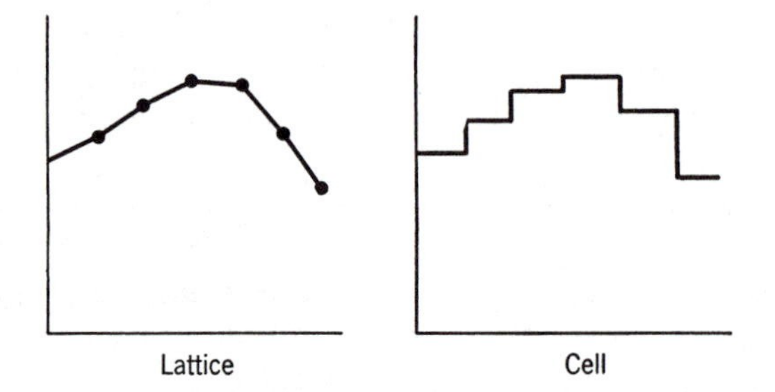

FIG. 12·18 Illustrating computation networks.

(2) The computation with waves is convenient only in *plane* flow, since it depends on the theorem that the strength of a wave does not change, even after intersection or reflection. In axially symmetric flow, and in three-dimensional flow generally, the strength of a wave varies continuously in space, and therefore cannot be conveniently used for computation. In this case it is simpler to use the basic characteristics method (Article 12·6).

(3) The wave method is more intuitive in *plane* flow than the characteristics method, and for this reason is usually preferred. In some problems it is also the more convenient one, for instance in applying the idea of wave cancellation to determine a boundary shape.

CHAPTER 13

Effects of Viscosity and Conductivity

13·1 Introduction

In almost all the problems treated so far we have dealt with a nonviscous, nonconducting fluid. Only in shock waves have the effects of viscosity and heat transfer appeared at all, and there only in a very indirect way, so that it was not necessary to treat these effects explicitly.

Since all real gases and fluids are, in fact, viscous and heat-conducting, it may seem surprising that in a book on gasdynamics only one chapter is devoted explicitly to the effects of viscous shear and heat transfer. The reason that the larger part of aerodynamics can be treated within the idealized concept of a perfect fluid is due to the fact that the viscosity and conductivity of gases are comparatively small.

The nondimensional parameter which measures the relative magnitude of the viscosity effects is the *Reynolds number* Re, defined by

$$Re = \frac{Ul\rho}{\mu} = \frac{Ul}{\nu} \tag{13·1}$$

U and l are a characteristic velocity and length, respectively, ρ is the density, and μ is the coefficient of viscosity, to be defined more precisely later. It is sometimes called the dynamic coefficient of viscosity, to distinguish it from the ratio $\nu = \mu/\rho$, which is called the kinematic coefficient of viscosity or, simply, the kinematic viscosity.

The relative magnitudes of viscosity and heat conductivity are measured by the *Prandtl number* Pr, which is defined by

$$Pr = \frac{c_p\mu}{k} \tag{13·2}$$

c_p is the specific heat at constant pressure and k is the coefficient of heat conductivity. The dimensions of μ and k are obvious from Eqs. 13·1 and 13·2.

Superficially, we can say that viscous effects in most aerodynamic problems are small since the Reynolds numbers are large. However, it is by no means obvious that viscous effects can be neglected. In fact, the problem of relating the solutions of perfect fluid flow and the corresponding solutions of real fluid flow represents one of the most difficult and intriguing problems of fluid mechanics.

Some phases of this problem are still unsolved. Fortunately, most (but not all) problems of high-speed gasdynamics come within the realm of boundary-layer theory. This concept, introduced by Prandtl in 1904, furnishes a method for relating perfect fluid flow and viscous flow past the same boundary. In the first place, boundary-layer theory furnishes a method for computing skin friction and heat transfer. Second, and possibly even more important, it states that *for flow at high Reynolds numbers past slender bodies viscosity does not influence the pressure field.* This statement justifies the extensive use of perfect fluid theory in aerodynamics. The pressure forces computed on the basis of a nonviscous fluid flow remain nearly unaffected by the effects of viscosity and heat conductivity, within the realm of applicability of boundary-layer theory.

The boundary layer shares with most other brilliant concepts the fate of having been at first disregarded and later accepted as obvious. Prandtl's boundary-layer theory, and its later extension to turbulent flow by von Kármán, have had an enormous impact on modern fluid dynamics, so much, in fact, that today it is sometimes forgotten that boundary-layer theory does not cover *all* aspects of viscosity effects in fluid dynamics.

We shall discuss first, in the next article, the simplest case of shear flow, the so-called Couette flow. The simplicity of the problem will enable us to exhibit the effects of compressibility upon shear flow, without the additional difficulties entailed in the boundary-layer problem. It will make the following study of boundary-layer flow much easier.

13·2 Couette Flow

The flow is two-dimensional, between two infinite, flat plates, distance d apart (Fig. 13·1). The coordinates are x, in the direction of flow, and y, normal to the plates. The space between the plates is filled with a gas. The upper plate slides in the x-direction with a constant velocity U. The problem is to compute the flow of the gas.

In perfect fluid theory the sliding motion of the upper wall would have no effect whatsoever on the gas, since the only boundary condition available applies to the velocity component normal to the surface. In real fluid theory it is necessary to add another boundary condition, for the velocity component parallel to the wall. This is the so-called no-slip condition: *the fluid at a solid boundary has the same velocity as the boundary.* In this problem, then, the gas next to the upper wall is moving with the wall at velocity U, while the gas at the lower wall is at rest. The no-slip condition permits the wall to transmit a shear force τ to the fluid.

Conditions are the same at every section x, and so the shear τ can depend only on y. For the same reason, there are no accelerations and no pressure gradients in the x-direction. Therefore, the equilibrium of forces on a fluid element shows that (Fig. 13·2)

$$\frac{d\tau}{dy} = 0 \tag{13·3}$$

Thus τ is constant throughout the flow, and must be equal to the shear stress, τ_w, on the walls.

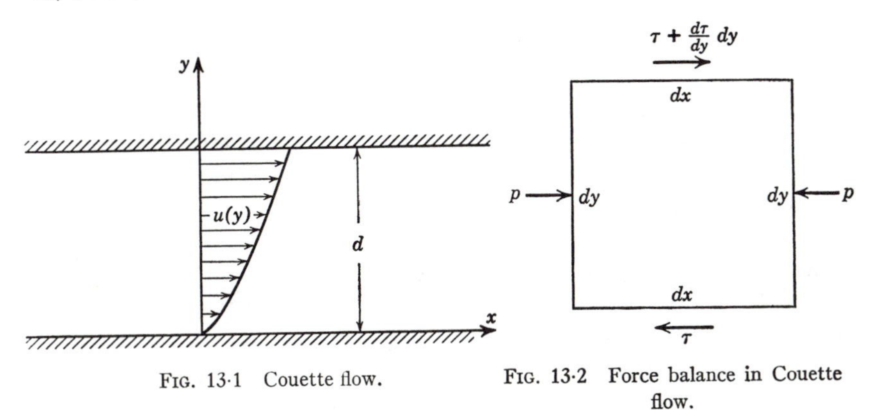

FIG. 13·1 Couette flow.

FIG. 13·2 Force balance in Couette flow.

Similarly, a momentum balance for the y-direction shows that $dp/dy = 0$, since $v = 0$. Thus the pressure is uniform throughout.

The shear stress is related to the velocity field $u(y)$ by Newton's form of the friction law

$$\tau = \mu \frac{du}{dy} \tag{13·4}$$

This equation defines the viscosity μ. For gases under a wide range of conditions,

$$\mu = \mu(T)$$

that is, μ does not depend on the pressure.

If the fluid is moving slowly, and if there is no heat transfer through the walls, the temperature will be nearly constant. Then $\mu =$ const., and since $\tau = \tau_w$ is also constant, Eq. 13·4 shows that u is a linear function of y. This is the example discussed in elementary physics textbooks; it is the case of *incompressible*, viscous flow.

If the velocity U of the upper wall is large enough so that the Mach number of the fluid there is appreciable, or if heat is transferred to the fluid through the walls, or if both occur, then T will depend on y. In such cases the problem is one of *compressible*, viscous flow.† In Couette flow, "*compressibility*" *appears only through temperature effects.* The pressure is constant, since there is no acceleration in the y-direction, and thus variations in ρ are due only to variations in T. For a *perfect gas*

$$\rho(y) = \frac{p}{RT(y)} = \frac{\text{const.}}{T} \tag{13·5}$$

The solution of Eq. 13·4 may be written, formally,

$$u = \tau_w \int_0^y \frac{dy}{\mu(T)} \tag{13·6}$$

but to find it explicitly one must know the temperature distribution $T(y)$ and the dependence of μ on T. For finding the temperature distribution we still have the energy equation at our disposal (the momentum equation (13·3) has already been used to set $\tau = \tau_w$ in Eq. 13·6).

To write the energy equation we shall have to compute the net flow of heat into a fluid element. The flow of heat, through unit area per unit time, is represented by q,‡ and is related to the temperature field by an expression analogous to the Newtonian friction law,

$$q = -k\frac{dT}{dy} \tag{13·7}$$

The negative sign is conventional, to show that q is positive when dT/dy is negative, that is, that heat flows from higher to lower temperature. Equation 13·7 defines the coefficient of heat conductivity k, which, like μ, is a function of T only,

$$k = k(T)$$

For all common gases the Prandtl number Pr is very nearly constant (independent of temperature and pressure),

$$Pr = \frac{c_p\mu}{k} = \text{const.} \tag{13·8}$$

The constant is of order 1 for the common gases. (In addition, for a fairly

†The extension to compressible flow was first given by C. R. Illingworth, "Some Solutions of the Flow of a Viscous, Compressible Fluid," *Cambridge Phil. Soc. Proc.*, *46* (1950), p. 469.

‡In this chapter we shall follow the usual convention and use q for the heat flux, *through unit area per unit time.* In Chapters 1, 2, and 7, q was used (also according to convention!) for the "external" or bulk addition of heat, *per unit mass.* In the latter usage, q is a scalar quantity, whereas in the usage of this chapter it is one component of the heat flux vector, $\mathbf{q}$ or q_i, which is introduced in Article 13·13. This ambiguous use of q appears to be well established in the literature; one does not usually encounter examples in which both quantities appear.

wide range of temperatures around room temperature, c_p is nearly constant, and therefore $\mu \sim k$.)

Since in this problem the state of a fluid element does not change as it flows along, the energy law requires simply that the rate of flow of heat into the element plus the rate of work on the element is zero. Here the only flow of heat is connected with q (Fig. 13·3), and the only work is that due to the shear. The energy equation, then, is

$$\left[q - \left(q + \frac{dq}{dy}\,dy\right)\right]dx + \left[\left(\tau + \frac{d\tau}{dy}\,dy\right)\left(u + \frac{du}{dy}\,dy\right) - \tau u\right]dx = 0 \qquad (13\cdot9)$$

or

$$\frac{d}{dy}(-q + \tau u) = 0 \qquad -q + \tau u = \text{const.} \qquad \blacktriangleright(13\cdot10)$$

The constant may be written in terms of the conditions at the lower wall, where $u = 0$. If the heat flux there is denoted by q_w, then

$$-q + \tau u = -q_w$$

q and τ may be replaced by the expressions in Eqs. 13·4, 13·7, and 13·8, giving

$$k\frac{dT}{dy} + \mu u\frac{du}{dy} = \mu\frac{d}{dy}\left(\frac{1}{Pr}\,c_pT + \frac{1}{2}\,u^2\right) = -q_w$$

where it has been assumed, in the second step, that $c_p = \text{const.}$ Integrated, this is

$$c_p(T - T_w) + \frac{1}{2}\,Pr\,u^2 = -Pr\,q_w\int_0^y \frac{dy}{\mu(T)} \qquad (13\cdot11)$$

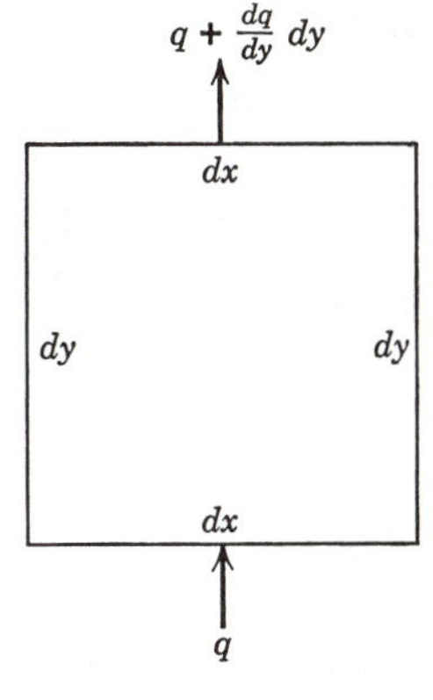

FIG. 13·3 Heat flow

where T_w is the temperature of the stationary wall. The integral on the right-hand side is given by Eq. 13·6. Substitution of that expression gives the so-called *energy integral*

$$c_p(T - T_w) + \frac{1}{2}\,Pr\,u^2 = -Pr\,\frac{q_w}{\tau_w}\,u \qquad \blacktriangleright(13\cdot12)$$

At the upper wall, where the velocity is U, we shall denote the temperature by T_∞. These are the boundary conditions that correspond most closely to the boundary-layer problem. The subscript on T_∞ is chosen for convenient comparison later. Putting these upper-wall conditions into Eq. 13·12 gives

$$c_pT_w = c_pT_\infty + Pr\left(\frac{U^2}{2} + \frac{q_w}{\tau_w}\,U\right) \qquad (13\cdot12a)$$

It is not necessary to have the restriction of constant specific heat c_p. Since p is constant, the enthalpy h is related to T by $dh = c_p\,dT$ (cf. Eq. 1·26). Hence Eq. 13·11 generalizes to

$$h - h_w + \frac{1}{2} Pr\, u^2 = -Pr\, q_w \int_0^y \frac{dy}{\mu(T)} \tag{13·11a}$$

or for a *single perfect gas* we have

$$h_w - h = \int_0^{T_w} c_p\, dT - \int_0^T c_p\, dT = \int_T^{T_w} c_p\, dT = \frac{Pr}{2} u^2 + Pr\, q_w \frac{u}{\tau_w}$$

and

$$\int_{T_\infty}^{T_w} c_p\, dT = Pr\left(\frac{U^2}{2} + \frac{q_w}{\tau_w} U\right) \tag{13·12b}$$

A further generalization to a dissociating gas is given in Article 13·18.

13·3 Recovery Temperature

If the lower wall is insulated, so that $q_w = 0$, what temperature does it attain? This special temperature is called the *recovery temperature*, and is denoted by T_r. From Eq. 13·12*a*, it has the value

$$T_r = T_\infty + \frac{Pr}{2c_p} U^2 \tag{13·13}$$

Introducing the Mach number $M_\infty = U/a_\infty$, this may be written

$$\frac{T_r}{T_\infty} = 1 + Pr \frac{\gamma - 1}{2} M_\infty{}^2 \tag{13·13a}$$

The recovery temperature is different from the *stagnation* temperature T_0, which, according to Eq. 2·30, is

$$\frac{T_0}{T_\infty} = 1 + \frac{\gamma - 1}{2} M_\infty{}^2$$

The expression

$$\frac{T_r - T_\infty}{T_0 - T_\infty} = r$$

is called the *recovery factor*. For our example of Couette flow of a perfect gas with constant c_p, the recovery factor is

$$\frac{T_r - T_\infty}{T_0 - T_\infty} = Pr$$

For air over a wide range of temperatures, $Pr = 0.73$. Sometimes this is approximated by $Pr = 1$. In that case, $T_r = T_0$ and the energy integral (for zero heat transfer) reduces to the familiar form of the one-dimensional energy equation.

From Eqs. 13·12a and 13·13 one also obtains the relation between heat transfer and shearing stress,

$$\frac{q_w}{\tau_w U} = \frac{c_p(T_w - T_r)}{Pr\, U^2} \tag{13·14}$$

This shows that the temperature of the wall, T_w, must be greater than T_r for heat transfer to the fluid (cf. Fig. 13·3). It is not sufficient to make $T_w > T_\infty$.

13·4 Velocity Distribution in Couette Flow

In Eq. 13·12 we now have the relation between T and u, which makes it possible to solve the momentum equation. It will be convenient to rewrite this expression in terms of T_∞,

$$c_p(T - T_\infty) = Pr\frac{q_w}{\tau_w}(U - u) + \frac{1}{2}Pr(U^2 - u^2)$$

or

$$\frac{T}{T_\infty} = 1 + Pr\frac{q_w}{U\tau_w}(\gamma - 1)M_\infty^2\left(1 - \frac{u}{U}\right) + Pr\frac{\gamma - 1}{2}M_\infty^2\left(1 - \frac{u^2}{U^2}\right) \tag{13·15}$$

The momentum equation is

$$\mu(T)\frac{du}{dy} = \tau_w = \text{const.}$$

The integral of this was given in one form in Eq. 13·6. Now that we may write $\mu = \mu(u)$, another form of the solution is

$$\int_0^u \mu(u)\, du = \tau_w y \tag{13·16}$$

This equation solves, in principle, the problem of the velocity distribution for any given $\mu(T)$, that is, $\mu(u)$. The dependence of μ on T, and thus on u through Eq. 13·15, can be obtained from empirical data or from kinetic theory (or both). The dependence is different for different gases, but *for all gases* μ increases with T. $\mu(T)$ can often be represented with sufficient accuracy by a power law,

$$\mu \sim T^\omega$$

or

$$\frac{\mu}{\mu_\infty} = \left(\frac{T}{T_\infty}\right)^\omega \tag{13·17}$$

For air, for example, $\omega = 0.76$ gives a good representation. With Eq. 13·17, Eq. 13·16 can be written explicitly

$$\int_0^u \left[1 + Pr\frac{q_w}{U\tau_w}(\gamma - 1)M_\infty^2\left(1 - \frac{u}{U}\right) + Pr\frac{\gamma - 1}{2}M_\infty^2\left(1 - \frac{u^2}{U^2}\right)\right]^\omega du$$

$$= \frac{\tau_w}{\mu_\infty}y \quad (13\cdot18)$$

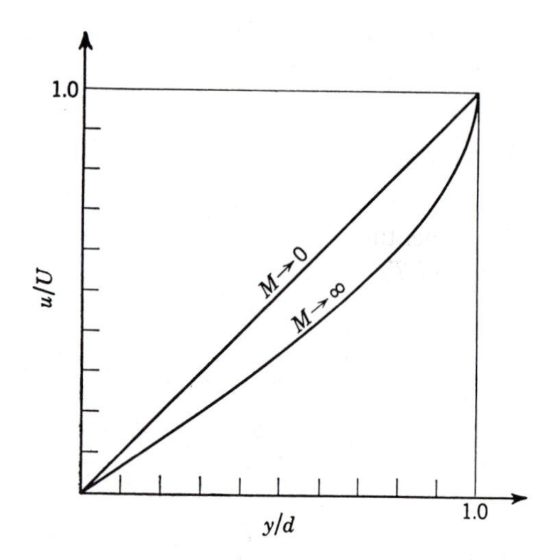

FIG. 13·4 Typical velocity profiles in Couette flow. ($\mu \sim T$.)

From Eq. 13·18, τ_w is obtained by extending the integral over the whole channel height, that is, up to $y = d$, $u = U$. This gives

$$\frac{\tau_w d}{\mu_\infty U}$$

$$= \int_0^1 \left[1 + Pr\frac{q_w}{U\tau_w}(\gamma - 1)M_\infty^2(1 - \xi) + Pr\frac{\gamma - 1}{2}M_\infty^2(1 - \xi^2)\right]^\omega d\xi \quad (13\cdot19)$$

where the variable $\xi = u/U$ has been introduced. For an arbitrary value of ω the integrals have to be evaluated numerically. As a simple, and significant, example, consider the case $\omega = 1$, that is, $\mu \sim T$, which is often used as an approximation in boundary-layer theory. Also, assume zero heat transfer, $q_w = 0$. In this case, Eq. 13·18 integrates easily, to give

$$\frac{\tau_w y}{\mu_\infty U} = \frac{u}{U} + Pr\frac{\gamma - 1}{2}M_\infty^2\left[\frac{u}{U} - \frac{1}{3}\left(\frac{u}{U}\right)^3\right] \quad (13\cdot20)$$

Equation 13·20 gives the velocity profile in implicit form. The velocity profile reduces to the linear profile if $M_\infty^2 \rightarrow 0$, as expected (Fig. 13·4). The shearing stress τ_w is given by

$$\frac{\tau_w d}{\mu_\infty U} = 1 + Pr\frac{\gamma - 1}{2} M_\infty^2 \left(1 - \frac{1}{3}\right)$$

and the friction coefficient, $C_f = \dfrac{\tau_w}{(\rho_\infty/2)U^2}$, is

$$C_f = 2\frac{\nu_\infty}{dU}\left(1 + \frac{Pr}{3}(\gamma - 1)M_\infty^2\right) = 2\frac{1 + Pr\dfrac{\gamma - 1}{3}M_\infty^2}{Re} \tag{13·21}$$

where *Re* is the Reynolds number based on the height.

This by no means exhausts the discussion of all the interesting features of compressible Couette flow. We shall discuss some further problems of Couette flow in Articles 13·18 and 14·10 and in the exercises.

13·5 Rayleigh's Problem. The Diffusion of Vorticity

In this article we shall come one step closer to the boundary-layer concept. Assume an infinite, flat plate in the x-z plane, in fluid at rest. At time $t = 0$ the plate is impulsively set into uniform motion in its own plane, say in the x-direction. Let U denote the velocity of the plate, and for simplicity *assume* $U \ll a$ *and no heat transfer through the plate.* This is the so-called problem of Rayleigh. (If another plate were placed at a height d above the sliding plate, we would have the initial state of the Couette flow problem.)

The velocity u at a distance y from the plate will now depend upon t, the time. The momentum equation for the x-direction is now

$$\rho\frac{\partial u}{\partial t} = \frac{\partial \tau}{\partial y} \tag{13·22}$$

that is, the rate of increase of u momentum of a fluid element is equal to the force acting on the element. The equation differs from the corresponding one for Couette flow in the time rate of change of the momentum, which appears here on the left-hand side (cf. Eq. 13·3). With the Newtonian expression (Eq. 13·4) for τ we have

$$\frac{\partial u}{\partial t} = \nu\frac{\partial^2 u}{\partial y^2} \tag{▶13·23}$$

Just as in Couette flow, the pressure is uniform throughout. Furthermore T = const. (since $U \ll a$ and $q_w = 0$) and hence μ, ρ, and ν are constants.

Equation 13·23 has the form of the heat equation. In fact, the corresponding heat transfer problem is simply that of a plate at rest, having its temperature suddenly raised.

The details of the computation are given in the Exercises. Here we are more interested in some specific features. First of all, Eq. 13·23 shows a very important *similarity property*. Namely, if $u(y, t)$ is a solution of the equation, we can get another solution from it, even for a different ν, by stretching of the coordinates in a specific way. The two flows are related by

$$u(y, t) = \tilde{u}(\tilde{y}, \tilde{t}) = \tilde{u}(Ay, Bt)$$

Placed in Eq. 13·23, this gives

$$B \frac{\partial \tilde{u}}{\partial \tilde{t}} = \nu A^2 \frac{\partial^2 \tilde{u}}{\partial \tilde{y}^2}$$

This is the Rayleigh equation for a fluid having kinematic viscosity

$$\nu' = \frac{\nu A^2}{B}$$

One of the stretching factors, say A, may now be eliminated, giving

$$u(y, t) = \tilde{u}\left(\sqrt{\frac{\nu'}{\nu} B}\, y; Bt\right) \quad \text{for arbitrary constant } B$$

In particular, if we keep $\nu' = \nu$, then $u(y, t)$ is equal to $\tilde{u}(\sqrt{B}\, y, Bt)$; that is, u remains constant along parabolas in the y, t plane.

This can be expressed conveniently by writing

$$u(y, t) = u\left(\frac{y}{\sqrt{\nu t}}\right)^{\dagger} \equiv u(\eta) \tag{13·24}$$

Since the boundary conditions are $u =$ const. at $y = 0$, and $u = 0$ at $y = \infty$, they fit the similarity, and hence we can expect our solution to be of the form of Eq. 13·24. Inserting Eq. 13·24 into Eq. 13·23 yields the total differential equation

$$u'' + \tfrac{1}{2}\eta u' = 0 \tag{13·25}$$

Hence

$$\frac{d}{d\eta}(\log u') = -\frac{\eta}{2}$$

and

$$u' = \text{const.} \cdot e^{-\eta^2/4} \tag{13·26}$$

Since

$$\frac{1}{\sqrt{\nu t}} \frac{du}{d\eta} = \frac{\partial u}{\partial y}$$

†This form is chosen in preference to $u(y/\sqrt{t})$, since the inclusion of ν makes the variable dimensionless. It would also be the appropriate one in case $\nu' \neq \nu$.

we obtain

$$\frac{\partial u}{\partial y} = \text{const.} \frac{1}{\sqrt{\nu t}} e^{-y^2/4\nu t} \tag{13·27}$$

For our discussion, the most important consequence of Eq. 13·27 is the fact that

$$\int_0^\infty \frac{\partial u}{\partial y} dy = \text{const.} \int_0^\infty \frac{e^{-y^2/4\nu t}}{\sqrt{\nu t}} dy = \text{const.} \quad \text{independent of } t$$

Now $\partial u/\partial y = \zeta$ is the vorticity (Article 7·10). Consequently we have

$$\frac{d}{dt} \int_0^\infty \zeta \, dy = 0 \tag{13·28}$$

The motion of the plate produces a certain amount of vorticity at $t = 0$. This vorticity spreads into the fluid but the total amount is preserved.

The analogy of this result to problems in heat transfer, and also mass diffusion, is evident. ζ corresponds to the total heat Q or the total mass m, both of which have to be considered in the analogous problems of heat or mass diffusion.

We can now define a certain velocity c which characterizes the spreading of the vorticity into the fluid. First we introduce a certain height, $\delta(t)$, which is defined by the equation

$$\zeta(0, t) \cdot \delta = \int_0^\infty \zeta \, dy$$

where $\zeta(0, t)$ is the vorticity next to the plate at any time t. It may be obtained from Eq. 13·27 by setting $y = 0$. Thus

$$\delta = \sqrt{\nu t} \int_0^\infty \frac{e^{-y^2/4\nu t}}{\sqrt{\nu t}} dy = 2\sqrt{\nu t} \int_0^\infty e^{-z^2} dz$$

or

$$\delta = \sqrt{\pi \nu t} \tag{13·29}$$

This may be regarded as a "vorticity thickness." The rate at which it grows defines the velocity c:

$$c = \frac{d\delta}{dt} = \frac{1}{2}\sqrt{\pi \frac{\nu}{t}} \tag{13·30}$$

The conclusions that we draw from this are the following:

Vorticity spreads like heat, with infinite signal velocity.† *Thus, strictly*

†This follows from Eq. 13·27: For any $t \neq 0$, ζ will differ from 0 for all y. But the effect at large y is obviously minute. For example, ζ/ζ_0 at 10 times the distance δ becomes: $\zeta/\zeta_0 = e^{-25\pi}$. Thus, the fact that the so-called "signal velocity" for diffusion problems

speaking, the influence of a source of vorticity is felt everywhere immediately, but one can define an effective region of influence δ and spreading velocity c.

It is possible, and highly interesting, to extend the discussion of Rayleigh's problem to compressible fluid flow, in which M_∞ is not necessarily small and heat transfer is not excluded. Problems like this have been worked out by Howarth†, Van Dyke‡, and others.

13·6 The Boundary-Layer Concept

A thin flat plate of length l is placed along the positive x-axis, with its leading edge at the origin. The fluid is assumed to have a uniform constant velocity U far upstream of the plate. In nonviscous fluid flow the plate has no influence upon the motion since it forms a stream surface. In a real fluid the plate will exert forces, and hence the flow field will be altered. *This problem, for large Reynolds numbers, is the classic boundary-layer problem.*

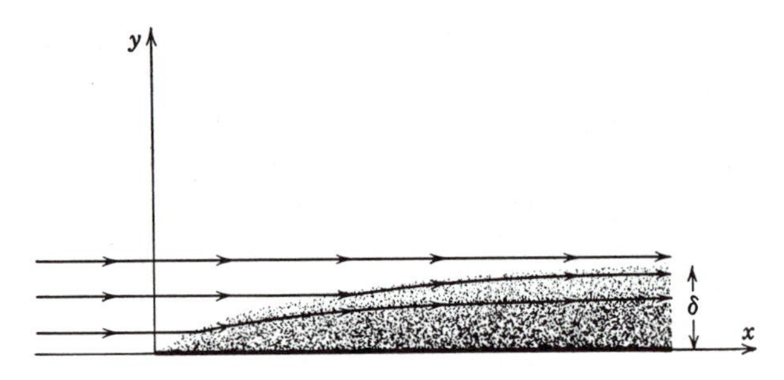

FIG. 13·5 Boundary layer on a flat plate.

To get an intuitive idea of the problem, assume the plate to be coated with a dye, soluble in the fluid. What will happen? Evidently the dye will diffuse into the moving fluid and will be carried downstream with the flow. The colored region of fluid will thus consist of a layer, starting at the leading edge of the plate, becoming thicker downstream, and leaving the plate as a colored wake. It is furthermore clear that this region will become narrower as the speed is increased. Now Rayleigh's problem has taught us that the diffusion of vorticity is similar to that of heat or matter. Consequently, the *vorticity* produced by the plate spreads like the dye, occupies a region which increases downstream, and finally continues into a wake (Fig. 13·5).

To estimate the order of magnitude of this region we can use the concepts

is infinite is no cause for concern, in fact, no more so than the small but finite probability given by a life insurance table for a man to reach an age of 1000 years.

†M. D. Van Dyke, "Impulsive Motion of an Infinite Flat Plate in a Viscous Compressible Fluid," *J. Appl. Math. and Phys.* (*Z.A.M.P.*), *3* (1952), p. 343.

‡ L. Howarth, "Some Aspects of Rayleigh's Problem for a Compressible Fluid," *Quart. J. Mech. Appl. Math.*, *4* (1951), p. 157.

of spreading velocity c and influence region δ developed with Rayleigh's problem. There we found that the spreading of vorticity proceeds at an effective speed c where

$$c \sim \sqrt{\nu/t}$$

We have to translate the concepts of c and δ into the present problem, which deals with *steady* flow. In this case the time t is simply related to the distance x from the leading edge and the mean velocity U:

$$t \sim \frac{x}{U} \tag{13·31}$$

To get a feeling for Eq. 13·31, imagine the events seen by an observer, moving with the stream, past the plate.† Hence, the spreading velocity c becomes

$$c \sim \sqrt{\frac{\nu U}{x}} \tag{13·32}$$

and the influence region

$$\delta \sim \sqrt{\frac{\nu x}{U}} \tag{13·33}$$

It is interesting, and often useful, to note that the Reynolds number can thus be written as the ratio of two velocities

$$\sqrt{Re} = \frac{U}{c} \tag{13·34}$$

which may be compared with the Mach number,

$$M = \frac{U}{a}$$

Hence, the *influence region of the vorticity spreading* bears a certain similarity to the Mach cone or Mach wedge, that is, the *influence region of the spreading of the pressure.* Furthermore, we may already note at this point that this definition of Reynolds number suggests an important combination of

†The transformation from t to x/U is very common in fluid dynamics. It is not always easy to make the statement precise. However, Eq. 13·31 becomes an exact equation (rather than a statement of order of magnitude only) if U is very large compared to the induced velocities and no forces act. That is, if

$$\frac{Du}{Dt} \equiv \frac{\partial u}{\partial t} + U\frac{\partial u}{\partial x} = 0$$

then

$$\frac{\partial}{\partial t} = -U\frac{\partial}{\partial x}$$

Reynolds number and Mach number, namely, that given by the ratio a/c. Thus

$$\frac{a}{c} = \frac{\sqrt{Re}}{M} \tag{13·35}$$

a parameter which becomes of great importance in hypersonic flow problems.

For our flat-plate problem we thus expect a region of influence starting from zero at the leading edge and reaching a maximum thickness

$$\delta \sim \sqrt{\frac{\nu l}{U}}$$

at the trailing edge. This region is called a *boundary layer* if $\delta/l \ll 1$, that is, if the Reynolds number is very large. If this condition is satisfied, the layer everywhere—except possibly in the neighborhood of the leading edge—will be very thin and will grow slowly. This leads to three important consequences, fundamental for boundary-layer theory:

(1) The pressure gradient across the boundary layer is zero. Hence, the presence of the plate affects only the momentum of the fluid in the x-direction, and this only through shear forces.

(2) Only one component of the stress tensor is of importance. The same is true for the heat transfer. Hence, the stresses and heat transfer terms are the same as those for Couette flow, for which there is no variation in x.

(3) The flow at any distance x from the leading edge is independent of the length l of the plate. Thus, a plate of length l is equivalent, within boundary-layer theory, to the first portion, of length l, of a semi-infinite plate. This very important effect (or lack of effect) results from the fact that, within boundary-layer theory, the layer is so thin that the diffusion of vorticity is normal to the surface.†

Statement (1) follows from the fact that the curvature of the streamlines, which is zero in the absence of viscosity, is still very small for a thin boundary layer, and hence the pressure differential that the centrifugal forces can support is very small indeed.

Statement (2) depends on the fact that the velocity and temperature gradients within the thin layer are very large normal to the plate, $\partial u/\partial y \sim U/\delta$, whereas these gradients are very small in the x-direction, $\partial u/\partial x \sim U/l$. The shear and heat transfer are proportional to these gradients. Statement (3) is closely related to statement (2).

†For small Reynolds number, that is, thick layers, vorticity diffuses *upstream and downstream* as well as normal to the plate, so that as $Re \rightarrow 1$ the flow becomes symmetrical. The trailing and leading edge become of equal importance.

13·7 Prandtl's Equations for a Flat Plate

On the basis of the discussion in Article 13·6, we can now set up the differential equations for the boundary layer on a flat plate. These will be quite general, for compressible flow with heat transfer, with only the (not very serious) restriction that the fluid be a perfect gas with constant specific heat. From the qualitative discussion in Article 13·6 we realize that the equations must be rather similar to those for Couette flow. The main point of difference is that now there are variations in the x-direction. In Couette flow $u = u(y)$, so that the continuity equation is automatically satisfied. Here $u = u(x, y)$, and so there must be a component $v(x, y)$, and the continuity equation must be included in the system of equations. Similarly, in Couette flow the transport terms for the momentum and energy were zero. Here, they have to be taken into account.

The system of boundary-layer equations for steady, two-dimensional flow is the following:

Continuity
$$\frac{\partial \rho u}{\partial x} + \frac{\partial \rho v}{\partial y} = 0 \qquad \blacktriangleright(13\cdot36a)$$

Momentum
$$\frac{\partial \rho u^2}{\partial x} + \frac{\partial \rho u v}{\partial y} = \frac{\partial \tau}{\partial y} \qquad \blacktriangleright(13\cdot36b)$$

Energy
$$\frac{\partial \rho u(h + \frac{1}{2}u^2)}{\partial x} + \frac{\partial \rho v(h + \frac{1}{2}u^2)}{\partial y} = \frac{\partial}{\partial y}(-q + \tau u) \qquad \blacktriangleright(13\cdot36c)$$

The terms on the left-hand sides are the transport terms, which are zero in Couette flow. They may be abstracted from the equations given in Chapter 7 for frictionless flow (e.g., Equations 7·10, 7·13, and 7·25). The viscous terms, which appear on the right-hand sides, are the same as in Couette flow. (In writing the energy equation, we neglect $\frac{1}{2}v^2$ compared to $\frac{1}{2}u^2$.)

It is convenient to rewrite the energy equation in terms of the familiar quantity†

$$J = \tfrac{1}{2}u^2 + h$$

Since we know, from one-dimensional flow and Couette flow, that $J =$ const. for some problems, the advantage of introducing this quantity is evident. The result is

$$\frac{\partial \rho u J}{\partial x} + \frac{\partial \rho v J}{\partial y} = \frac{\partial}{\partial y}(\tau u - q) \qquad \blacktriangleright(13\cdot36d)$$

which is a more convenient form of the energy equation.‡

†It is conventional to use J here rather than h_0 (cf. Chapters 2 and 7).

‡The left-hand side of this equation can also be obtained directly from Eq. 7·25, for steady flow.

The shearing stress and heat flux in the equations may be written in terms of the velocity and temperature gradients, that is,

$$\tau = \mu \frac{\partial u}{\partial y}$$

$$q = -k \frac{\partial T}{\partial y} = -\frac{1}{Pr} \mu c_p \frac{\partial T}{\partial y}$$

where

$$Pr = \frac{c_p \mu}{k}$$

is the Prandtl number.

13·8 Characteristic Results from the Boundary-Layer Equation

The general solution of the flat-plate problem, that is, the solution of the system of Eqs. 13·36, for given values of U and specified temperature distribution or heat transfer at the wall, is fairly involved. Before outlining explicit methods of solution we shall discuss a few results that are obtained without much computation.†

(1) *The shearing stress near the wall is constant.* That is, $\partial\tau/\partial y = 0$ as $y \to 0$. This follows immediately from the momentum equation (13·36*b*). The left-hand side of the equation contains the squares or products of u and v. As $y \to 0$ these terms become arbitrarily small, leaving

$$\left(\frac{\partial \tau}{\partial y}\right)_{y=0} = 0 \tag{13·37}$$

(2) *For Prandtl number* $Pr = 1$, *there exists a simple energy integral, similar to the one obtained in the case of Couette flow.* For $Pr = 1$ we have simply

$$q = -k \frac{\partial T}{\partial y} = -\mu \frac{\partial c_p T}{\partial y}$$

Thus the energy equation (13·36*d*) becomes

$$\frac{\partial \rho u J}{\partial x} + \frac{\partial \rho v J}{\partial y} = \frac{\partial}{\partial y} \mu \left(u \frac{\partial u}{\partial y} + \frac{\partial c_p T}{\partial y}\right) = \frac{\partial}{\partial y}\left(\mu \frac{\partial J}{\partial y}\right) \tag{13·38}$$

Equation 13·38 is certainly satisfied if

$$J = \text{const.}$$

that is,

$$\tfrac{1}{2}u^2 + c_p T = \text{const.} = \tfrac{1}{2}U^2 + c_p T_\infty$$

†For convenience in writing the equation we shall assume constant c_p. However, it is easily generalized. Since p is constant, $dh = c_p\, dT$, and the results are still valid if $c_p T$ is replaced by $h = \int c_p\, dT$. (Cf. the generalization of Eqs. 13·11 to 13·11*a*.)

where T_∞ is the temperature of the free stream. But

$$\tfrac{1}{2}U^2 + c_pT_\infty = c_pT_0$$

where T_0 is the stagnation temperature. Hence we have

$$\tfrac{1}{2}u^2 + c_pT = c_pT_0 \qquad \blacktriangleright(13{\cdot}39)$$

as a possible integral of the boundary-layer energy equation. It may be shown that Eq. 13·39 corresponds to zero heat transfer at the wall, that is, to the case of an insulated plate, for if Eq. 13·39 is differentiated with respect to y, then

$$u\frac{\partial u}{\partial y} + c_p\frac{\partial T}{\partial y} = 0$$

Since $u \to 0$ as $y \to 0$, so also $\partial T/\partial y \to 0$ and $q \to 0$. *Consequently the temperature of the insulated plate for Prandtl number unity is equal to* T_0.

(3) For $Pr = 1$ the energy integral may easily be extended to the case with heat transfer at the wall, in a fashion similar to that for Couette flow. Put

$$J = c_pT_w + \text{const. } u \qquad (13{\cdot}40)$$

where T_w now denotes a uniform wall temperature. We see that Eq. 13·38 is still satisfied because the first term of Eq. 13·40 makes both sides of Eq. 13·38 zero, and the second reduces Eq. 13·38 to the momentum equation (13·36*b*). Therefore, Eq. 13·40 is another solution, in which the constant depends simply on the heat transfer at the wall.

From

$$\tfrac{1}{2}u^2 + c_pT = c_pT_w + \text{const. } u$$

we calculate the temperature gradient at the wall and obtain

$$-\frac{c_p}{k}q_w = \frac{\text{const.}}{\mu}\tau_w$$

and since

$$\frac{c_p\mu}{k} \equiv Pr = 1$$

$$\text{const.} = -\frac{q_w}{\tau_w}$$

Thus the boundary-layer equations for $Pr = 1$ have an energy integral

$$\frac{1}{2}u^2 + c_pT = c_pT_w - \frac{q_w}{\tau_w}u \qquad (13{\cdot}41)$$

corresponding to the uniform wall temperature T_w. The heat transfer from the wall is given by

$$\frac{q_w}{\tau_w U} = \frac{c_p(T_w - T_0)}{U^2} \tag{13·42}$$

(4) *Shearing stress (in general) and heat transfer (for uniform T_w) vary as $x^{-1/2}$.* This result stems from the similarity property of boundary-layer flow. It may be expected that the velocity profiles $u = u(y)$, at various distances x from the leading edge, will be reduced to a *single* curve when plotted against the parameter y/δ, where $\delta = \delta(x)$ is given in Eq. 13·33. That is, we may expect

$$\frac{u}{U} = f\left(\frac{y}{\delta}\right) = f(\eta) \tag{13·43}$$

where

$$\eta = y\sqrt{\frac{U}{\nu x}}$$

That this similarity does exist may be verified by inserting Eq. 13·43 into the equations of motion and showing that they reduce to an ordinary differential equation in η. (This is straightforward for incompressible flow, but for the compressible case it is more troublesome, because of the dependence of ν on T.) If we admit that the similarity exists, it follows that

$$\left(\frac{\partial u}{\partial y}\right)_{y=0} = U\sqrt{\frac{U}{\nu x}}f'(0)$$

and consequently

$$\tau_w = f'(0)\frac{\mu U^{3/2}}{\sqrt{\nu x}} \sim \frac{1}{\sqrt{x}} \tag{13·44a}$$

It follows from Eq. 13·42 that for $Pr = 1$ and uniform wall temperature,

$$q_w \sim \tau_w \sim \frac{1}{\sqrt{x}} \tag{13·44b}$$

The skin friction coefficient

$$C_f = \frac{\tau_w}{(\rho_\infty/2)U^2}$$

is thus inversely proportional to the square root of the Reynolds number Re_x. The constant of proportionality has to be computed by solving the boundary-layer equations. For *incompressible* flow the well-known result is

$$C_f = \frac{0.664}{\sqrt{Re_x}}$$

For compressible flow the effect of Mach number on the value of this constant is briefly discussed in Article 13·10.

13·9 The Displacement Effect of the Boundary Layer. Momentum and Energy Integrals

Very useful results may be obtained by integrating the equations of motion over the y-coordinate. This gives relations between certain mean values of the boundary-layer properties. The best known of these integral relations is von Kármán's momentum integral; the integral of the continuity equation, giving the displacement effect, and the integral of the energy equation are, however, of almost equal importance.

We know that there exists a distance δ from the wall beyond which the component u is effectively equal to the free-stream velocity U, and the temperature T is equal to the free-stream temperature T_∞. We shall integrate the continuity, momentum, and energy equations from 0 to δ and interpret the results.

(1) *Continuity Equation.*

$$\frac{\partial \rho u}{\partial x} + \frac{\partial \rho v}{\partial y} = 0$$

$$\int_0^\delta \frac{\partial \rho v}{\partial y}\, dy = (\rho v)_\delta = -\int_0^\delta \frac{\partial \rho u}{\partial x}\, dy \tag{13·45}$$

Adding to Eq. 13·45 the integral

$$\int_0^\delta \frac{\partial}{\partial x} (\rho_\infty U)\, dy$$

which is zero since $\rho_\infty U$ are constants, we have

$$(\rho v)_\delta = \int_0^\delta \frac{\partial}{\partial x} (\rho_\infty U - \rho u)\, dy$$

or

$$(\rho v)_\delta = \frac{d}{dx} \int_0^\delta (\rho_\infty U - \rho u)\, dy\dagger$$

Since u remains equal to U for $y > \delta$ and also $\rho = \rho_\infty$ for $y > \delta$, we can extend the integration to ∞ and write

$$\frac{v_\infty}{U} = \frac{d}{dx} \int_0^\infty \left(1 - \frac{\rho u}{\rho_\infty U}\right) dy \tag{13·46}$$

The integral on the right-hand side has the dimension of a length. It is

†Differentiation before the integral is possible because according to Leibnitz' rule

$$\frac{d}{dx} \int_0^{\delta(x)} f(x, y)\, dy = \int_0^\delta \frac{\partial f}{\partial x}\, dy + f(x, \delta) \frac{d\delta}{dx}$$

In our case, f vanishes for $y = \delta$ since $\rho u = \rho_\infty U$ there, and thus the second term is zero.

called the *displacement thickness* of the boundary layer and is usually denoted by $\delta^*(x)$. Thus

$$v_\infty = U \frac{d\delta^*}{dx} \tag{13·46a}$$

Equation 13·46*a* states that *the effect of the presence of the boundary layer upon the external flow is equivalent to a source distribution of strength* $U(d\delta^*/dx)$. The boundary layer gives the flat plate an apparent thickness by "displacing" the external flow.

(2) *Momentum Equation.* The same procedure applied to the momentum equation yields the following

$$\frac{\partial \rho u^2}{\partial x} + \frac{\partial \rho v u}{\partial y} = \frac{\partial \tau}{\partial y}$$

$$\int_0^\delta \frac{\partial \rho u^2}{\partial x}\, dy + \rho u v \Big|_0^\delta = \tau \Big|_0^\delta$$

or

$$\int_0^\delta \frac{\partial \rho u^2}{\partial x}\, dy + U(\rho v)_\delta = -\tau_w$$

since $\tau_\delta = 0$. But from Eq. 13·45

$$(\rho v)_\delta = -\int_0^\delta \frac{\partial \rho u}{\partial x}\, dy$$

and hence

$$\int_0^\delta \left(\frac{\partial \rho u^2}{\partial x} - U \frac{\partial \rho u}{\partial x} \right) dy = -\tau_w$$

Again the differentiation can be taken before the integral and the limits extended to infinity. Thus, von Kármán's momentum integral for compressible fluid flow past a flat plate is obtained

$$\frac{d}{dx} \int_0^\infty \frac{\rho u}{\rho_\infty U} \left(1 - \frac{u}{U} \right) dy = \frac{\tau_w}{\rho_\infty U^2} \tag{13·47}$$

The length ϑ defined by the integral is called the *momentum thickness* of the boundary layer. Thus

$$\frac{\tau_w}{\rho_\infty U^2} = \frac{d\vartheta}{dx} \tag{13·47a}$$

Equation 13·47 expresses the shearing stress at the wall, that is, the skin friction, in terms of the "momentum defect" in the layer.

(3) *Energy Equation.*

$$\frac{\partial \rho u J}{\partial x} + \frac{\partial \rho v J}{\partial y} = \frac{\partial}{\partial y}(\tau u - q)$$

hence

$$\int_0^\delta \frac{\partial \rho u J}{\partial x}\, dy + (\rho v)_\delta J_\infty = +q_w$$

The shear term drops out because $\tau_\delta = 0$ and $u(0) = 0$. Thus

$$\frac{d}{dx}\int_0^\infty \frac{\rho u}{\rho_\infty U}\left(1 - \frac{J}{J_\infty}\right) dy = \frac{-q_w}{\rho_\infty J_\infty U} = \frac{-q_w}{\rho_\infty U C_p T_0} \qquad (13\cdot48)$$

The length θ defined by the integral can be called *energy thickness*. Equation 13·48 states that the change of θ with x is due only to heat transfer at the wall.

Specifically, for zero heat transfer θ is constant. Thus, J can be a constant equal to J_∞ throughout the layer, as we found for $Pr = 1$, or else the variation of J from J_∞ can be such as to make the integral zero. In the latter case, if the plate recovery temperature is less than stagnation temperature T_0, there must exist regions in the layer where the total temperature is larger than T_0.

13·10 Change of Variables

So far the velocity component u was considered the dependent variable and x, y the independent ones. For compressible flow this is not a very fortunate choice, since it tends to cover up some of the physically significant effects, and one has to introduce other more suitable variables. There are various possibilities, but here we shall give only one example, which will indicate the usefulness of change of variables. This will be the application of Crocco's choice of variables to the momentum integral equation.

Crocco chooses as dependent variable the shearing stress τ, and as independent variables u, or u/U, and x. The momentum integral equation (13·47) is

$$\frac{d}{dx}\int_0^\infty \frac{\rho u}{\rho_\infty U}\left(1 - \frac{u}{U}\right) dy = \frac{\tau_w}{\rho_\infty U^2}$$

Now

$$\tau = \mu \frac{\partial u}{\partial y}$$

and, since the integral is taken at constant x, we may write

$$dy = \mu \frac{du}{\tau}$$

Furthermore, similarity requires that the distribution of shear depends on y/δ, that is,

$$\tau = \tau_w(x)\tilde{g}\left(\frac{y}{\delta}\right)$$

or, since $y/\delta = f(u/U)$, we may write

$$\tau = \tau_w(x) g\left(\frac{u}{U}\right) = \tau_w(x) g(\xi)$$

in the new variables. Thus the momentum integral can be rewritten in terms of g and ξ

$$\frac{d}{dx} \frac{\mu_\infty U}{\tau_w} \int_0^1 \frac{\rho\mu}{\rho_\infty \mu_\infty} \xi(1 - \xi) \frac{d\xi}{g(\xi)} = \frac{\tau_w}{\rho_\infty U^2} \tag{13·49}$$

Now for constant wall temperature, $\rho\mu/\rho_\infty\mu_\infty$ is independent of x, and hence the integral is independent of x. Call it α. Equation 13·49 can then be written

$$\frac{1}{\tau_w} \frac{d}{dx}\left(\frac{1}{\tau_w}\right) = \frac{1}{\mu_\infty \rho_\infty U^3 \alpha}$$

or

$$\tau_w = \sqrt{\frac{\alpha}{2}} \sqrt{\frac{\mu_\infty \rho_\infty U^3}{x}}$$

$$C_f = \frac{\tau_w}{\frac{1}{2}\rho_\infty U^2} = \sqrt{2\alpha} \sqrt{\frac{\nu_\infty}{Ux}} \tag{13·50}$$

This form for C_f is nothing new; it followed already from similarity (Eq. 13·44*a*). What is important here is that Eq. 13·49 shows that the Mach number effect on the constant α is entirely through the temperature variation of the product $\rho\mu$. Now, ρ/ρ_∞ in the boundary layer is equal to T_∞/T, since the pressure is a constant. We thus find that there exists a special viscosity-temperature relation *for which the skin friction is independent of Mach number*, namely, for $\mu/\mu_\infty = T/T_\infty$.

A good approximation to the viscosity-temperature relation is given by

$$\frac{\mu}{\mu_\infty} = \left(\frac{T}{T_\infty}\right)^\omega$$

where ω is constant, and it follows from Eqs. 13·49 and 13·50 that the skin friction coefficient C_f increases or decreases with Mach number, depending on whether ω is larger or smaller than unity. For air $\omega \doteq 0.76$ and C_f *decreases, but only a little, with increasing Mach number. For all common gases the effect of Mach number on the skin friction is fairly small.* Figure 13·6 shows the effect for air. Figure 13·7 shows a typical set of velocity profiles.

13·11 Boundary Layers on Profiles Other than a Flat Plate

The flat plate demonstrates the essence of boundary-layer theory very clearly, but it is obviously not a practical case. That is, in general, the

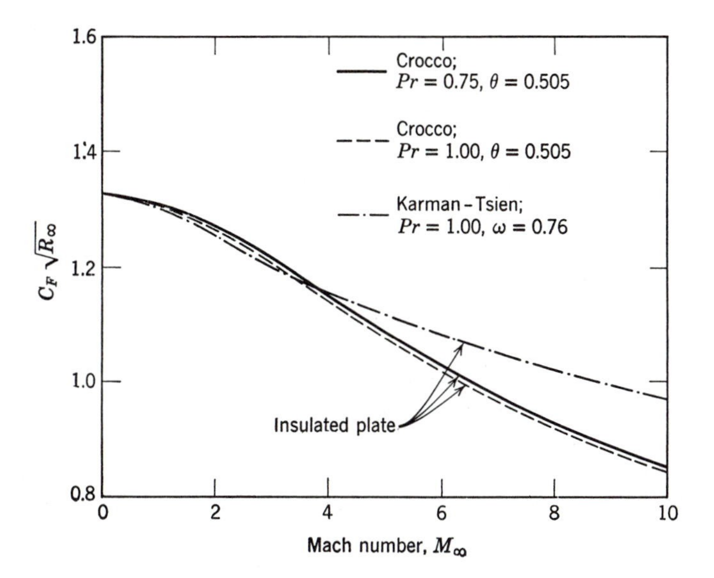

FIG. 13·6 Comparison of mean skin-friction coefficients, using Sutherland law $\dfrac{\mu}{\mu_\infty} = \left(\dfrac{T}{T_\infty}\right)^{\frac{1}{2}} \dfrac{1+\theta}{1+\theta(T_\infty/T)}$ and power law $\dfrac{\mu}{\mu_\infty} = \left(\dfrac{T}{T_\infty}\right)^{\omega}$. [From E. R. Van Driest, "Investigation of Laminar Boundary Layers in Compressible Fluids Using the Crocco Method," *NACA Tech. Note* 2597 (1952).]

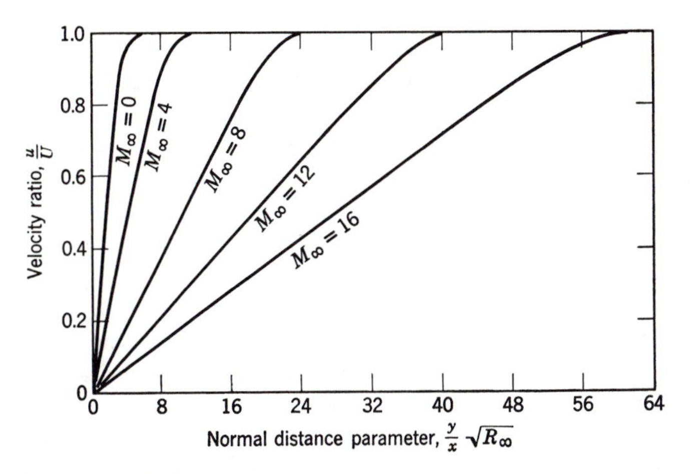

FIG. 13·7 Velocity distributions across laminar boundary layer on insulated flat plate. $Pr = 0.75$. R_∞ is based on x and U. [From E. R. Van Driest, *NACA Tech. Note* 2597 (1952).]

fluid flow problems of practical interest deal with flow past boundaries that are often slender, but still not flat plates. A complete discussion of boundary layers on arbitrary bodies goes far beyond the scope of this book and, in many aspects, even beyond present knowledge. But a few comments, applicable to thin or slender bodies, should be added here.

The essence of classical boundary-layer theory is that the presence of the layer has very small effect on the external flow field. Thus the pressure field which will exist in the flow past a body is almost undisturbed by the boundary layer, and can be computed by potential theory. It acts upon the boundary layer only as a given impressed force. Since the boundary-layer momentum equation involves only the component of the momentum in the direction of the free stream, the effect of body shape is felt essentially through the component of the pressure gradient along the body, that is, $\partial p/\partial x$ in the case of slender bodies.

Thus the equations of motion of the boundary layer are modified by the addition of the term $\partial p/\partial x$ in the momentum equation, where $\partial p/\partial x$ is a function of x known from the potential flow solution. The energy equation as written for J remains unaltered. The boundary-layer equations then are

$$\frac{\partial \rho u^2}{\partial x} + \frac{\partial \rho u v}{\partial y} = \frac{\partial \tau}{\partial y} - \frac{\partial p}{\partial x} \qquad \blacktriangleright(13\cdot51a)$$

$$\frac{\partial \rho u J}{\partial x} + \frac{\partial \rho v J}{\partial y} = \frac{\partial}{\partial y}\,(u\tau - q) \qquad \blacktriangleright(13\cdot51b)$$

Thus the shearing stress τ_w is altered by the pressure gradient, according to the first equation, and so is the heat transfer, which is always related to the shearing stress through the second equation. For $\partial p/\partial x > 0$, that is, pressure increasing downstream, or so-called "adverse" pressure gradient, the shearing stress decreases even faster than in flat-plate flow, whereas for "favorable" gradient, $\partial p/\partial x < 0$, it decreases less rapidly, and may even increase if the gradient is favorable enough. The heat transfer follows a similar behavior.

The case $\partial p/\partial x > 0$ soon leads to conditions for which the classic theory of laminar boundary layers ceases to be valid: a laminar boundary layer in a region of increasing pressure becomes turbulent or separates; in both cases the classical approach of boundary-layer theory breaks down. Since for laminar layers these effects occur for very small pressure gradients, it is safe to state that for all practical applications no laminar boundary layer will exist in a region of increasing pressure. On the other hand, the case of decreasing pressure has many applications to practical problems. It is fortunately the simpler one, and even relatively simple approximate methods give reliable results for skin friction and heat transfer coefficients in regions of falling pressure.

13·12 Flow through a Shock Wave

As another typical case of the effects of viscosity and heat transfer we shall briefly consider the flow through a shock wave. The details of the structure of a shock wave are usually less important than those for a boundary layer, but they are highly interesting and instructive. In a shock wave the changes of velocity, pressure, etc., occur in the direction of motion

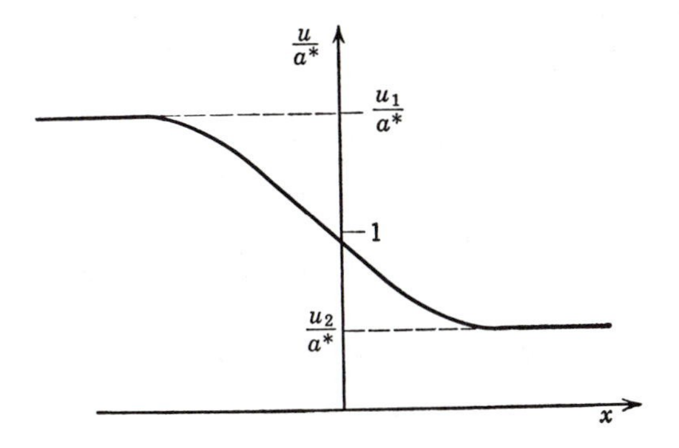

FIG. 13·8 Velocity distribution within a shock wave.

(Fig. 13·8). Thus a shock wave is a *longitudinal wave.* In the case of Couette flow the gradients of velocity, temperature, etc., are exactly *transversal;* in boundary-layer flow they are approximately so. Thus the flow through a shock wave gives us another typical example of viscous, compressible flow of essentially different nature from the Couette and boundary-layer flows.

The essential difference between a boundary layer and a shock wave may be picturesquely described as follows. A streamline that enters a shock wave comes out again downstream, whereas a streamline that enters a boundary layer remains within a region of shear. Hence in flowing through a shock wave a fluid particle passes from one state of thermodynamical equilibrium to another state of equilibrium. In boundary-layer flow the particle does not reach a state of final thermodynamical equilibrium. The losses due to flow through a shock wave can therefore be described in terms of an entropy difference. Since the entropy is a function of state, the path of the fluid particle does not matter and *the drag due to shock wave is independent of the viscosity and heat transfer coefficients.* This is the deeper reason why it is possible to compute the wave drag in supersonic flow without explicit reference to viscosity and heat conductivity.

The equations of motion for viscous, compressible flow through a steady shock wave are easy to write down: the flow is one-dimensional, with variations only in the x-direction, along the flow. The continuity equation is simply

$$\frac{d\rho u}{dx} = 0 \tag{13·52a}$$

To account for the viscosity of the fluid we have to add a stress term in the momentum equation. This *compressive* stress will be denoted by $\tilde{\tau}$ to distinguish it from the *shear* stress τ. In the energy equation there is a similar term for the heat flux $\tilde{q}$. Thus the momentum and energy equations are

$$\frac{d\rho u^2}{dx} = -\frac{dp}{dx} + \frac{d\tilde{\tau}}{dx} \tag{13·52b}$$

$$\frac{d\rho uJ}{dx} = \frac{d}{dx}(\tilde{\tau}u - \tilde{q}) \tag{13·52c}$$

The equations can immediately be integrated once:

$$\rho u \equiv m = \text{const.} \tag{13·53a}$$

$$\rho u^2 - \rho_1 u_1{}^2 = -(p - p_1) + \tilde{\tau} \tag{13·53b}$$

$$\rho uJ - \rho_1 u_1 J_1 = \tilde{\tau}u - \tilde{q} \tag{13·53c}$$

The subscript 1 denotes the conditions upstream of the shock zone, in the region where both the shear and the heat transfer vanish. From Eq. 13·53 we can obtain the shock equations used in our earlier discussion, that is, the so-called "jump condition." Namely, if the integration is extended across the shock wave, to the downstream region where $\tilde{\tau}$ and $\tilde{q}$ again vanish, Eqs. 13·53*b* and 13·53*c* give

$$m(u_1 - u_2) = p_2 - p_1 \tag{13·54a}$$

$$m(J_1 - J_2) = 0 \tag{13·54b}$$

The second equation is

$$\tfrac{1}{2}u_1{}^2 + h_1 = \tfrac{1}{2}u_2{}^2 + h_2$$

which was previously obtained from purely thermodynamic reasoning.

Hence the stress $\tilde{\tau}$ and the heat flux $\tilde{q}$ are, as expected, of no influence on the jump conditions (13·54*a*) and (13·54*b*). But the effects of $\tilde{\tau}$ and $\tilde{q}$ do account for the entropy difference across the shock wave. This is seen from the differential form of the second law of thermodynamics written per unit mass

$$\frac{ds}{dt} = \frac{1}{T}\left(\frac{dh}{dt} - \frac{1}{\rho}\frac{dp}{dt}\right) \tag{13·55}$$

where d/dt is for the rate of change following a particle. Hence

$$\frac{ds}{dt} = m\frac{ds}{dx} = \frac{1}{T}\left(m\frac{dh}{dx} - \frac{m}{\rho}\frac{dp}{dx}\right)$$

By using Eq. 13·52, this may be rewritten in the form

$$m\frac{ds}{dx} = \frac{\tilde{\tau}}{T}\frac{du}{dx} - \frac{1}{T}\frac{d\tilde{q}}{dx}$$

or

$$m(s_2 - s_1) = \int_{(1)}^{(2)} \frac{\tilde{\tau}}{T}\frac{du}{dx}\,dx - \int_{(1)}^{(2)} \frac{1}{T}\frac{d\tilde{q}}{dx}\,dx \tag{13·56}$$

The stress and heat flux are related to the velocity and temperature gradients, respectively, by expressions similar to Newton's law for the shearing stress τ and the law for heat flux q. These are

$$\tilde{\tau} = \tilde{\mu}\frac{du}{dx}, \qquad \tilde{q} = -\tilde{k}\frac{dT}{dx} \tag{13·57}$$

Here $\tilde{\mu} \neq \mu$, but $\tilde{k} = k$. (The difference is due to the fact that momentum is a vector quantity whereas heat is a scalar quantity.)

Equation 13·56, with Eq. 13·57 inserted, now becomes

$$m(s_2 - s_1) = \int_{(1)}^{(2)} \frac{\tilde{\mu}}{T}\left(\frac{du}{dx}\right)^2 dx + \int_{(1)}^{(2)} \frac{1}{T}\frac{d}{dx}\left(k\frac{dT}{dx}\right) dx$$

The second integral can be integrated once by parts and the final result becomes

$$m(s_2 - s_1) = \int_{(1)}^{(2)} \frac{\tilde{\mu}}{T}\left(\frac{du}{dx}\right)^2 dx + \int_{(1)}^{(2)} \frac{k}{T^2}\left(\frac{dT}{dx}\right)^2 dx \qquad \blacktriangleright(13·58)$$

Equation 13·58 relates the entropy increase to the dissipation and heat transfer within the shock zone. The expressions $\frac{\tilde{\mu}}{T}\left(\frac{du}{dx}\right)^2$ and $\frac{k}{T^2}\left(\frac{dT}{dx}\right)^2$ are thus sources of entropy, a fact already mentioned in Chapter 1. It will be noted that they are both positive.

With the expressions for $\tilde{\tau}$ and $\tilde{q}$, the equations of motion can be integrated, and $u(x)$, $T(x)$, etc., can be found. From these a shock-wave thickness ϵ can be defined. In general the integration of the equation requires numerical computation. However, in two special cases the results are more easily obtained. The first is the case where

$$\widetilde{Pr} = \frac{\tilde{\mu} c_p}{k} = 1$$

that is, where the Prandtl number based on $\tilde{\mu}$ is equal to unity. It is clear, by comparing with Couette or boundary-layer flow, that in this case there is a simple energy integral

$$J = \tfrac{1}{2}u^2 + h = \text{const.}$$

In this particular case J not only has the same value on either side of the shock wave but also is constant throughout. The second case which can be easily worked out is the weak shock, where $u_1/a^* - 1 \ll 1$. The details of these computations are left as an exercise.

As regards the shock thickness, it may be noted that $u(x)$, $T(x)$, etc., are functions which fair asymptotically into the initial and final states, $u = u_1$ and $u = u_2$. Hence the problem of defining a thickness ϵ is similar to that for the boundary layer. It can be done in many different ways, of which several examples are given in the Exercises.

In all cases the thickness ϵ is proportional to $\tilde{\mu}$ and k. Furthermore, for all common gases

$$\frac{\epsilon\, \Delta u \rho^*}{\tilde{\mu}^*} \approx 1 \tag{13·59}$$

This is a Reynolds number based on the velocity jump across the shock, the shock thickness, and values of ρ and $\tilde{\mu}$ evaluated at sonic conditions, that is, at $T = T^*$. This Reynolds number is of order unity. For a weak shock, for example, Eq. 13·59 gives

$$\epsilon \sim \frac{\tilde{\mu}^*}{\rho^* a^* (M_1 - 1)} \tag{13·60}$$

Since a^* is quite large, the shock thickness ϵ is in general very small indeed.

Shock-wave profiles and thicknesses have been measured by Sherman in a low-density flow. Figure 13·9 shows measurements of the temperature of a fine wire at various positions in the shock wave. (Temperature and distance are normalized.) The measured profile compares well with a theoretical profile based on Navier-Stokes equations, that is, on continuum theory. Shown for comparison is a profile based on the method of Mott-Smith, which assumes a certain model for the molecular velocity distribution function.

13·13★ The Navier-Stokes Equations

The Euler equations for the motion of a frictionless fluid have been derived in Chapter 7. Here we shall give the set of equations that describe

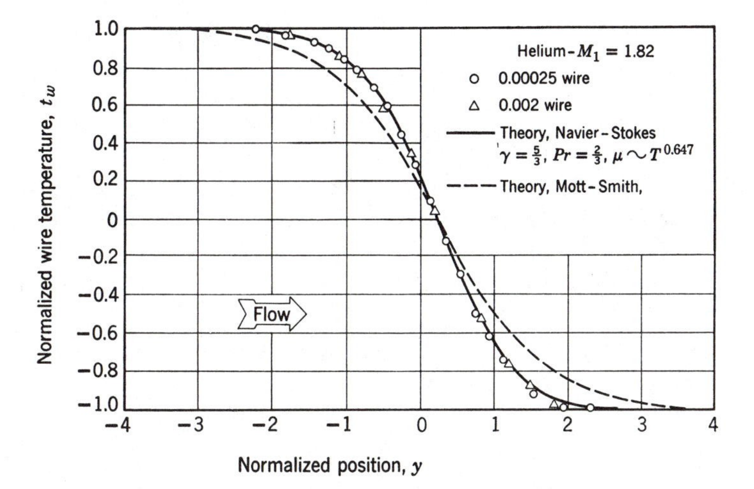

FIG. 13·9 Temperature profile through a normal shock wave. $M_1 = 1.82$ in helium. Temperatures measured with a resistance wire thermometer. [From F. S. Sherman, "A Low-Density Wind Tunnel Study of Shock Wave Structure and Relaxation Phenomena in Gases," *NACA Tech. Note* 3298 (1955).]

the motion of a viscous, heat-conducting, compressible fluid. These equations are usually called the Navier-Stokes equations. In a viscous fluid the surface forces acting on a particular mass of fluid are not necessarily normal to the surface element. Thus the forces in the momentum equation are different from the corresponding terms in the Euler equations. Furthermore, there may now be heat flow in the fluid, as well as irreversible transformation of kinetic energy into heat due to the action of the viscous stresses. The energy equation has to be rewritten to take account of the exchange of energy between kinetic energy, internal energy, and heat. The continuity equation remains unchanged, since it does not involve forces or energy.

As in Chapter 7, consider a volume V enclosed by a surface A. The surface force acting on dA can be written

$$\mathbf{P}\, dA$$

where $\mathbf{P}$ denotes a stress vector. For a frictionless fluid, $\mathbf{P}$ is parallel to $\mathbf{n}$, the vector of dA, and the factor of proportionality is $-p$. For a viscous fluid $\mathbf{P}$ is not necessarily parallel or proportional to $\mathbf{n}$, *but is a linear*† *function of* $\mathbf{n}$. Hence

$$P_i = T_{ik} n_k \tag{13·61}$$

The T_{ik} are the components of a stress tensor. It is more convenient to

†This is related to the requirement that the force per unit volume, due to surface stresses, be finite.

separate the viscous and nonviscous terms in T_{ik} and write

$$T_{ik} = -p\,\delta_{ik} + \tau_{ik} \tag{13·62}$$

where the τ_{ik} vanish for a frictionless fluid and the T_{ik} thus reduce to the proper stress for the Euler equations.

Using Eq. 13·62 we can easily generalize the momentum equations (7·12). We have for the equilibrium of forces on the volume V the equation:

$$\int_V \frac{\partial \rho u_i}{\partial t}\,dV + \int_A (\rho u_i)u_j n_j\,dA = \int_V \rho f_i\,dV + \int_A P_i\,dA \tag{13·63}$$

Only the last term differs from the corresponding one in the derivation of the Euler equations (7·13 to 7·19). This term is

$$\int_A P_i\,dA = -\int_A p n_i\,dA + \int_A \tau_{ik} n_k\,dA \tag{13·64}$$

If Eq. 13·64 is transformed into a volume integral by applying Gauss's theorem, we find:

$$\int_A P_i\,dA = -\int_V \frac{\partial p}{\partial x_i}\,dV + \int_V \frac{\partial \tau_{ik}}{\partial x_k}\,dV \tag{13·65}$$

Hence the momentum equations for a viscous fluid become:

$$\frac{\partial \rho u_i}{\partial t} + \frac{\partial \rho u_i u_k}{\partial x_k} = -\frac{\partial p}{\partial x_i} + \frac{\partial \tau_{ik}}{\partial x_k} + \rho f_i \tag{13·66a}$$

or using the continuity equation, as in Eq. 7·20,

$$\rho\left[\frac{\partial u_i}{\partial t} + u_k \frac{\partial u_i}{\partial x_k}\right] = -\frac{\partial p}{\partial x_i} + \frac{\partial \tau_{ik}}{\partial x_k} + \rho f_i \tag{▶13·66b}$$

To obtain the energy equation we have to apply the law of conservation of energy to the volume V. The energy content of the fluid in V consists of internal energy e and kinetic energy $\frac{1}{2}u^2$, both referred to unit mass. Hence, the rate of change of the energy in V can be written:

$$\int_V \frac{\partial}{\partial t}[\rho(e + \tfrac{1}{2}u^2)]\,dV + \int_A \rho(e + \tfrac{1}{2}u^2)u_j n_j\,dA \tag{13·67}$$

This rate of change of energy in V is due to the addition of heat through the boundary and due to the work done on V by the stresses. The rate of work done by the stresses is given by

$$\int_A \mathbf{P}\cdot\mathbf{u}\,dA = \int_A P_j u_j\,dA$$

or

$$\int_A \mathbf{P} \cdot \mathbf{u}\, dA = -\int_A p u_j n_j\, dA + \int_A \tau_{jk} u_j n_k\, dA \tag{13·68}$$

If external volume forces f_i are present, an additional term

$$\int_V f_i u_i\, dV$$

has to be added.

The heat transfer through the surface is described in terms of a heat flux vector $\mathbf{q}$, which denotes the quantity of heat that flows per unit time through unit area.† Hence, the heat *transferred* to V can be written:

$$-\int_A \mathbf{q} \cdot \mathbf{n}\, dA = -\int_A q_k n_k\, dA \tag{13·69}$$

Using Gauss's theorem to convert the surface integrals into volume integrals, we find that the energy equation becomes:

$$\frac{\partial}{\partial t} \rho \left(e + \frac{1}{2} u^2 \right) + \frac{\partial}{\partial x_j} \rho u_j \left(e + \frac{1}{2} u^2 \right) = -\frac{\partial p u_j}{\partial x_j} + \frac{\partial \tau_{jk} u_j}{\partial x_k} - \frac{\partial q_k}{\partial x_k} + \rho f_i u_i \tag{13·70}$$

It is often more convenient to rewrite Eq. 13·70 in terms of the quantity

$$J = \frac{1}{2} u^2 + h = \frac{1}{2} u^2 + e + \frac{p}{\rho}$$

This is easily done by adding $(\partial p / \partial t)$ to both sides of Eq. 13·70 and combining the first term on the right side of Eq. 13·70 with the second term on the left. Thus,

$$\frac{\partial}{\partial t} \rho \left[e + \frac{1}{2} u^2 + \frac{p}{\rho} \right] + \frac{\partial}{\partial x_j} \rho u_j \left(e + \frac{1}{2} u^2 + \frac{p}{\rho} \right) = \frac{\partial p}{\partial t} + \frac{\partial}{\partial x_k} (u_j \tau_{jk} - q_k) + \rho f_i u_i$$

or

$$\frac{\partial \rho J}{\partial t} + \frac{\partial \rho u_j J}{\partial x_j} = \frac{\partial p}{\partial t} + \frac{\partial}{\partial x_k} (u_j \tau_{jk} - q_k) + \rho f_i u_i \tag{13·71}$$

Equations 13·66 and 13·71 together with the continuity equation and an equation of state form a complete set of equations describing the motion of a viscous, compressible fluid. So far, however, these equations are incom-

†In this derivation we are omitting the *bulk heat addition* of Chapter 7, which was also denoted by q (scalar), though it has different dimensions (cf. footnote in Article 13·2).

plete, since τ_{jk} and q_k are not expressed in terms of the velocity gradients and temperature gradients, respectively. To do this we have to identify the fluid more specifically: We assume that *the stress tensor is a linear function of the rate of strain tensor.* This assumption is closely satisfied in most cases of interest with possible deviation only in the flow through very strong shock waves. Similarly, *we assume a linear relation between* **q** *and the gradient of the temperature.* Finally, we restrict ourselves to an isotropic fluid, that is, a fluid without distinguished axis; this last assumption is hardly restrictive at all for gasdynamics since all gases and pure liquids have this property.

The rate of strain tensor ϵ_{ij} is related to the velocity field by:

$$\epsilon_{ij} = \frac{1}{2}\left(\frac{\partial u_i}{\partial x_j} + \frac{\partial u_j}{\partial x_i}\right) \tag{13·72}$$

that is, it is the symmetrical part of the so-called deformation tensor $\partial u_i/\partial x_j$ (Eq. 7·43).

Linearity between τ_{ij} and ϵ_{ij} means that

$$\tau_{ij} = \alpha_{ijlm}\epsilon_{lm} \tag{13·73}$$

and linearity between **q** and grad T,

$$q_i = \beta_{ij}\frac{\partial T}{\partial x_j} \tag{13·74}$$

α_{ijlm} and β_{ij} are thus generalized coefficients of viscosity and heat conductivity, respectively. Fortunately, for isotropic fluids α_{ijlm} and β_{ij} must be of such a form that Eqs. 13·73 and 13·74 remain invariant under rotation of the coordinate system.

This means that **q** must be parallel to grad T and that the two symmetrical tensors τ_{ij} and ϵ_{lm} must have the same directions for their principal axes. With these restrictions α_{ijlm} must have the form:

$$\alpha_{ijlm} = \lambda\,\delta_{ij}\,\delta_{lm} + \mu(\delta_{il}\,\delta_{jm} + \delta_{im}\,\delta_{jl}) \tag{13·75}$$

Thus there are only two coefficients of viscosity, λ and μ (just as there are two coefficients of elasticity for isotropic solids).

The β_{ij} become simply

$$\beta_{ij} = -k\,\delta_{ij} \tag{13·76}$$

where k denotes the coefficient of heat conductivity and the negative sign is commonly chosen to have **q** positive in the direction of falling temperature.

Thus the stress tensor τ_{ij} becomes

$$\tau_{ij} = \lambda\frac{\partial u_l}{\partial x_l}\delta_{ij} + \mu\left(\frac{\partial u_i}{\partial x_j} + \frac{\partial u_j}{\partial x_i}\right) \tag{13·77}$$

The trace of τ_{ij}, that is, the sum of the diagonal terms, becomes

$$\tau_{ii} = (3\lambda + 2\mu)\frac{\partial u_i}{\partial x_i} = (3\lambda + 2\mu)\operatorname{div}\mathbf{u}$$

The coefficient $\kappa = 3\lambda + 2\mu$ is called bulk viscosity or second viscosity coefficient. Sometimes in the literature it is assumed to be zero, in order to make $\frac{1}{3}\tau_{ii} = -p$ (cf. Eq. 13·62). However, except for very special conditions, for example, monatomic gases, there is no reason to assume $3\lambda = -2\mu$.

Finally Eq. 13·74 together with Eq. 13·76 takes the familiar form

$$q_i = -k\frac{\partial T}{\partial x_i} \tag{13·78}$$

or

$$\mathbf{q} = -k \operatorname{grad} T$$

Thus, summarizing the rather sketchy derivation in this paragraph, we have for the description of the flow of a viscous, compressible fluid, the set of equations:

Continuity:

$$\frac{\partial \rho}{\partial t} + \frac{\partial \rho u_j}{\partial x_j} = 0$$

Momentum:

$$\frac{\partial \rho u_i}{\partial t} + \frac{\partial \rho u_i u_j}{\partial x_j} = -\frac{\partial p}{\partial x_i} + \frac{\partial \tau_{ik}}{\partial x_k} + \rho f_i$$

Energy:

$$\frac{\partial \rho J}{\partial t} + \frac{\partial \rho u_j J}{\partial x_j} = \frac{\partial p}{\partial t} + \frac{\partial}{\partial x_k}[u_j\tau_{jk} - q_k] + \rho f_i u_i$$

State:

$$f(p, \rho, T) = 0 \tag{13·79}$$

where

$$\tau_{ij} = \lambda\,\delta_{ij}\frac{\partial u_l}{\partial x_l} + \mu\left(\frac{\partial u_i}{\partial x_j} + \frac{\partial u_j}{\partial x_i}\right)$$

$$q_j = -k\frac{\partial T}{\partial x_j}$$

Finally, we may obtain a "continuity equation" for the specific entropy s, which shows how entropy is produced irreversibly by the action of viscosity and conduction. By subtracting the continuity equation from the energy equation, we obtain

$$\frac{D}{Dt}(h + \tfrac{1}{2}u^2) = \frac{1}{\rho}\frac{\partial p}{\partial t} + f_i u_i + \frac{1}{\rho}\frac{\partial}{\partial x_k}(u_j\tau_{jk} - q_k)$$

which is similar to Eq. 7·25, except for the viscous term. Similarly, by subtracting the continuity equation from the momentum equation we obtain

$$\frac{Du_i}{Dt} = -\frac{1}{\rho}\frac{\partial p}{\partial x_i} + f_i + \frac{1}{\rho}\frac{\partial \tau_{ik}}{\partial x_k}$$

which differs from Euler's equation (7·19) by the viscous term. If this equation is multiplied by u_i and subtracted from the preceding form of the energy equation, the result is

$$\frac{Dh}{Dt} - \frac{1}{\rho}\frac{Dp}{Dt} = \frac{1}{\rho}\tau_{jk}\frac{\partial u_j}{\partial x_k} - \frac{1}{\rho}\frac{\partial q_k}{\partial x_k}$$

Now the entropy is related to the other variables of state by $T\,ds = dh - dp/\rho$. Hence the rate of change of entropy of a fluid particle is given by

$$T\frac{Ds}{Dt} = \frac{1}{\rho}\tau_{jk}\frac{\partial u_j}{\partial x_k} - \frac{1}{\rho}\frac{\partial q_k}{\partial x_k}$$

Using Eq. 13·78, this may also be written

$$\rho\frac{Ds}{Dt} + \frac{\partial}{\partial x_i}\left(\frac{q_i}{T}\right) = \frac{1}{T}\tau_{ik}\frac{\partial u_i}{\partial x_k} + \frac{k}{T^2}\left(\frac{\partial T}{\partial x_i}\frac{\partial T}{\partial x_i}\right) \qquad \blacktriangleright(13·80)$$

which has the form of a continuity equation with "source terms" appearing on the right-hand side. These terms, which are positive, are sometimes called the irreversible entropy production, or dissipation, terms. (With bulk addition of heat there would be an additional term like that on the right-hand side of Eq. 7·24. Cf. footnote above. This additional term is "reversible," that is, it is positive or negative, depending on whether the bulk term is positive or negative.) This equation may also be compared with the result obtained for a shock-wave layer, in Eq. 13·58.

13·14 The Turbulent Boundary Layer

Any laminar shear flow at a sufficiently high Reynolds number becomes unstable and then *turbulent*. That is, the flow ceases to be steady (even if the boundary conditions do not depend on time) and the velocity components fluctuate in a random manner. In Article 13·17 a little more will be said about turbulence in general. Here it suffices to say that the existence of turbulent fluctuations results in pronounced "*mixing*." Shear, heat transfer, and diffusion are greatly increased by this exchange mechanism, which may be described as *molar*, in contrast to the laminar exchange mechanism, which is *molecular*. A laminar boundary layer becomes turbulent at a certain, sufficiently high Reynolds number and continues as a *turbulent boundary layer*.

In general, the turbulent layer will still be thin compared to the characteristic body dimensions, and the boundary-layer approximations may still be applied to the mean motion. Indeed, these equations can be formally written in exactly the same form as for the laminar layer, except that now the shearing stress τ represents the sum of viscous shear and *turbulent shear.* The latter is sometimes called apparent shear. Similarly q, the heat flux, now represents the sum of the contributions from molecular transfer and molar transfer. One may also define an "effective Prandtl number."

Thus the equations for a turbulent boundary layer along a flat plate are formally identical with the set of equations (13·36):

Continuity
$$\frac{\partial \rho u}{\partial x} + \frac{\partial \rho v}{\partial y} = 0 \tag{13·80a}$$

Momentum
$$\frac{\partial \rho u^2}{\partial x} + \frac{\partial \rho u v}{\partial y} = \frac{\partial \tau}{\partial y} \tag{13·80b}$$

Energy
$$\frac{\partial \rho u J}{\partial x} + \frac{\partial \rho v J}{\partial y} = \frac{\partial}{\partial y}(\tau u - q) \tag{13·80c}$$

where ρ, u, v, T, stand for the *mean values*, that is, for the quantities observed by a "slow reading" instrument. For example, the *instantaneous* velocity component in the x-direction will be $u + u'$, where $u'(x, y, t)$ denotes a *turbulent fluctuation.* It is possible to actually measure u', for example with a hot-wire anemometer. If a large number of such *instantaneous observations* of u' are added, the sum tends to zero. This obviously does not occur for quantities like $(u')^2$, which are always positive, nor does it occur, in general, for products like $u'v'$. This fact, that quantities like $u'v'$, etc., do not vanish in the average, accounts in fact for the existence of "apparent" or molar shear and heat transfer.

To solve the *laminar* boundary-layer equations we introduced the expressions that relate τ and q to the velocity and temperature fields,

$$\tau = \mu \frac{\partial u}{\partial y}$$

$$q = -k \frac{\partial T}{\partial y}$$

There does not exist at present a theory of turbulent motion which allows us to make the same step for the turbulent shear and turbulent heat transfer. To find the relations between τ, q on the one hand and u, v, T, etc., on the other is indeed the main aim of a future theory of turbulent motion. At present we have only some general laws of similarity and semi-empirical formulae for τ and q. However, a discussion of these would lead far beyond the scope of this book. Even without a detailed theory, one can still obtain some general ideas that are helpful in understanding the experimental observations. In particular, we are interested in the effects of compressibility on the skin friction and heat transfer coefficients. These may be summarized as follows.

(*a*) An increase in Mach number of the outer flow affects the layer essentially only through the circumstance that dissipation, and hence the temperature, in the layer increase. This behavior is a consequence of the validity of the boundary-layer approximation, that the mean pressure remains constant across the layer. In the *laminar* layer, an increase in temperature decreases ρ and (for a gas) increases μ. The first effect decreases the skin friction (because it increases the thickness and hence decreases the gradients), and the second increases it (because τ is proportional to μ). We have seen previously that as a consequence of these two effects the shearing stress on the wall depends only on the ratio $\rho\mu/\rho_\infty\mu_\infty$, and hence the effect of Mach number on the skin friction is slight.

On the other hand, for a *turbulent* layer, much the greater contribution to the shear comes from molar mixing, and this process does not depend on temperature as directly and strongly as does viscous shear. Thus the effect of temperature upon the density predominates, and the turbulent skin friction and heat transfer coefficient decrease appreciably with increasing Mach number of the free stream. Figure 13·10 shows the most recent experimental results for the Mach number dependency of skin friction.

(*b*) The turbulent mixing affects both skin friction and heat transfer. It is quite reasonable to assume that the effective Prandtl number is about unity. In this case, as for laminar flow, the same simple energy integral exists; that is, across the boundary layer of a flat plate we have, for zero heat transfer,

$$\tfrac{1}{2}u^2 + c_pT = \text{const.} = c_pT_0 \tag{13·81}$$

whereas, with heat transfer,

$$\tfrac{1}{2}u^2 + c_pT = c_pT_w - \frac{q_w}{\tau_w}u \tag{13·82}$$

corresponding to Eqs. 13·39 and 13·41 for the laminar layer.

Hence heat transfer and shearing stress on the wall are again related approximately by

$$\frac{q_w}{\tau_wU} = \frac{c_p(T_w - T_0)}{U^2} \tag{13·83}$$

13·15 Boundary-Layer Effects on the External Flow Field

Within boundary-layer theory the *pressure field* of a body in a viscous fluid flow is nearly the same as in inviscid flow.† The effect of the viscous

†A difficulty arises here: Potential flow past a solid body usually leads to more than one

shear layer near the surface upon the pressure field is small and can be accounted for by the *displacement effect* of the layer: The pressure field corresponds to a slightly thickened body; the displacement thickness δ^* (Eq. 13·46*a*) has to be added to the thickness distribution of the body.

In order for the boundary layer concepts to apply at all, δ^* has to be relatively small. Even so, there exist conditions where the boundary-layer induced pressure effects are very important. That is, a small change in the effective body dimensions may have a comparatively large effect upon the pressure field, as in transonic and hypersonic flow. In some cases, an unusual accuracy in the pressure field is required, as in the design of supersonic wind tunnel nozzles for uniform flow.

To discuss all possible cases in detail here would lead far beyond the scope of this book. Using the similarity laws of high-speed flow (Chapter 10), one can demonstrate the expected effects for the classic flat plate and then qualitatively at least extend them to other cases.

The relation between the body thickness ratio τ,† the free-stream Mach number M, and the pressure coefficient C_p at a point of a thin body follows the similarity laws (Eqs. 10·18 and 10·43):

$$C_p = \frac{\tau^{2/3}}{[M^2(\gamma + 1)]^{1/3}} fn\left(\frac{M^2 - 1}{[(\gamma + 1)M^2\tau]^{2/3}}\right) \tag{13·84}$$

in the transonic range, and

$$C_p = \tau^2 fn(\tau\sqrt{M^2 - 1}) \tag{13·85}$$

in the hypersonic range.

To apply Eqs. 13·84 and 13·85 to the boundary-layer displacement effect on a flat plate, we have to replace τ by δ^*/l, where l denotes the length of the plate or the first l units of a semi-infinite plate. For an insulated flat plate, δ^* has the form

$$\delta^* = \alpha\sqrt{\frac{\nu l}{U}}\left[1 + \beta\frac{\gamma - 1}{2}M^2\right] \tag{13·86}$$

where α and β are constants depending on the viscosity-temperature law of the fluid, the Prandtl number, etc. For example, for an incompressible fluid, α is 1.73 and $\beta = 0$. For our similarity consideration the numerical values of α and β are of little importance.

---

solution. Boundary-layer theory is a perturbation procedure about one of these. For slender, symmetrical bodies, especially the classic flat plate at zero angle of attack, the relevant potential solution is easily chosen. For lifting bodies, and even more so for bluff bodies, the proper choice of the potential solution is a formidable task and represents a still largely unsolved problem in fluid mechanics.

†There should be no confusion with the shearing stress.

The characteristic difference in δ^*/l for small and large values of M is apparent from Eq. 13·86. For small Mach numbers, that is $M \approx 1$,

$$\frac{\delta^*}{l} \approx \sqrt{\frac{\nu}{lU}} = \frac{1}{\sqrt{Re}} \tag{13·87}$$

For large Mach numbers, that is $M \gg 1$,

$$\frac{\delta^*}{l} \approx (\gamma - 1)\frac{M^2}{\sqrt{Re}} \tag{13·88}$$

The increase of δ^* with M^2 expresses, of course, the predominant effect of dissipative heating at high Mach numbers.

Inserting Eq. 13·87 into Eq. 13·84, and Eq. 13·88 into Eq. 13·85, yields the similarity laws:

$$C_p[(\gamma + 1)M^2 Re]^{1/3} = fn\left(\frac{(M^2 - 1)Re^{1/3}}{[(\gamma + 1)M^2]^{2/3}}\right), \quad M \sim 1 \tag{13·89}$$

and

$$\frac{C_p Re}{(\gamma - 1)^2 M^4} = fn\left(\frac{(\gamma - 1)M^3}{\sqrt{Re}}\right), \qquad M \gg 1 \tag{13·90}$$

The form of the functions in Eq. 13·89 and 13·90 can be obtained from solutions of transonic and hypersonic flow, respectively, past a parabolic body.

Some measurements of the effect in hypersonic flow are shown in Fig. 13·11. The induced pressure near the leading edge (large values of the abscissa) is considerably higher than the free-stream value, but decays downstream practically as $x^{-1/2}$.

The boundary layers encountered in supersonic nozzle problems are usually turbulent, not laminar as assumed in deriving the similarity laws (Eqs. 13·89 and 13·90). In supersonic flow at high Reynolds numbers, the layers are usually quite thin and a sufficiently accurate correction is often obtained by using values of δ^* computed for flat-plate flow and altering the area ratio in the nozzle accordingly; or else δ^* is measured and the nozzle readjusted. For low Reynolds number flow and (or) hypersonic Mach numbers, the boundary-layer effects become much more pronounced and an empirical iteration scheme involving a sequence of nozzle changes is sometimes necessary. For the latter cases, the flow in the nozzle may also be strongly altered when a model is introduced in the test section. This occurs through the influence of the model's pressure field on the nozzle boundary layer, which in turn affects the nozzle flow.

13·16 Shock-Wave Boundary-Layer Interaction

In the previous article the influence of a boundary layer upon the external flow field was discussed. In supersonic and hypersonic flows this influence can be called "shock-wave boundary-layer interaction" since the presence of the viscous layer results in the appearance of new shock waves (as on the leading edge of a perfect flat plate) or alters existing ones (as on a wedge or cone).

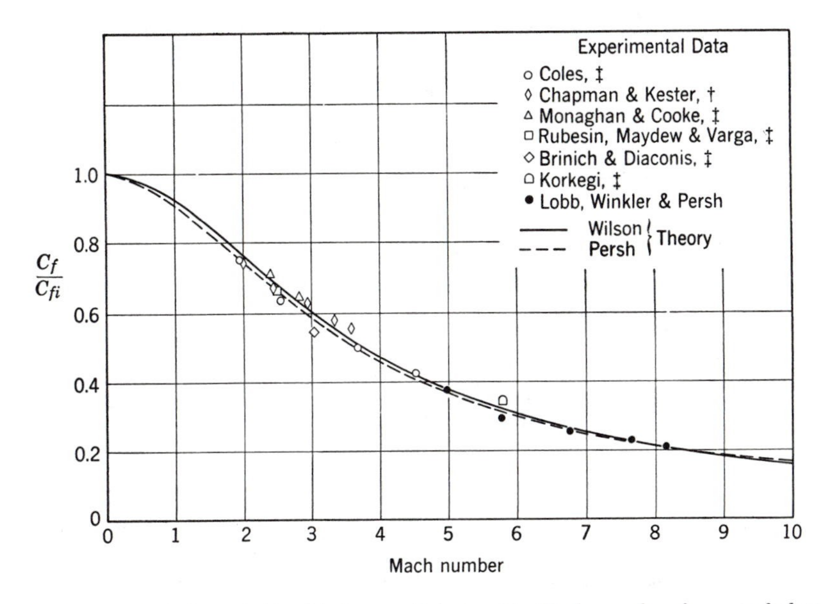

FIG. 13·10 Variation of skin friction coefficient with Mach number for a turbulent boundary layer at zero heat transfer. C_{fi} is the value for incompressible flow. From R. K. Lobb, E. M. Winkler, and J. Persh, U.S. Naval Ordnance Laboratory, *Rept. 3880* (1955). †Mean skin friction values. ‡Local skin friction values, for Reynolds number $Re_\theta = 8000$ based on momentum thickness.

Besides the leading edge problem, there exist a great number of other interactive effects. In all cases where a shock wave either originates or reflects from a solid boundary, shock-wave boundary-layer interaction must occur. The presence of a boundary layer on the wall results in an altered boundary condition for the shock wave, and, on the other hand, the very large pressure gradients due to the shock wave strongly alter the boundary-layer flow. In general, the problem is a nonlinear one, in which neither the simplification of boundary-layer theory nor the simplification of simple shock-wave theory apply.

Experimentally the effects were first observed near the trailing edge of supersonic airfoils and in transonic flow. Essential features of the problem can be seen from two typical configurations: (1) flow along a straight wall followed by a corner (Fig. 13·12) and (2) the reflection of a shock wave from a wall with boundary layer (Fig. 13·13).

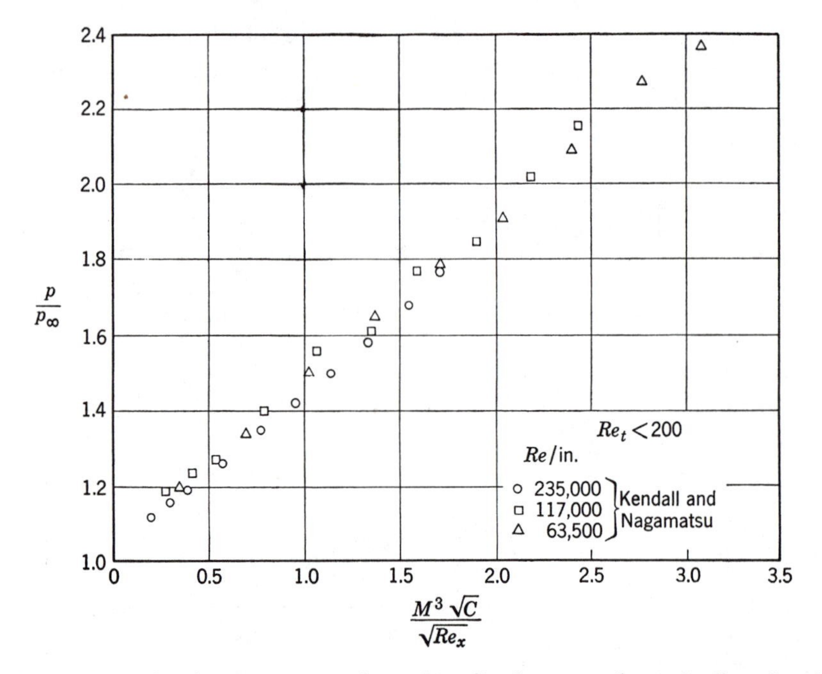

FIG. 13·11 Induced surface pressure due to boundary layer growth near leading edge of a flat plate. Re_x is based on distance from leading edge. Re_t is based on leading edge thickness. C is from the viscosity-temperature law, $\mu/\mu_\infty = C(T/T_\infty)$. From J. M. Kendall, "Experimental Investigation of Leading Edge Shock-Wave Boundary-Layer Interaction at Hypersonic Speeds," *J. Aeronaut. Sci.*, *23* (1956).

In both cases one finds an essential difference between laminar and turbulent boundary layers. This is not surprising in view of the fact that the influence of a pressure gradient upon a laminar layer is much more pronounced than on a turbulent layer. Indeed, the difference is much reduced if one chooses for comparison not the *same* shock strengths but *different* strengths such that $\Delta_{p\,\text{laminar}}/\Delta_{p\,\text{turbulent}}$ is in the ratio $\tau_{w\,\text{laminar}}/\tau_{w\,\text{turbulent}}$.

The presence of the shear layer near the wall spreads the pressure jump of the shock over a considerable region, sometimes as much as 50 boundary-layer thicknesses. Hence the shock wave is felt *upstream*. This effect can cause much trouble in supersonic wind tunnels, especially at low Reynolds numbers with thick laminar layers; the shock wave from a model can actually alter the flow field ahead of the model by separating the wall boundary layer.

A similar effect is observed in flow past blunt bodies with a thin spike (Fig. 13·14). In this case, shock-wave boundary-layer interaction creates a conical, separated region that provides a reduction of drag of the configuration, compared to the body without needle.

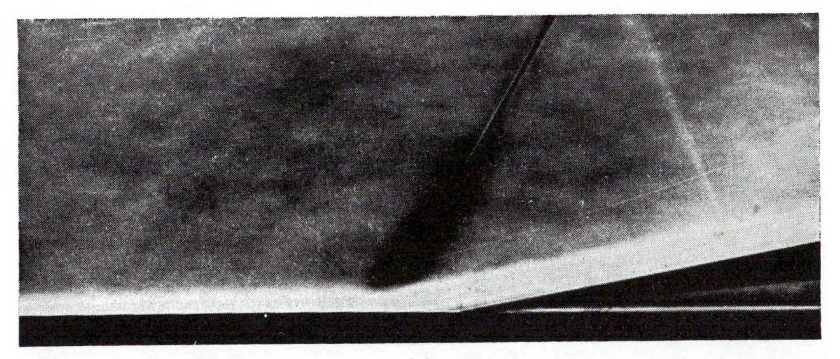

(a)

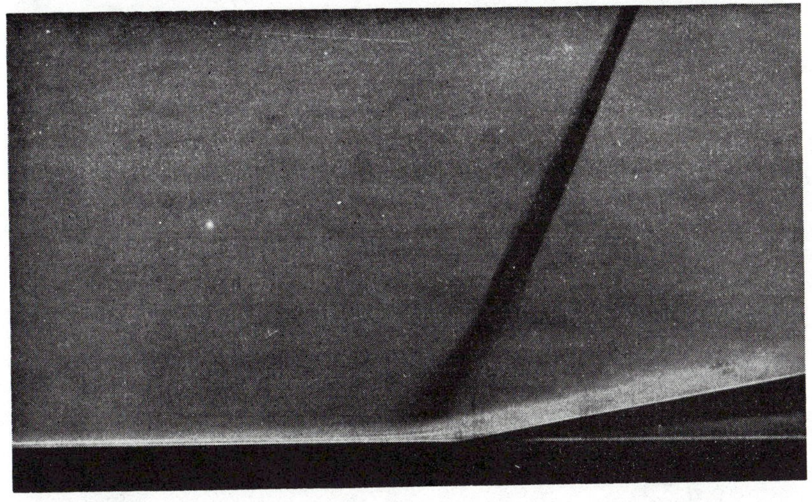

(b)

FIG. 13·12 Schlieren photographs of shock-wave boundary-layer interaction at a corner. (a) Boundary layer turbulent ahead of corner, $M_1 = 1.38$, corner angle $\theta = 10°$; (b) boundary layer laminar ahead of corner, $M_1 = 1.38$, $\theta = 10°$.

The effects due to shock-wave boundary-layer interaction upon the drag of various aerodynamic configurations are a rather fascinating subject. From subsonic flow one is used to thinking that boundary-layer separation will always increase the drag. This is not necessarily so in transonic and supersonic flow because of the interdependence of viscous drag and wave drag. Contrary to earlier belief, for example, the drag rise of thin profiles in transonic flow is not much altered by boundary-layer separation. Here, the interaction between the shear layer and the shock often results mainly in transfer of the energy loss from the shear layer to the shock, and vice versa.

It is not possible to describe or even to list here all the various interactive

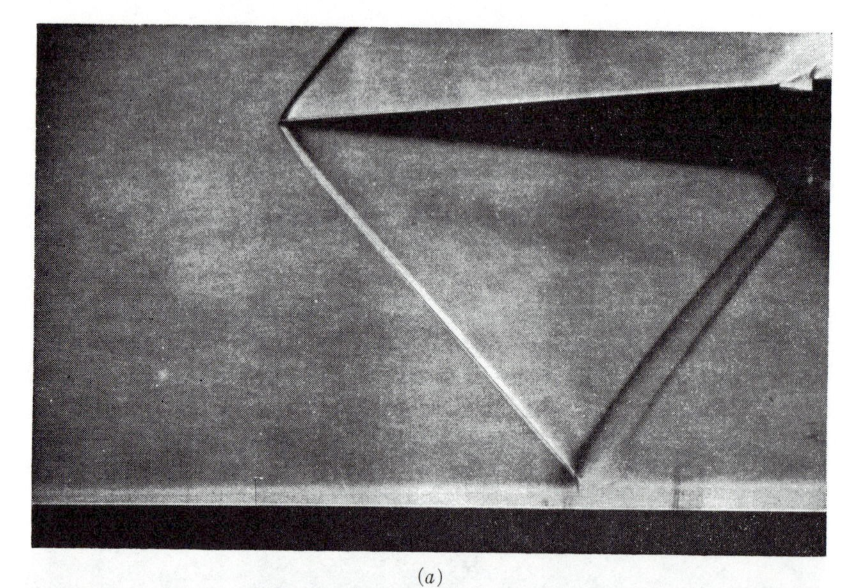

(a)

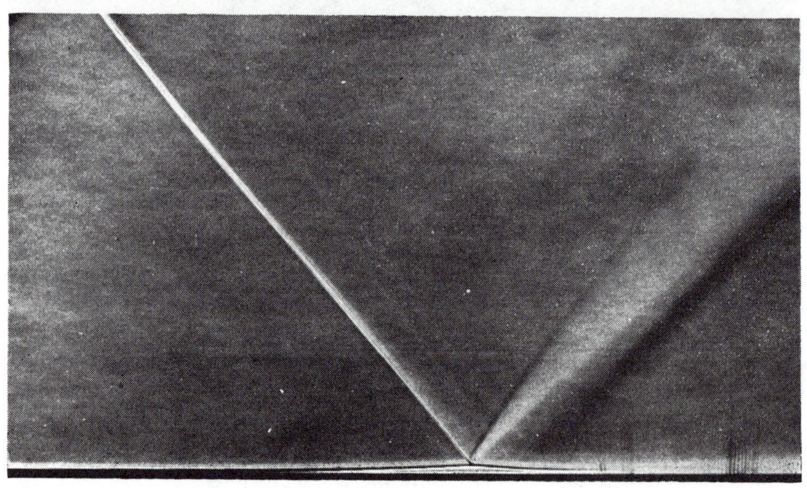

(b)

FIG. 13·13 Shock-wave boundary-layer interaction. (a) Boundary layer turbulent, $M_1 = 1.45$, $\theta = 4.5°$ for incident shock; (b) boundary layer laminar, $M_1 = 1.40$, $\theta = 3°$.

effects that may occur between shear and compression. Another example, aerodynamic noise, is mentioned in the following article.

13·17 Turbulence

It is a well-known experimental result that flow at high Reynolds numbers is always nonstationary. Even with stationary boundary conditions it is impossible to keep the velocity field steady. It is possible to split the velocity field into a mean field and a fluctuating field:

$$w_i(x_i, t) = u_i(x_i, t) + u'_i(x_i, t)$$

with

$$u_i(x_i, t) = \overline{w_i(x_i, t)}$$

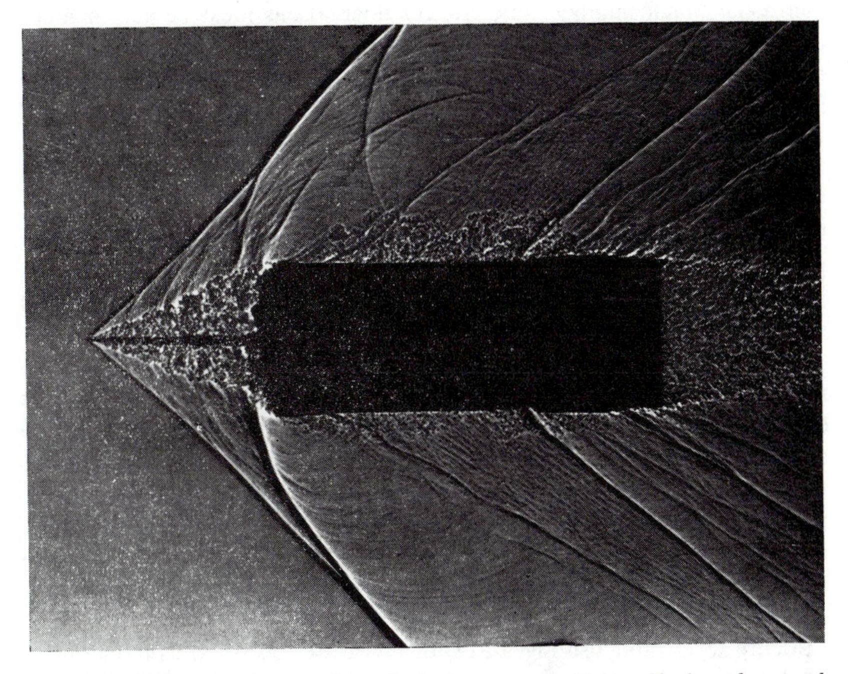

Fig. 13·14 Effect of spike on a blunt body in supersonic flight. Shadow photograph at $M_1 = 1.72$. (Courtesy Ballistics Research Laboratory, Aberdeen Proving Ground U.S.A., and G. Birkhoff, *Hydrodynamics*, Princeton University Press, 1950.)

If the $u'(x_i, t)$ have a random, statistical character, we call the flow *turbulent*.

For low speeds there is a large body of experimental results and measurements of turbulent flows. A complete theory does not exist, but semi-empirical theoretical studies have been applied successfully to turbulent shear flows and some progress has been achieved with statistical theories in simpler cases.

Turbulence is a phenomenon that is contained in the Navier-Stokes equations. It is not necessary or useful to consider turbulence from the molecular point of view. The best evidence for this is in the similarity of turbulent flows in water and in air. The real difficulty in developing a theory of turbulence lies in the nonlinear character of the Navier-Stokes equations;

there are no general methods for investigating the ensemble of possible solutions.

Experience has shown that turbulence, as expected, also exists in high-speed and supersonic flow. In fact, the investigations of turbulent boundary layers in supersonic flow have shown that qualitatively, up to Mach numbers of the order 4, at least, the effects are similar to those in incompressible flow. However, it is clear from general considerations that, with increasing Mach number, any velocity fluctuations must be accompanied by density, temperature, and pressure fluctuations. Thus at high speeds we expect new effects due to the coupling of the turbulent velocity or vorticity field with fluctuations in the variables of state. The most spectacular of these new effects is the coupling of turbulent and sound fields.

At subsonic Mach numbers this effect is largely one way: *sound* is produced from turbulent fluctuations but the energy in the sound field, although large in acoustical terms, is small compared to the energy in the turbulence. The problem of aerodynamic noise has attracted much interest because of its immediate practical importance. Lighthill has pioneered the research into sound created by turbulence at subsonic speeds.†

At very high velocities it is to be expected that the relative energy transmitted from the turbulence to the sound becomes appreciable, and turbulent energy is radiated away. The effect is quite similar to heat loss by radiation at high temperatures. Furthermore intense random noise may *induce* turbulence. Some observations in supersonic wind tunnels can be thus interpreted. It is likely that the fluctuation level in high-speed wind tunnels is largely governed by acoustic noise radiated from the wall boundary layers, and not by turbulence convected downstream from the settling chamber.

Figure 13·15 (and also Fig. 13·14) demonstrates plainly the production of random noise from boundary-layer turbulence. In cases like these, the production of acoustic radiation is most easily visualized by considering fluctuation in the displacement effect of the boundary layer. Figure 13·15 is quite characteristic of this effect. This instantaneous picture shows two kinds of waves in the region between the cone and the shock wave. There are stationary Mach waves, produced by irregularities in the surface, and "noise" waves, which are fragmented and irregular, and which are actually moving but have been "stopped" in this picture.

13·18 Couette Flow of a Dissociating Gas

So far we have discussed shear flow problems for a perfect gas with constant specific heat. As a simple and highly instructive problem we shall

†M. J. Lighthill, "On Sound Generated Aerodynamically," *Proc. Roy. Soc. A*, *211* (1952). p. 564, and *222* (1954), p. 1.

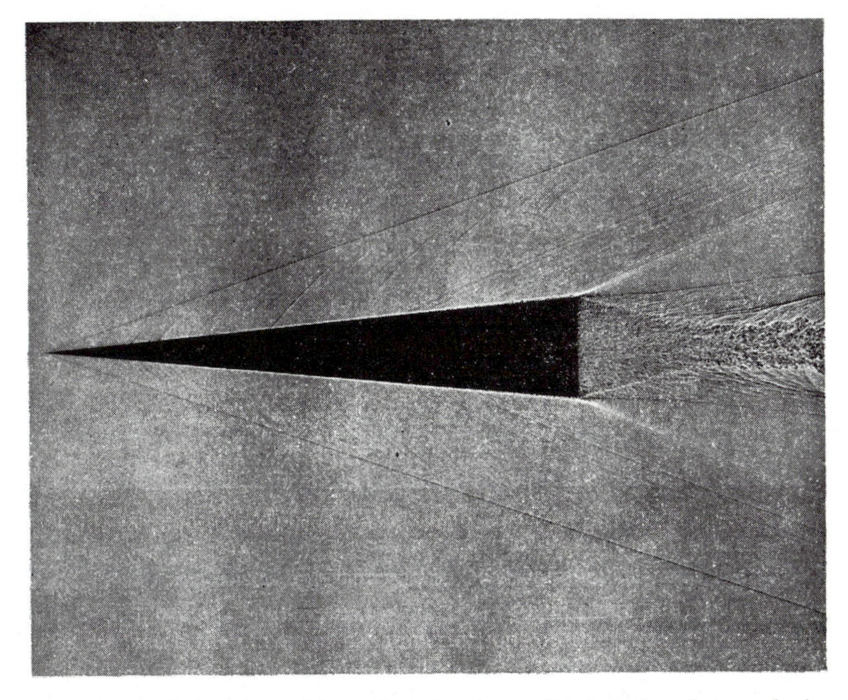

FIG. 13·15 Radiation of sound from a boundary layer. Spark shadow photograph of a cone in flight in a pressurized ballistics range. ($M_1 = 3.2$, half cone angle $\theta = 5°$, $p_1 = 3$ atmospheres.) (Courtesy U.S. Naval Ordnance Laboratory.)

briefly sketch the effects of gas imperfections on the Couette flow problem. Specifically we shall deal with dissociation. However, the same approach applies to ionization and condensation effects. Of special interest is the wall recovery temperature T_r (cf. Article 13·3), and we shall study dissociation effects upon T_r.

The energy equation for Couette flow reads (13·10)

$$u\tau - q = -q_w \tag{13·91}$$

For an insulated wall $q_w = 0$, and, introducing the Newtonian expression for q and τ, we have from Eq. 13·91

$$\mu \frac{d}{dy}\left(\frac{u^2}{2}\right) + k\frac{dT}{dy} = 0 \tag{13·92}$$

Now, *since p = const. throughout*, we have

$$dh = c_p\, dT$$

in spite of the fact that for a dissociating gas $h = h(p, T)$ (cf. Eq. 1·23).

Thus Eq. 13·92 becomes:

$$\frac{d}{dy}\left(\frac{u^2}{2}\right) + \frac{1}{Pr}\frac{dh}{dy} = 0$$

The Prandtl number Pr is not exactly constant, i.e., independent of T and hence y. But the variation appears to be quite small.† Consequently we obtain, neglecting the small variation in Pr,

$$\frac{u^2}{2} + \frac{h}{Pr} = \text{const.}$$

or

$$h_r - h_\infty = \frac{Pr}{2} U^2 \qquad \blacktriangleright(13\cdot93)$$

Thus the enthalpy difference equals $\frac{Pr}{2} U^2$ just as for the perfect gas (cf. Eq. 13·12), but now h is *not directly proportional to* T. For a dissociating gas, h can be expressed in terms of l_D, the heat of the reaction, and α the degree of dissociation by Eq. 1·75

$$h = h_2 + \alpha l_D$$

h_2 is the enthalpy of the molecular gas; hence we put, approximately,

$$h_2 = c_p T$$

where c_p is the (constant) value of specific heat for a perfect gas, and we also neglect the small variation of l_D with T.

Equation 13·93 yields then the simple and very instructive result

$$c_p(T_r - T_\infty) = \frac{Pr}{2} U^2 - [\alpha(p, T_r) - \alpha(p, T_\infty)]l_D \qquad \blacktriangleright(13\cdot94)$$

Since $T_r > T_\infty$, $\alpha(p, T_r) > \alpha(p, T_\infty)$ and consequently the recovery temperature is *reduced by dissociation*; i.e., part of the dissipated energy goes into breaking of the molecular bond and is thus not available for heating the surface. (The same equation applies to a droplet-loaded gas stream for which evaporization effects appear.)

As a limiting case take T_∞ and T_r such that the gas is completely undissociated for $T = T_\infty$ and completely dissociated for $T = T_r$. Then:

$$\frac{T_r}{T_\infty} = 1 + \frac{PrU^2}{2c_pT_\infty} - \frac{l_D}{c_pT_\infty}$$

†Cf. C. F. Hansen, "Note on the Prandtl Number for Dissociated Air," *J. Aeronaut. Sci.* *20* (1953), p. 789.

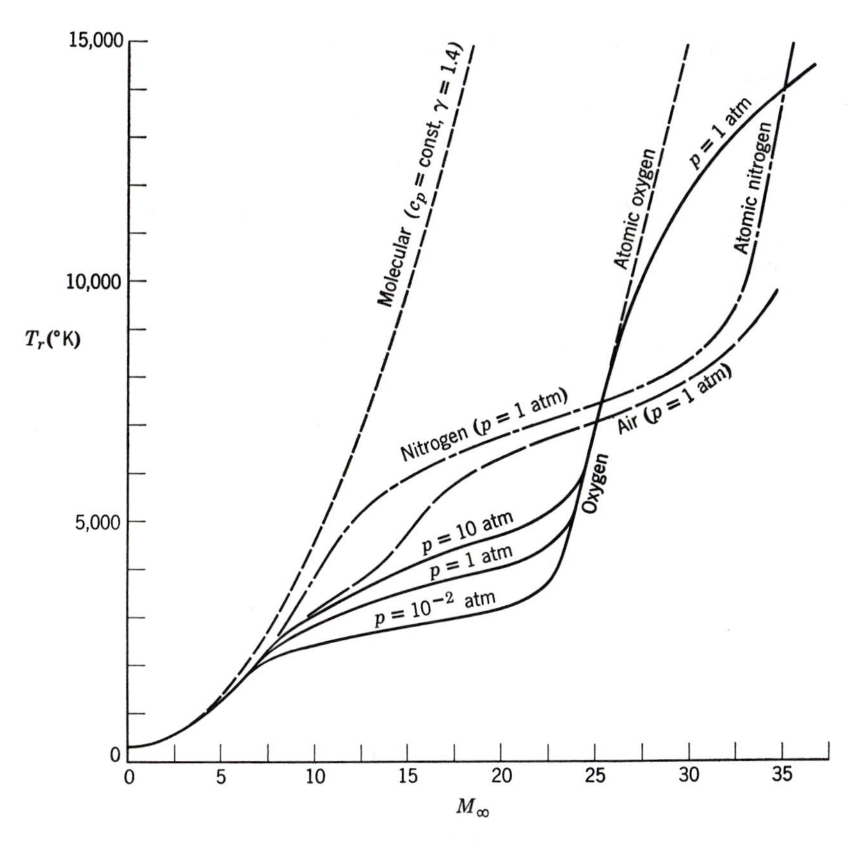

Fig. 13·16 Effects of dissociation and ionization on recovery temperature in Couette flow.

or

$$\frac{T_r}{T_\infty} = 1 + Pr\frac{\gamma - 1}{2}M_\infty^2\left(1 - \frac{2l_D}{PrU^2}\right)$$

or

$$\frac{T_r}{T_\infty} = 1 + \frac{\gamma - 1}{2}\left(PrM_\infty^2 - \frac{2\theta_D}{\gamma T_\infty}\right)$$

where γ and a_∞ refer to the undissociated gas. The characteristic temperature θ_D, for example, for O_2 is 59,000° K! (Cf. Article 1·16 and Table I at the end of the book.) Consequently it is evident that this limiting case requires Mach numbers M_∞ in excess of 20.

Figure 13·16 shows the results of some calculations of T_r versus M_∞, using Eq. 13·93.† If the specific heat were constant, T_r would follow the

†H. W. Liepmann and Z. O. Bleviss, "The Effects of Dissociation and Ionization on Compressible Couette Flow," Douglas Aircraft Co., *Rept. SM–19831* (1956).

curve marked "molecular." The actual results for pure oxygen and pure nitrogen are calculated from the equation of state for a dissociating diatomic gas, shown in Figs. 1.6 and 1.7. After complete dissociation, T_r would follow the curves marked "atomic" if there were no ionization. The actual results with ionization are shown for oxygen at a pressure of 1 atmosphere. Results are also presented for air at a pressure of 1 atmosphere. These are calculated from tabulated values of the thermodynamic properties of air.†

†F. R. Gilmore, "Equilibrium Composition and Thermodynamic Properties of Air to 24,000° K," RAND Corporation, *Memo. RM-1543* (1955).

CHAPTER 14

Concepts from Gaskinetics

14·1 Introduction

In this chapter we shall briefly examine some ideas from the kinetic theory of gases, or *gaskinetics*. We want to see how the results of this theory are related to the continuum theory of gases, or *gasdynamics*, treated in the preceding chapters, and, in particular, to see what modifications are required in the latter case if the gas is highly rarefied or moving at very high speed. For this reason it will be helpful first to review the basic assumptions of gasdynamics.

Gasdynamics deals with *macroscopic* quantities, such as the force exerted on a body by a moving gas, the heat transferred into or out of the gas, and so on. These quantities are "perceived" or measured by the body or by instruments, which *average* over a certain volume, or area, or interval of time. The quantities are described and related to each other by the equations of gasdynamics, which are *field equations*. Thus there is a velocity field, $\mathbf{w}(\mathbf{r}, t)$, a pressure field, $p(\mathbf{r}, t)$, a temperature field, $T(\mathbf{r}, t)$, from which it is possible to calculate such macroscopic quantities as mass flow, forces, heat transfer, and so on. The equations themselves, in their most primitive form, are expressions of very broad laws of experience—the law of conservation of mass, Newton's law, and the law of conservation of energy.

The equations that express these laws are incomplete, however, until the *working fluid* is specified. In perfect fluid theory it suffices to specify the fluid by its equations of state. For a real gas it is necessary to specify also the relations between the rate of strain and the stress, and between the flux of heat and the temperature gradient. Finally, the boundary conditions must also be specified.

If the relations between stress and rate of strain and between heat flux and temperature gradient are assumed to be *linear*, one obtains the *Navier-Stokes equations*. Within the frame of these equations, a gas is specified by its equations of state and by the transport parameters, k, λ, and μ.

In contrast to the treatment of gasdynamics on the basis of empirically determined *bulk* properties, it is the aim of *kinetic theory* to predict the dynamics of a gas on the basis of *molecular* mechanics. From a knowledge of forces between molecules, within the gas and at the boundaries, kinetic theory is capable, in principle, of computing the complete motion of the gas. Thus it furnishes the link between the properties of matter in bulk and the properties of individual molecules and atoms. It "explains" the equations of state, and the concepts of temperature and of heat. Furthermore, kinetic theory is expected to describe transport processes, that is, to furnish the *general relations* between stress and rate of strain and between heat flux and temperature gradient. From these it should be possible to infer the range of applicability of the *linear relations* and hence of the Navier-Stokes equations. One would especially like to obtain the *corrections* to these equations.

In dealing explicitly with the molecular structure of the gas, the task of kinetic theory is to solve a typical problem in particle mechanics, that is, to solve the equations of motion for given forces and initial conditions. However, in the multibody problem of kinetic theory, the initial conditions are unknown, and in fact are random. It would be impossible to predict the motion of a given molecule. As a matter of fact, this is not at all interesting, for the aim of kinetic theory is to predict the average, bulk behavior, due to the cooperative effect of all molecules. To accomplish this, kinetic theory applies *the laws of probability and the methods of statistics* to the mechanical problem. For example, one is interested in the *probability* that a molecule has certain coordinates and velocities. From such probability distributions one can compute the averages or mean values of various parameters, such as the velocity, temperature, etc. These mean values are then related to the macroscopically measureable quantities.

It turns out to be very easy to obtain the *equilibrium* equations of state of a monatomic, perfect gas. Multiatomic perfect gases and van der Waals' gases in thermodynamic equilibrium can still be handled without great difficulty. *Nonequilibrium* or transport processes, however, are much more difficult. For instance, to derive the Navier-Stokes equations and to predict the coefficients of viscosity and conductivity, λ, μ, and k, is very laborious. To develop the theory beyond the Navier-Stokes equations is a formidable task. None of the so-called higher-order equations obtained so far is free of objections, and the problem of improving on the Navier-Stokes equations cannot yet be considered solved.

Kinetic theory was developed in the second half of the last century, by Maxwell, Boltzmann, Clausius, and others. At that time the very existence of molecules was a matter of argument. The present intense interest in kinetic theory, as applied to gasdynamics, is due to the possibility of flight at extreme altitudes and speeds. Much has been written in the aeronautical literature on the breakdown of continuum theory and on the necessity of replacing it by kinetic theory. These statements, or their interpretation, are not always correct. It must be remembered that *the general field equations of gasdynamics are statements of laws which hold also in kinetic theory;*

these equations will always apply and will always relate the mean quantities. It is possible that one may be interested in small regions of the flow, or in comparatively short times. Measurements may then not yield the mean values which are related by the field equations. *Fluctuations* about these mean values will then be noticeable and possibly decisive. Hence, in rarefied gas flow, fluctuations about the average values have to be expected, and these do restrict the use of the continuum equations. Furthermore the Navier-Stokes equations may become insufficiently accurate, but so far there seems to be no clear-cut example of this occurrence. Finally the boundary conditions, for instance, the "no-slip condition," may have to be modified. In any event, the problem as to whether gasdynamics has to be replaced by gaskinetics is not easy to answer, and should be approached with much caution.

The following articles give a brief and incomplete discussion of concepts from the kinetic theory of gases, mainly to introduce the reader to the concepts and terminology, and to emphasize the warning given above.

14·2 Probability Concepts

The concepts of probability used in kinetic theory can be illustrated by the problem of throwing dice. Assume that two dice are thrown repeatedly and that the result of each throw is represented graphically in a plane (Fig. 14·1), that is, the number that comes up on the first die is the x-coordinate and that on the second die is the y-coordinate. The outcome of a throw is recorded, or tallied, in the appropriate square, or "cell." For instance, a throw that gives a 5 on the first die and a 2 on the second die gives a tally, or "sample point," in the (5, 2) cell.

After a very large number of throws, we examine the "score" registered by the tallies in the various cells. If the dice are *perfect,* that is, not loaded, no cell has a preference over any other, so that, given a sufficient number of experiments for a valid statistical sample, the distribution of tallies over the x-y plane is *uniform.* We say that every cell has the same *weight.* There are 36 cells, all of equal weight. Thus the probability of throwing a specific combination, say (5, 2) is

$$\text{Pr}(5, 2) = \tfrac{1}{36}$$

The probability of any combination (x, y) is the same. It may be expressed in more general terms as the ratio of the area of the x-y cell to the total area, that is,

$$\text{Pr}(x, y) = \frac{A_{xy}}{A} = \frac{A_{xy}}{\sum\limits_{x}\sum\limits_{y} A_{xy}} \tag{14·1}$$

From this geometrical interpretation of probability we obtain two basic laws of probability:

(1) The probability $\text{Pr}(5)$ of throwing a 5 with the first die, regardless of the outcome on the other one, is equal to the area of the shaded strip in Fig. 14·1, divided by the total area, that is,

$$\text{Pr}(5) = \frac{A_{51} + A_{52} + \cdots + A_{56}}{A} = \frac{1}{6}$$

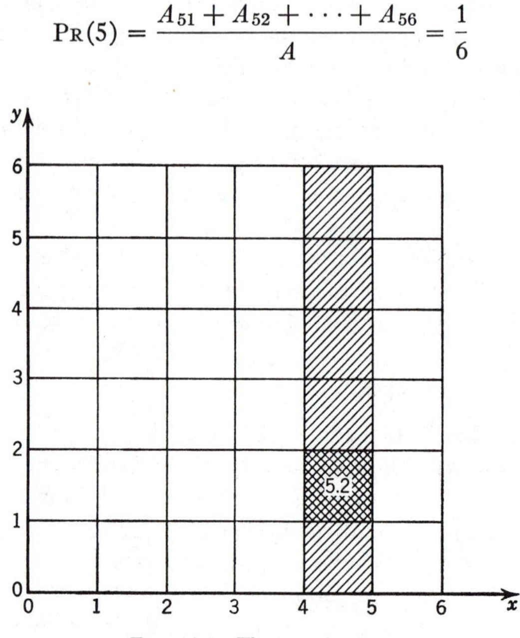

Fig. 14·1 The sample plane.

Written more generally,

$$\text{Pr}(x) = \sum_{y=1}^{6} \text{Pr}(x, y) \tag{14·2}$$

This is the *law of addition.*

(2) The "simple probabilities" $\text{Pr}(x)$ and $\text{Pr}(y)$ are each 1/6, the "joint probability" $\text{Pr}(x, y)$ is 1/36. This is an example of the *law of multiplication,*

$$\text{Pr}(x, y) = \text{Pr}(x)\text{Pr}(y) \tag{14·3}$$

It states that the joint probability is the product of the simple probabilities. It applies only if the simple probabilities are *independent,* that is, the outcome of x is independent of the outcome of y. For the throwing of two dice this is a natural assumption, but it is easy to imagine cases for which it is not so, for example, two dice connected by a rigid rod.

If the dice are *loaded*, the problem is somewhat altered. The sample point now has a preference for certain cells. The cells now have different weights. Instead of Eq. 14·1 we now have for the probability,

$$\text{Pr}(x, y) = \frac{\phi(x, y)A_{xy}}{\sum_{x=1}^{6}\sum_{y=1}^{6} \phi(x, y)A_{xy}} = \frac{\phi(x, y)}{\sum_x \sum_y \phi(x, y)} \tag{14·4}$$

the second step being obtained by dividing through by the fixed area A_{xy}. $\phi(x, y)$ denotes the *weight* of the cell (x, y). Equation 14·4 reduces to Eq. 14·1 if ϕ = const., say $\phi = 1$.

If $\phi(x, y)$ is known, we can compute the probability of any throw, and also the relation between probabilities, as before. But how is ϕ obtained? In principle, its determination is a mechanical problem. Given the distribution of weight in a cube (die), the laws of mechanics determine its motion for given initial conditions. However, the shaking of the dice in the cup and the throwing introduce statistical elements into the initial conditions, quite analogous to those in the problem of molecular motions. One has then, in general, to combine the laws of mechanics and of statistics.†

From our definition of probability we can proceed to compute averages. For instance, we may ask for the average x-coordinate after very many repeated throws. This average value, $\bar{x}$, is given by

$$\bar{x} = \sum \sum x\text{Pr}(x, y) = \frac{\sum_x \sum_y x\phi(x, y)}{\sum_x \sum_y \phi(x, y)} \tag{14·5}$$

It will be seen that those values of x which are expected to appear more often are "weighted" more in forming the average.

$\bar{x}$ may also be imagined to be the x-coordinate of the center of mass for a distribution of mass points with mass ϕ. It is often convenient to think of ϕ as a mass distribution. Then $\overline{x^2}$, $\overline{xy}$, etc., correspond to the components of the moment of inertia.

In general, the mean value of *any function*, $F(x, y)$, associated with the coordinates (x, y), is obtained from

$$\bar{F} = \frac{\sum_x \sum_y F(x, y)\phi(x, y)}{\sum_x \sum_y \phi(x, y)} \tag{14·6}$$

In our example of throwing dice, x and y are not continuous, that is, they

†For a die to be loaded its weight must be distributed in such a way that the center of gravity does not coincide with the geometrical center. One can see fairly easily that the essential quantity in determining $\phi(x)$ is the differences in potential energy of the die in various positions.

can have only discrete values. It is not difficult to generalize our definitions to the case of *continuous* distributions,† like those used in kinetic theory. The cell "area" then becomes $dx\,dy$ instead of unity, the summations are replaced by integrals, and $\phi(x, y)$ becomes a continuous weight distribution, or "probability density." Thus 14·6 generalizes to

$$\bar{F} = \frac{\iint F(x, y)\phi(x, y)\,dx\,dy}{\iint \phi(x,y)\,dx\,dy} \tag{14·7}$$

The law of addition becomes

$$\phi(x) = \frac{\int \phi(x, y)\,dy}{\iint \phi(x, y)\,dx\,dy} \tag{14·7a}$$

and the law of multiplication, for independent probabilities, is the same as before.

Extension of the definition to more than two variables is also straightforward. For example, if $F = F(x, y, z)$, then

$$\bar{F} = \frac{\iiint F(x, y, z)\phi(x, y, z)\,dx\,dy\,dz}{\iiint \phi(x, y, z)\,dx\,dy\,dz} \tag{14·8}$$

Finally we must make a remark about the so-called *ergodic property*. In the dice experiment, instead of *repeating the throws N times*, we expect to obtain the same result by *throwing an "ensemble" of N pairs* of similar dice simultaneously. That is, we expect each throw in the repeated experiment to be independent of the previous ones. For the dice problem this seems evident. In the molecular problem, however, the "mixing" and "throwing" are continuous processes, so that such independence is not evident a priori. We now have to observe the properties of one molecule or a group of molecules *over a length of time, t*. This corresponds to the experiment of repeated throws. Alternately, we may have N systems simultaneously, and observe the properties of the group of molecules in each system. That is, we have an "ensemble" consisting of N replicas of the gas system in question. We expect that the results of both experiments will be the same if N and t are large. This is called an ergodic property of the system. It is equivalent to the assumption that there exists a time $t = t^*$ such that the system, for

†To extend the dice throwing example to continuous distributions, imagine a spherical die on which the longitude and latitude define the coordinates of a throw.

example, the molecule, has lost all memory of, and has a subsequent motion independent of, the motion prior to $t = 0$. Thus a single system observed over a time $t > t^*$ is equivalent to the observation of N simultaneous systems, the relation between the two being given by

$$N \approx \frac{t}{t^*}$$

Thus a *time average* for time t over a single system is equivalent to an *ensemble average* over N systems. In this sense, $\phi(x, y)\, dx\, dy \Big/ \iint \phi(x, y)\, dx\, dy$ is equal to the average fraction of the time the molecule spends in the cell (x, y).

In the discussion of the ergodic property for a continuous system, there was an implicit assumption, that, during the time averaging, the "conditions" of the system did not change. More precisely, the *statistical properties* of the system should not change with time. This is called a *macroscopically stationary system.* It means that the result obtained by observing at a given time is the same as that obtained at any other time, provided that the intervals of observation are greater than t^* in each case.

It is difficult to imagine a gas in a physical container to which the ergodic property would not apply, for the particles of a solid boundary also have random, thermal motions and hence exert random forces on any molecule colliding with the walls.

14·3 Distribution Functions

Consider a gas enclosed in a container of volume V. The gas is in thermodynamic equilibrium, so that its state may be specified by a few variables of state, p, ρ, T, etc.

From the point of view of kinetic theory, the gas consists of very many, N, molecules that move in V and collide with each other and with the walls of the vessel. The mechanical state of this system is specified by the *space coordinates* and the *velocity components* of each of the N molecules; the number of variables is very large! The few thermodynamic variables are *average values* of certain quantities associated with these N molecules.

For example, consider the density ρ, which is the average mass per unit volume. From the molecular point of view, ρ is equal to the average number of molecules per unit volume times the mass, m, of each one. Since ρ is uniform, the average number of molecules per unit volume is everywhere the same, and so the statistical weight $\phi(x, y, z)$ of each volume element dV is unity. The probability of finding a given molecule in a given volume element dV is simply dV/V, corresponding exactly to the problem of throwing unloaded dice. With this simple weighting function, we can easily compute other averages associated with the molecular mass, e.g., the center of mass of the gas in V, etc. The reader can easily check that the center of mass corresponds to the average position of any molecule!

So far we have not considered the *velocity components*, which are also required, in addition to the mass coordinates, to specify the mechanical state of the molecules. We shall see presently that certain averages of the velocity components are related to the pressure p, and the temperature T. But how does one compute these averages? We cannot use the space weighting function $\phi(x, y, z)$, because all the molecules have different velocities. We need another weighting function to tell us how the velocities are distributed. We must imagine then a "velocity space" (u, v, w), with a corresponding weighting *distribution function* $\phi(u, v, w)$. The definitions introduced earlier apply directly. For instance, the probability of finding a molecule in the "volume" element $du\,dv\,dw$ (i.e., a molecule having velocity components between u and $u + du$, v and $v + dv$, w and $w + dw$) is

$$\frac{\phi(u, v, w)\,du\,dv\,dw}{\displaystyle\iiint_{-\infty}^{\infty} \phi(u, v, w)\,du\,dv\,dw}$$

As another example, the mean square of u is

$$\overline{u^2} = \frac{\displaystyle\iiint_{-\infty}^{\infty} u^2\phi(u, v, w)\,du\,dv\,dw}{\displaystyle\iiint_{-\infty}^{\infty} \phi(u, v, w)\,du\,dv\,dw}$$

The integrals must extend over all possible values of each velocity component, from $-\infty$ to $+\infty$. We can see then that $\phi(u, v, w)$ cannot be unity, or constant, for this would result in an infinite value for the mean kinetic energy, $\frac{1}{2}m(\overline{u^2} + \overline{v^2} + \overline{w^2})$, whereas the mean energy of the gas is finite. In fact, $\phi(u, v, w)$ must decrease rapidly to zero, for large values of u, v, w. Thus the molecular problem in velocity space corresponds to the problem of *loaded* dice, that is with a nonconstant ϕ.

One problem of kinetic theory is that of determining $\phi(u, v, w)$ from mechanical and statistical considerations. For a *gas in equilibrium* the problem is relatively simple; it leads to the distribution law of *Maxwell* and *Boltzmann*,

$$\phi(u, v, w) = Ae^{-\beta(u^2+v^2+w^2)} \tag{14·9}$$

Maxwell first obtained this function by a very simple and intuitive argument:

(1) Since no direction in velocity space is preferred, $\phi(u, v, w)$ must be independent of any rotation of axes. Hence $\phi(u, v, w)$ depends only on $u^2 + v^2 + w^2$. That is, $\phi(u, v, w) = fn(u^2 + v^2 + w^2)$.

(2) Assuming that u, v, w are statistically independent,† i.e., that $\phi(u, v, w) = \phi(u)\phi(v)\phi(w)$, it follows that

$$\phi(u)\phi(v)\phi(w) = fn(u^2 + v^2 + w^2) \tag{14·10}$$

The only nonsingular function ϕ which satisfies this relation has the form of Eq. 14·9.

For our purposes, it is important to remember that, once $\phi(u, v, w)$ is known, we can obtain mean values of all functions of u, v, w. Furthermore, these mean values can be interpreted as *time averages* or as *ensemble averages*.

14·4 The Virial Theorem of Clausius

One way to obtain the equation of state of a perfect gas is to compute the total momentum transferred to a portion of a wall. It is very instructive, however, to consider also the problem from a somewhat different point of view which is due to Clausius, and in which the mechanical equations of motion are explicitly used.

Consider the motion of a single molecule of a gas. Let m be its mass, $\mathbf{r}$ its position vector measured from an arbitrary origin, and $\mathbf{F}(\mathbf{r}, t)$ the forces acting on it. These forces are due to the other molecules and to the wall of the vessel containing the gas. The equation of motion is

$$m \frac{d^2\mathbf{r}}{dt^2} = \mathbf{F} \tag{14·11}$$

We now form the scalar product with $\mathbf{r}$ and slightly transform the left side of the resulting equation,

$$m\mathbf{r} \cdot \frac{d^2\mathbf{r}}{dt^2} = m \frac{d}{dt}\left(\mathbf{r} \cdot \frac{d\mathbf{r}}{dt}\right) - m \left|\frac{d\mathbf{r}}{dt}\right|^2 = \mathbf{F} \cdot \mathbf{r} \tag{14·12}$$

or, with $d\mathbf{r}/dt = \mathbf{c}$,

$$m \frac{d}{dt} (\mathbf{r} \cdot \mathbf{c}) - mc^2 = \mathbf{F} \cdot \mathbf{r} \tag{14·13}$$

We shall now *average* this equation. To do this we imagine a large number of replicas of our gas. In each of these systems, we pick a molecule and write Eq. 14·13 for it. We then add all these equations and divide by the

†This is certainly the *simplest* assumption, and may be legitimately tried for comparison with experience. In the application of classic mechanics to kinetic theory, the assumption of independence is rigorously justified from a discussion of collision processes.

number of systems. The result is an equation between *ensemble* averages

$$m\frac{d}{dt}\overline{(\mathbf{r}\cdot\mathbf{c})} - m\overline{c^2} = \overline{\mathbf{F}\cdot\mathbf{r}} \tag{14·14}$$

In the first term, differentiation with respect to t and summation over all the members of the ensemble are independent linear operations and can be interchanged. But we are considering a *macroscopically stationary system, and hence all mean values must be independent of time.* The first term then vanishes, and we obtain one form of the virial theorem of Clausius:

$$m\overline{c^2} = -\overline{\mathbf{F}\cdot\mathbf{r}} \tag{14·15}$$

$\overline{\mathbf{F}\cdot\mathbf{r}}$ is called the *virial* of the forces. The virial theorem was obtained by averaging over an ensemble. We can now use the ergodic property and consider the mean values to be *time averages.* Thus $m\overline{c^2}$ is twice the average of the kinetic energy of a single molecule observed over a long time, and $\overline{\mathbf{F}\cdot\mathbf{r}}$ is the time average of the force times displacement in the *direction of the force* for this single molecule.

14·5 The Equation of State of a Perfect Gas

We shall now apply the virial theorem to the molecules of a gas enclosed in a vessel of volume V. For simplicity, we may choose a spherical vessel of radius r_0 and use its center as origin.

The size of each molecule and the range of its force field are assumed to be vanishingly small. This implies the assumption of *point masses* for the molecules, and leads, as we shall see, to the perfect gas equation.

For point molecules, the contribution to the virial of the intermolecular forces is zero. This is so because the force vector $\mathbf{F}$ *at any point* is randomly distributed, in direction, and thus the average $\mathbf{F}\cdot\mathbf{r}$ is zero. Thus the only contribution comes from the *force at the wall.* To find this we multiply the virial equation by N, the number of molecules in V, and write

$$Nm\overline{c^2} = -N\overline{\mathbf{F}\cdot\mathbf{r}}$$

At the wall, the force $\mathbf{F}$ is related to the *pressure.* In magnitude it is equal to the pressure times the area of the wall, and is directed radially inward. Furthermore, $|\mathbf{r}| = r_0$. Thus,

$$Nm\overline{c^2} = p4\pi r_0^2\cdot r_0 = 3pV$$

or

$$\frac{p}{\rho} = \tfrac{1}{3}\overline{c^2} \tag{14·16}$$

This becomes identical with the thermal equation of state for the gas

$$\frac{p}{\rho} = RT$$

if we put

$$RT = \tfrac{1}{3}\overline{c^2} \tag{14·17}$$

Now $\frac{1}{2}\overline{c^2} = e_T$ is the mean kinetic energy in translation per unit mass of gas. Thus

$$\frac{p}{\rho} = RT = \tfrac{2}{3}e_T \tag{14·18}$$

We have thus succeeded in expressing the temperature in mechanical terms.

Furthermore, if the gas consists of mass points without internal structure, then e_T accounts for its *total* energy per unit mass and hence is equal to the *internal energy* of thermodynamics. Thus

$$e = \tfrac{3}{2}RT \tag{14·19}$$

Equation 14·19 is the caloric equation of state of a *monatomic* gas. The specific heat c_v and the ratio of specific heats γ are determined from

$$c_v = \frac{de}{dT} = \frac{3}{2}R \tag{14·20}$$

and

$$\gamma = \frac{c_p}{c_v} = \frac{R + c_v}{c_v} = \frac{5}{3} \tag{14·21}$$

These values agree with observations on the noble gases, e.g., He, A, etc., over a fairly large temperature range.

Multiatomic gases will have other energies besides the kinetic energy of translation, for example, rotational and vibrational energies. In this case

$$e = \tfrac{3}{2}RT + e_{\text{rot}} + e_{\text{vib}} \quad \text{etc.} \tag{14·22}$$

Thus c_v for multiatomic gases is larger than $\frac{3}{2}R$, and γ is smaller than $\frac{5}{3}$.

14·6 The Maxwell-Boltzmann Distribution

The virial theorem has led us to a relation between the mean square of the absolute velocity of a molecule $\overline{c^2}$, the thermodynamic pressure p, and the temperature T. We can now combine these results with the distribution function $\phi(u, v, w)$ and bring the latter into its final form.

$\phi(u, v, w)$ was given by:

$$\phi(u, v, w)\, du\, dv\, dw = Ae^{-\beta(u^2+v^2+w^2)}\, du\, dv\, dw \tag{14·23}$$

where A and β are constants, independent of u, v, w. We first relate A and β by normalizing ϕ, i.e., by choosing A so that

$$\iiint_{-\infty}^{\infty} Ae^{-\beta(u^2+v^2+w^2)}\, du\, dv\, dw = 1 \tag{14·24}$$

The integral splits into the product of three well-known integrals of the form:

$$\int_{-\infty}^{\infty} e^{-\beta u^2}\, du = \sqrt{\frac{\pi}{\beta}} \tag{14·25}$$

Thus Eq. 14·24 becomes

$$A\left(\frac{\pi}{\beta}\right)^{3/2} = 1$$

$$A = \left(\frac{\beta}{\pi}\right)^{3/2} \tag{14·26}$$

Next we relate β to the *mean square velocity*. Such a relation is suggested by dimensional reasoning; for we see from Eq. 14·23 that β must have the dimensions of 1 over velocity squared. For example,

$$\overline{u^2} = \iiint_{-\infty}^{\infty} u^2 \phi(u, v, w)\, du\, dv\, dw$$

$$= \left(\frac{\beta}{\pi}\right)^{3/2} \iiint_{-\infty}^{\infty} u^2 e^{-\beta(u^2+v^2+w^2)}\, du\, dv\, dw = \left(\frac{\beta}{\pi}\right)^{1/2} \int_{-\infty}^{\infty} u^2 e^{-\beta u^2}\, du$$

The integral here is easily related to Eq. 14·25 by the very useful method of "differentiation with respect to a parameter":

$$\int_{-\infty}^{\infty} e^{-\beta u^2} u^2\, du = -\frac{d}{d\beta}\int_{-\infty}^{\infty} e^{-\beta u^2}\, du = -\frac{d}{d\beta}\sqrt{\frac{\pi}{\beta}} = \frac{1}{2}\sqrt{\frac{\pi}{\beta^3}} \tag{14·27}$$

We thus obtain

$$\overline{u^2} = \frac{1}{2\beta} \tag{14·28}$$

and identical expressions for $\overline{v^2}$ and $\overline{w^2}$. We have, in fact, also computed $\overline{c^2}$ because

$$\overline{c^2} = \overline{(u^2 + v^2 + w^2)} = \overline{u^2} + \overline{v^2} + \overline{w^2}$$

Thus

$$\overline{c^2} = \frac{3}{2\beta} \tag{14·29}$$

But we have found previously (Eq. 14·17) the relation:

$$\overline{c^2} = 3RT$$

Hence

$$\beta = \frac{1}{2RT}$$

and the Maxwell-Boltzmann distribution law can be written in the form

$$\phi(u, v, w)\, du\, dv\, dw = (2\pi RT)^{-3/2} e^{-(u^2+v^2+w^2)/2RT}\, du\, dv\, dw \quad ▶(14\cdot30)$$

It is often more convenient to rewrite Eq. 14·30 in terms of the absolute molecular velocity c and two angles Φ and Ψ. That is, one introduces polar coordinates in velocity space. In this case one simply has to transform the "cell volume" $du\, dv\, dw$. For polar coordinates in three dimensions we have:

$$du\, dv\, dw = c^2 \sin \Psi\, dc\, d\Psi\, d\Phi$$

and Eq. 14·30 becomes

$$\phi(u, v, w)\, du\, dv\, dw = (2\pi RT)^{-3/2} c^2 e^{-c^2/2RT} \sin \Psi\, dc\, d\Psi\, d\Phi \quad (14\cdot31)$$

Equation 14·31 can be written

$$\phi(u, v, w)\, du\, dv\, dw = f(c, \Psi, \Phi)\, dc\, d\Psi\, d\Phi$$

where $f(c, \Psi, \Phi)$ is the weighting function for the cell, $dc\, d\Psi\, d\Phi$,

$$f(c, \Psi, \Phi) = (2\pi RT)^{-3/2} c^2 e^{-c^2/2RT} \sin\Psi \quad (14\cdot32)$$

as compared to

$$\phi(u, v, w) = (2\pi RT)^{-3/2} e^{-(u^2+v^2+w^2)/2RT} \quad (14\cdot33)$$

From Eq. 14·31 we can also obtain the probability of finding a molecule with an absolute velocity between c and $c + dc$, *regardless* of its direction of motion. In this case we have simply to integrate over all values of Ψ and Φ. Thus

$$\int_0^{2\pi} d\Phi \int_0^{\pi} d\Psi f(c, \Psi, \Phi) = 4\pi (2\pi RT)^{-3/2} c^2 e^{-c^2/2RT} = \zeta(c)$$

and hence

$$\zeta(c) dc = \frac{4\pi}{(2\pi RT)^{3/2}} c^2 e^{-c^2/2RT} dc \quad ▶(14\cdot34)$$

denotes the probability of finding a molecule with an absolute velocity between c and $c + dc$, regardless of its direction of motion. These examples illustrate how the function may be transformed. The clue always lies in the transformation of the unit cell.

Finally a few useful relations are here collected:

$$\overline{c^2} = \int_0^{\infty} c^2 \zeta(c)\, dc$$

This integral can again be reduced by differentiation with respect to a parameter†

$$\int_0^\infty c^4 e^{-\beta c^2}\, dc = \frac{d^2}{d\beta^2} \int_0^\infty e^{-\beta c^2}\, dc$$

Thus we recover our old result

$$\overline{c^2} = 3RT \tag{14·35}$$

The mean speed $\bar{c}$, on the other hand, becomes

$$\bar{c} = \int_0^\infty c\zeta(c)\, dc$$

The integral can be reduced to the elementary one

$$\int_0^\infty c e^{-\beta c^2}\, dc = \frac{1}{2\beta}$$

and hence

$$\bar{c} = \sqrt{\frac{8RT}{\pi}} \tag{14·36}$$

Note the relation of $\bar{c}$ and $\sqrt{\overline{c^2}}$ to the speed of sound a:

$$a = \sqrt{\gamma RT} = \bar{c}\sqrt{\frac{\pi\gamma}{8}} = \sqrt{\overline{c^2}}\sqrt{\frac{\gamma}{3}} \tag{14·37}$$

The mean molecular velocities are thus of the same order as the speed of sound.

14·7★ The Specific Heats of Gases

In the preceding article we dealt only with the velocities of the gas molecules. In the Maxwell-Boltzmann distribution law (Eq. 14·23) we found the statistical weight $\phi(u, v, w)$ of a cell $du\, dv\, dw$ in velocity space and were then able to compute all mean values related to the *translational degrees of freedom*. In general, a molecule is a complex mechanical system capable of rotation and vibration as well as translation. To compute the mean energy of a molecule we need therefore the distribution function of all degrees of freedom. This is the task of statistical mechanics.

A thorough discussion of this problem far surpasses the scope of this book; here we shall only outline the way in which one derives the expressions for $c_p(T)$ and $c_v(T)$ (cf. Eqs. 1·91, 1·92).

(*a*) We are interested here only in perfect gases at high temperatures. In this case the *statistical* approach is the same in classic and in quantum statistical mechanics: each molecule can be considered a member of a sta-

†In comparing with Eq. 14·27 note that the limits have been changed since c varies from 0 to ∞ only!

tistical ensemble, i.e., each molecule corresponds to one of the dice in our simplest example.

The energy ϵ_m of a molecule will consist of translational energy ϵ_T and a contribution from the internal degrees of freedom (e.g., rotation and vibration) which we shall call simply ϵ. Hence we have

$$\epsilon_m = \epsilon_T + \epsilon$$

and the mean values

$$\overline{\epsilon_m} = \overline{\epsilon_T} + \bar{\epsilon}$$

An expression for the translational energy was already obtained in Eq. 14·19, but there it was written for unit mass,

$$e_T\,(T) = \tfrac{3}{2}RT$$

The corresponding energy per molecule, of mass m, is

$$\overline{\epsilon_T} = me_T = \tfrac{3}{2}\mathbf{k}T$$

where $\mathbf{k} = mR$ is the Boltzmann constant. (It may be seen from Eq. 14·17 that $\overline{\epsilon_T} = \frac{1}{2}m\overline{c^2}$.) We now have to discuss the contribution $\bar{\epsilon}$, from the internal degrees of freedom.

(*b*) In *classic* mechanics, ϵ is a continuous function of the so-called generalized coordinates q_j and momenta p_j.† As a case of special interest for us, take a spring-mass system, the harmonic oscillator. The energy ϵ of such a system oscillating in the x-direction, say, is

$$\epsilon = \tfrac{1}{2}m\dot{x}^2 + \tfrac{1}{2}m\omega_0{}^2x^2$$

if m denotes the mass, ω_0 the proper angular frequency. The generalized coordinates here are simply $q \equiv x$ and $p = m\dot{x}$. Hence we have

$$\epsilon(p, q) = \frac{p^2}{2m} + \frac{m\omega_0{}^2q^2}{2} \tag{14·38}$$

In *quantum* mechanics, ϵ can take only discrete values ϵ_j, the so-called *eigenvalues*. For the harmonic oscillator, these are given by

$$\epsilon_j = \hbar\omega_0(j + \tfrac{1}{2}) \quad j = 0, 1, 2, \cdots \tag{14·38a}$$

$\hbar$ is Planck's constant, h, divided by 2π.

(*c*) The statistical weight ϕ is given by the so-called canonical distribution,

$$\phi = \text{const.}\ e^{-\epsilon/\mathbf{k}T} \tag{14·39}$$

Thus in classical statistical mechanics $\phi(p_j, q_j)$ is the weight of a cell in

†q_j and p_j are standard notation. This should cause no confusion with the components of the heat flux vector and the pressure.

"phase space," i.e., in a space with p_j, q_j as coordinates. For example, the mean energy of the oscillator is given by

$$\bar{\epsilon} = \frac{\iint \epsilon\phi \, dp \, dq}{\iint \phi \, dp \, dq} = \frac{\iint \left[\frac{p^2}{2m} + \frac{m\omega_0^2 q^2}{2}\right] e^{-\left(\frac{p^2}{2m\mathbf{k}T} + \frac{m\omega_0^2 q^2}{2\mathbf{k}T}\right)} dp \, dq}{\iint e^{-(1/2\mathbf{k}T)(p^2/m + m\omega_0^2 q^2)} \, dp \, dq}$$

Therefore,

$$\bar{\epsilon} = \frac{\int \frac{p^2}{2m} e^{-p^2/2m\mathbf{k}T} \, dp}{\int e^{-p^2/2m\mathbf{k}T} \, dp} + \frac{\int \frac{m\omega_0^2 q^2}{2} e^{-m\omega_0^2 q^2/\mathbf{k}T} \, dq}{\int e^{-m\omega_0^2 q^2/\mathbf{k}T} \, dq}$$

Evaluation of these integrals, which are of the same type as occurred previously (cf. Eqs. 14·25 and 14·27), gives

$$\bar{\epsilon} = \tfrac{1}{2}\mathbf{k}T + \tfrac{1}{2}\mathbf{k}T = \mathbf{k}T$$

This equation gives the famous *law of equipartition of energy:* Any term in ϵ which is quadratic in either p or q contributes $\frac{1}{2}\mathbf{k}T$ to $\bar{\epsilon}$.

A classic "degree of freedom" is essentially defined by the fact that it contributes a quadratic term to the energy function. Thus every degree of freedom contributes $\frac{1}{2}\mathbf{k}T$ to the mean energy of the molecule or $\frac{1}{2}RT$ to the mean specific energy. Thus, if the number of (classic) degrees of freedom is z, we have

$$\frac{e}{RT} = \frac{c_v}{R} = \frac{z(\frac{1}{2}RT)}{RT} = \frac{z}{2} \tag{14·40a}$$

or

$$\gamma = \frac{c_p}{c_v} = \frac{c_v + R}{c_v} = 1 + \frac{2}{z} \tag{14·40b}$$

An alternate form was given in Eq. 1·91.

In quantum statistical mechanics, Eq. 14·39 has to be written

$$\phi_j = \text{const. } e^{-\epsilon_j/\mathbf{k}T} \tag{▶14·39a}$$

and represents the statistical weight of the jth state. The mean energy $\bar{\epsilon}$ has now to be obtained by

$$\bar{\epsilon} = \frac{\sum_j \epsilon_j \phi_j}{\sum_j \phi_j}$$

For the harmonic oscillator ϵ_j is given by Eq. 14·38a and $\bar{\epsilon}$ can be easily obtained. For the computation one can drop the additive $\frac{1}{2}$ in Eq. 14·38a, since it contributes only to the energy constant, and write $\epsilon_j = \hbar\omega_0 j$. Thus

$$\bar{\epsilon} = \frac{\sum \hbar\omega_0 j e^{-(\hbar\omega_0/\mathbf{k}T)j}}{\sum e^{-(\hbar\omega_0/\mathbf{k}T)j}}$$

The denominator is a geometrical series

$$\mathcal{P} \equiv \sum (e^{-\hbar\omega_0/\mathbf{k}T})^j = (1 - e^{-\hbar\omega_0/\mathbf{k}T})^{-1}$$

and the numerator can be expressed as $-\,d\mathcal{P}/d(\mathbf{k}T)^{-1}$. Hence

$$\bar{\epsilon} = -\frac{1}{\mathcal{P}}\frac{d\mathcal{P}}{d(\mathbf{k}T)^{-1}} = -\frac{d}{d(\mathbf{k}T)^{-1}}\ln \mathcal{P}$$

or

$$\bar{\epsilon} = \frac{\hbar\omega_0}{e^{\hbar\omega_0/\mathbf{k}T} - 1}$$

If we write θ_v for $\hbar\omega_0/\mathbf{k}$, we obtain

$$(c_v)_{\text{vib}} = \frac{de}{dT} = \frac{1}{m}\frac{d\bar{\epsilon}}{dT} = \frac{\mathbf{k}}{m}\left(\frac{\theta_v}{T}\right)^2 e^{\theta_v/T}(e^{\theta_v/T} - 1)^{-2}$$

or

$$\frac{(c_v)_{\text{vib}}}{R} = \frac{(\theta_v/2T)^2}{[\sinh\,(\theta_v/2T)]^2} \qquad (14\cdot41)$$

A diatomic molecule contains translational, rotational, and vibrational energies. In the temperature range of interest here, the rotational degrees of freedom are "fully excited," i.e., they have reached their classical equipartition value. The vibrational part can be well approximated by a harmonic oscillator, and its contribution to c_v or c_p is given by Eq. 14·41. Consequently we obtain Eq. 1·92 for $c_p(T)$, i.e.,

$$c_p = R + c_v = R + (c_v)_T + (c_v)_{\text{rot}} + (c_v)_{\text{vib}}$$

A diatomic molecule has two rotational degrees of freedom. Thus

$$\frac{c_p}{R} = 1 + \frac{3}{2} + \frac{2}{2} + \frac{(c_v)_{\text{vib}}}{R}$$

or

$$\frac{c_p}{R} = \frac{7}{2} + \frac{(\theta_v/2T)^2}{[\sinh\,(\theta_v/2T)]^2}$$

14·8 Molecular Collisions. Mean Free Path and Relaxation Times

The moving molecules collide with each other and with the walls of the container. In each collision the molecule is acted on by a force **F**, which is randomly distributed (cf. Article 14·5). Even at the wall, where there is a preferred direction inward, there is an element of randomness due to the thermal motion of the wall molecules and the imperfections of the surface. Owing to the *randomness* of the forces on it, the motion of a molecule must, after a time, be independent of its previous history. We denote this characteristic time, for "loss of memory," by t^*. The concept can be made more precise by introducing a "correlation function," but for the discussion here this will not be necessary.

Evidently t^* will depend on the number, n, of collisions that the molecule experiences per unit time. The simplest assumption is to take t^* inversely proportional to n, and write

$$t^* = \frac{\alpha}{n}$$

where α denotes a factor of proportionality. Evidently *α is not necessarily the same for collisions between molecules and for collisions with the wall.* In a *mixture* of gases it will also be different for collisions with the same kind and with another kind of molecule. Furthermore, α may be different for different effects, even with molecules of the same kind. For instance, to change the vibrational energy may require more collisions than to change the translational energy. *t^* is called a relaxation time.*

To obtain α, a detailed discussion and knowledge of the collision process are required. Hence α depends on the gas and, for collisions with the wall, on the wall material.

To obtain n, the mean number of collisions per unit time, we may use a crude model since we are interested primarily in t^*; it is useless to refine the computation of n without a detailed computation of α. First, we note that in a gas made up of point molecules the probability of a collision *between molecules* is zero. Thus to compute collisions it is necessary to bring in the "size of the molecule," or "range of influence." One defines a *collision cross-section* A of a molecule, this being the target area which a molecule presents to another one moving toward it. The average number of collisions that a molecule experiences per unit time depends on A, on the mean speed $\bar{c}$, and on the number of molecules per unit volume, N. In unit time a molecule sweeps out a "corridor" of volume $\bar{c}A$, which encloses approximately $NA\bar{c}$ molecules. Thus the expected number of collisions with other molecules is

$$n = NA\bar{c}$$

The same result may also be obtained by dimensional reasoning. The time for collision between molecules is, then,

$$t = \frac{1}{NA\bar{c}} \tag{14·42}$$

The average distance traveled during this time is called the *mean free path,*

$$t\bar{c} = \Lambda = \frac{1}{NA} \tag{14·43}$$

The relaxation time, then, is

$$t^* = \alpha t = \frac{\alpha}{NA\bar{c}} \tag{14·42a}$$

and the *relaxation length* is

$$\Lambda^* = \alpha\Lambda = \frac{\alpha}{NA} \tag{14·43a}$$

The number of collisions, n, with a *solid boundary* will depend on the setup, that is, on the probability that a molecule hits the boundary before it terminates its mean free path. Hence collisions with the wall are important in a border zone of thickness Λ.

In the flow of a rarefied gas between walls, as in the example of Couette flow of Article 13·2, or in gas in equilibrium at low pressure in a vessel of volume V, it is possible that the border zone becomes of the same order as, or larger than, the dimensions of the setup. The molecules then collide with the walls more frequently than with each other, and the relaxation time is then simply of order

$$t^* = \frac{\tilde{\alpha} d}{\bar{c}} \tag{14·44}$$

where d is a typical dimension, for example, the distance between the walls in Couette flow, or in the vessel. In this case $\tilde{\alpha}$ depends on both the gas and the wall material.

The concept of relaxation time is very important, and usually is more convenient than the concept of mean free path. A relaxation time t^* can be assigned to any property which is carried by a molecule and which may be altered by a collision, for example, the momentum and the kinetic energy. In addition, the relaxation times of rotational energy modes might be different from those of the translational energy modes. Furthermore, a multiatomic molecule has internal degrees of freedom or energy modes; that is, the atoms may vibrate relative to each other. These internal modes may also be disturbed in a collision, the relaxation time being in general different from that of the translational and rotational modes.

To get an idea of the magnitude of t^*, consider the collision rate n in air at standard conditions, that is, room temperature and pressure.

$$N = 2.69 \times 10^{19}\ \text{cm}^{-3}$$

$$\bar{c} \doteq 4.5 \times 10^{4}\ \text{cm sec}^{-1}$$

$$A \doteq 10^{-15}\ \text{cm}^2$$

Hence

$$n \doteq 10^9 \text{ sec}^{-1}$$

$$t^* \doteq \alpha \times 10^{-9} \text{ sec}$$

$$\Lambda^* \doteq 4\alpha \times 10^{-5} \text{ cm}$$

An analysis of the collision processes shows that for the translational degrees of freedom, α is of order unity. For the internal degrees of freedom α may be quite large, as large as 10^6 in some cases!

The relaxation times evidently decrease with increasing density; the temperature dependence is more complicated since α and A depend on the temperature, in addition to $\bar{c}$.

The relaxation times are a measure of the rate at which small deviations from equilibrium subside. These deviations may be differences in momentum or energy, in the degree of dissociation or ionization, or, in general, in the distribution of energy over the various degrees of freedom. Thus the relaxation times determine the *currents* which are set up in a nonequilibrium state. They are intimately related to the *transport parameters* in the macroscopic point of view (e.g. μ and k).

14·9 Shear Viscosity and Heat Conduction

We have seen that it is possible to derive the thermodynamic equations of state and also the velocity distribution law without a detailed consideration of the collision processes. On the other hand, for gases that are *not in thermodynamic equilibrium* the collision processes become of primary importance. The rigorous theory of transport processes is, in fact, very difficult; the velocity distribution function has to be constructed for a given physical situation. To do this, Boltzmann derived an integrodifferential equation for the distribution function $\phi(x, y, z, u, v, w)$. In spite of considerable and excellent work since then, the problem of solving this Boltzmann equation is still not completely settled.

Here we shall consider only small deviations from equilibrium, and assume linear relations between stress and rate of strain and between heat flux and temperature gradient. Thus the main aim will be to obtain gaskinetic expressions for the shear viscosity μ and heat conductivity k. From dimensional reasoning, we expect μ/ρ and $k/c_p\rho$ to be of the form,

$$\frac{\mu}{\rho} = (\bar{c})^2 t^*_1 = \bar{c}\Lambda^*_1 = \alpha_1 \bar{c}\Lambda$$

$$\frac{k}{c_p\rho} = (\bar{c})^2 t^*_2 = c\Lambda^*_2 = \alpha_2 \bar{c}\Lambda$$

where t^*_1 and t^*_2 are relaxation times for momentum and energy, respectively. We have written an equality sign since the t^* may include an arbitrary constant. Introducing the expressions for t^* (Eqs. 14·42 and 14·43), we have

$$\nu = \frac{\mu}{\rho} = \alpha_1 \frac{\bar{c}}{NA}$$

$$\frac{k}{c_p \rho} = \alpha_2 \frac{\bar{c}}{NA}$$

(It is also possible to absorb the constant α in the collision cross-section A.) N is the number of molecules per unit volume and thus mN is the density ρ. Consequently

$$\mu = \nu\rho = \alpha_1 \frac{m}{A} \bar{c}$$

$$k = \alpha_2 \frac{m}{A} c_p \bar{c}$$

Thus, μ and k do not depend explicitly upon ρ or p.

It turns out experimentally that viscosity and heat conductivity for gases indeed depend only on T for a fairly wide range of pressures or densities. This shows that α and A do not depend strongly on p. However, they do vary with T, and so does their ratio α/A. Thus μ and k are not simply proportional to $\bar{c} \sim \sqrt{RT}$, but follow a more complicated temperature law.

In small deviations from equilibrium it is plausible that the transport coefficients are related to the corresponding relaxation times. For example, in the Rayleigh problem (Article 13·5) a deviation from thermodynamic equilibrium is set up initially by giving an excess momentum to the fluid near the boundary. This momentum excess then diffuses into the gas. Its time rate of change is *measured macroscopically in terms of the viscosity, microscopically in terms of the corresponding relaxation time.* Heat conduction consists of a similar diffusion of an excess of energy.

The two corresponding transport parameters for shear and heat diffusion are ν and $k/c_p\rho$. Both have the same dimension, namely *velocity* $\times$ *length* or better *velocity squared* $\times$ *time.* The characteristic velocity is a mean molecular velocity, e.g., $\bar{c}$ or $\sqrt{\bar{c^2}}$; it does not matter which one we choose since we shall in either event have an undetermined constant. The characteristic time is the relaxation time t^*, and the characteristic length is the relaxation length Λ^*.

14·10 Couette Flow of a Highly Rarefied Gas

Couette flow represents one of the few cases for which the solution of the Navier-Stokes equations is known for all Mach numbers and Reynolds numbers. Furthermore it is a case of pure shear flow. Thus it is very convenient for demonstrating the application of the concepts of the preceding article.

The shearing stress τ_w in Couette flow is related to the viscosity μ, the velocity U of the moving wall, and the channel height d by Eq. 13·16,

$$\frac{\tau_w d}{\mu_\infty U} = \int_0^1 \frac{\mu}{\mu_\infty}\, d\left(\frac{u}{U}\right) \tag{14·45}$$

μ_∞ is the value of the viscosity at the temperature of the moving wall.

For *low-speed flow*, $\mu/\mu_\infty \to 1$, and then

$$\tau_w = \mu \frac{U}{d} \tag{14·46}$$

or, in terms of skin friction coefficient,

$$C_f = \frac{2}{Re} \tag{14·46a}$$

We may now express μ in terms of the molecular quantities introduced in the last article. For pressures large enough so that collisions between molecules determine the relaxation time t^*, we had

$$\mu = \rho(\bar{c})^2 t^*_1 = \frac{\alpha m}{A}\bar{c}$$

which gives

$$\tau_w = \alpha_1 \frac{m}{A}\frac{\bar{c}U}{d}$$

This equation does not give any really new result. It is important, rather, as the relation from which α/A may be determined experimentally.

If the pressure of the gas is so low that t^*_1 is determined mainly by collisions with the wall, then

$$t^*_1 = \tilde{\alpha}\frac{d}{\bar{c}}$$

$$\mu = \rho(\bar{c})^2 t^*_1 = \tilde{\alpha}_1 \rho \bar{c} d$$

and then Eq. 14·45 becomes

$$\tau_w = \tilde{\alpha}_1 \rho \bar{c} U \tag{14·47}$$

For low pressures τ_w is independent of the spacing between the walls!

Equation 14·47 can be rewritten in terms of the pressure and temperature. For a perfect gas,

$$\frac{p}{\rho} = RT$$

and

$$\bar{c} = \sqrt{\frac{8}{\pi}}\sqrt{RT}$$

Hence

$$\tau_w = \tilde{\alpha}_1 \sqrt{\frac{8}{\pi}} \frac{pU}{\sqrt{RT}} \qquad \blacktriangleright(14\cdot48)$$

The corresponding skin friction coefficient is

$$C_f = \frac{\tau_w}{\frac{1}{2}\rho U^2} = 2\tilde{\alpha}_1 \sqrt{\frac{8}{\pi\gamma}} \frac{1}{M}$$

as compared to Eq. 14·46*a*, for incompressible flow,

$$C_f = \frac{2}{Re}$$

The characteristic parameter which determines whether the gas flow is highly rarefied can be expressed in terms of mean free path Λ or in terms of macroscopic quantities as follows:

$$\frac{d}{\Lambda} \sim \frac{d}{\bar{c}t^*} \sim \frac{Ud}{\nu}\frac{a}{U} = \frac{Re}{M} \qquad \blacktriangleright$$

Thus, if $M/Re \ll 1$, the gas is to be considered *dense* and the shearing stress determined from Eq. 14·45, whereas, if $M/Re \gg 1$, the gas is *highly rarefied* and the shearing stress is determined from Eq. 14·48.

Finally we may consider Couette flow of a rarefied gas without the restriction of small U. We have then to use Eq. 13·16 with

$$\frac{\mu}{\mu_\infty} = \frac{\bar{c}}{\bar{c}_\infty} = \sqrt{\frac{T_\infty}{T}}$$

For zero heat transfer, T/T_∞ is related to $u/U = \xi$ by Eq. 13·15,

$$T/T_\infty = 1 + Pr\frac{\gamma - 1}{2}M_\infty^2(1 - \xi^2)$$

Let us call $Pr\frac{\gamma - 1}{2}M_\infty^2 = \Gamma$ for short; the shearing stress is then obtained from

$$\frac{\tau_w d}{\mu_\infty U} = \int_0^1 \frac{d\xi}{\sqrt{1 + \Gamma(1 - \xi^2)}} = \frac{1}{\sqrt{\Gamma}} \sin^{-1} \sqrt{\frac{\Gamma}{1 + \Gamma}} \qquad (14\cdot49)$$

For small M_∞, i.e., small Γ,

$$\frac{1}{\sqrt{\Gamma}} \sin^{-1} \sqrt{\frac{\Gamma}{1+\Gamma}} \to 1$$

For large M_∞, i.e., large Γ,

$$\frac{1}{\sqrt{\Gamma}} \sin^{-1} \sqrt{\frac{\Gamma}{1+\Gamma}} \to \frac{\pi}{2\sqrt{\Gamma}}$$

Thus we have

$$\tau_w = \tilde{\alpha}_1 \sqrt{\frac{8}{\pi}} \frac{pU}{\sqrt{RT}} \qquad \text{(for small } M_\infty\text{)}$$

i.e., we recover our previous result in Eq. 14·48. In the opposite limit of rarefied gas flow at high speed, we find

$$\tau_w = 2\tilde{\alpha}_1 \sqrt{\frac{\pi}{(\gamma-1)PrM_\infty^2}} \frac{pU}{\sqrt{RT}} \qquad \text{(for large } M_\infty\text{)} \qquad (14\cdot50)$$

The corresponding skin friction coefficients are:

$$C_f = 2\tilde{\alpha}_1 \sqrt{\frac{8}{\pi\gamma}} \frac{1}{M_\infty}$$

and

$$C_f = 4\tilde{\alpha}_1 \sqrt{\frac{\pi}{\gamma(\gamma-1)Pr}} \frac{1}{M_\infty^2}$$

respectively.

Similar results can be obtained for the heat transfer, which for low pressures becomes, like τ_w, independent of the channel height d. This result is typical and forms the basis for measuring low pressures by heat transfer or force measurement (Pirani gage, Knudsen gage).

14·11 The Concepts of Slip and of Accommodation

A molecule that collides with a solid boundary may not acquire the momentum and energy corresponding to the state of the boundary after *a single collision.* For a gas in equilibrium this effect is immaterial because the boundaries and the gas have the same mean momentum (namely zero) and the same temperature. Under nonequilibrium conditions the effect can become important. If there exists a relative motion between the gas and the boundaries, momentum is transferred between them. If there exists a temperature difference, heat or energy is transferred between them. If, in average, a gas molecule does not acquire the momentum of the wall at one collision, one says that there exists "slip." If, in average, a molecule does not acquire the energy corresponding to the temperature of the boundary after one collision, one says that there exists a lack of complete "accommodation." Therefore in the case of slip the molecules near the boundary will have a mean velocity different from the velocity of the wall. In the case of a lack of accommodation there will exist a temperature jump.

Since these are effects connected with the collisions with the wall, they are important only in a zone near the wall, of thickness of the order Λ. Hence for the dynamics of dense gases these effects are immaterial and can always be neglected. In rarefied gas flow they may become important. The value of the constant $\tilde{\alpha}$, introduced in the definition of t^*, the relaxation time for collision with the wall, obviously depends on the magnitude of slip and accommodation.

The effects of slip and lack of accommodation are usually expressed in terms of discontinuities v, in velocity, and θ, in temperature, between the boundary and the gas. For instance, in Couette flow, with the upper wall moving with velocity U, the difference between the gas velocities at upper and lower walls is only $U - 2v$, due to the slip (v) at each wall. Thus the shearing stress in the gas is:

$$\tau = \mu \frac{U - 2v}{d}$$

Close to the wall the flow behaves like free molecular flow (cf. Article 14·10), so that we may also write (Eq. 14·47),

$$\tau = \tilde{\alpha}_1 \rho \bar{c} v$$

Consequently we have, with $\mu = \alpha_1 \rho \bar{c} \Lambda$ (cf. Article 14·9),

$$v = \frac{U}{2 + \dfrac{\tilde{\alpha}_1}{\alpha_1} \dfrac{d}{\Lambda}} = \frac{U}{d} \frac{\Lambda}{2 \dfrac{\Lambda}{d} + \dfrac{\tilde{\alpha}_1}{\alpha_1}}$$

For small Λ/d this becomes

$$v = \frac{\alpha_1 \Lambda}{\tilde{\alpha}_1} \frac{U}{d} = \xi \frac{dU}{dy}$$

The coefficient $\xi = \alpha_1 \Lambda / \tilde{\alpha}_1$ is called the slip coefficient.

Similarly, for the temperature, one can run through an analogous argument to introduce θ. This gives

$$\theta = \frac{\alpha_2 \Lambda}{\tilde{\alpha}_2} \frac{T_1 - T_2}{d} = g \frac{dT}{dy}$$

where $g = \dfrac{\alpha_2}{\tilde{\alpha}_2} \Lambda$ is the corresponding coefficient for the temperature jump.

The ratio $\alpha/\tilde{\alpha}$ is ordinarily of order unity and hence ξ and g are of order Λ, a result first obtained by Maxwell.

14·12 Relaxation Effects of the Internal Degrees of Freedom

The concept of relaxation time can be applied to *any* transfer of energy by collision. As noted earlier, the relaxation times for vibrational and rotational modes may be different from those for translation. The relaxation times of the internal degrees of freedom do not enter directly into the shear viscosity, which is related only to the translational degrees of freedom, but they do appear in cases of rapid heating and cooling, or rapid compression and expansion.

For example, assume that the molecules of a gas have z_1 degrees of freedom with *short* relaxation times t^*_1, and z_2 degrees with comparatively *long* relaxation times t^*_2. In the best-known case, t^*_1 is for translation and rotation whereas t^*_2 is for vibration. The ratio of specific heats, γ, is related to z (Eq. 14·40*b*) by

$$\gamma = \frac{z+2}{z} = \frac{z_1 + z_2 + 2}{z_1 + z_2}$$

If a flow process is such that heating or cooling occurs in a time $t \ll t^*_2$, then the degrees of freedom corresponding to t^*_2 are unable to reach their equilibrium energy, and *the gas behaves as if γ were determined by the z_1 degrees of freedom, alone.* This gives a higher value for γ than if all the degrees of freedom were brought to equilibrium. The classic example is the propagation of sound waves of high frequency. If f is the frequency, the velocity of the sound wave is different for $\frac{1}{f} \gtrless t^*_2$. The values are

$$a_1 = \sqrt{\gamma RT} = \sqrt{\frac{z_1 + z_2 + 2}{z_1 + z_2} RT} \quad \text{for } ft^*_2 \ll 1$$

$$a_2 = \sqrt{\frac{z_1 + 2}{z_1} RT} \quad \text{for } ft^*_2 \gg 1$$

The change-over is accompanied by strong absorption of sound, since in the region where $ft^*_2 \sim 1$ there are currents of energy in the gas and consequently entropy is produced.

These effects are even more pronounced in shock waves. The slow degrees of freedom cannot follow the initial rapid change of pressure, density, and temperature, but adjust more slowly. It is easy to see qualitatively the result of this lag by examining the equilibrium relations for a shock wave (Eqs. 2·41). These may be put in the form

$$h_2 - h_1 = \tfrac{1}{2}u_1^2[1 - (u_2/u_1)^2]$$

$$p_2 - p_1 = \rho_1 u_1^2[1 - u_2/u_1]$$

For *strong* shock waves, $(u_2/u_1)^2 \ll 1$ and hence

$$h_2 - h_1 \doteq \tfrac{1}{2}u_1^2\dagger$$

That is, the enthalpy difference depends mainly on the shock speed, not on the gas. If the specific heat were constant, that is, if T increased linearly with h, this would give

$$T_2 = T_1 + \tfrac{1}{2}u_1^2/c_p$$

But with variable specific heat (particularly with dissociation, ionization, etc.) the increase of T with h is slower than linear (for example, see Fig. 1·7), and hence a given enthalpy difference $h_2 - h_1$ corresponds to a smaller

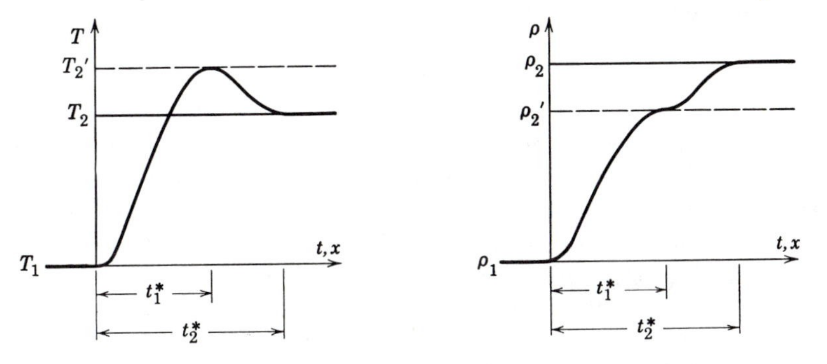

FIG. 14·2 The effect of molecular relaxation on the temperature and density profile in a shock wave.

temperature difference $T_2 - T_1$. The pressure difference $p_2 - p_1$ is practically unaffected; the term u_2/u_1, though not always negligible, has only a small influence, and a slight change in it due to variable specific heat leaves $p_2 - p_1$ practically unchanged.

Thus, if the specific heat were constant, we would have a *higher temperature* T'_2, practically the *same pressure* p_2, and hence *lower density* ρ'_2, than with variable specific heat. Now the effect of the relaxation time is to delay the adjustments that lead to the "true" specific heat, that is, to make the gas behave initially as if specific heat were constant. Hence it reaches the temperature T'_2 and density ρ'_2 before coming to the final equilibrium values T_2 and ρ_2. This is illustrated schematically in Fig. 14·2. Observations on the density distribution in shock waves can thus be used to determine internal relaxation times. Figure 6·15*d* shows a typical interferogram of a shock wave with relaxation effect.

In *subsonic* flow, steep gradients can be produced by flow past *small* obstacles. This was first pointed out by Kantrowitz,‡ who has discussed

†Note the similarity with Couette flow (Eq. 13·93).
‡A. Kantrowitz, *J. Chem. Phys.*, *14* (1946), p. 150.

in detail the flow near the mouth of an impact tube. Here the fluid is brought to rest in a time $t \sim d/U$, where d denotes the diameter of the tube. Consequently we can expect relaxation effects if

$$\frac{d}{U} \sim t^*$$

The appearance of relaxation effects is noted by the loss of total head measured by a small tube, as compared with a large one. The explanation is again to be found in the currents that exist in the nonequilibrium zone, where the energy distribution between the degrees of freedom adjusts itself. The resultant entropy change ΔS is related to the actual and the measured pressures by the relation (Eq. 2·15*a*)

$$\frac{p_0}{p_0'} = e^{\Delta S/R}$$

Kantrowitz has demonstrated how relaxation times can be obtained from pitot tube measurements. It is evident that similar results are obtained for reactions like condensation, dissociation, and ionization. Here too, relaxation times (or reaction rates) can be defined, and phenomena similar to the previous ones can be studied. Indeed ideas like these were first introduced in Einstein's theory of sound propagation in dissociating gases.

In these examples, the effects are essentially due to longitudinal, compression waves, and the effect of internal relaxation times is related to the so-called bulk viscosity or second viscosity coefficient, κ. In Article 13·12 we wrote $\tilde{\tau} = \tilde{\mu}\, du/dx$ for the viscous compressive stress. Referring to the linear stress relation (Eq. 13·77) we see that the "viscosity coefficient" for compressive stress is

$$\tilde{\mu} = \lambda + 2\mu$$

and that it is, in general, different from the shear viscosity coefficient μ. In terms of the bulk viscosity coefficient, $\kappa = 3\lambda + 2\mu$, we may also write

$$\tilde{\mu} = \tfrac{4}{3}\mu + \tfrac{1}{3}\kappa$$

As noted in Article 13·13, the so-called "Stokes assumption," $\kappa = 0$, is not valid, except for monatomic gases. Thus $\tilde{\mu}$ is affected, in general, by the internal relaxation times.

14·13 The Limit of Continuum Theory

The discussion of viscosity from a molecular point of view and its application to Couette flow have shown us that there exist two limiting cases for the flow: one in which the density is so high that the intermolecular collisions determine the relaxation times, and the other in which the gas is so rarefied that the collisions with the boundaries determine the relaxation times.

For flow similar to Couette flow, i.e., flow that is confined between walls, one can easily define these two limiting cases by the ratio of mean free path Λ to channel diameter d. Λ/d is often called the Knudsen number; if $\Lambda/d \ll 1$ intermolecular collisions dominate; if $\Lambda/d \gg 1$, collisions with the boundaries dominate. Since $\nu \sim \Lambda\bar{c}$ and $\bar{c} \sim a$ one can express the Knudsen number in terms of Mach number and Reynolds number by

$$\frac{\Lambda}{d} \sim \frac{M}{Re}$$

Consequently, flow with $M/Re \geq 1$ can be properly called rarefied gas flow.

It is often stated that continuum theory breaks down for rarefied gas flow and that flow problems in this regime have to be computed by kinetic theory. Such a statement is much too strong and usually results from a comparison of an experiment in rarefied gas flow with a solution of the Navier-Stokes equations which does not apply to the problem. For example, it is not possible to consider rarefied gas flow as *incompressible* flow, because this would mean $M \to 0$ and then M/Re could not, in general, be large. Thus, it is not possible to compare the solution of the Navier-Stokes equations for incompressible flow through a pipe (Poiseuille flow) with rarefied gas flow. Furthermore rarefied gas flow is always a low Reynolds number flow. Consequently we cannot compare, for example, the results of skin friction measurements in rarefied gas flow with boundary layer theory, which applies only to large Reynolds number. Similarly the use of the ratio of mean free path and boundary-layer thickness as a characteristic parameter makes no sense.

It should also be kept in mind that the simple dependence of k and μ on T, i.e., $\mu = \mu(T)$ and $k = k(T)$, applies only to a limited range of densities. At very low (and very high) density, $\mu = \mu(\rho, T)$ and $k = k(\rho, T)$. Again care has to be exercised in comparing solutions of the Navier-Stokes equations with experience or with kinetic theory in the rarefied gas regime.

In principle, continuum theory will become inapplicable only when the fluctuations due to the molecular structure become so large that measurements do not furnish the proper mean values. *This depends not only on the ratio of mean free path to the channel dimensions but, because of the ergodic property, also on the length of the time average.* With an instrument that averages over a very long time it is possible in principle to make measurements, say of the velocity profile in pipe flow, at very low pressures. Thus, even in rarefied gas flows, it is possible to define mean values of quantities such as pressure and force, and to apply the continuum equations of motion to these mean values. However, for rarefied gases it is often *simpler and more direct* to apply kinetic considerations even if it is the mean values that are required. Similarly, at normal conditions it is simpler and more direct to apply gasdynamical considerations, and to take μ and k from experience or from a single kinetic computation. Thus gasdynamics and gaskinetics actually overlap, and the choice of the approach is dictated by simplicity rather than by necessity.

For the most interesting case, namely flow of unlimited extent past a body, both methods encounter difficulties concerning the proper boundary conditions at infinity. For instance, there is difficulty obtaining a solution of the Navier-Stokes equations for *compressible*, viscous flow past a body, such as a sphere, which is valid for all M and Re. On the other hand, the solutions of kinetic theory usually apply only to the limit of so-called "free molecular flow," corresponding to $\Lambda \rightarrow \infty$. Just how this limit is approached in the case of an unbounded fluid is difficult to ascertain.

In general, the field of rarefied gas flow problems is still largely unclarified.

Exercises

Exercises, Chapter 1

1·1 Show that

$$\left(\frac{\partial p}{\partial \rho}\right)_s = \gamma \left(\frac{\partial p}{\partial \rho}\right)_T$$

Verify this for a perfect gas.

1·2 Show that

$$c_p - c_v = T \left(\frac{\partial v}{\partial T}\right)_p \left(\frac{\partial p}{\partial T}\right)_v$$

and

$$\gamma(c_p - c_v) = T(a \cdot \alpha)^2$$

where

$$a^2 = \left(\frac{dp}{d\rho}\right)_s, \qquad \alpha = \frac{1}{v}\left(\frac{\partial v}{\partial T}\right)_p$$

1·3 At the critical point, $(\partial p/\partial v) = (\partial^2 p/\partial v^2) = 0$. Compute $\kappa = (p_c v_c)/(RT_c)$ for the Dieterici equation of state:

$$p = \frac{RT}{v - \beta} \exp\left[-\frac{\alpha}{RTv}\right]$$

Compare κ with the values given in Table I.

1·4 Use the results of (1·1) and (1·2) to find the expression for the velocity of sound $a^2 = (\partial p/\partial \rho)_s$ for a gas which follows the Dieterici equation of state.

1·5 Consider an adiabatic throttling process. Compute dT/dp for a gas, using the equation of state in the form (Eq. 1·85) or (Eq. 1·87).

1·6 Find the canonical equation of state $E = E(V, S)$ corresponding to Dieterici's equation and to Eq. 1·85.

1·7 A simple pendulum is placed in a heat-insulated, rigid container. At $t = 0$ the pendulum is allowed to begin to oscillate from an initial displacement. What is the change of entropy of the system between the initial state and the final state, i.e., the state after motion has subsided?

1·8 A body whose drag depends on its speed, $D = D(U)$, moves along an arbitrary trajectory in an insulated closed box containing a perfect gas. If the body is at rest and the gas is in equilibrium at times t_1 and t_2, show that

$$\frac{T_2}{T_1} = e^{\Delta s/c_v} \quad \text{where} \quad \Delta S = \int_{t_1}^{t_2} \frac{DU}{T}\, dt$$

If T_2 is not much larger than T_1 and the motion is steady, show that the rate of entropy production is approximately given by

$$\frac{dS}{dt} = \frac{DU}{T_1}$$

1·9 (*a*) A perfect gas, enclosed by an insulated (upright) cylinder and piston, is at equilibrium at conditions p_1, v_1, T_1. A weight is placed on the piston. After a number of oscillations, the motion subsides and the gas reaches a new equilibrium at conditions p_2, v_2, T_2. Find the temperature ratio T_2/T_1 in terms of the pressure ratio $\lambda = p_2/p_1$. Show that the change of entropy is given by

$$s_2 - s_1 = R \log \left[\frac{1 + (\gamma - 1)\lambda}{\gamma}\right]^{\gamma/(\gamma-1)} \frac{1}{\lambda}$$

Show that, if the initial disturbance is small, so that $\lambda = 1 + \epsilon$, $\epsilon \ll 1$, then

$$(s_2 - s_1)/R \doteq \epsilon^2/2\gamma$$

(*b*) Show that the frequency of small oscillations about the equilibrium position, using isentropic relations, is

$$n = \frac{1}{2\pi}\sqrt{\gamma\frac{g}{l}}\sqrt{\left(1 + \frac{p_1 A}{W}\right)}$$

where l is the height of the cylinder. Compare with the frequency of a simple pendulum. Justify the use of the isentropic relations.

1·10 Consider an insulated (closed) system made up of unit mass of a dissociating, diatomic gas, e.g., N and N_2. Let s_1, e_1 and x denote the specific entropy, energy and mass fraction of N, and let s_2, e_2 and $(1 - x)$ be the corresponding variables for N_2. Thus

$$s = xs_1 + (1 - x)s_2$$
$$e = xe_1 + (1 - x)e_2$$

For a closed system s has a maximum in equilibrium. Hence $\delta s = 0$. Since no heat is added and no external work is done, E does not change and thus $\delta E = 0$.

Apply these two conditions to the system here; i.e., let $\delta s = (\partial s/\partial x)\,\delta x + (\partial s/\partial T)\,\delta T = 0$ and similarly $\delta e = (\partial e/\partial x)\,\delta x + (\partial e/\partial T)\,\delta T = 0$. For simplicity assume constant specific heats. Show that:

(*a*) $$s_1 T - (e_1 + R_1 T) = s_2 T - (e_2 + R_2 T)$$

i.e., $$g_1 = g_2$$

(*b*) $$\rho_1^2/\rho_2 = \text{const.}\ T^{\Delta c_v/R} \exp\left(-\Delta e_0/RT\right)$$

where $\Delta c_v = c_{v1} - c_{v2}$; $\Delta e_0 = e_{01} - e_{02}$ and $R \equiv R_2 = \frac{1}{2}R_1$.

Compare the results with Eq. 1·78 of the text and with the general law of mass action (Eq. 1·74). Note that $\dfrac{\Delta c}{R} = \dfrac{7}{2}$ as follows from Eqs. 1·90 and 1·91 of the text.

1·11 Apply the law of mass action (Eq. 1·74) (or a consideration similar to the one of Exercise 1·10) to an ionizing, monatomic gas, i.e., to a reaction of the type

$$A^+ + e = A$$

Argon ion + electron = neutral argon

The electron gas can be assumed to be perfect. The masses and hence the gas constants of A^+ and A are approximately equal since the mass of an electron is very small. Assume constant specific heat of all three gases.

Let x denote the degree of ionization, i.e., x is the mass fraction of A^+. Show that:

$$\frac{x^2}{1 - x^2} = \text{const.}\ \frac{T^{5/2}}{p} \exp\left(-e_0/RT\right)$$

(The constant cannot be determined from thermodynamics. Statistical mechanics yields for the constant the expression $(2\pi m)^{3/2}k^{5/2}h^{-3}$, where m is the electron mass, and h Planck's constant.)

The expression

$$\frac{x^2}{1-x^2} = \frac{(2\pi m)^{3/2}}{ph^3} (\mathrm{k}T)^{5/2} \exp(-e_0/\mathrm{k}T)$$

is the famous formula of Eggert-Saha.

1·12 Consider unit mass of a two phase system, e.g., a fraction x of water vapor and $(1-x)$ of water.

(*a*) Show that in equilibrium

$$g_1 = g_2$$

(*b*) Apply this equilibrium condition to two neighboring points in the p, T diagram, i.e.,

$$g_1(p, T) = g_2(p, T)$$

$$g_1(p + dp, T + dT) = g_2(p + dp, T + dT)$$

and show that:

$$\frac{dp}{dT} = \frac{s_1 - s_2}{v_1 - v_2}$$

(*c*) Show that $T(s_1 - s_2) = l \equiv$ the latent heat, i.e., the heat of transformation. Thus

$$\frac{dp}{dT} = \frac{l}{T(v_1 - v_2)}$$

This is the general form of the Clapeyron-Clausius equation.

If l is assumed independent of T and the vapor treated as a perfect gas, one finds:

$$p = \text{const.}\ e^{-l/RT}$$

Compare with Eq. 1·83 of the text.

1·13 Consider a cavity of volume V from which all material has been evacuated. The walls of the cavity are kept at a temperature T. The electromagnetic radiation which fills the cavity can be considered a thermodynamic system with an energy

$$E = \sigma V T^4 \qquad \text{(Stefan's law; } \sigma = \text{constant)}$$

Show that the radiation exerts a pressure

$$p = \frac{1}{3}\frac{E}{V}$$

on the walls of the cavity and that its entropy is given by

$$S = \frac{4}{3}\frac{E}{T}$$

(The application of thermodynamics to this so-called "black body radiation" is due to Boltzmann.) *Hint:* Use the reciprocity relations.

Exercises, Chapter 2

2·1 Show that for a perfect gas

$$M = \frac{u}{a_0}\left[1 - \frac{\gamma - 1}{2}\left(\frac{u}{a_0}\right)^2\right]^{-1/2}$$

At low Mach number this gives

$$M \doteq \frac{u}{a_0}\left[1 + \frac{\gamma - 1}{4}\left(\frac{u}{a_0}\right)^2\right]$$

Also obtain T/T_0, p/p_0, ρ/ρ_0 as functions of $(u/a_0)^2$.

2·2 Show that the Bernoulli equation (2·38) reduces to the incompressible form (2·18*c*) when the Mach number is low. *Hint:* In this case p differs only slightly from p_0, that is, $p = p_0(1 - \epsilon)$, $\epsilon \ll 1$. Show that

$$\frac{1}{2}\rho_0 u^2 = (p_0 - p)\left(1 + \frac{\epsilon}{2\gamma} + \cdots\right)$$

2·3 Show that local and reference (free-stream) conditions are related by

$$\frac{T}{T_1} = 1 - \frac{\gamma - 1}{2} M_1^2\left[\left(\frac{u}{U}\right)^2 - 1\right]$$

and obtain corresponding expressions for p/p_1, ρ/ρ_1. Obtain simple, approximate expressions for the case of small perturbations, i.e., $u' = u - U \ll U$.

2·4 Show that in flow from a reservoir the maximum velocity that may be reached is given by

$$u_m^2 = 2h_0$$

$$= \frac{2}{\gamma - 1} a_0^2 \qquad \text{for a perfect gas}$$

What are the corresponding values of T and M? Interpret this result.

2·5 The pressure ratio across a normal shock is sometimes a more convenient parameter than the Mach number. Show that

$$\frac{\rho_2}{\rho_1} = \frac{u_1}{u_2} = \frac{1 + \dfrac{\gamma + 1}{\gamma - 1}\dfrac{p_2}{p_1}}{\dfrac{\gamma + 1}{\gamma - 1} + \dfrac{p_2}{p_1}} = \frac{p_2}{p_1}\frac{T_1}{T_2}$$

These are called the *Rankine-Hugoniot* relations.

2·6 Obtain the following expressions for a weak normal shock $\left(\dfrac{\Delta p}{p_1} = \dfrac{p_2 - p_1}{p_1} \ll 1\right)$:

$$\frac{\Delta\rho}{\rho_1} \doteq -\frac{\Delta u}{u_1} \doteq \frac{1}{\gamma}\frac{\Delta p}{p_1}$$

$$M_1^2 = 1 + \frac{\gamma + 1}{2\gamma}\frac{\Delta p}{p_1} \qquad \text{(exact)}$$

$$M_2^2 \doteq 1 - \frac{\gamma + 1}{2\gamma}\frac{\Delta p}{p_1}$$

$$\frac{\Delta p_0}{p_0} \doteq -\frac{\gamma + 1}{12\gamma^2}\left(\frac{\Delta p}{p_1}\right)^3$$

Exercises, Chapter 3

3·1 Obtain the normal shock-wave relations by applying the conservation equations directly to the case of a propagating shock wave (cf. Article 3·2). *Hint:* Consider the fluid between an advancing piston and a reference section ahead of the shock.

3·2 Show that the equation of continuity is

$$\frac{\partial \rho}{\partial t} + \frac{\partial}{\partial r}(\rho u) + N\frac{\rho u}{r} = 0$$

where $N = 0$, 1, and 2, for plane flow, cylindrical flow, and spherical flow, respectively. Euler's equation (3·8) is the same for all three cases. Obtain the corresponding acoustic equations for cylindrical and spherical waves.

Show that in all three cases the motion is irrotational (Article 7·10) so that the velocity may be obtained from a velocity potential, and the acoustic equations may be written in the form

$$\frac{1}{a_1^2}\frac{\partial^2 \phi}{\partial t^2} = \nabla^2 \phi \equiv \frac{\partial^2 \phi}{\partial r^2} + \frac{N}{r}\frac{\partial \phi}{\partial r}$$

Show that for spherical waves ($N = 2$) the general solution is

$$\phi = \frac{1}{r}[\phi_1(r - a_1 t) + \phi_2(r + a_1 t)]$$

Compare this solution with that for plane waves ($N = 0$, Article 3·4).

What of the case $N = 1$? Plane and spherical waves are similar, except for the differences in attenuation, but cylindrical waves are fundamentally different in structure. For a discussion of the wave equation see Ref. B·1. The difference between solutions of the one- and two-dimensional wave equation is also noted in connection with steady supersonic flow (e.g., Articles 9·7 and 9·11).

3·3 (*a*) Show that when a plane acoustic wave of strength Δp is reflected from a closed end of a tube the pressure there becomes $2\Delta p$; at an open end it is zero, and so the reflected wave must be an expansion wave.

(*b*) Discuss the reflection of a shock and of a centered expansion wave from a closed end. Sketch the *x-t* diagrams (cf. Exercises 12·5 and 3·7).

3·4 Show that the one-dimensional equations of motion may be written in the form

$$\frac{2}{\gamma - 1}\left(\frac{\partial a}{\partial t} + u\frac{\partial a}{\partial x}\right) + a\frac{\partial u}{\partial x} = 0$$

$$\frac{\partial u}{\partial t} + u\frac{\partial u}{\partial x} + \frac{2a}{\gamma - 1}\frac{\partial a}{\partial x} = 0$$

Adding and subtracting, respectively, gives the equations

$$\left[\frac{\partial}{\partial t} + (u + a)\frac{\partial}{\partial x}\right]\left(u + \frac{2a}{\gamma - 1}\right) = 0$$

$$\left[\frac{\partial}{\partial t} + (u - a)\frac{\partial}{\partial x}\right]\left(u - \frac{2a}{\gamma - 1}\right) = 0$$

which show that the quantities $P = u + \dfrac{2a}{\gamma - 1}$ and $Q = u - \dfrac{2a}{\gamma - 1}$ are constant on curves which have slopes $dx/dt = u + a$ and $dx/dt = u - a$, respectively. These curves are the characteristics; P and O are the Riemann invariants. Set up a computing method based on the above properties. (*Hint:* Compare Article 12·4.) Show that in a *simple* expansion (or isentropic compression) wave one of the Riemann invariants is constant. Thus check Eq. 3·23.

3·5 Obtain the following shock-tube equations:

$$M_s = \frac{c_s}{a_1} = \left(\frac{\gamma_1 - 1}{2\gamma_1} + \frac{\gamma_1 + 1}{2\gamma_1}\frac{p_2}{p_1}\right)^{1/2} \quad \text{Shock speed}$$

$$M_2 = \frac{1}{\gamma_1}\left(\frac{p_2}{p_1} - 1\right)\left[\frac{p_2}{p_1}\left(\frac{\gamma_1 + 1}{2\gamma_1} + \frac{\gamma_1 - 1}{2\gamma_1}\frac{p_2}{p_1}\right)\right]^{-1/2} \quad \text{Mach number behind shock}$$

$$M_3 = \frac{2}{\gamma_4 - 1}\left[\left(\frac{p_4/p_1}{p_2/p_1}\right)^{(\gamma_4 - 1)/2\gamma_4} - 1\right] \quad \text{Mach number behind contact surface}$$

3·6 A shock tube may be used as a short-duration wind tunnel by utilizing the flow behind the shock wave. Show that, in terms of the shock speed $M_s = c_s/a_1$, the density ratio $\eta = \rho_2/\rho_1$, and conditions in the expansion chamber (1), the flow conditions behind the shock in region (2) are given by the following:

$$\frac{p_2}{p_1} = 1 + \gamma_1 M_s^2\left(1 - \frac{1}{\eta}\right)$$

$$\frac{h_2}{h_1} = 1 + \frac{\gamma_1 - 1}{2} M_s^2\left(1 - \frac{1}{\eta^2}\right)$$

$$\frac{u_2}{a_1} = M_s\left(1 - \frac{1}{\eta}\right)$$

$$\frac{h_{02}}{h_1} = 1 + \frac{c_s u_2}{h_1} = 1 + (\gamma_1 - 1)M_s^2\left(1 - \frac{1}{\eta}\right)$$

The above expressions are general. Given the equation of state, they may be solved by an iteration procedure starting with an assumption for η. Work out the above expressions for shocks which are weak enough so that the specific heat remains constant. *Hint:* For this case η is given by Eq. 2·47. Calculate $M_2 = u_2/a_2$. Show that M_2 has the limiting value

$$M_2 \rightarrow \sqrt{2/\gamma_1(\gamma_1 - 1)}$$

The real gas effects actually give somewhat higher values. To obtain still higher flow Mach numbers, the flow must be expanded in a nozzle. (In region 3, there is no limit on M_3, but h_{02} is low.)

3·7 If the conditions behind the shock after its reflection from the end of the tube are denoted by (5) and the shock speed relative to the tube is U_R, show that, in terms of the density ratios $\eta = \rho_2/\rho_1$ and $\zeta = \rho_5/\rho_1$,

$$\frac{U_R}{c_s} = \frac{\eta - 1}{\zeta - \eta}$$

$$\frac{p_5}{p_1} = 1 + \gamma_1 M_s^2 \frac{(\eta - 1)(\zeta - 1)}{(\zeta - \eta)}$$

$$\frac{h_5}{h_1} = 1 + (\gamma_1 - 1)M_s^2 \frac{(\eta - 1)(\zeta - 1)}{(\zeta - \eta)}\frac{1}{\eta}$$

For a given incident shock, M_s and η are first calculated (Exercise 3·6); the above equations are then solved by assuming values for ζ and finding consistent values of p_5 and h_5.

3·8 Show that the maximum shock velocity attainable in a (uniform) shock tube, for $p_4/p_1 \to \infty$, is given by

$$M_{s\infty} = \frac{\gamma_1 + 1}{\gamma_4 - 1} \frac{a_4}{a_1}$$

Hint: From Eqs. 3·26 and 3·2, respectively, obtain

$$\sqrt{p_2/p_1} \doteq (a_4/a_1)\sqrt{2\gamma_1(\gamma_1 + 1)}/(\gamma_4 - 1)$$

$$M_s \doteq \sqrt{(\gamma_1 + 1)/2\gamma_1}\sqrt{p_2/p_1}$$

Calculate the limiting values of M_s for combinations of air-air, helium-air, hydrogen-air. What is the effect of temperature ratio, T_4/T_1?

Exercises, Chapter 4

4·1 Show that the pressure coefficient corresponding to a small flow deflection may be written, to second order,

$$C_p = \frac{2}{\sqrt{M_1^2 - 1}}\theta + \frac{(\gamma + 1)M_1^4 - 4(M_1^2 - 1)}{2(M_1^2 - 1)^2}\theta^2$$

Show that the same result is obtained from the oblique shock relations and from the Prandtl-Meyer relation, θ being positive for a compression. Why can one expect the coefficients of θ and θ^2 (called the Busemann coefficients) to be the same in the two cases? *Hint:* Entropy changes are of order θ^3.

4·2 A weak oblique shock makes an angle ϵ with Mach lines ahead of it (Eq. 4·18). Show that it makes the same angle with Mach lines downstream of it. That is, the shock position is the "average" of the Mach line positions on either side. Use this result to prove that the shape of the shock in the region of attenuation by an expansion (Fig. 4·17*a*) is parabolic. *Hint:* Use the well-known optical result that a parabolic "reflector" focuses a parallel beam of rays.

4·3 Calculate $\beta' - \beta$ (Fig. 4·11*a*), assuming the shocks to be weak enough so that the approximate expressions of Article 4·7 may be used. Find the locus of M_1 versus θ for the condition $\beta' - \beta = 0$. Also obtain a few points for stronger shocks by using the shock charts.

4·4 A "free" surface is defined as one along which the pressure is constant, for example, the edge of a jet issuing into the atmosphere. Show that the "reflection" of an oblique shock wave from a free surface is an expansion wave. Compute its strength for the case of a weak incident shock.

4·5 Assuming the interacting shocks in Fig. 4·12 to be weak, apply the approximate expressions of Article 4·7, and show that, approximately, $\delta = \theta' - \theta$. (To this order of accuracy, the "turning strength" of a shock is the same after intersection as it is before; cf. Article 12·10). Calculate the "strength" of the slipstream in terms of $(u_3' - u_3)/u_3$ or $(\rho_3' - \rho_3)/\rho_3$.

4·6 Describe the motion set up by a wedge sliding along a wall, with velocity v_1, into a fluid at rest. Show that the shock speed, normal to its front, is $v_1 \sin\beta$ and that the flow behind it is set into motion in the same direction. Trace the path of a particle, and thus show that the instantaneous streamlines are perpendicular to the shock front. *Hint:* Apply a uniform velocity transformation to the flow in Fig. 4·3.

4·7 Show that the energy equation may be written in the form

$$\frac{\gamma - 1}{\gamma + 1}\cos^2\mu + \sin^2\mu = \frac{a^{*2}}{w^2} \quad \text{where } \sin\mu = 1/M$$

Use this to obtain the relation

$$\sqrt{M^2 - 1}\,\frac{dw}{w} = \frac{b^2 - 1}{b^2 + \tan^2 \mu}\, d\mu \quad \text{where } b^2 = \frac{\gamma - 1}{\gamma + 1}$$

and so evaluate the integral

$$\int \sqrt{M^2 - 1}\,\frac{dw}{w} = \mu - \frac{1}{b}\tan^{-1}\left(\frac{1}{b}\tan \mu\right) + \text{const.}$$

This result may be converted to the form given in Eq. 4·21*b*.

4·8 Write the equations of a streamline in a Prandtl-Meyer expansion. *Hints:* Use the corner as origin and the length of a "ray" as parameter. Note that the mass flow between the wall and the streamline is fixed.

4·9 Show that for a thin, symmetrical airfoil, whose profile is a "lens" defined by two circular arcs, the drag coefficient is

$$C_D = \frac{16}{3\sqrt{M_1^2 - 1}}\left(\frac{t}{c}\right)^2$$

Prove that for a given thickness ratio, t/c, the profile for minimum drag is a symmetrical diamond profile.

4·10 Prove that on a supersonic swept-back wing of infinite span the thin-airfoil pressure coefficient (Eq. 4·26) is multiplied by the sweepback factor, $\dfrac{1}{\sqrt{1 - n^2}}$, where $n = \dfrac{\tan \Lambda}{\sqrt{M_1^2 - 1}}$ and Λ is the sweepback angle (Fig. 4·21). This result also applies to finite wings, in regions not affected by the "tips." (*Hint:* Busemann suggested that one may resolve the free-stream Mach number into components normal and parallel to the leading edge, M_n and M_p. The flow may then be studied in planes normal to the leading edge, using the thin-airfoil theory to calculate the pressure ratio $\Delta p/p_1$. From this, the pressure coefficient referred to M_1 may be calculated.)

4·11 Write the expression for the drag of a Busemann biplane (Fig. 4·23*a*) at off-design Mach number. *Hint:* The escaping waves (Fig. 4·22*c*) are closer together than they would be with no interference. Thus they carry away the same momentum as the wave system from a monoplane of smaller chord.

4·12 Show that the equation of the shock polar, in the hodograph plane (Fig. 4·24*a*) is

$$\left(\frac{v_2}{a^*}\right)^2 = \left(\frac{u_1}{a^*} - \frac{u_2}{a^*}\right)^2 \frac{u_1 u_2 - a^{*2}}{\dfrac{2}{\gamma + 1} u_1^2 - u_1 u_2 + a^{*2}}$$

Another form is

$$\left(\frac{v_2}{u_1}\right)^2 = \left(1 - \frac{u_2}{u_1}\right)^2 \frac{(M_1^2 - 1) - \dfrac{\gamma + 1}{2x} M_1^2 \left(1 - \dfrac{u_2}{u_1}\right)}{1 + \dfrac{\gamma + 1}{2} M_1^2 \left(1 - \dfrac{u_2}{u_1}\right)}$$

Show that for $M_1 \to \infty$ the shock polar is a circle, which *touches* the circle

$$u/a^* = u_m/a^*$$

4·13 Using Eq. 4·9, show that

$$\tan \beta = [(\eta - 1) \pm \sqrt{(\eta - 1)^2 - 4\eta \tan^2 \theta}]/2 \tan \theta$$

where $\eta = \rho_2/\rho_1$ is the density ratio across the shock. Show how this may be used to calculate oblique shock solutions for the case of variable specific heat (cf. Exercise 3·6). Show that for small values of θ the "weak solution" (using the minus sign above) reduces to

$$\tan \beta = \frac{\eta}{\eta - 1} \theta$$

Compare with Eq. 4·11*b* for large M_1 and constant specific heat.

Show that at detachment

$$\tan \theta = (\eta - 1)/2\sqrt{\eta}$$

What effect does departure from constant specific heat have on detachment? (Real gas effects tend to increase η.) (Ivey and Cline, *NACA Tech. Note* 2196, 1950.)

4·14 On Fig. 4·15, plot the two curves which are the loci of detachment Mach number for wedges and cones, respectively. (For each value of δ/d there is a corresponding wedge or cone angle θ, for which the detachment Mach number may be found in Chart 1 at the back of the book or on Fig. 4·27*a*.) For a wedge or cone of given θ, the experimental measurements of δ/d versus M_1 fall on a curve which starts at the detachment locus and is asymptotic to the appropriate limiting curve in Fig. 4·15.

4·15 $F(x, y, z) = 0$ represents a surface in space. Let it be a shock front, and show that the continuity, momentum, and energy equations require that the following three expressions have the same value on either side of the shock wave:

$$\rho \mathbf{w} \cdot \operatorname{grad} F \tag{1}$$

$$\rho(\mathbf{w} \cdot \operatorname{grad} F)^2 + p(\operatorname{grad} F)^2 \tag{2}$$

$$\frac{1}{2}(\mathbf{w} \cdot \operatorname{grad} F)^2 + \frac{\gamma}{\gamma - 1} \frac{p}{\rho} (\operatorname{grad} F)^2 \tag{3}$$

Exercises, Chapter 5

5·1 The mass flow through a Laval nozzle may be expressed in terms of conditions in the reservoir and throat. Show that

$$m = \rho_0 a_0 A^* M^2 \left(1 + \frac{\gamma - 1}{2} M^2\right)^{-\gamma/(\gamma-1)}$$

and

$$(m)_{\max} = m^* = \left(\frac{2}{\gamma + 1}\right)^{\gamma/(\gamma-1)} \rho_0 a_0 A^*$$

5·2 Plot the distribution of Mach number along a conical, supersonic nozzle, which expands from area A^* at the throat to $A_E = 4A^*$ at the exit. (If the divergence is gradual, the one-dimensional results give fairly good accuracy.)

Show that the minimum pressure ratio p_0/p_E to maintain supersonic flow up to the exit is

$$\frac{p_0}{p_E} = \left(1 + \frac{\gamma - 1}{2} M_1^2\right)^{\gamma/(\gamma-1)} \left(\frac{2\gamma}{\gamma + 1} M_1^2 - \frac{\gamma - 1}{\gamma + 1}\right)$$

For values of pressure ratio less than this minimum, find the positions of a normal shock wave in the conical nozzle described above.

5·3 For a duct of constant area, the adiabatic, frictionless, one-dimensional equations give two possible solutions: (*a*) uniform flow; (*b*) transition through a normal shock.

Flows with friction and nonuniformities may be approximately treated by using the average values, $\bar{u}$, $\bar{T}$, etc. Show that

$$\bar{h} + \left(\frac{m}{2A}\right)^2 \frac{1}{\bar{\rho}^2} = h_0$$

where m is the mass flow. Since $\rho = \rho(h, s)$ this defines a curve in the h-s plane, for given m/A and h_0. This is called a *Fanno line.* Sketch a Fanno line, assuming a perfect gas to compute $\rho = \rho(h, s)$. Show that the entropy has a maximum value at sonic conditions. Discuss the possible changes of velocity, pressure, etc., in a constant area duct, for supersonic and subsonic flows, respectively.

5·4 Show that for constant-area, *frictionless, diabatic* flow

$$\bar{p} + \left(\frac{m}{A}\right)^2 \frac{1}{\bar{\rho}} = \text{const.}$$

This equation, called the *Rayleigh line,* is useful in cases of constant-area flow with external heat addition. Sketch a Rayleigh line in the h-s plane, and show that there is a maximum value of enthalpy and of entropy, the latter again occurring at sonic conditions. Discuss supersonic and subsonic flows with heat addition.

5·5 Calculate the running time of an intermittent wind tunnel in terms of the storage tank volume and pressure, and the test section area and Mach number. Compare the "pressure" and "vacuum" methods of operation. See Fig. 5·8 for the minimum pressure ratio required at each Mach number.

5·6 Write the "ideal power" (Eqs. 5·7*a*, 5·9) per square foot of *test section area,* for atmospheric stagnation conditions. Using Fig. 5·8 for the pressure ratio, plot this power as a function of M.

5·7 Show that a close estimate of the ideal power is obtained by writing

$$Q = \tfrac{1}{2}(T_0 + T_c)\,\Delta S$$

that is, by choosing an average temperature for the entropy transfer. The resulting expression is

$$P' = 2490 \log \lambda$$

horsepower per square foot of throat area.

5·8 Plot the basic wind tunnel cycle (Fig. 5·9) on p-v and h-s (or T-s) diagrams.

5·9 Using one-dimensional flow theory, study the velocity and pressure distributions in the throat of a nozzle. In particular, describe the branch point which occurs when the throat becomes sonic (Fig. 5·3). *Hint:* Integrate the area-velocity relation (Eq. 2·27), using the approximations, $A = A^* + cx^2$, $u = a^* + u'$. Show that

$$u'^2 - \frac{2ca^{*2}}{(\gamma + 1)A^*} x^2 = \text{const.}$$

5·10 At one of the design points of the Southern California Cooperative Wind Tunnel the conditions are as follows: $M_1 = 1.8$, $p_0 = 1420$ lb/ft^2, $T_0 = 120°$ F, test section area = 96 sq ft. The compressor power is 40,000 hp. Compare this with the ideal power for these conditions, using Eq. 5·9 and assuming normal shock recovery.

Exercises, Chapter 6

6·1 If the velocity indicated by a standard airspeed indicator is u_i, the true speed may be obtained from $u/u_i = \sqrt{1/\sigma}$ where $\sigma = \rho_1/\rho_s$ is the ratio of actual air density to standard, sea-level density, and compressibility is neglected. Show that at subsonic speeds the compressibility correction may be obtained from

$$\left(\frac{u}{u_i}\right)^2 = \frac{\gamma}{2\sigma} M_1^2 \left[\left(1 + \frac{\gamma - 1}{2} M_1^2\right)^{\gamma/(\gamma-1)} - 1\right]^{-1}$$

which may be approximated by

$$\frac{u}{u_i} = \frac{1}{\sqrt{\sigma}}\left[1 - \frac{M_1^2}{4} + \frac{2\gamma - 1}{48} M_1^4 + \cdots\right]$$

Estimate the Mach number range for which the approximate formula may be used with the first two terms to give 1% accuracy.

6·2 What pressure is indicated by a pitot tube at $M = 2$ in the test section of a blow-down wind tunnel operating into a vacuum tank? What pressure is indicated if the pitot tube is on the nose of an airplane flying at $M = 2$?

6·3 A pitot tube is mounted on a supersonic wing with a sharp (wedge) leading edge. Show that the reading of the pitot tube is higher when its mouth is downstream of the leading edge shock wave than when it is ahead in the free stream.

6·4 In a boundary layer, what measurements in addition to the surface static pressure are needed to obtain the Mach number profile? Compare subsonic and supersonic cases. What measurements are needed for the velocity profile? For the density profile? Etc.

6·5 If q_0 is the (constant) rate of heat input to unit surface area of a semi-infinite body, the temperature distribution in the body is given by

$$T - T_1 = \frac{2q_0}{k}\left[\sqrt{\frac{Kt}{\pi}} \exp\left(-\frac{x^2}{4Kt}\right) - \frac{x}{2}\, erfc \frac{x}{2\sqrt{Kt}}\right]^{\dagger}$$

where T_1 is the initial temperature of the body, at $t = 0$, k is the thermal conductivity, $K = k/\rho c$, ρ = density, c = specific heat of body, x is the distance from the surface. Show how a thin metallic film attached to an insulating surface may be used as a resistance thermometer to measure the heat transfer. Discuss the following limiting cases: (*a*) the film is so thin that its temperature is essentially the surface temperature (at $x = 0$); (*b*) the film is thick enough so that during some time interval it retains all the heat transferred into it. Show that the film thickness h for which these limiting cases are good approximations is given by

$$h \lessgtr \sqrt{Kt}\ddagger$$

6·6 Compute typical ray deflections for a schlieren system on wind tunnels with test sections of 2-inch and 12-inch widths. Use typical density gradients in a boundary layer (Exercise 13·4); a Prandtl-Meyer expansion at the shoulder of an airfoil (Article 4·10); a circular arc airfoil (Exercise 4·9); a weak shock wave (Exercise 13·7). Compare the last computation with the result obtained by applying Snell's law of refraction to the case of glancing incidence.

6·7 If the ray deflection in a schlieren system is too large, the corresponding image at the knife edge may fall completely on or completely off the cutoff, in which case any further deflection produces no corresponding effect at the screen. Show that the limiting ray deflection ϵ_m is proportional to $1/s$, and that its maximum value $2/s$ is obtained when the

† Carslaw and Jaeger, *Conduction of Heat in Solids*, Oxford, 1947.

‡ J. Rabinowicz, M. E. Jessey, and C. A. Bartsch, "Resistance Thermometer for Transient High-Temperature Studies," *J. Appl. Phys.*, *27* (1956), p. 97.

knife edge is set so that half the primary image is cut off. (s is the sensitivity.) Such "nonlinear" effects are sometimes helpful!

6·8 Show that the sensitivity of a schlieren system cannot be changed by magnifying the source (e.g., with a condenser lens) but that the illumination of the screen may be increased.

6·9 Show that the Mach number increment between two adjacent interference fringes is

$$\Delta M \doteq \frac{1}{M}\left(1 + \frac{\gamma - 1}{2} M^2\right)^{\gamma/(\gamma-1)} \left(\frac{\rho_s}{\rho_0}\right) \frac{\lambda}{\beta L}$$

At what Mach number is ΔM a minimum, that is, the sensitivity a maximum? Do the same for the pressure coefficient increment ΔC_p (Bryson, *NACA Tech. Note* 2560, 1951).

6·10 Show that for axially symmetric flow the fringe-shift relation (6·18c) becomes

$$N = \frac{2\beta}{\lambda \rho_s} \int_{r_0}^{r_L} \frac{[\rho(r) - \rho_1]}{\sqrt{r^2 - r_0^2}}\, r\, dr$$

for a light path perpendicular to the axis of symmetry and at distance r_0 from it.

6·11 (a) Show how to use the calibration curves of Fig. 6·18 to determine the local mass flow ρu and the total temperature T_0 from measurements of i, R_w, and R_e (the latter is obtained by using a very low current). *Hint:* It may be assumed as a first step in an iteration procedure that $T_2 = T_e$. A second step is in fact usually not necessary.

(b) Show how to use these measurements, together with a surface pressure measurement, to determine profiles of ρ, u, M, T, etc., in a boundary layer. *Hint:* $(\rho u)/(\gamma p a_0)$ may be written as a function of M.

Exercises, Chapter 7

7·1 Show that for axially symmetric flow the continuity equation in natural coordinates is

$$\rho u(2\pi r\, \Delta n) = \text{const.}$$

where r is the distance from the axis of symmetry and Δn is the distance between streamlines, in a meridian plane. By combining with Euler's equation, obtain the equation of motion

$$\frac{M^2 - 1}{u} \frac{\partial u}{\partial s} - \frac{\partial \theta}{\partial n} - \frac{1}{r} \frac{\partial r}{\partial s} = 0$$

Compare with the case of plane flow. The equation of vorticity is the same as in plane flow.

7·2 Show that the variation of total pressure across streamlines is given, for a perfect gas, by

$$-\frac{1}{\rho_0} \frac{dp_0}{dn} = \left(1 + \frac{\gamma - 1}{2} M^2\right) u\zeta + \frac{1}{2} C_p M^2 \frac{dT_0}{dn}$$

and hence, for incompressible flow, the total-pressure gradient is related to the vorticity by

$$\frac{dp_0}{dn} = -\rho_0 u \zeta$$

7·3 A fluid at rest at the conditions p_1, ρ_1, T_1 is slightly disturbed. Show that for this case Eqs. 7·45 and 7·46 may be simplified and combined to give

$$a_1^2 \frac{\partial^2 u_j}{\partial x_j\, \partial x_i} - \frac{\partial^2 u_i}{\partial t^2} = 0$$

and that in terms of the velocity potential, $u_i = \partial\phi/\partial x_i$, this gives the general *acoustic equation* (cf. Exercise 3·2)

$$\frac{\partial^2\phi}{\partial x_j\,\partial x_j} \equiv \nabla^2\phi = \frac{1}{a_1{}^2}\frac{\partial^2\phi}{\partial t^2}$$

Hint: Follow the linearizing procedure outlined in Article 3·4. Show that the pressure is obtained from

$$p - p_1 = -\rho_1\frac{\partial\phi}{\partial t}$$

Show that, if the disturbance is produced by a body that is moving at constant speed U, the Galilean transformation, $x'_1 = x_1 + Ut$, $x'_2 = x_2$, $x'_3 = x_3$, transforms the above to the steady-state equations

$$\nabla^2\phi = \frac{U^2}{a_1{}^2}\frac{\partial^2\phi}{\partial x'_1{}^2}$$

$$C_p = -\frac{2}{U}\frac{\partial\phi}{\partial x'_1}$$

Compare with the results of Articles 8·2, 8·3 and 9·19.

Exercises, Chapter 8

8·1 Study the wave-shaped wall in a subsonic flow of finite lateral extent.

(*a*) Assume the flow is bounded at $x_2 = b$ by a solid wall, i.e., impose the boundary condition $v = 0$ at $x_2 = b$. Show that

$$\phi = \frac{\epsilon U}{\sqrt{1 - M_\infty{}^2}}\cos\alpha x_1\frac{\cosh\left[\alpha\sqrt{1 - M_\infty{}^2}(b - x_2)\right]}{\sinh\left[\alpha\sqrt{1 - M_\infty{}^2}\,b\right]}$$

(*b*) Assume the flow is bounded at $x_2 = b$ by a free surface ($p = p_\infty = \text{const.}$). Obtain ϕ. Discuss the solution, expecially with respect to the tunnel inteference problem. Check the limit $b \to \infty$ and the similarity behavior. Discuss the "mixed conditions" (slotted tunnel walls), which may be used to simulate free flight conditions.

8·2 Use the technique of the Fourier integral to build up the solution past an arbitrary, nonlifting two-dimensional body from the solution for the wave-shaped wall. Apply the technique specifically to the flow past an aerodynamic shape in a tunnel or free jet, using the solutions obtained in Exercise 8·1. (J. Cole, 1947.)

8·3 Consider flow past an infinite, "corrugated" cylinder of radius $R = R_0 + \epsilon\sin\alpha x_1$. The problem corresponds to the wave-shaped wall for the case of axially symmetrical flow (cf. Article 9·5). Find the velocities and pressures for subsonic and supersonic motion. (von Kármán, 1935.) *Hint:* The solution involves Bessel functions of order zero. The proper combination of functions in supersonic flow may be found by examining the asymptotic behavior of the solution for large r. It must have the two-dimensional form $f(x - r\sqrt{M_\infty{}^2 - 1})$.

8·4 Consider small perturbations of a plane flow for which $U = U(y)$ is a given function and the undisturbed pressure p_∞ is a constant. Find the differential equation for the pressure perturbation, i.e., put $p = p_\infty + p'$ where $p' = p'(x, y) \ll p_\infty$, and derive the equation for p':

$$\frac{\partial^2 p'}{\partial x^2} + \frac{M^2}{1 - M^2}\left(\frac{\partial}{\partial y}\frac{1}{M^2}\right)\frac{\partial p'}{\partial y} + \frac{1}{1 - M^2}\frac{\partial^2 p'}{\partial y^2} = 0\dagger$$

†Lighthill, 1950.

Try to apply the equation to the reflection and refraction of a weak shock wave by a shear layer for which U varies linearly from $U = U_1$ at $y = b$ to $U = U_2 < U_1$ at $y = 0$, with U_1, $U_2 > a_1$, a_2, i.e., supersonic flow throughout. (Use the equations of motion in natural coordinates, Eqs. 7·53 to 7·56, to derive the pressure equation.)

8·5 Find the lift and drag of a thin diamond section airfoil in supersonic flow by applying the momentum integrals (Eq. 7·16) to a control surface consisting of a rectangular box. *Hint:* Within linearized theory the disturbances are propagated unchanged along Mach lines. The integration need be carried out over only those portions of control surface which lie between its intersections with leading and trailing edge Mach waves. Sketch the momentum flux through the surface.

Exercises, Chapter 9

9·1 The condition $f(0) = 0$ for the applicability of Eq. 9·33 implies $dA/dx = 0$ at $x = 0$. Show that for meridian profiles which behave like $r =$ const. x^n, near the nose, this condition is fulfilled if $n > \frac{1}{2}$.

9·2 Show that the solution for flow over a slender axially symmetric body of profile section

$$R = \frac{2t}{L^2} x(L - x)$$

gives

$$u = \frac{\partial \phi}{\partial x} = -\frac{4Ut^2}{L^4}\left[3x(L - 3x) + (L^2 - 6Lx + 6x^2) \log \frac{2x}{R\sqrt{M^2 - 1}}\right]$$

(Ref. C·19). Compare the distribution of pressure and of drag with that of a two-dimensional airfoil having the same biconvex profile (cf. Eq. 4·9).

9·3 Von Kármán noted that the wave drag integral (Eq. 9·35*b*) is analogous to the classical incompressible formula for the induced drag of a lifting wing, of span b:

$$D_i = -\frac{\rho_\infty}{4\pi} \int_{-b/2}^{-b/2} \int_{-b/2}^{b/2} \Gamma'(\xi)\Gamma'(x) \log |x - \xi| \, dx \, d\xi$$

Identify the corresponding terms in the analogy. The result that minimum D_i for given lift occurs for elliptic lift distribution may be used to find the ogive profile whose wave drag is minimum for a given base area. Sketch the profile.

9·4 The nose of a cylindrical body has the profile $R = \epsilon x^{3/2}$, $0 \leq x \leq 1$. Show that the pressure distribution is given by

$$\frac{C_p}{\epsilon^2} = 6x \log \frac{2}{\epsilon\sqrt{M^2 - 1}} - 3x \log x - \frac{33}{4} x$$

Estimate the drag for $M = \sqrt{2}$ and $\epsilon = 0.1$. Compare with the value for a conical nose and for the optimum von Kármán ogive (Exercise 9·3).

9·5 Show that, if a cylindrical control surface, alined parallel to the flow, is used in the momentum theorem (Eqs. 7·16), and if the velocity perturbations u_x, u_r, u_θ are assumed small, then

$$\frac{D}{q} = \int_{A_2} \frac{u_r^2 + u_\theta^2 + (M^2 - 1)u_x^2}{U^2} \, dA_2 - \int_{A_3} \frac{u_x u_r}{U^2} \, dA_3$$

$$\frac{L}{q} = -2 \int_{A_2} \frac{u_r \sin\theta + u_\theta \cos\theta}{U} \, dA_2 + 2 \int_{A_3} \frac{u_x}{U} \sin\theta \, dA_3$$

where A_2 is the area of the cylindrical surface and A_3 is the area of its (downstream) base (cf. Exercise 8·5). Write these out for axially symmetric flow, and use them to check the results of Articles 9·12 and 9·17.

9·6 Use the ideas of slender body theory to compute the distribution of pressure on a flat wing of arbitrary plan form† and low aspect ratio (slender wing). *Hint:* The potential for the incompressible cross-flow past a two-dimensional flat plate is

$$\phi = U_c z \pm U_c \sqrt{\left(\frac{b}{2}\right)^2 - y^2}$$

Show that, if the chordwise distribution of span is $b = b(x)$, the chordwise lift loading is given by

$$\frac{dL}{dx} = \pi q \alpha b \frac{db}{dx}$$

and hence

$$C_L = \frac{\pi}{2} \alpha \mathit{R}$$

where $\mathit{R} = b_m^2/A$ is the aspect ratio, b_m the maximum span, at the trailing edge. Compare with $C_L = 4\alpha/\sqrt{M^2 - 1}$ for two-dimensional supersonic wings. The criterion for slender wings is that $\mathit{R}\sqrt{M^2 - 1} \ll 1$.

Show that for a slender delta wing the *spanwise* distribution of lift is

$$\frac{dL}{dy} = 4q\alpha \sqrt{\left(\frac{b_m}{2}\right)^2 - y^2}$$

This is an elliptic lift distribution, and hence the corresponding (vortex) drag due to lift is given by

$$C_{Di} = C_L^2/\pi \mathit{R}$$

Thus $D_i = \frac{1}{2}L\alpha$. One might have expected to find that the resultant force is normal to the flat wing, that is, $D_i \doteq L\alpha$. Why is this not so? *Hint:* The infinite suction at the leading edge provides enough thrust to cancel part of the streamwise component (R. T. Jones, *NACA Rep.* 835, 1946). Compare with the two-dimensional lifting flat wing, which has zero drag, and with the wing of large aspect ratio.

Exercise, Chapter 10

10·1 Consider flow at very high Mach number M_∞ past a slender body, characterized by a thickness ratio τ and a shape given by $B(x, y, z) = 0$. Let the shock surfaces present be described by $F(x, y, z) = 0$ (cf. Exercise 4·15). Introduce the independent variables

$$\bar{x} = x$$

$$\bar{y} = \frac{1}{\tau} y$$

$$\bar{z} = \frac{1}{\tau} z$$

† The span must be monotonically increasing downstream, so that there are no trailing side edges.

and the dependent variables

$$u = U[1 + \tau^2 \bar{u}(\bar{x}, \bar{y}, \bar{z})]$$

$$v = U\tau\bar{v}$$

$$w = U\tau\bar{w}$$

$$p = p_\infty \gamma M^2 \tau^2 \bar{p}$$

$$\rho = \rho_\infty \bar{\rho}$$

and boundary conditions

$$B = \bar{B}$$

$$F = \bar{F}$$

where $\bar{u}$, $\bar{v}$, etc., depend on $\bar{x}$, $\bar{y}$, $\bar{z}$ only. Insert these variables into the equations of motion, the boundary conditions, and the shock relations. Neglecting all terms in τ^2, we obtain the hypersonic, small-perturbation equations.

Derive the equations. Note specifically that only the y and z momentum equations are needed. Discuss the similarity laws and the meaning of the choice of variables. (Van Dyke, *NACA Tech. Note* 3173, 1954.)

Exercise, Chapter 11

11·1 (*a*) In flow over an airfoil the first appearance of sonic speed occurs locally at the point of minimum pressure. Show that this critical pressure coefficient is given by

$$C_{pc} = \frac{2}{\gamma M_\infty^2}\left[\left(\frac{2}{\gamma+1} + \frac{\gamma-1}{\gamma+1} M_{\infty c}^2\right)^{\gamma/(\gamma-1)} - 1\right]$$

Plot C_{pc} against $M_{\infty c}$.

(*b*) Find the critical Mach number M_c for an airfoil which in incompressible flow has a minimum pressure coefficient of -1.0. *Hint:* Plot the variation of C_p with M_∞, using the Prandtl-Glauert rule, on the plot of (*a*).

11·2 (*a*) Show that an approximate expression for the shock polar (Exercise 4.12), valid near sonic conditions, is

$$\left(\frac{v}{a^*}\right)^2 = \frac{\gamma+1}{2} \frac{(u' + u'')^2(u' - u'')}{a^{*3}}$$

where v, u', and u'' are perturbation velocities defined by $u_1 = a^* + u'$, $u_2 = a^* - u''$, $v_2 = v$.

(*b*) Show that an approximate expression for the Prandtl-Meyer relation (Eq. 4·21), valid near sonic conditions, is

$$\nu = \tfrac{2}{3}\sqrt{\gamma+1}\left(\frac{w'}{a^*}\right)^{3/2}$$

where w' is defined by $w = a^* + w'$.

(*c*) Use these relations to obtain an approximate expression for the pressures on a diamond airfoil at slightly supersonic speeds (cf. Fig. 11·3).

Exercises, Chapter 12

12·1 A simple expansion (or compression) wave produces a certain flow deflection $\Delta\theta$. Show that after interaction with another simple wave of the opposite family it still produces the same deflection $\Delta\theta$. This result for isentropic waves is not limited to *weak* waves (cf. Exercise 4·5).

12·2 Show that the last term in Eq. 12·13 may be written, for a perfect gas, in the form,

$$\frac{\cot\mu}{w^2}[T\,\Delta S - \Delta h_0] = \frac{\sin\mu\cos\mu}{\gamma}\left[\frac{\Delta S}{R} - \frac{\Delta h_0}{RT}\right]$$

12·3 Show that in the expansion over the shoulder of an axially symmetric body the pressure change is given by the two-dimensional, Prandtl-Meyer theory. *Hint:* Just at the shoulder, $\frac{1}{w}\frac{\partial w}{\partial s} \gg \frac{\sin\theta}{r}$.

12·4 Show how to design a supersonic nozzle by specifying the distribution of Mach number along the axis. The distribution may be arbitrary but should fair smoothly into the test section Mach number and should have the correct slope at the throat. An estimate of the latter may be obtained from Exercise 5·9; a more exact value may be obtained from a solution of the transonic equation for a particular throat.

12·5 Set up a computing procedure for one-dimensional nonsteady flow, following the method outlined in Article 12·4 (cf. Article 3·3 and Exercise 3·4). Compute the reflection of a centered expansion wave at the end of a closed tube.

Exercises, Chapter 13

13·1 Compute the skin friction for Couette flow, using the Sutherland viscosity law

$$\frac{\mu}{\mu_1} = \left(\frac{T}{T_1}\right)^{1/2}\frac{1+\theta}{1+\theta(T/T_1)}$$

where $\theta = 0.505$. Compare the result with that obtained for $\mu/\mu_1 = T/T_1$ and $\mu/\mu_1 = (T/T_1)^\omega$ where $\omega = 0.76$.

13·2 Compute and plot the temperature distribution in Couette flow for various values of M_∞ and q_w. For simplicity use $\mu \sim T$.

13·3 (*a*) Compute the velocity distribution $u/U = f(\eta)$ for Rayleigh's problem.

(*b*) Does the fluid become heated? What is the parameter that measures the increase of temperature of the fluid and hence determines the applicability of incompressible flow theory?

13·4 A good approximation for $g(\xi)$ in Eq. 13·49 is

$$g(\xi) = \sqrt{1-\xi^2}$$

(*a*) Using this expression and Eq. 13·49, find C_f for the case $\mu/\mu_\infty = T/T_\infty$, and also for the case with $Pr = 1$ and $\mu/\mu_\infty = (T/T_\infty)^\omega$. Compare the result with the exact solution, $C_f = 0.664/\sqrt{Re}$ for incompressible flow.

(*b*) Find the velocity profile $u/U = f\left(y\Big/\sqrt{\frac{\nu x}{U}}\right)$ for the case with $Pr = 1$ for various M_∞. *Hint:* Integrate the relation $dy = \mu(T)\,du/\tau$.

(*c*) Compute and plot the density profile.

13·5 Transform the expression for the displacement thickness

$$\delta^* = \int\left(1 - \frac{\rho u}{\rho_\infty U}\right)dy$$

in Crocco's variables. Show that if M_∞ is large, and $\mu \sim T$, then $\delta^* \sim M_\infty{}^2/\sqrt{Re}$.

13·6 Consider a flat-plate boundary layer for the cases

(*a*) $$c_p = c_p(T)$$
(*b*) $$h = h(p, T) \quad \text{(e.g., dissociation occurs)}$$

Are the equations of motion and the energy integral different from the simple case where $h = c_pT$? What are the essential differences in the flow? (Compare with Article 13·18.)

13·7 Integrate Eqs. 13·53 for a weak shock, for which

$$u = a^* + u', \quad u' \ll a^*$$

(*a*) Find the velocity profile $u' = u'(x)$.

(*b*) Compute the increase in entropy, using Eq. 13·58, and compare the result with the expression obtained from the usual shock wave "jump" equations (Article 2·13).

Within the above approximation, $\mu(T)$ can be replaced by $\mu(T^*) = \mu^* = \text{const.}$ The energy integral $\frac{1}{2}u^2 + c_pT = \text{const.}$ can also be used. Prove that the maximum shear $\tilde{\tau}_{\max}$ occurs at the point where $u = a^*$, i.e., at the sonic point.

13·8 Show that the thickness of a shock wave is of order

$$\epsilon = \frac{\tilde{\mu}^*}{\rho^* \Delta u}$$

where Δu is the "jump" in velocity across the shock. *Hint:* The rate of entropy production per unit mass is $(\tilde{\mu}/T)(du/dx)^2 \sim (u^*/T^*)(\Delta u/\epsilon)^2$. Compare this with the rate of entropy production determined by the jump equations (Article 2·13), using the mass flow and Eq. 2·51. Strictly, there is an additional contribution from $(k/T^2)(dT/dx)^2$. Show how this modifies the above result.

Estimate shock-wave thicknesses for $M_1 = 1.1$ and 1.01.

13·9 Define a shock-wave thickness ϵ by

$$\epsilon = \int_{-\infty}^{\infty} \tilde{\tau}/\tilde{\tau}_{\max}\, dx$$

Find ϵ explicitly for the weak wave of Exercise 13·7. Plot ϵ as a function of the Mach number.

13·10 Compare ϵ as defined in Exercise 13·9 with the following alternate definitions of shock thickness

$$\epsilon_1 = \frac{u_1 - u_2}{(du/dx)_{\max}}, \quad \epsilon_2 = \int_{-\infty}^{\infty} \left(\frac{u_1 - u}{u_1 - a^*} + \frac{u - u_2}{a^* - u_2}\right) dx$$

13·11 Show that for boundary-layer flow with pressure gradient Eq. 13·37 generalizes to

$$\left(\frac{\partial \tau}{\partial y}\right)_{\text{wall}} = \frac{dp}{dx}$$

(*a*) Discuss the curvature of the velocity profile near the wall for $M_\infty = 0$ and arbitrary M_∞, but $Pr = 1$ and $q_w = 0$.

(*b*) For incompressible flow $dp/dx = -\rho U\, dU/dx$. Assume that $U = \text{const.}\ x^n$ and that the velocity profiles are similar, i.e., that

$$\frac{u}{U} = f\left(\frac{y}{\delta}\right)$$

Show that for such a case

$$\delta \sim x^{(1-n)/2}$$

$$\tau_w \sim x^{(3n-1)/2}$$

(c) Show that the conditions near the wall for $q_w = 0$ and $dp/dx \neq 0$ are the same as for another case where $q_w \neq 0$ and $dp/dx = 0$. Determine the necessary relation between q_w and dp/dx.

13·12 A special solution of the laminar boundary-layer equation for low-speed flow past a flat plate is

$$\frac{u}{U} = f\left(\frac{y}{\delta}\right) \quad \text{with } \textit{constant } \delta$$

Show that this solution corresponds to uniform suction on the plate with a velocity v_0. Determine $f\left(\frac{y}{\delta}\right)$ and the relation between δ and v_0.

13·13 The velocity distribution of a turbulent boundary layer near a smooth wall has the form

$$\frac{u}{U} = f\left(\frac{u_\tau y}{\nu}\right); \quad u_\tau = \sqrt{\frac{\tau_w}{\rho}} = u_\tau(x)$$

Use the continuity equation to find v near the wall, and plot the streamline direction (D. Coles, 1955).

13·14 In the absence of a pressure gradient, the velocity of a laminar boundary layer near the wall has the form:

$$u = \frac{\tau_w}{\mu} y + \cdots$$

Write the boundary-layer equation, using this linear law as first approximation. (The equations have many applications, e.g., to the computation of τ_w (Weyl, 1941) and heat transfer (Lighthill, 1950).)

13·15 The boundary-layer equations for axially symmetrical flow are:

$$\frac{\partial \rho u r}{\partial x} + \frac{\partial \rho v r}{\partial y} = 0$$

$$\frac{\partial \rho u^2 r}{\partial x} + \frac{\partial \rho u v r}{\partial y} = -r\frac{\partial p}{\partial x} + \frac{\partial r\tau}{\partial y}$$

$$\frac{\partial \rho u J r}{\partial x} + \frac{\partial \rho v J r}{\partial y} = \frac{\partial}{\partial y}(\tau u r - r q)$$

where r is the radial coordinate, y the normal distance from the surface of the body; x lies along the body axis.

(a) Consider the laminar boundary layer on a straight cone at supersonic velocities. Here p = const. Find the momentum integral relation.

(b) Transform the momentum integral into Crocco's variables, and show that, for given x,

$$(\tau_w)_{\text{cone}} = \sqrt{3}(\tau_w)_{\text{flat plate}}$$

where both are evaluated for the same free-stream conditions. *Hint:* Bring the momentum integral relation into the form $\frac{1}{r\tau_w}\frac{d}{dx}\frac{r}{\tau_w}$ = const., integrate for the case of the cone, $r = \theta x$, and compare with the case $r \to \infty$.

Exercises, Chapter 14

14·1 To obtain an appreciation for the magnitudes involved, compute the fraction of molecules which at standard conditions have velocities in excess of one-tenth of the velocity of light.

14·2 Let N denote the number of perfect gas molecules in a volume V. Show that the number of pairs whose intermolecular distance is between s and $s + ds$ is $2\pi N^2 s^2\, ds/V$. Compute the mean distance.

14·3 Show that the number of molecules n which escape per unit time from a vessel, through a very small orifice of area ΔA, is given by

$$n = \Delta A \bar{c}/4$$

Use the result to show that the pressures p_1, p_2 and temperatures T_1, T_2 in two vessels connected by a small orifice are related by:

$$\frac{p_1}{p_2} = \sqrt{\frac{T_1}{T_2}}$$

("Small" implies that $\Delta A \ll \Lambda^2$.)

14·4 Let q_i and p_i be the generalized coordinates and momenta of a classical mechanical system. The probability $\phi(p_i, q_i)$ of finding the system in the element

$$d\Omega = dq_1 \cdots dq_n\, dp_1 \cdots dp_n$$

is given by the canonical distribution:

$$\phi(p_i, q_i)\, d\Omega = A \exp\left[-\frac{\epsilon(p_i, q_i)}{kT}\right] d\Omega$$

where ϵ is the energy of the system, A a constant to be determined by normalizing ϕ.

(1) Consider a system of perfect gas molecules in a gravitational field, for which

$$\epsilon = \frac{1}{2m}(p_1{}^2 + p_2{}^2 + p_3{}^2) + mgq_3$$

mgq_3 is the gravitational potential. Find the distribution of particles with altitude (q_3), and identify the result with the barometric formula obtained from hydrostatics.

(2) Consider molecules that exert forces upon each other. Let the potential of these forces be $x(s)$, where s denotes the distance between the centers. Show that the number of pairs of molecules at distance s apart now becomes (cf. Exercise 14·2)

$$\frac{2\pi N^2 s^2\, ds e^{-x/kT}}{V}$$

14·5 Use the result of Exercise 14·4(2) to compute the contribution to the virial from the intermolecular forces.

Consider a spherical container, as in Article 14·5, and show that $\overline{\mathbf{F} \cdot \mathbf{s}}$ due to the intermolecular forces differs from zero only in a zone near the container walls. Try to compute the second virial coefficient (cf. Eq. 1·85). The result is

$$b(T) = -\frac{2\pi}{3RT}\int_0^\infty s^3 \frac{\partial x}{\partial s} e^{-x(s)/RT}\, ds$$

or, after partial integration

$$b(T) = \frac{2\pi}{RT}\int_0^\infty s^2(e^{-x/RT} - 1)\, ds$$

Discuss $b(T)$ for rigid molecules, for which $x = 0$ for $s > D$ and $x = \infty$ for $s \leq D$.

Selected References

A. Thermodynamics and Physics of Gases

1. Born, M., *Natural Philosophy of Cause and Chance,* Oxford University Press, 1949.
2. Epstein, P. S., *Thermodynamics,* John Wiley & Sons, New York, 1937.
3. Fowler, R. H., and E. A. Guggenheim, *Statistical Thermodynamics,* Cambridge, 1952.
4. Guggenheim, E. A., *Thermodynamics,* North Holland Publishing Co., Amsterdam, 1950.
5. Hall, Newman A., *Thermodynamics of Fluid Flow,* Prentice-Hall, New York, 1951.
6. Jeans, J., *An Introduction to the Kinetic Theory of Gases,* Cambridge, 1946.
7. Rossini, F. D. (Editor), *Thermodynamics and Physics of Matter.* Vol. I of *High Speed Aerodynamics and Jet Propulsion,* Princeton, 1955.
8. Sommerfeld, A., *Lectures on Theoretical Physics,* Vol. V, *Thermodynamics and Statistical Mechanics,* Academic Press, New York, 1956.

B. Wave Motion

1. Courant, R., and K. O. Friedrichs, *Supersonic Flow and Shock Waves,* Interscience, New York, 1948.
2. Oswatitsch, K., *Gasdynamik,* Springer, Vienna, 1952.
3. Rayleigh, J. W. S., *The Theory of Sound,* Dover, New York, 1945.
4. Rudinger, G., *Wave Diagrams for Nonsteady Flow in Ducts,* Van Nostrand, New York, 1950.
5. Sommerfeld, A., *Lectures on Theoretical Physics,* Vol. II, *Mechanics of Deformable Bodies,* Academic Press, New York, 1950.

C. High-Speed Aerodynamics

1. Ackeret, J., *Gasdynamik, Handbuch der Physik,* Vol. 7, Chapter 5, Springer, Berlin, 1927.
2. Ames Research Staff, "Equations, Tables and Charts for Compressible Flow," *NACA Report* 1135 (1953).
3. Busemann, A., *Gasdynamik, Handbuch der Experimentalphysik,* Vol. 4, Part I, Akademischer Verlag, Leipzig, 1931.
4. Carrier, G. F. (Editor), *Foundations of High Speed Aerodynamics,* Dover, New York, 1951. (A collection of original papers. Also an extensive bibliography.)
5. Courant, R., and K. O. Friedrichs, *Supersonic Flow and Shock Waves,* Interscience Publishers, Inc., New York, 1948.
6. Emmons. H. W. (Editor), *Foundations of Gas Dynamics,* Vol. III of *High Speed Aerodynamics and Jet Propulsion,* Princeton, 1956.
7. Ferri, A., *Elements of Aerodynamics of Supersonic Flows,* Macmillan, New York, 1949.
8. Frankl, F. I., and E. A. Karpovich, *Gasdynamics of Thin Bodies,* Interscience Publishers, Inc., New York, 1953.
9. Guderley, G., *Advances in Applied Mechanics,* III (von Kármán and von Mises, Editors), Academic Press, New York, 1953.

10. Howarth, L. (Editor), *Modern Developments in Fluid Dynamics, High Speed Flow* (2 volumes), Oxford, 1953.
11. von Kármán, T., "The Problem of Resistance in Compressible Fluids," *Proc. 5th Volta Congress*, Rome (1935), pp. 255–264.
12. von Kármán, T., "Compressibility Effects in Aerodynamics," *J. Aeronaut. Sci., 8*, (1941), pp. 337–356.
13. von Kármán, T., "Supersonic Aerodynamics," *J. Aeronaut. Sci., 14* (1947), p. 373.
14. Kuethe, A. M., and J. D. Schetzer, *Foundations of Aerodynamics*, John Wiley & Sons, 1950.
15. Massachusetts Institute of Technology, Department of Electrical Engineering, Center of Analysis, "Tables of Supersonic Flow around Cones by the Staff of the Computing Section, under the direction of Zdenek Kopal," *Technical Report* 1, Cambridge (1947).
16. Oswatitsch, K., *Gasdynamik*, Springer, Vienna, 1952; Academic Press, New York, 1956.
17. Prandtl, L., "General Considerations on the Flow of Compressible Fluids," *NACA Tech. Mem.* No. 805 (1936) (translated from *Proc. 5th Volta Congress*, 1935).
18. Sauer, R., *Theoretische Einfuehrung in die Gasdynamik*, Springer, Berlin, 1943. (Reprinted in English by J. W. Edwards, Ann Arbor, Mich., 1947.)
19. Sears, W. R. (Editor), *Theory of High Speed Aerodynamics*, Vol. VI of *High Speed Aerodynamics and Jet Propulsion*, Princeton, 1954.
20. Shapiro, A. H., *The Dynamics and Thermodynamics of Compressible Fluid Flow*, 2 vols., Ronald, New York, 1953.
21. Tables of Compressible Airflow, Oxford, 1952.
22. Taylor, G. I., and J. W. Maccoll, *The Mechanics of Compressible Fluids* (Durand, *Aerodynamic Theory*, Vol. 3), California Institute of Technology, 1943.
23. Ward, G. N., *Linearized Theory of Steady High-Speed Flow*, Cambridge, 1955.

D. Viscous, Compressible Flow. Turbulence

1. Howarth, L. (Editor), *Modern Developments in Fluid Dynamics, High Speed Flow*, 2 volumes, Oxford, 1953.
2. Lin, C. C. (Editor), *Laminar Flows and Transition to Turbulence*, Vol. IV of *High Speed Aerodynamics and Jet Propulsion*, Princeton, 1957.
3. Prandtl, L., *Essentials of Fluid Dynamics*, Blackie, London; Hafner, New York, 1952.
4. Schlichting, H., *Boundary Layer Theory*, McGraw-Hill, New York, 1955.

E. Experimental Methods and Facilities

1. Howarth, L. (Editor), *Modern Developments in Fluid Dynamics, High Speed Flow*, Vol. II, Oxford, 1953.
2. Ladenburg, Lewis, Pease, and Taylor (Editors), *Physical Measurements in Gas Dynamics and Combustion.* Vol. IX of *High Speed Aerodynamics and Jet Propulsion*, Princeton, 1954.
3. Newell, Homer E., *High Altitude Rocket Research*, Academic Press, New York, 1953.
4. Pankhurst, R. C., and D. W. Holder, *Wind Tunnel Technique*, Pitman, London, 1952.
5. Pope, A., *Wind Tunnel Testing*, John Wiley & Sons, New York, 1954.

Tables

TABLE I

CRITICAL DATA AND CHARACTERISTIC TEMPERATURES FOR SEVERAL GASES

	p_c (atm)	T_c (°K)	R $\left(\frac{\text{atm cm}^3}{\text{deg gr}}\right)$	$\frac{p_c v_c}{RT_c}$	θ_v (°K)	θ_D (°K)
O_2	49.7	154.3	2.56	.292	2230	59,000
N_2	33.5	126.0	2.93	.292	3340	113,300
NO	65.0	179.1	2.73	.255	2690	75,500
H_2	12.8	33.2	40.7	.306	6100	52,400
He	2.26	5.2	20.5	.306	—	—
A	48.0	151.1	2.05	.291	—	—
CO_2	73.0	304.2	1.86	.280	954†	40,000‡

	θ_i (°K)
O	158,000
N	168,800
H	157,800
He	285,400
A	182,900
C	130,800

Boltzmann's constant $\mathbf{k} = 1.380 \times 10^{-16}$ erg deg^{-1}
Planck's constant $h = 6.625 \times 10^{-27}$ erg sec

† Lowest value.
‡ Approximate value.

TABLE II

FLOW PARAMETERS VERSUS M FOR SUBSONIC FLOW

M	p/p_0	ρ/ρ_0	T/T_0	a/a_0	A^*/A
.00	1.0000	1.0000	1.0000	1.0000	.00000
.01	.9999	1.0000	1.0000	1.0000	.01728
.02	.9997	.9998	.9999	1.0000	.03455
.03	.9994	.9996	.9998	.9999	.05181
.04	.9989	.9992	.9997	.9998	.06905
.05	.9983	.9988	.9995	.9998	.08627
.06	.9975	.9982	.9993	.9996	.1035
.07	.9966	.9976	.9990	.9995	.1206
.08	.9955	.9968	.9987	.9994	.1377
.09	.9944	.9960	.9984	.9992	.1548
.10	.9930	.9950	.9980	.9990	.1718
.11	.9916	.9940	.9976	.9988	.1887
.12	.9900	.9928	.9971	.9986	.2056
.13	.9883	.9916	.9966	.9983	.2224
.14	.9864	.9903	.9961	.9980	.2391
.15	.9844	.9888	.9955	.9978	.2557
.16	.9823	.9873	.9949	.9974	.2723
.17	.9800	.9857	.9943	.9971	.2887
.18	.9776	.9840	.9936	.9968	.3051
.19	.9751	.9822	.9928	.9964	.3213
.20	.9725	.9803	.9921	.9960	.3374
.21	.9697	.9783	.9913	.9956	.3534
.22	.9668	.9762	.9904	.9952	.3693
.23	.9638	.9740	.9895	.9948	.3851
.24	.9607	.9718	.9886	.9943	.4007
.25	.9575	.9694	.9877	.9938	.4162
.26	.9541	.9670	.9867	.9933	.4315
.27	.9506	.9645	.9856	.9928	.4467
.28	.9470	.9619	.9846	.9923	.4618
.29	.9433	.9592	.9835	.9917	.4767
.30	.9395	.9564	.9823	.9911	.4914
.31	.9355	.9535	.9811	.9905	.5059
.32	.9315	.9506	.9799	.9899	.5203
.33	.9274	.9476	.9787	.9893	.5345
.34	.9231	.9445	.9774	.9886	.5486

TABLE II (*Continued*)

FLOW PARAMETERS VERSUS M FOR SUBSONIC FLOW

M	p/p_0	ρ/ρ_0	T/T_0	a/a_0	A^*/A
.35	.9188	.9413	.9761	.9880	.5624
.36	.9143	.9380	.9747	.9873	.5761
.37	.9098	.9347	.9733	.9866	.5896
.38	.9052	.9313	.9719	.9859	.6029
.39	.9004	.9278	.9705	.9851	.6160
.40	.8956	.9243	.9690	.9844	.6289
.41	.8907	.9207	.9675	.9836	.6416
.42	.8857	.9170	.9659	.9828	.6541
.43	.8807	.9132	.9643	.9820	.6663
.44	.8755	.9094	.9627	.9812	.6784
.45	.8703	.9055	.9611	.9803	.6903
.46	.8650	.9016	.9594	.9795	.7019
.47	.8596	.8976	.9577	.9786	.7134
.48	.8541	.8935	.9560	.9777	.7246
.49	.8486	.8894	.9542	.9768	.7356
.50	.8430	.8852	.9524	.9759	.7464
.51	.8374	.8809	.9506	.9750	.7569
.52	.8317	.8766	.9487	.9740	.7672
.53	.8259	.8723	.9468	.9730	.7773
.54	.8201	.8679	.9449	.9721	.7872
.55	.8142	.8634	.9430	.9711	.7968
.56	.8082	.8589	.9410	.9701	.8063
.57	.8022	.8544	.9390	.9690	.8155
.58	.7962	.8498	.9370	.9680	.8244
.59	.7901	.8451	.9349	.9669	.8331
.60	.7840	.8405	.9328	.9658	.8416
.61	.7778	.8357	.9307	.9647	.8499
.62	.7716	.8310	.9286	.9636	.8579
.63	.7654	.8262	.9265	.9625	.8657
.64	.7591	.8213	.9243	.9614	.8732
.65	.7528	.8164	.9221	.9603	.8806
.66	.7465	.8115	.9199	.9591	.8877
.67	.7401	.8066	.9176	.9579	.8945
.68	.7338	.8016	.9153	.9567	.9012
.69	.7274	.7966	.9131	.9555	.9076

TABLE II (*Continued*)

FLOW PARAMETERS VERSUS M FOR SUBSONIC FLOW

M	p/p_0	ρ/ρ_0	T/T_0	a/a_0	A^*/A
.70	.7209	.7916	.9107	.9543	.9138
.71	.7145	.7865	.9084	.9531	.9197
.72	.7080	.7814	.9061	.9519	.9254
.73	.7016	.7763	.9037	.9506	.9309
.74	.6951	.7712	.9013	.9494	.9362
.75	.6886	.7660	.8989	.9481	.9412
.76	.6821	.7609	.8964	.9468	.9461
.77	.6756	.7557	.8940	.9455	.9507
.78	.6690	.7505	.8915	.9442	.9551
.79	.6625	.7452	.8890	.9429	.9592
.80	.6560	.7400	.8865	.9416	.9632
.81	.6495	.7347	.8840	.9402	.9669
.82	.6430	.7295	.8815	.9389	.9704
.83	.6365	.7242	.8789	.9375	.9737
.84	.6300	.7189	.8763	.9361	.9769
.85	.6235	.7136	.8737	.9347	.9797
.86	.6170	.7083	.8711	.9333	.9824
.87	.6106	.7030	.8685	.9319	.9849
.88	.6041	.6977	.8659	.9305	.9872
.89	.5977	.6924	.8632	.9291	.9893
.90	.5913	.6870	.8606	.9277	.9912
.91	.5849	.6817	.8579	.9262	.9929
.92	.5785	.6764	.8552	.9248	.9944
.93	.5721	.6711	.8525	.9233	.9958
.94	.5658	.6658	.8498	.9218	.9969
.95	.5595	.6604	.8471	.9204	.9979
.96	.5532	.6551	.8444	.9189	.9986
.97	.5469	.6498	.8416	.9174	.9992
.98	.5407	.6445	.8389	.9159	.9997
.99	.5345	.6392	.8361	.9144	.9999
1.00	.5283	.6339	.8333	.9129	1.0000

Numerical values taken from NACA TN 1428, courtesy of the National Advisory Committee for Aeronautics. Setup from A. M. Kuethe and J. D. Schetzer, *Foundations of Aerodynamics*, John Wiley & Sons, New York, 1950.

TABLE III

Flow Parameters versus M for Supersonic Flow

M	$\frac{p}{p_0}$	$\frac{\rho}{\rho_0}$	$\frac{T}{T_0}$	$\frac{a}{a_0}$	$\frac{A^*}{A}$	$\frac{\frac{\rho}{2}V^2}{p_0}$	θ
1.00	.5283	.6339	.8333	.9129	1.0000	.3698	0
1.01	.5221	.6287	.8306	.9113	.9999	.3728	.04473
1.02	.5160	.6234	.8278	.9098	.9997	.3758	.1257
1.03	.5099	.6181	.8250	.9083	.9993	.3787	.2294
1.04	.5039	.6129	.8222	.9067	.9987	.3815	.3510
1.05	.4979	.6077	.8193	.9052	.9980	.3842	.4874
1.06	.4919	.6024	.8165	.9036	.9971	.3869	.6367
1.07	.4860	.5972	.8137	.9020	.9961	.3895	.7973
1.08	.4800	.5920	.8108	.9005	.9949	.3919	.9680
1.09	.4742	.5869	.8080	.8989	.9936	.3944	1.148
1.10	.4684	.5817	.8052	.8973	.9921	.3967	1.336
1.11	.4626	.5766	.8023	.8957	.9905	.3990	1.532
1.12	.4568	.5714	.7994	.8941	.9888	.4011	1.735
1.13	.4511	.5663	.7966	.8925	.9870	.4032	1.944
1.14	.4455	.5612	.7937	.8909	.9850	.4052	2.160
1.15	.4398	.5562	.7908	.8893	.9828	.4072	2.381
1.16	.4343	.5511	.7879	.8877	.9806	.4090	2.607
1.17	.4287	.5461	.7851	.8860	.9782	.4108	2.839
1.18	.4232	.5411	.7822	.8844	.9758	.4125	3.074
1.19	.4178	.5361	.7793	.8828	.9732	.4141	3.314
1.20	.4124	.5311	.7764	.8811	.9705	.4157	3.558
1.21	.4070	.5262	.7735	.8795	.9676	.4171	3.806
1.22	.4017	.5213	.7706	.8778	.9647	.4185	4.057
1.23	.3964	.5164	.7677	.8762	.9617	.4198	4.312
1.24	.3912	.5115	.7648	.8745	.9586	.4211	4.569
1.25	.3861	.5067	.7619	.8729	.9553	.4223	4.830
1.26	.3809	.5019	.7590	.8712	.9520	.4233	5.093
1.27	.3759	.4971	.7561	.8695	.9486	.4244	5.359
1.28	.3708	.4923	.7532	.8679	.9451	.4253	5.627
1.29	.3658	.4876	.7503	.8662	.9415	.4262	5.898
1.30	.3609	.4829	.7474	.8645	.9378	.4270	6.170
1.31	.3560	.4782	.7445	.8628	.9341	.4277	6.445
1.32	.3512	.4736	.7416	.8611	.9302	.4283	6.721
1.33	.3464	.4690	.7387	.8595	.9263	.4289	7.000
1.34	.3417	.4644	.7358	.8578	.9223	.4294	7.279

TABLE III (*Continued*)

FLOW PARAMETERS VERSUS M FOR SUPERSONIC FLOW

M	$\frac{p}{p_0}$	$\frac{\rho}{\rho_0}$	$\frac{T}{T_0}$	$\frac{a}{a_0}$	$\frac{A^*}{A}$	$\frac{\frac{\rho}{2}V^2}{p_0}$	θ
1.35	.3370	.4598	.7329	.8561	.9182	.4299	7.561
1.36	.3323	.4553	.7300	.8544	.9141	.4303	7.844
1.37	.3277	.4508	.7271	.8527	.9099	.4306	8.128
1.38	.3232	.4463	.7242	.8510	.9056	.4308	8.413
1.39	.3187	.4418	.7213	.8493	.9013	.4310	8.699
1.40	.3142	.4374	.7184	.8476	.8969	.4311	8.987
1.41	.3098	.4330	.7155	.8459	.8925	.4312	9.276
1.42	.3055	.4287	.7126	.8442	.8880	.4312	9.565
1.43	.3012	.4244	.7097	.8425	.8834	.4311	9.855
1.44	.2969	.4201	.7069	.8407	.8788	.4310	10.15
1.45	.2927	.4158	.7040	.8390	.8742	.4308	10.44
1.46	.2886	.4116	.7011	.8373	.8695	.4306	10.73
1.47	.2845	.4074	.6982	.8356	.8647	.4303	11.02
1.48	.2804	.4032	.6954	.8339	.8599	.4299	11.32
1.49	.2764	.3991	.6925	.8322	.8551	.4295	11.61
1.50	.2724	.3950	.6897	.8305	.8502	.4290	11.91
1.51	.2685	.3909	.6868	.8287	.8453	.4285	12.20
1.52	.2646	.3869	.6840	.8270	.8404	.4279	12.49
1.53	.2608	.3829	.6811	.8253	.8354	.4273	12.79
1.54	.2570	.3789	.6783	.8236	.8304	.4266	13.09
1.55	.2533	.3750	.6754	.8219	.8254	.4259	13.38
1.56	.2496	.3710	.6726	.8201	.8203	.4252	13.68
1.57	.2459	.3672	.6698	.8184	.8152	.4243	13.97
1.58	.2423	.3633	.6670	.8167	.8101	.4235	14.27
1.59	.2388	.3595	.6642	.8150	.8050	.4226	14.56
1.60	.2353	.3557	.6614	.8133	.7998	.4216	14.86
1.61	.2318	.3520	.6586	.8115	.7947	.4206	15.16
1.62	.2284	.3483	.6558	.8098	.7895	.4196	15.45
1.63	.2250	.3446	.6530	.8081	.7843	.4185	15.75
1.64	.2217	.3409	.6502	.8064	.7791	.4174	16.04
1.65	.2184	.3373	.6475	.8046	.7739	.4162	16.34
1.66	.2151	.3337	.6447	.8029	.7686	.4150	16.63
1.67	.2119	.3302	.6419	.8012	.7634	.4138	16.93
1.68	.2088	.3266	.6392	.7995	.7581	.4125	17.22
1.69	.2057	.3232	.6364	.7978	.7529	.4112	17.52

TABLE III (*Continued*)

Flow Parameters versus M for Supersonic Flow

M	$\frac{p}{p_0}$	$\frac{\rho}{\rho_0}$	$\frac{T}{T_0}$	$\frac{a}{a_0}$	$\frac{A^*}{A}$	$\frac{\frac{\rho}{2}V^2}{p_0}$	θ
1.70	.2026	.3197	.6337	.7961	.7476	.4098	17.81
1.71	.1996	.3163	.6310	.7943	.7423	.4086	18.10
1.72	.1966	.3129	.6283	.7926	.7371	.4071	18.40
1.73	.1936	.3095	.6256	.7909	.7318	.4056	18.69
1.74	.1907	.3062	.6229	.7892	.7265	.4041	18.98
1.75	.1878	.3029	.6202	.7875	.7212	.4026	19.27
1.76	.1850	.2996	.6175	.7858	.7160	.4011	19.56
1.77	.1822	.2964	.6148	.7841	.7107	.3996	19.86
1.78	.1794	.2932	.6121	.7824	.7054	.3980	20.15
1.79	.1767	.2900	.6095	.7807	.7002	.3964	20.44
1.80	.1740	.2868	.6068	.7790	.6949	.3947	20.73
1.81	.1714	.2837	.6041	.7773	.6897	.3931	21.01
1.82	.1688	.2806	.6015	.7756	.6845	.3914	21.30
1.83	.1662	.2776	.5989	.7739	.6792	.3897	21.59
1.84	.1637	.2745	.5963	.7722	.6740	.3879	21.88
1.85	.1612	.2715	.5936	.7705	.6688	.3862	22.16
1.86	.1587	.2686	.5910	.7688	.6636	.3844	22.45
1.87	.1563	.2656	.5884	.7671	.6584	.3826	22.73
1.88	.1539	.2627	.5859	.7654	.6533	.3808	23.02
1.89	.1516	.2598	.5833	.7637	.6481	.3790	23.30
1.90	.1492	.2570	.5807	.7620	.6430	.3771	23.59
1.91	.1470	.2542	.5782	.7604	.6379	.3753	23.87
1.92	.1447	.2514	.5756	.7587	.6328	.3734	24.15
1.93	.1425	.2486	.5731	.7570	.6277	.3715	24.43
1.94	.1403	.2459	.5705	.7553	.6226	.3696	24.71
1.95	.1381	.2432	.5680	.7537	.6175	.3677	24.99
1.96	.1360	.2405	.5655	.7520	.6125	.3657	25.27
1.97	.1339	.2378	.5630	.7503	.6075	.3638	25.55
1.98	.1318	.2352	.5605	.7487	.6025	.3618	25.83
1.99	.1298	.2326	.5580	.7470	.5975	.3598	26.10
2.00	.1278	.2300	.5556	.7454	.5926	.3579	26.38
2.01	.1258	.2275	.5531	.7437	.5877	.3559	26.66
2.02	.1239	.2250	.5506	.7420	.5828	.3539	26.93
2.03	.1220	.2225	.5482	.7404	.5779	.3518	27.20
2.04	.1201	.2200	.5458	.7388	.5730	.3498	27.48

TABLE III (*Continued*)

FLOW PARAMETERS VERSUS M FOR SUPERSONIC FLOW

M	$\frac{p}{p_0}$	$\frac{\rho}{\rho_0}$	$\frac{T}{T_0}$	$\frac{a}{a_0}$	$\frac{A^*}{A}$	$\frac{\frac{\rho}{2}V^2}{p_0}$	θ
2.05	.1182	.2176	.5433	.7371	.5682	.3478	27.75
2.06	.1164	.2152	.5409	.7355	.5634	.3458	28.02
2.07	.1146	.2128	.5385	.7338	.5586	.3437	28.29
2.08	.1128	.2104	.5361	.7322	.5538	.3417	28.56
2.09	.1111	.2081	.5337	.7306	.5491	.3396	28.83
2.10	.1094	.2058	.5313	.7289	.5444	.3376	29.10
2.11	.1077	.2035	.5290	.7273	.5397	.3355	29.36
2.12	.1060	.2013	.5266	.7257	.5350	.3334	29.63
2.13	.1043	.1990	.5243	.7241	.5304	.3314	29.90
2.14	.1027	.1968	.5219	.7225	.5258	.3293	30.16
2.15	.1011	.1946	.5196	.7208	.5212	.3272	30.43
2.16	.09956	.1925	.5173	.7192	.5167	.3252	30.69
2.17	.09802	.1903	.5150	.7176	.5122	.3231	30.95
2.18	.09650	.1882	.5127	.7160	.5077	.3210	31.21
2.19	.09500	.1861	.5104	.7144	.5032	.3189	31.47
2.20	.09352	.1841	.5081	.7128	.4988	.3169	31.73
2.21	.09207	.1820	.5059	.7112	.4944	.3148	31.99
2.22	.09064	.1800	.5036	.7097	.4900	.3127	32.25
2.23	.08923	.1780	.5014	.7081	.4856	.3106	32.51
2.24	.08785	.1760	.4991	.7065	.4813	.3085	32.76
2.25	.08648	.1740	.4969	.7049	.4770	.3065	33.02
2.26	.08514	.1721	.4947	.7033	.4727	.3044	33.27
2.27	.08382	.1702	.4925	.7018	.4685	.3023	33.53
2.28	.08252	.1683	.4903	.7002	.4643	.3003	33.78
2.29	.08123	.1664	.4881	.6986	.4601	.2982	34.03
2.30	.07997	.1646	.4859	.6971	.4560	.2961	34.28
2.31	.07873	.1628	.4837	.6955	.4519	.2941	34.53
2.32	.07751	.1609	.4816	.6940	.4478	.2920	34.78
2.33	.07631	.1592	.4794	.6924	.4437	.2900	35.03
2.34	.07512	.1574	.4773	.6909	.4397	.2879	35.28
2.35	.07396	.1556	.4752	.6893	.4357	.2859	35.53
2.36	.07281	.1539	.4731	.6878	.4317	.2839	35.77
2.37	.07168	.1522	.4709	.6863	.4278	.2818	36.02
2.38	.07057	.1505	.4688	.6847	.4239	.2798	36.26
2.39	.06948	.1488	.4668	.6832	.4200	.2778	36.50

TABLE III (*Continued*)

FLOW PARAMETERS VERSUS M FOR SUPERSONIC FLOW

M	$\frac{p}{p_0}$	$\frac{\rho}{\rho_0}$	$\frac{T}{T_0}$	$\frac{a}{a_0}$	$\frac{A^*}{A}$	$\frac{\frac{\rho}{2}V^2}{p_0}$	θ
2.40	.06840	.1472	.4647	.6817	.4161	.2758	36.75
2.41	.06734	.1456	.4626	.6802	.4123	.2738	36.99
2.42	.06630	.1439	.4606	.6786	.4085	.2718	37.23
2.43	.06527	.1424	.4585	.6771	.4048	.2698	37.47
2.44	.06426	.1408	.4565	.6756	.4010	.2678	37.71
2.45	.06327	.1392	.4544	.6741	.3973	.2658	37.95
2.46	.06229	.1377	.4524	.6726	.3937	.2639	38.18
2.47	.06133	.1362	.4504	.6711	.3900	.2619	38.42
2.48	.06038	.1347	.4484	.6696	.3864	.2599	38.66
2.49	.05945	.1332	.4464	.6681	.3828	.2580	38.89
2.50	.05853	.1317	.4444	.6667	.3793	.2561	39.12
2.51	.05762	.1302	.4425	.6652	.3757	.2541	39.36
2.52	.05674	.1288	.4405	.6637	.3722	.2522	39.59
2.53	.05586	.1274	.4386	.6622	.3688	.2503	39.82
2.54	.05500	.1260	.4366	.6608	.3653	.2484	40.05
2.55	.05415	.1246	.4347	.6593	.3619	.2465	40.28
2.56	.05332	.1232	.4328	.6579	.3585	.2446	40.51
2.57	.05250	.1218	.4309	.6564	.3552	.2427	40.75
2.58	.05169	.1205	.4289	.6549	.3519	.2409	40.96
2.59	.05090	.1192	.4271	.6535	.3486	.2390	41.19
2.60	.05012	.1179	.4252	.6521	.3453	.2371	41.41
2.61	.04935	.1166	.4233	.6506	.3421	.2353	41.64
2.62	.04859	.1153	.4214	.6492	.3389	.2335	41.86
2.63	.04784	.1140	.4196	.6477	.3357	.2317	42.09
2.64	.04711	.1128	.4177	.6463	.3325	.2298	42.31
2.65	.04639	.1115	.4159	.6449	.3294	.2280	42.53
2.66	.04568	.1103	.4141	.6435	.3263	.2262	42.75
2.67	.04498	.1091	.4122	.6421	.3232	.2245	42.97
2.68	.04429	.1079	.4104	.6406	.3202	.2227	43.19
2.69	.04362	.1067	.4086	.6392	.3172	.2209	43.40
2.70	.04295	.1056	.4068	.6378	.3142	.2192	43.62
2.71	.04229	.1044	.4051	.6364	.3112	.2174	43.84
2.72	.04165	.1033	.4033	.6350	.3083	.2157	44.05
2.73	.04102	.1022	.4015	.6337	.3054	.2140	44.27
2.74	.04039	.1010	.3998	.6323	.3025	.2123	44.48

TABLE III (*Continued*)

Flow Parameters versus M for Supersonic Flow

M	$\frac{p}{p_0}$	$\frac{\rho}{\rho_0}$	$\frac{T}{T_0}$	$\frac{a}{a_0}$	$\frac{A^*}{A}$	$\frac{\frac{\rho}{2}V^2}{p_0}$	θ
2.75	.03978	.09994	.3980	.6309	.2996	.2106	44.69
2.76	.03917	.09885	.3963	.6295	.2968	.2089	44.91
2.77	.03858	.09778	.3945	.6281	.2940	.2072	45.12
2.78	.03799	.09671	.3928	.6268	.2912	.2055	45.33
2.79	.03742	.09566	.3911	.6254	.2884	.2039	45.54
2.80	.03685	.09463	.3894	.6240	.2857	.2022	45.75
2.81	.03629	.09360	.3877	.6227	.2830	.2006	45.95
2.82	.03574	.09259	.3860	.6213	.2803	.1990	46.16
2.83	.03520	.09158	.3844	.6200	.2777	.1973	46.37
2.84	.03467	.09059	.3827	.6186	.2750	.1957	46.57
2.85	.03415	.08962	.3810	.6173	.2724	.1941	46.78
2.86	.03363	.08865	.3794	.6159	.2698	.1926	46.98
2.87	.03312	.08769	.3777	.6146	.2673	.1910	47.19
2.88	.03263	.08675	.3761	.6133	.2648	.1894	47.39
2.89	.03213	.08581	.3745	.6119	.2622	.1879	47.59
2.90	.03165	.08489	.3729	.6106	.2598	.1863	47.79
2.91	.03118	.08398	.3712	.6093	.2573	.1848	47.99
2.92	.03071	.08307	.3696	.6080	.2549	.1833	48.19
2.93	.03025	.08218	.3681	.6067	.2524	.1818	48.39
2.94	.02980	.08130	.3665	.6054	.2500	.1803	48.59
2.95	.02935	.08043	.3649	.6041	.2477	.1788	48.78
2.96	.02891	.07957	.3633	.6028	.2453	.1773	48.98
2.97	.02848	.07872	.3618	.6015	.2430	.1758	49.18
2.98	.02805	.07788	.3602	.6002	.2407	.1744	49.37
2.99	.02764	.97705	.3587	.5989	.2384	.1729	49.56
3.00	.02722	.07623	.3571	.5976	.2362	.1715	49.76
3.01	.02682	.07541	.3556	.5963	.2339	.1701	49.95
3.02	.02642	.07461	.3541	.5951	.2317	.1687	50.14
3.03	.02603	.07382	.3526	.5938	.2295	.1673	50.33
3.04	.02564	.07303	.3511	.5925	.2273	.1659	50.52
3.05	.02526	.07226	.3496	.5913	.2252	.1645	50.71
3.06	.02489	.07149	.3481	.5900	.2230	.1631	50.90
3.07	.02452	.07074	.3466	.5887	.2209	.1618	51.09
3.08	.02416	.06999	.3452	.5875	.2188	.1604	51.28
3.09	.02380	.06925	.3437	.5862	.2168	.1591	51.46

TABLE III (*Continued*)

FLOW PARAMETERS VERSUS M FOR SUPERSONIC FLOW

M	$\frac{p}{p_0}$	$\frac{\rho}{\rho_0}$	$\frac{T}{T_0}$	$\frac{a}{a_0}$	$\frac{A^*}{A}$	$\frac{\frac{\rho}{2}V^2}{p_0}$	θ
3.10	.02345	.06852	.3422	.5850	.2147	.1577	51.65
3.11	.02310	.06779	.3408	.5838	.2127	.1564	51.84
3.12	.02276	.06708	.3393	.5825	.2107	.1551	52.02
3.13	.02243	.06637	.3379	.5813	.2087	.1538	52.20
3.14	.02210	.06568	.3365	.5801	.2067	.1525	52.39
3.15	.02177	.06499	.3351	.5788	.2048	.1512	52.57
3.16	.02146	.06430	.3337	.5776	.2028	.1500	52.75
3.17	.02114	.06363	.3323	.5764	.2009	.1487	52.93
3.18	.02083	.06296	.3309	.5752	.1990	.1475	53.11
3.19	.02053	.06231	.3295	.5740	.1971	.1462	53.29
3.20	.02023	.06165	.3281	.5728	.1953	.1450	53.47
3.21	.01993	.06101	.3267	.5716	.1934	.1438	53.65
3.22	.01964	.06037	.3253	.5704	.1916	.1426	53.83
3.23	.01936	.05975	.3240	.5692	.1898	.1414	54.00
3.24	.01908	.05912	.3226	.5680	.1880	.1402	54.18
3.25	.01880	.05851	.3213	.5668	.1863	.1390	54.35
3.26	.01853	.05790	.3199	.5656	.1845	.1378	54.53
3.27	.01826	.05730	.3186	.5645	.1828	.1367	54.71
3.28	.01799	.05671	.3173	.5633	.1810	.1355	54.88
3.29	.01773	.05612	.3160	.5621	.1793	.1344	55.05
3.30	.01748	.05554	.3147	.5609	.1777	.1332	55.22
3.31	.01722	.05497	.3134	.5598	.1760	.1321	55.39
3.32	.01698	.05440	.3121	.5586	.1743	.1310	55.56
3.33	.01673	.05384	.3108	.5575	.1727	.1299	55.73
3.34	.01649	.05329	.3095	.5563	.1711	.1288	55.90
3.35	.01625	.05274	.3082	.5552	.1695	.1277	56.07
3.36	.01602	.05220	.3069	.5540	.1679	.1266	56.24
3.37	.01579	.05166	.3057	.5529	.1663	.1255	56.41
3.38	.01557	.05113	.3044	.5517	.1648	.1245	56.58
3.39	.01534	.05061	.3032	.5506	.1632	.1234	56.75
3.40	.01513	.05009	.3019	.5495	.1617	.1224	56.91
3.41	.01491	.04958	.3007	.5484	.1602	.1214	57.07
3.42	.01470	.04908	.2995	.5472	.1587	.1203	57.24
3.43	.01449	.04858	.2982	.5461	.1572	.1193	57.40
3.44	.01428	.04808	.2970	.5450	.1558	.1183	57.56

TABLE III (*Continued*)

FLOW PARAMETERS VERSUS M FOR SUPERSONIC FLOW

M	$\frac{p}{p_0}$	$\frac{\rho}{\rho_0}$	$\frac{T}{T_0}$	$\frac{a}{a_0}$	$\frac{A^*}{A}$	$\frac{\frac{\rho}{2}V^2}{p_0}$	θ
3.45	.01408	.04759	.2958	.5439	.1543	.1173	57.73
3.46	.01388	.04711	.2946	.5428	.1529	.1163	57.89
3.47	.01368	.04663	.2934	.5417	.1515	.1153	58.05
3.48	.01349	.04616	.2922	.5406	.1501	.1144	58.21
3.49	.01330	.04569	.2910	.5395	.1487	.1134	58.37
3.50	.01311	.04523	.2899	.5384	.1473	.1124	58.53
3.60	.01138	.04089	.2784	.5276	.1342	.1033	60.09
3.70	9.903×10^{-3}	.03702	.2675	.5172	.1224	.09490	61.60
3.80	8.629×10^{-3}	.03355	.2572	.5072	.1117	.08722	63.04
3.90	7.532×10^{-3}	.03044	.2474	.4974	.1021	.08019	64.44
4.00	6.586×10^{-3}	.02766	.2381	.4880	.09329	.07376	65.78
4.10	5.769×10^{-3}	.02516	.2293	.4788	.08536	.06788	67.08
4.20	5.062×10^{-3}	.02292	.2208	.4699	.07818	.06251	68.33
4.30	4.449×10^{-3}	.02090	.2129	.4614	.07166	.05759	69.54
4.40	3.918×10^{-3}	.01909	.2053	.4531	.06575	.05309	70.71
4.50	3.455×10^{-3}	.01745	.1980	.4450	.06038	.04898	71.83
4.60	3.053×10^{-3}	.01597	.1911	.4372	.05550	.04521	72.92
4.70	2.701×10^{-3}	.01464	.1846	.4296	.05107	.04177	73.97
4.80	2.394×10^{-3}	.01343	.1783	.4223	.04703	.03861	74.99
4.90	2.126×10^{-3}	.01233	.1724	.4152	.04335	.03572	75.97
5.00	1.890×10^{-3}	.01134	.1667	.4082	.04000	.03308	76.92
6.00	6.334×10^{-4}	5.194×10^{-3}	.1220	.3492	.01880	.01596	84.96
7.00	2.416×10^{-4}	2.609×10^{-3}	.09259	.3043	9.602×10^{-3}	8.285×10^{-3}	90.97

TABLE III (*Continued*)

FLOW PARAMETERS VERSUS M FOR SUPERSONIC FLOW

M	$\frac{p}{p_0}$	$\frac{\rho}{\rho_0}$	$\frac{T}{T_0}$	$\frac{a}{a_0}$	$\frac{A^*}{A}$	$\frac{\frac{\rho}{2}V^2}{p_0}$	θ
8.00	1.024×10^{-4}	1.414×10^{-3}	.07246	.2692	5.260×10^{-3}	4.589×10^{-3}	95.62
9.00	4.739×10^{-5}	8.150×10^{-4}	.05814	.2411	3.056×10^{-3}	2.687×10^{-3}	99.32
10.00	2.356×10^{-5}	4.948×10^{-4}	.04762	.2182	1.866×10^{-3}	1.649×10^{-3}	102.3
100.00	2.790×10^{-12}	5.583×10^{-9}	4.998×10^{-4}	.02236	2.157×10^{-8}	1.953×10^{-8}	127.6
∞	0	0	0	0	0	0	130.5

Numerical values taken from NACA TN 1428, courtesy of the National Advisory Committee for Aeronautics. Setup from A. M. Kuethe and J. D. Schetzer, *Foundations of Aerodynamics*, John Wiley & Sons, New York, 1950.

TABLE IV

PARAMETERS FOR SHOCK FLOW

M_{1n}	p_2/p_1	ρ_2/ρ_1	T_2/T_1	a_2/a_1	p_2^0/p_1^0	M_2 for Normal Shocks Only
1.00	1.000	1.000	1.000	1.000	1.0000	1.0000
1.01	1.023	1.017	1.007	1.003	1.0000	.9901
1.02	1.047	1.033	1.013	1.007	1.0000	.9805
1.03	1.071	1.050	1.020	1.010	1.0000	.9712
1.04	1.095	1.067	1.026	1.013	.9999	.9620
1.05	1.120	1.084	1.033	1.016	.9999	.9531
1.06	1.144	1.101	1.039	1.019	.9998	.9444
1.07	1.169	1.118	1.046	1.023	.9996	.9360
1.08	1.194	1.135	1.052	1.026	.9994	.9277
1.09	1.219	1.152	1.059	1.029	.9992	.9196
1.10	1.245	1.169	1.065	1.032	.9989	.9118
1.11	1.271	1.186	1.071	1.035	.9986	.9041
1.12	1.297	1.203	1.078	1.038	.9982	.8966
1.13	1.323	1.221	1.084	1.041	.9978	.8892
1.14	1.350	1.238	1.090	1.044	.9973	.8820
1.15	1.376	1.255	1.097	1.047	.9967	.8750
1.16	1.403	1.272	1.103	1.050	.9961	.8682
1.17	1.430	1.290	1.109	1.053	.9953	.8615
1.18	1.458	1.307	1.115	1.056	.9946	.8549
1.19	1.485	1.324	1.122	1.059	.9937	.8485
1.20	1.513	1.342	1.128	1.062	.9928	.8422
1.21	1.541	1.359	1.134	1.065	.9918	.8360
1.22	1.570	1.376	1.141	1.068	.9907	.8300
1.23	1.598	1.394	1.147	1.071	.9896	.8241
1.24	1.627	1.411	1.153	1.074	.9884	.8183
1.25	1.656	1.429	1.159	1.077	.9871	.8126
1.26	1.686	1.446	1.166	1.080	.9857	.8071
1.27	1.715	1.463	1.172	1.083	.9842	.8016
1.28	1.745	1.481	1.178	1.085	.9827	.7963
1.29	1.775	1.498	1.185	1.088	.9811	.7911
1.30	1.805	1.516	1.191	1.091	.9794	.7860
1.31	1.835	1.533	1.197	1.094	.9776	.7809
1.32	1.866	1.551	1.204	1.097	.9758	.7760
1.33	1.897	1.568	1.210	1.100	.9738	.7712
1.34	1.928	1.585	1.216	1.103	.9718	.7664

TABLE IV (*Continued*)

PARAMETERS FOR SHOCK FLOW

M_{1n}	p_2/p_1	ρ_2/ρ_1	T_2/T_1	a_2/a_1	p_2^0/p_1^0	M_2 for Normal Shocks Only
1.35	1.960	1.603	1.223	1.106	.9697	.7618
1.36	1.991	1.620	1.229	1.109	.9676	.7572
1.37	2.023	1.638	1.235	1.111	.9653	.7527
1.38	2.055	1.655	1.242	1.114	.9630	.7483
1.39	2.087	1.672	1.248	1.117	.9606	.7440
1.40	2.120	1.690	1.255	1.120	.9582	.7397
1.41	2.153	1.707	1.261	1.123	.9557	.7355
1.42	2.186	1.724	1.268	1.126	.9531	.7314
1.43	2.219	1.742	1.274	1.129	.9504	.7274
1.44	2.253	1.759	1.281	1.132	.9476	.7235
1.45	2.286	1.776	1.287	1.135	.9448	.7196
1.46	2.320	1.793	1.294	1.137	.9420	.7157
1.47	2.354	1.811	1.300	1.140	.9390	.7120
1.48	2.389	1.828	1.307	1.143	.9360	.7083
1.49	2.423	1.845	1.314	1.146	.9329	.7047
1.50	2.458	1.862	1.320	1.149	.9298	.7011
1.51	2.493	1.879	1.327	1.152	.9266	.6976
1.52	2.529	1.896	1.334	1.155	.9233	.6941
1.53	2.564	1.913	1.340	1.158	.9200	.6907
1.54	2.600	1.930	1.347	1.161	.9166	.6874
1.55	2.636	1.947	1.354	1.164	.9132	.6841
1.56	2.673	1.964	1.361	1.166	.9097	.6809
1.57	2.709	1.981	1.367	1.169	.9061	.6777
1.58	2.746	1.998	1.374	1.172	.9026	.6746
1.59	2.783	2.015	1.381	1.175	.8989	.6715
1.60	2.820	2.032	1.388	1.178	.8952	.6684
1.61	2.857	2.049	1.395	1.181	.8914	.6655
1.62	2.895	2.065	1.402	1.184	.8877	.6625
1.63	2.933	2.082	1.409	1.187	.8838	.6596
1.64	2.971	2.099	1.416	1.190	.8799	.6568
1.65	3.010	2.115	1.423	1.193	.8760	.6540
1.66	3.048	2.132	1.430	1.196	.8720	.6512
1.67	3.087	2.148	1.437	1.199	.8680	.6485
1.68	3.126	2.165	1.444	1.202	.8640	.6458
1.69	3.165	2.181	1.451	1.205	.8599	.6431

TABLE IV (*Continued*)

PARAMETERS FOR SHOCK FLOW

M_{1n}	p_2/p_1	ρ_2/ρ_1	T_2/T_1	a_2/a_1	p_2^0/p_1^0	M_2 for Normal Shocks Only
1.70	3.205	2.198	1.458	1.208	.8557	.6405
1.71	3.245	2.214	1.466	1.211	.8516	.6380
1.72	3.285	2.230	1.473	1.214	.8474	.6355
1.73	3.325	2.247	1.480	1.217	.8431	.6330
1.74	3.366	2.263	1.487	1.220	.8389	.6305
1.75	3.406	2.279	1.495	1.223	.8346	.6281
1.76	3.447	2.295	1.502	1.226	.8302	.6257
1.77	3.488	2.311	1.509	1.229	.8259	.6234
1.78	3.530	2.327	1.517	1.232	.8215	.6210
1.79	3.571	2.343	1.524	1.235	.8171	.6188
1.80	3.613	2.359	1.532	1.238	.8127	.6165
1.81	3.655	2.375	1.539	1.241	.8082	.6143
1.82	3.698	2.391	1.547	1.244	.8038	.6121
1.83	3.740	2.407	1.554	1.247	.7993	.6099
1.84	3.783	2.422	1.562	1.250	.7948	.6078
1.85	3.826	2.438	1.569	1.253	.7902	.6057
1.86	3.870	2.454	1.577	1.256	.7857	.6036
1.87	3.913	2.469	1.585	1.259	.7811	.6016
1.88	3.957	2.485	1.592	1.262	.7765	.5996
1.89	4.001	2.500	1.600	1.265	.7720	.5976
1.90	4.045	2.516	1.608	1.268	.7674	.5956
1.91	4.089	2.531	1.616	1.271	.7628	.5937
1.92	4.134	2.546	1.624	1.274	.7581	.5918
1.93	4.179	2.562	1.631	1.277	.7535	.5899
1.94	4.224	2.577	1.639	1.280	.7488	.5880
1.95	4.270	2.592	1.647	1.283	.7442	.5862
1.96	4.315	2.607	1.655	1.287	.7395	.5844
1.97	4.361	2.622	1.663	1.290	.7349	.5826
1.98	4.407	2.637	1.671	1.293	.7302	.5808
1.99	4.453	2.652	1.679	1.296	.7255	.5791
2.00	4.500	2.667	1.688	1.299	.7209	.5773
2.01	4.547	2.681	1.696	1.302	.7162	.5757
2.02	4.594	2.696	1.704	1.305	.7115	.5740
2.03	4.641	2.711	1.712	1.308	.7069	.5723
2.04	4.689	2.725	1.720	1.312	.7022	.5707

TABLE IV (*Continued*)

PARAMETERS FOR SHOCK FLOW

M_{1n}	p_2/p_1	ρ_2/ρ_1	T_2/T_1	a_2/a_1	p_2^0/p_1^0	M_2 for Normal Shocks Only
2.05	4.736	2.740	1.729	1.315	.6975	.5691
2.06	4.784	2.755	1.737	1.318	.6928	.5675
2.07	4.832	2.769	1.745	1.321	.6882	.5659
2.08	4.881	2.783	1.754	1.324	.6835	.5643
2.09	4.929	2.798	1.762	1.327	.6789	.5628
2.10	4.978	2.812	1.770	1.331	.6742	.5613
2.11	5.027	2.826	1.779	1.334	.6696	.5598
2.12	5.077	2.840	1.787	1.337	.6649	.5583
2.13	5.126	2.854	1.796	1.340	.6603	.5568
2.14	5.176	2.868	1.805	1.343	.6557	.5554
2.15	5.226	2.882	1.813	1.347	.6511	.5540
2.16	5.277	2.896	1.822	1.350	.6464	.5525
2.17	5.327	2.910	1.831	1.353	.6419	.5511
2.18	5.378	2.924	1.839	1.356	.6373	.5498
2.19	5.429	2.938	1.848	1.359	.6327	.5484
2.20	5.480	2.951	1.857	1.363	.6281	.5471
2.21	5.531	2.965	1.866	1.366	.6236	.5457
2.22	5.583	2.978	1.875	1.369	.6191	.5444
2.23	5.635	2.992	1.883	1.372	.6145	.5431
2.24	5.687	3.005	1.892	1.376	.6100	.5418
2.25	5.740	3.019	1.901	1.379	.6055	.5406
2.26	5.792	3.032	1.910	1.382	.6011	.5393
2.27	5.845	3.045	1.919	1.385	.5966	.5381
2.28	5.898	3.058	1.929	1.389	.5921	.5368
2.29	5.951	3.071	1.938	1.392	.5877	.5356
2.30	6.005	3.085	1.947	1.395	.5833	.5344
2.31	6.059	3.098	1.956	1.399	.5789	.5332
2.32	6.113	3.110	1.965	1.402	.5745	.5321
2.33	6.167	3.123	1.974	1.405	.5702	.5309
2.34	6.222	3.136	1.984	1.408	.5658	.5297
2.35	6.276	3.149	1.993	1.412	.5615	.5286
2.36	6.331	3.162	2.002	1.415	.5572	.5275
2.37	6.386	3.174	2.012	1.418	.5529	.5264
2.38	6.442	3.187	2.021	1.422	.5486	.5253
2.39	6.497	3.199	2.031	1.425	.5444	.5242

TABLE IV (*Continued*)

PARAMETERS FOR SHOCK FLOW

M_{1n}	p_2/p_1	ρ_2/ρ_1	T_2/T_1	a_2/a_1	p_2^0/p_1^0	M_2 for Normal Shocks Only
2.40	6.553	3.212	2.040	1.428	.5401	.5231
2.41	6.609	3.224	2.050	1.432	.5359	.5221
2.42	6.666	3.237	2.059	1.435	.5317	.5210
2.43	6.722	3.249	2.069	1.438	.5276	.5200
2.44	6.779	3.261	2.079	1.442	.5234	.5189
2.45	6.836	3.273	2.088	1.445	.5193	.5179
2.46	6.894	3.285	2.098	1.449	.5152	.5169
2.47	6.951	3.298	2.108	1.452	.5111	.5159
2.48	7.009	3.310	2.118	1.455	.5071	.5149
2.49	7.067	3.321	2.128	1.459	.5030	.5140
2.50	7.125	3.333	2.138	1.462	.4990	.5130
2.51	7.183	3.345	2.147	1.465	.4950	.5120
2.52	7.242	3.357	2.157	1.469	.4911	.5111
2.53	7.301	3.369	2.167	1.472	.4871	.5102
2.54	7.360	3.380	2.177	1.476	.4832	.5092
2.55	7.420	3.392	2.187	1.479	.4793	.5083
2.56	7.479	3.403	2.198	1.482	.4754	.5074
2.57	7.539	3.415	2.208	1.486	.4715	.5065
2.58	7.599	3.426	2.218	1.489	.4677	.5056
2.59	7.659	3.438	2.228	1.493	.4639	.5047
2.60	7.720	3.449	2.238	1.496	.4601	.5039
2.61	7.781	3.460	2.249	1.500	.4564	.5030
2.62	7.842	3.471	2.259	1.503	.4526	.5022
2.63	7.903	3.483	2.269	1.506	.4489	.5013
2.64	7.965	3.494	2.280	1.510	.4452	.5005
2.65	8.026	3.505	2.290	1.513	.4416	.4996
2.66	8.088	3.516	2.301	1.517	.4379	.4988
2.67	8.150	3.527	2.311	1.520	.4343	.4980
2.68	8.213	3.537	2.322	1.524	.4307	.4972
2.69	8.275	3.548	2.332	1.527	.4271	.4964
2.70	8.338	3.559	2.343	1.531	.4236	.4956
2.71	8.401	3.570	2.354	1.534	.4201	.4949
2.72	8.465	3.580	2.364	1.538	.4166	.4941
2.73	8.528	3.591	2.375	1.541	.4131	.4933
2.74	8.592	3.601	2.386	1.545	.4097	.4926

TABLE IV (*Continued*)

PARAMETERS FOR SHOCK FLOW

M_{1n}	p_2/p_1	ρ_2/ρ_1	T_2/T_1	a_2/a_1	p_2^0/p_1^0	M_2 for Normal Shocks Only
2.75	8.656	3.612	2.397	1.548	.4062	.4918
2.76	8.721	3.622	2.407	1.552	.4028	.4911
2.77	8.785	3.633	2.418	1.555	.3994	.4903
2.78	8.850	3.643	2.429	1.559	.3961	.4896
2.79	8.915	3.653	2.440	1.562	.3928	.4889
2.80	8.980	3.664	2.451	1.566	.3895	.4882
2.81	9.045	3.674	2.462	1.569	.3862	.4875
2.82	9.111	3.684	2.473	1.573	.3829	.4868
2.83	9.177	3.694	2.484	1.576	.3797	.4861
2.84	9.243	3.704	2.496	1.580	.3765	.4854
2.85	9.310	3.714	2.507	1.583	.3733	.4847
2.86	9.376	3.724	2.518	1.587	.3701	.4840
2.87	9.443	3.734	2.529	1.590	.3670	.4833
2.88	9.510	3.743	2.540	1.594	.3639	.4827
2.89	9.577	3.753	2.552	1.597	.3608	.4820
2.90	9.645	3.763	2.563	1.601	.3577	.4814
2.91	9.713	3.773	2.575	1.605	.3547	.4807
2.92	9.781	3.782	2.586	1.608	.3517	.4801
2.93	9.849	3.792	2.598	1.612	.3487	.4795
2.94	9.918	3.801	2.609	1.615	.3457	.4788
2.95	9.986	3.811	2.621	1.619	.3428	.4782
2.96	10.06	3.820	2.632	1.622	.3398	.4776
2.97	10.12	3.829	2.644	1.626	.3369	.4770
2.98	10.19	3.839	2.656	1.630	.3340	.4764
2.99	10.26	3.848	2.667	1.633	.3312	.4758
3.00	10.33	3.857	2.679	1.637	.3283	.4752
3.10	11.05	3.947	2.799	1.673	.3012	.4695
3.20	11.78	4.031	2.922	1.709	.2762	.4643
3.30	12.54	4.112	3.049	1.746	.2533	.4596
3.40	13.32	4.188	3.180	1.783	.2322	.4552
3.50	14.13	4.261	3.315	1.821	.2129	.4512
3.60	14.95	4.330	3.454	1.858	.1953	.4474
3.70	15.80	4.395	3.596	1.896	.1792	.4439
3.80	16.68	4.457	3.743	1.935	.1645	.4407
3.90	17.58	4.516	3.893	1.973	.1510	.4377

TABLE IV (*Continued*)

PARAMETERS FOR SHOCK FLOW

M_{1n}	p_2/p_1	ρ_2/ρ_1	T_2/T_1	a_2/a_1	p_2^0/p_1^0	M_2 for Normal Shocks Only
4.00	18.50	4.571	4.047	2.012	.1388	.4350
5.00	29.00	5.000	5.800	2.408	.06172	.4152
6.00	41.83	5.268	7.941	2.818	.02965	.4042
7.00	57.00	5.444	10.47	3.236	.01535	.3974
8.00	74.50	5.565	13.39	3.659	8.488×10^{-3}	.3929
9.00	94.33	5.651	16.69	4.086	4.964×10^{-3}	.3898
10.00	116.5	5.714	20.39	4.515	3.045×10^{-3}	.3876
100.00	11,666.5	5.997	1945.4	44.11	3.593×10^{-8}	.3781
∞	∞	6	∞	∞	0	.3780

Data taken from NACA TN 1428, courtesy of National Advisory Committee of Aeronautics. Setup from A. M. Kuethe and J. D. Schetzer, *Foundations of Aerodynamics*, John Wiley & Sons, New York, 1950.

TABLE V

MACH NUMBER AND MACH ANGLE VERSUS PRANDTL-MEYER FUNCTION

ν (deg)	M	μ (deg)	ν (deg)	M	μ (deg)
0.0	1.000	90.000	17.5	1.689	36.293
0.5	1.051	72.099	18.0	1.706	35.874
1.0	1.082	67.574	18.5	1.724	35.465
1.5	1.108	64.451	19.0	1.741	35.065
2.0	1.133	61.997	19.5	1.758	34.673
2.5	1.155	59.950	20.0	1.775	34.290
3.0	1.177	58.180	20.5	1.792	33.915
3.5	1.198	56.614	21.0	1.810	33.548
4.0	1.218	55.205	21.5	1.827	33.188
4.5	1.237	53.920	22.0	1.844	32.834
5.0	1.256	52.738	22.5	1.862	32.488
5.5	1.275	51.642	23.0	1.879	32.148
6.0	1.294	50.619	23.5	1.897	31.814
6.5	1.312	49.658	24.0	1.915	31.486
7.0	1.330	48.753	24.5	1.932	31.164
7.5	1.348	47.896	25.0	1.950	30.847
8.0	1.366	47.082	25.5	1.968	30.536
8.5	1.383	46.306	26.0	1.986	30.229
9.0	1.400	45.566	26.5	2.004	29.928
9.5	1.418	44.857	27.0	2.023	29.632
10.0	1.435	44.177	27.5	2.041	29.340
10.5	1.452	43.523	28.0	2.059	29.052
11.0	1.469	42.894	28.5	2.078	28.769
11.5	1.486	42.287	29.0	2.096	28.491
12.0	1.503	41.701	29.5	2.115	28.216
12.5	1.520	41.134	30.0	2.134	27.945
13.0	1.537	40.585	30.5	2.153	27.678
13.5	1.554	40.053	31.0	2.172	27.415
14.0	1.571	39.537	31.5	2.191	27.155
14.5	1.588	39.035	32.0	2.210	26.899
15.0	1.605	38.547	32.5	2.230	26.646
15.5	1.622	38.073	33.0	2.249	26.397
16.0	1.639	37.611	33.5	2.269	26.151
16.5	1.655	37.160	34.0	2.289	25.908
17.0	1.672	36.721	34.5	2.309	25.668

TABLE V (*Continued*)

MACH NUMBER AND MACH ANGLE VERSUS PRANDTL-MEYER FUNCTION

ν (deg)	M	μ (deg)	ν (deg)	M	μ (deg)
35.0	2.329	25.430	52.5	3.146	18.532
35.5	2.349	25.196	53.0	3.174	18.366
36.0	2.369	24.965	53.5	3.202	18.200
36.5	2.390	24.736	54.0	3.230	18.036
37.0	2.410	24.510	54.5	3.258	17.873
37.5	2.431	24.287	55.0	3.287	17.711
38.0	2.452	24.066	55.5	3.316	17.551
38.5	2.473	23.847	56.0	3.346	17.391
39.0	2.495	23.631	56.5	3.375	17.233
39.5	2.516	23.418	57.0	3.406	17.076
40.0	2.538	23.206	57.5	3.436	16.920
40.5	2.560	22.997	58.0	3.467	16.765
41.0	2.582	22.790	58.5	3.498	16.611
41.5	2.604	22.585	59.0	3.530	16.458
42.0	2.626	22.382	59.5	3.562	16.306
42.5	2.649	22.182	60.0	3.594	16.155
43.0	2.671	21.983	60.5	3.627	16.005
43.5	2.694	21.786	61.0	3.660	15.856
44.0	2.718	21.591	61.5	3.694	15.708
44.5	2.741	21.398	62.0	3.728	15.561
45.0	2.764	21.207	62.5	3.762	15.415
45.5	2.788	21.017	63.0	3.797	15.270
46.0	2.812	20.830	63.5	3.832	15.126
46.5	2.836	20.644	64.0	3.868	14.983
47.0	2.861	20.459	64.5	3.904	14.840
47.5	2.886	20.277	65.0	3.941	14.698
48.0	2.910	20.096	65.5	3.979	14.557
48.5	2.936	19.916	66.0	4.016	14.417
49.0	2.961	19.738	66.5	4.055	14.278
49.5	2.987	19.561	67.0	4.094	14.140
50.0	3.013	19.386	67.5	4.133	14.002
50.5	3.039	19.213	68.0	4.173	13.865
51.0	3.065	19.041	68.5	4.214	13.729
51.5	3.092	18.870	69.0	4.255	13.593
52.0	3.119	18.701	69.5	4.297	13.459

TABLE V (*Continued*)

MACH NUMBER AND MACH ANGLE VERSUS PRANDTL-MEYER FUNCTION

ν (deg)	M	μ (deg)	ν (deg)	M	μ (deg)
70.0	4.339	13.325	87.5	6.390	9.003
70.5	4.382	13.191	88.0	6.472	8.888
71.0	4.426	13.059	88.5	6.556	8.774
71.5	4.470	12.927	89.0	6.642	8.660
72.0	4.515	12.795	89.5	6.729	8.546
72.5	4.561	12.665	90.0	6.819	8.433
73.0	4.608	12.535	90.5	6.911	8.320
73.5	4.655	12.406	91.0	7.005	8.207
74.0	4.703	12.277	91.5	7.102	8.095
74.5	4.752	12.149	92.0	7.201	7.983
75.0	4.801	12.021	92.5	7.302	7.871
75.5	4.852	11.894	93.0	7.406	7.760
76.0	4.903	11.768	93.5	7.513	7.649
76.5	4.955	11.642	94.0	7.623	7.538
77.0	5.009	11.517	94.5	7.735	7.428
77.5	5.063	11.392	95.0	7.851	7.318
78.0	5.118	11.268	95.5	7.970	7.208
78.5	5.174	11.145	96.0	8.092	7.099
79.0	5.231	11.022	96.5	8.218	6.989
79.5	5.289	10.899	97.0	8.347	6.881
80.0	5.348	10.777	97.5	8.480	6.772
80.5	5.408	10.656	98.0	8.618	6.664
81.0	5.470	10.535	98.5	8.759	6.556
81.5	5.532	10.414	99.0	8.905	6.448
82.0	5.596	10.294	99.5	9.055	6.340
82.5	5.661	10.175	100.0	9.210	6.233
83.0	5.727	10.056	100.5	9.371	6.126
83.5	5.795	9.937	101.0	9.536	6.019
84.0	5.864	9.819	101.5	9.708	5.913
84.5	5.935	9.701	102.0	9.885	5.806
85.0	6.006	9.584			
85.5	6.080	9.467			
86.0	6.155	9.350			
86.5	6.232	9.234			
87.0	6.310	9.119			

Numerical values taken from *Publication No.* 26, Jet Propulsion Laboratory, California Institute of Technology.

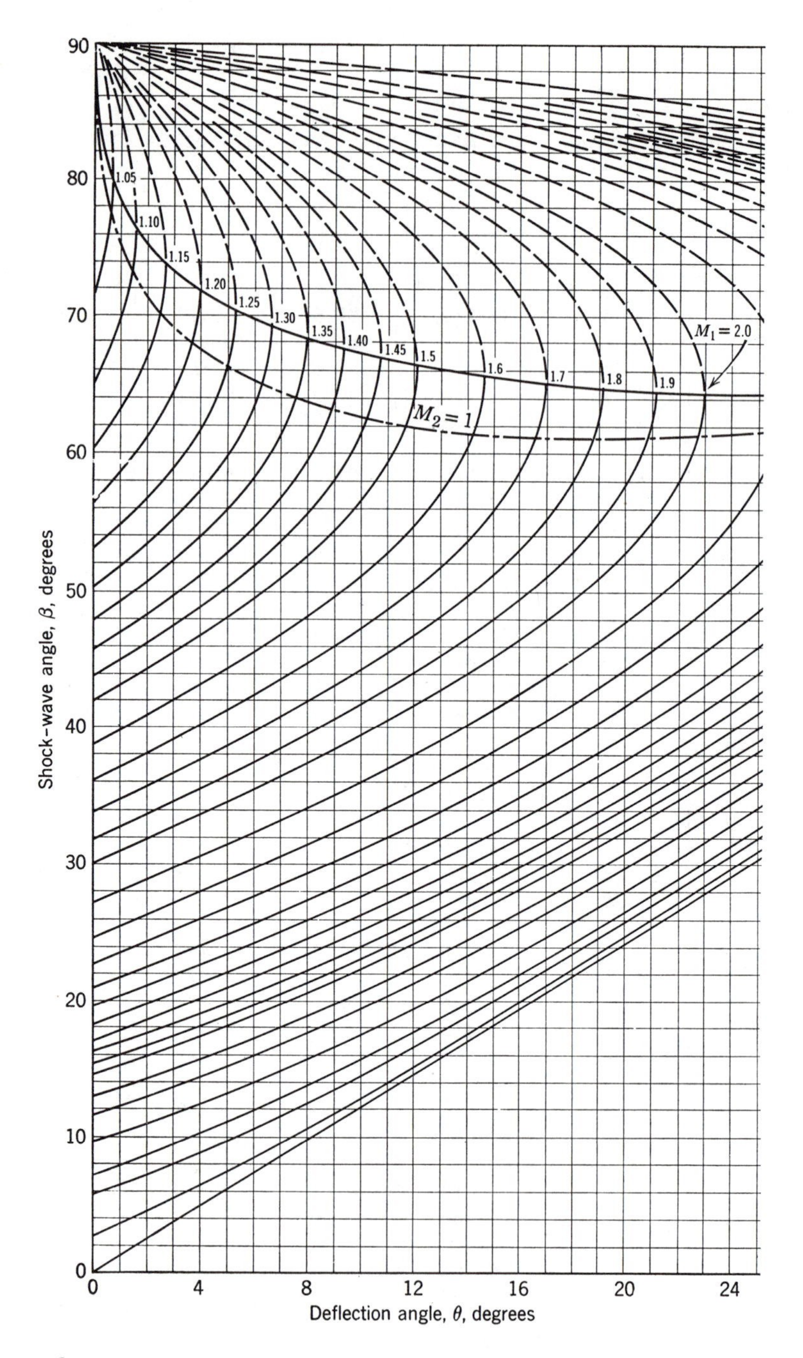

OBLIQUE SHOCK CHART 1 Variation of shock-wave angle with flow-deflection angle for various upstream Mach numbers. Perfect gas, $\gamma = 1.40$. (From *NACA Report* 1135.)

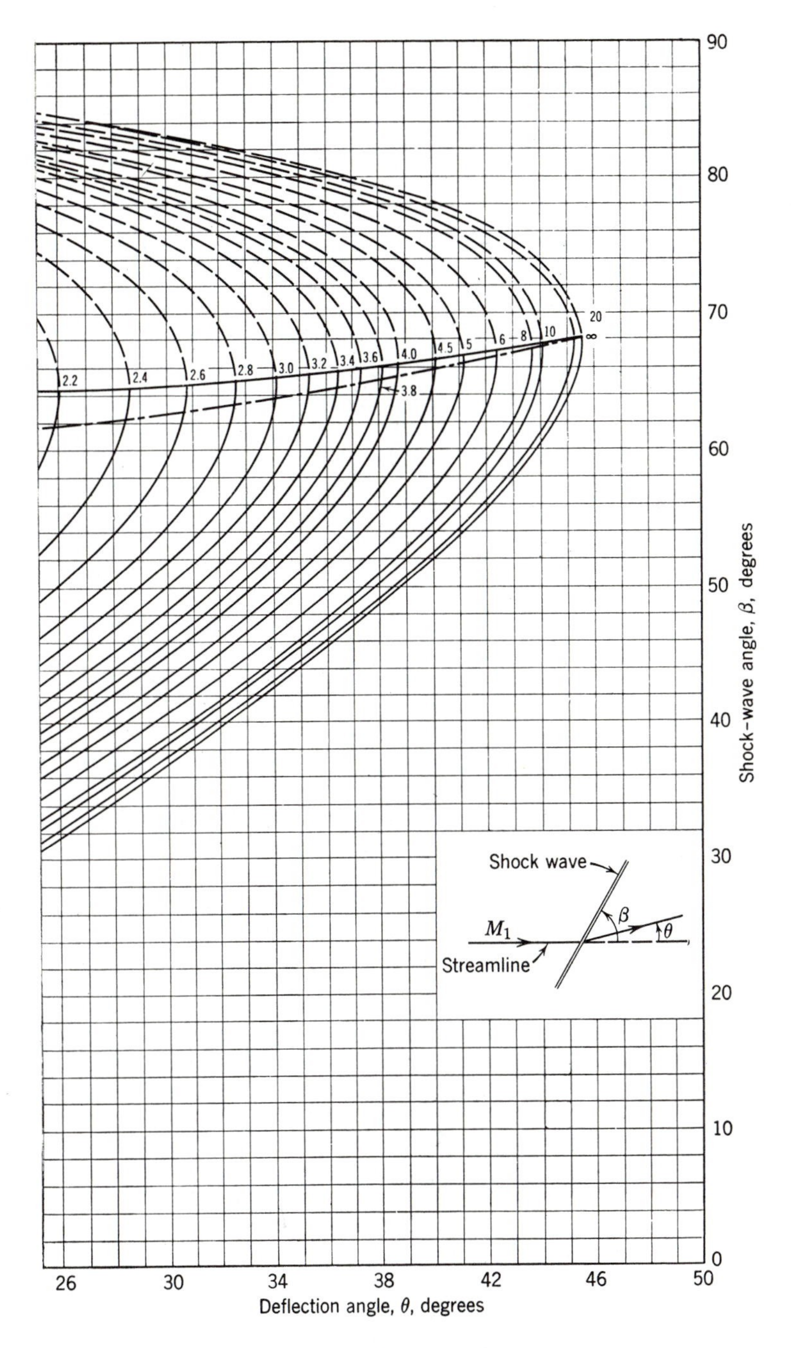

CHART 1 (continued)

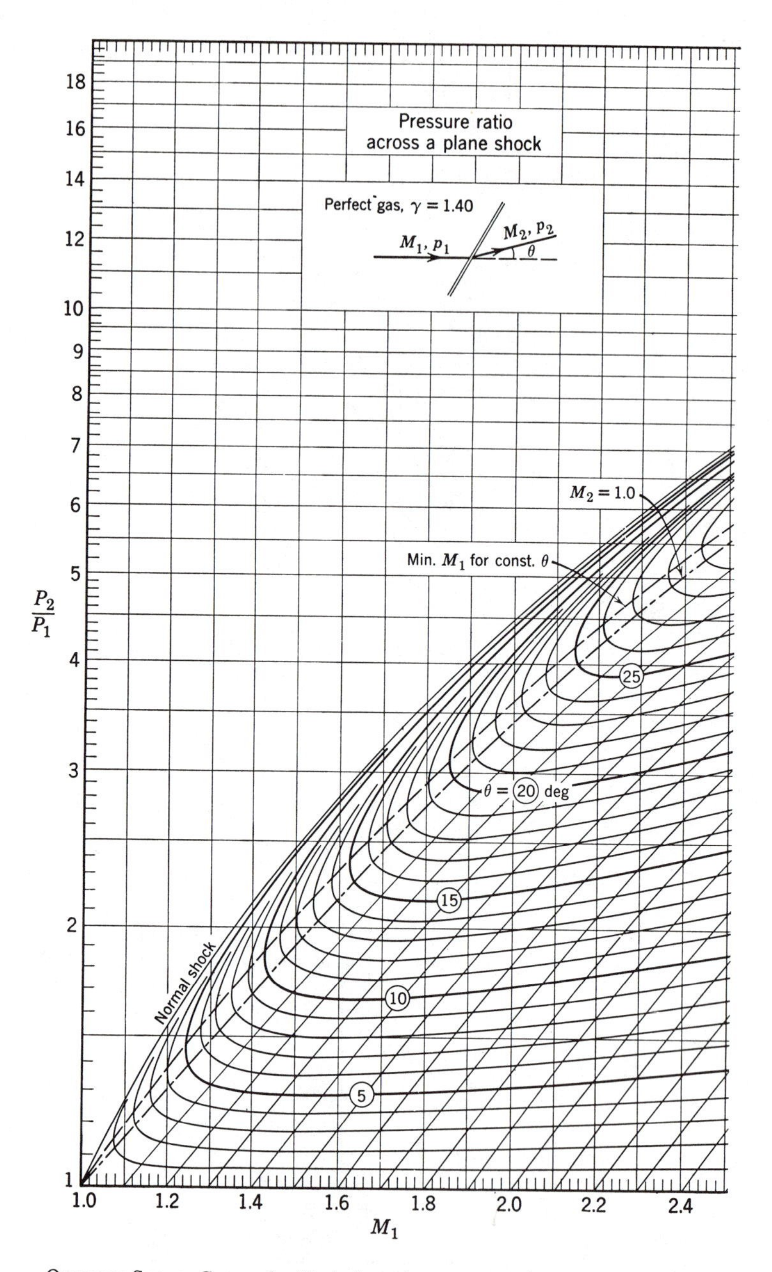

Oblique Shock Chart 2 Variation of pressure ratio and downstream Mach number with flow-deflection angle and upstream Mach number. (Data from C. L. Dailey and F. C. Wood, *Computation Curves for Compressible Flow Problems*, Wiley, 1949.)

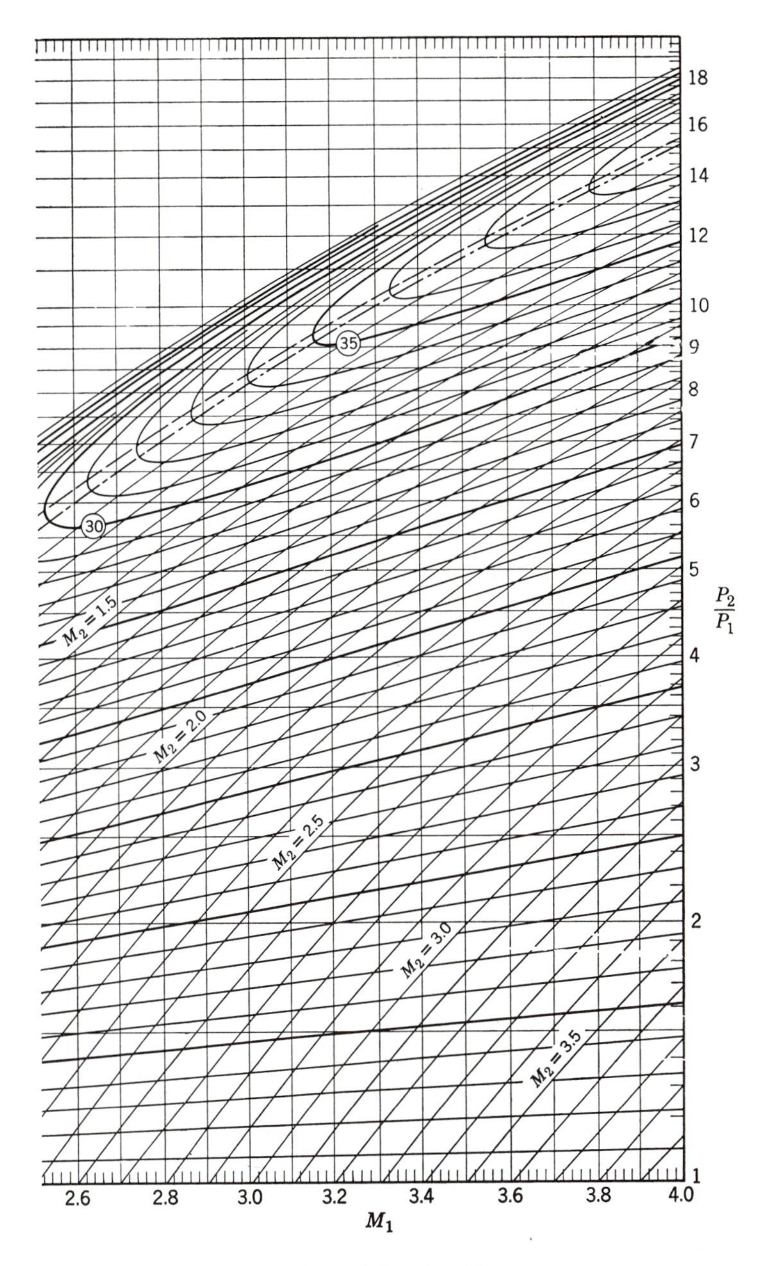

CHART 2 (continued)

Index